AF746967

Recent Perspectives in Pyrolysis Research

Edited by Mattia Bartoli and Mauro Giorcelli

Published in London, United Kingdom

IntechOpen

Supporting open minds since 2005

Recent Perspectives in Pyrolysis Research
http://dx.doi.org/10.5772/intechopen.95170
Edited by Mattia Bartoli and Mauro Giorcelli

Contributors
Vesna Mislej, Barbara Novosel, Viktor Grilc, Brice H. Leeds, Jamiu Mosebolatan Jabar, Asfaw Gezae Daful, Meegalla R. Chandraratne, Marie Loridon, Arturo Aguirre Gómez, Laura Virginia Nuñez-Balderas, Claudia Hidalgo Moreno, Jorge Dionisio Etchevers Barra, Lidija Ržek, Mojca Ravnikar Turk, Marjan Tušar, Marimuthu Karunakaran, Kasinathan Kasirajan, Murugesan Rajini, Karuppasamy Radhi Devi, Sergey Vasilevich, Dmitry Degterov, Marco Frediani, Piero Frediani, Mzukisi Matandabuzo, Delford Dovorogwa, Ngonidzashe L. Shangwa, Wilson R. Nyemba, Simon Chinguwa, Tien-Chien Jen, M. Jahiding, Mashuni, Sonia Huertas de la Torre, Deodatus Kazawadi, Justin Ntalikwa, Godlisten Kombe, William Gong, Michiaki Matsukawa, Georgia Kastrinaki, George Ganas, George Karagiannakis, Miguel Bengoechea, Aitor Eguia-Barrio, Iratxe De Meatza, Vittor Rodrigues Santos Alves, Jin Hu, Jorge Reyes-Rivera, Tatsuya Senzaki, Takanori Yonai, Haruka Taniguchi, Akiyuki Matsushita, Takahiko Sasaki, Mokoto Hagiwara

© The Editor(s) and the Author(s) 2022
The rights of the editor(s) and the author(s) have been asserted in accordance with the Copyright, Designs and Patents Act 1988. All rights to the book as a whole are reserved by INTECHOPEN LIMITED. The book as a whole (compilation) cannot be reproduced, distributed or used for commercial or non-commercial purposes without INTECHOPEN LIMITED's written permission. Enquiries concerning the use of the book should be directed to INTECHOPEN LIMITED rights and permissions department (permissions@intechopen.com).
Violations are liable to prosecution under the governing Copyright Law.

Individual chapters of this publication are distributed under the terms of the Creative Commons Attribution 3.0 Unported License which permits commercial use, distribution and reproduction of the individual chapters, provided the original author(s) and source publication are appropriately acknowledged. If so indicated, certain images may not be included under the Creative Commons license. In such cases users will need to obtain permission from the license holder to reproduce the material. More details and guidelines concerning content reuse and adaptation can be found at http://www.intechopen.com/copyright-policy.html.

Notice
Statements and opinions expressed in the chapters are these of the individual contributors and not necessarily those of the editors or publisher. No responsibility is accepted for the accuracy of information contained in the published chapters. The publisher assumes no responsibility for any damage or injury to persons or property arising out of the use of any materials, instructions, methods or ideas contained in the book.

First published in London, United Kingdom, 2022 by IntechOpen
IntechOpen is the global imprint of INTECHOPEN LIMITED, registered in England and Wales, registration number: 11086078, 5 Princes Gate Court, London, SW7 2QJ, United Kingdom
Printed in Croatia

British Library Cataloguing-in-Publication Data
A catalogue record for this book is available from the British Library

Additional hard and PDF copies can be obtained from orders@intechopen.com

Recent Perspectives in Pyrolysis Research
Edited by Mattia Bartoli and Mauro Giorcelli
p. cm.
Print ISBN 978-1-83969-914-6
Online ISBN 978-1-83969-915-3
eBook (PDF) ISBN 978-1-83969-916-0

For EU product safety concerns:
IN TECH d.o.o., Prolaz Marije Krucifikse Kozulić 3, 51000 Rijeka, Croatia, info@intechopen.com or visit our website at intechopen.com.

We are IntechOpen, the world's leading publisher of Open Access books

Built by scientists, for scientists

5,700+
Open access books available

141,000+
International authors and editors

180M+
Downloads

156
Countries delivered to

Our authors are among the
Top 1%
most cited scientists

12.2%
Contributors from top 500 universities

CLARIVATE ANALYTICS
BOOK CITATION INDEX
INDEXED

WEB OF SCIENCE™

Selection of our books indexed in the Book Citation Index (BKCI) in Web of Science Core Collection™

Interested in publishing with us?
Contact book.department@intechopen.com

Numbers displayed above are based on latest data collected.
For more information visit www.intechopen.com

Meet the editor

Dr. Mattia Bartoli has always performed at the highest levels throughout his education and training, as proved by research outputs with several published peer-reviewed papers in top international journals. After obtaining his Ph.D., Dr. Bartoli moved to the Biorefinery Conversion Network, University of Alberta, Canada, where he contributed to developing new materials and new technologies. In 2018, he joined the Carbon Group, Polytechnic University of Turin, Italy, where he studied both the production and use of carbon from thermochemical conversion of waste streams for material science applications. Since 2021, Dr. Bartoli has been working on CO2 electrochemical and thermochemical conversion at the Center for Sustainable Future Technologies (CSFT@POLITO). He is also a member of several journal editorial boards and international societies.

Dr. Mauro Giorcelli is co-founder of the Carbon Group, Polytechnic University of Turin, Italy, where he was also a post-doctoral researcher in the Material and Science Technology Department. He obtained a degree in Electronic Engineering and a Ph.D. in Physics in 2009. Currently, he is a researcher in composite materials. He is a carbon materials specialist, with more than fifteen years of experience in the field. He has published more than eighty articles in international journals. His main expertise is low-cost carbon materials derived from recycled materials, in particular carbon materials derived from biomasses (biochar). He recently published a book dedicated to innovative biochar applications. Dr. Giorcelli has a widespread collaboration network in Europe, Asia, and Canada for biochar production and applications.

Contents

Preface XIII

Section 1
Pyrolysis for Energy Applications 1

Chapter 1 3
Features of Pyrolysis of Plant Biomass at Excessive Pressure
by Sergey Vasilevich and Dmitry Degterov

Chapter 2 23
Catalytic Fast Pyrolysis of Lignocellulosic Biomass to Benzene, Toluene, and Xylenes
by William H. Gong

Chapter 3 49
On the Feasibility of the Closed Cycle Local Economy Based on Pyrolysis
by Brice H. Leeds

Chapter 4 65
Advances in the Pyrolysis Process and the Generation of Bioenergy
by Vittor Rodrigues Santos Alves

Chapter 5 85
Cheminformatics Applied to Analytical Pyrolysis of Lignocellulosic Materials
by Jorge Reyes-Rivera

Chapter 6 107
Biowastes as a Potential Energy Source in Africa
by Deodatus Kazawadi, Justin Ntalikwa and Godlisten Kombe

Chapter 7 129
Pyrolysis: A Convenient Route for Production of Eco-Friendly Fuels and Precursors for Chemical and Allied Industries
by Jamiu Mosebolatan Jabar

Chapter 8 145
Recent Advances in Thermochemical Conversion of Biomass
by Meegalla R. Chandraratne and Asfaw G. Daful

Section 2
Pyrolysis For Material Science Applications 177

Chapter 9 179
Functional Materials Synthesis and Physical Properties
by Tatsuya Senzaki, Michiaki Matsukawa, Takanori Yonai, Haruka Taniguchi, Akiyuki Matsushita, Takahiko Sasaki and Mokoto Hagiwara

Chapter 10 197
Aerosol Spray Pyrolysis Synthesis of Doped $LiNi_{0.5}Mn_{1.5}O_4$ Cathode Materials for Next Generation Lithium-Ion Batteries
by George Ganas, George Karagiannakis, Aitor Eguia-Barrio, Miguel Bengoechea, Iratxe de Meatza and Georgia Kastrinaki

Chapter 11 213
Accelerate the Aging of Polymer as Energy-Saving Method Prior to the Pyrolysis Process
by Jin Hu

Chapter 12 233
The Use of Biochar of High Growth Rate Plants to Agriculturally Remediate Heavy Metal Polluted Acidic Mine Wastes
by Arturo Aguirre Gómez, Laura Virginia Nuñez Balderas, Claudia Hidalgo Moreno and Jorge Dionisio Etchevers Barra

Chapter 13 261
Rejuvenator Obtained by Pyrolysis of Waste Tires for Use in Asphalt Mixtures with Added Reclaimed Asphalt
by Lidija Ržek, Mojca Ravnikar Turk and Marjan Tušar

Chapter 14 279
The Biomass Waste Pyrolysis for Biopesticide Application
by Mashuni and M. Jahiding

Chapter 15 293
Pyrolysis-GC/MS, A Powerful Analytical Tool for Additives and Polymers Characterization
by Sonia Huertas de la Torre

Chapter 16 313
Structural, Morphological and Optical Properties of Perfume Atomizer Spray Pyrolysis CdO Thin Films: Effect of Solution Volume
by Kasinathan Kasirajan, Karuppasamy Radhi Devi, Murugesan Rajini and Marimuthu Karunakaran

Chapter 17 327
Recent Perspectives in Biochar Production, Characterization and Applications
by Asfaw Gezae Daful, Meegalla R. Chandraratne and Marie Loridon

Section 3
Pyrolysis for Waste Management 345

Chapter 18 347
Mixed or Contaminated Waste Plastic Recycling through Microwave - Assisted Pyrolysis
by Piero Frediani and Marco Frediani

Chapter 19 371
Basic Morphological, Thermal and Physicochemical Properties of Sewage Sludge for Its Sustainable Energy and Material Use in the Circular Economy
by Barbara Novosel, Vesna Mislej and Viktor Grilc

Chapter 20 411
Activated Carbons from Waste Tyre Pyrolysis: Application
by Mzukisi Matandabuzo and Delford Dovorogwa

Chapter 21 423
Conceptualization and Design of a Small Pyrolysis Plant for the Sustainable Production of Flexible Bricks and Bituminous Concrete from Polyethylene Terephthalate Waste
by Ngonidzashe L. Shangwa, Wilson R. Nyemba, Simon Chinguwa and Tien-Chien Jen

Preface

Reactivity is a vast realm with many specific sub-fields. Among them, reactivity under high-temperature conditions is one of the most attractive due to the complex chemical routes active in the high-temperature regime. Pyrolysis is the most studied high-temperature conversion procedure. It breaks down polymeric macromolecules into compounds with low molecular weight in a free-oxygen atmosphere [1]. Pyrolytic treatment leads to three classes of products: gas, liquid, and carbonaceous residue. Several technologies have been developed to efficiently convert polymeric structures into useful products. Some of these technologies include heating technology [2], apparatus design [3–6], and the use of catalysts [7–9]. Furthermore, a pyrolytic approach can be used for the inorganic synthesis of chemical species produced by the thermal degradation of template precursors.

The flexibility of the pyrolytic process has contributed to its spread across several research and industrial projects. Pyrolytic conversions have potential applications in producing advanced materials and sustainable fuels. Furthermore, pyrolysis of waste streams has attracted attention worldwide as a possible solution to convert waste materials into eco-friendly, usable materials. The assembling of multi-product platforms built around pyrolysis units has become a reality in the world of next-generation biorefineries.

This book paints a complete and multidisciplinary picture of the many possible applications of pyrolysis. Chapters report on the use of pyrolytic methods for the production of fuels, chemicals, and nanostructured materials for high-tech applications and to convert waste streams (both biomass and plastics) into value-added products. The book provides several key lab and field examples to further understanding of the topic. It also discusses whether pyrolysis processes are convenient and economically sound considering both more traditional and unconventional approaches. The book is a useful resource for students, researchers, company advisors, and other interested readers.

Mattia Bartoli
Center for Sustainable Future Technologies (CSFT),
Istituto Italiano di Tecnologia (IIT),
Torino, Italy

Dr. Mauro Giorcelli
Polytechnic University of Turin,
Torino, Italy

References

[1] Wampler, T.P. *Applied pyrolysis handbook*; CRC Press: 2006.

[2] Bridgwater, A.; Peacocke, G. Fast pyrolysis processes for biomass. *Renewable and sustainable energy reviews* **2000**, 4, 1-73.

[3] Ragucci, R.; Giudicianni, P.; Cavaliere, A. Cellulose slow pyrolysis products in a pressurized steam flow reactor. *Fuel* **2013**, 107, 122-130, doi:http://dx.doi.org/10.1016/j.fuel.2013.01.057.

[4] Tsai, W.; Lee, M.; Chang, Y. Fast pyrolysis of rice straw, sugarcane bagasse and coconut shell in an induction-heating reactor. *Journal of analytical and applied pyrolysis* **2006**, 76, 230-237.

[5] Karaduman, A.; Şimşek, E.H.; Çiçek, B.; Bilgesü, A.Y. Flash pyrolysis of polystyrene wastes in a free-fall reactor under vacuum. *Journal of Analytical and Applied Pyrolysis* **2001**, 60, 179-186, doi:http://dx.doi.org/10.1016/S0165-2370(00)00169-8.

[6] Li, D.; Briens, C.; Berruti, F. Improved lignin pyrolysis for phenolics production in a bubbling bed reactor – Effect of bed materials. *Bioresource Technology* **2015**, 189, 7-14, doi:http://dx.doi.org/10.1016/j.biortech.2015.04.004.

[7] Shen, Y.; Yoshikawa, K. Recent progresses in catalytic tar elimination during biomass gasification or pyrolysis—A review. *Renewable and Sustainable Energy Reviews* **2013**, 21, 371-392, doi:http://dx.doi.org/10.1016/j.rser.2012.12.062.

[8] Bu, Q.; Lei, H.; Zacher, A.H.; Wang, L.; Ren, S.; Liang, J.; Wei, Y.; Liu, Y.; Tang, J.; Zhang, Q.; et al. A review of catalytic hydrodeoxygenation of lignin-derived phenols from biomass pyrolysis. *Bioresource Technology* **2012**, 124, 470-477, doi:http://dx.doi.org/10.1016/j.biortech.2012.08.089.

[9] Bridgwater, A.V. Review of fast pyrolysis of biomass and product upgrading. *Biomass and Bioenergy* **2012**, 38, 68-94, doi:http://dx.doi.org/10.1016/j.biombioe.2011.01.048.

Section 1

Pyrolysis for Energy Applications

Chapter 1

Features of Pyrolysis of Plant Biomass at Excessive Pressure

Sergey Vasilevich and Dmitry Degterov

Abstract

The paper describes methods for producing charcoal (highly porous carbon materials) based on plant (wood) raw materials, and the equipment used to implement these processes, the use of activated carbons. The paper describes results of an experimental study of the effect of pressure on the formation of charcoal in the pyrolysis of birch chips. The experimental investigation was carried out at pressures of 0.1, 0.3, 0.5, 0.7 MPa. To investigate the effect of pressure on the pyrolysis process, a laboratory bench was designed and constructed. It was found that increasing the pressure from 0.1 MPa to 0.7 MPa increases the yield of charcoal from 25.1 to 32.4% by weight (relative to the dry weight of the starting material) and the carbon content from 89.1% by weight at 0.1 MPa to 96.4% by weight at 0.7 MPa. The calorific value of charcoal decreases from 34.86 MJ/kg at a pressure of 0.1 MPa to 30.23 MJ/kg at a pressure of 0.7 MPa. This is due to the release of oxygen-containing components, which have a higher calorific value than pure carbon, from the porous coal structure. Reduction of the charcoal heat combustion with a decrease in the amount of oxygen-containing components confirms conclusion that their calorific value exceeds the calorific value of pure carbon.

Keywords: pyrolysis, birch chips, pressure, laboratory reactor, charcoal

1. Introduction

Charcoal (**Figure 1**) is a solid porous pyrolysis product of plant biomass (thermal decomposition without oxygen access), which is formed along with liquid products (resins) and combustible gases. Charcoal is used as a high quality renewable solid fuel (calorific value 30–35 MJ/kg), as well as in industry, for example, in the production of technical silicon [1].

Currently, the use of charcoal in the production of carbon sorbents is expanding, which are used in various technologies of adsorption purification and separation [2, 3].

The density of wood has a significant effect on the strength of coal, which can vary significantly both from breed to breed, and in different parts of the wood. **Table 1** shows the average physical characteristics of oak and birch wood, which are most often used in Belarus for the production of high-quality charcoal.

The yield of charcoal during pyrolysis increases as the size of the pieces of wood increases, while the yield of almost all liquid products decreases. Depending on the size of the wood raw material, it is possible to achieve different densities of its packing in the working chamber. Depending on the size of the wood raw material, it is possible to achieve different densities of its packing in the working chamber.

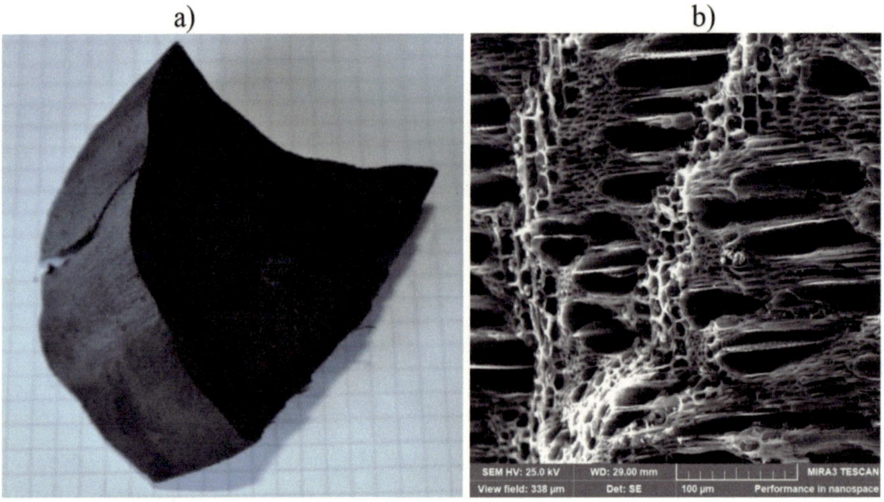

Figure 1.
General view (a) and micrograph (b) of charcoal.

Wood species	Birch	Oak
Density in an absolutely dry state, kg/m^3	520	550
Cellulose content, %	45,8	37,1
Content of hemicellulose, %	26,0	28,8
Lignin content, %	21,2	22,5
Heat capacity, J/kg K	1700	
Calorific value, kWh/kg	4,0	
Thermal conductivity across the fibers, mW/m K	150	200
Charcoal output, kg/m^3	150	160

Table 1.
Physical parameters of raw materials for the production of charcoal [2].

Type of firewood	Coniferous firewood, m			Deciduous firewood, m		
	0,25	1,0	3,0	0,25	1,0	3,0
Round 3–10 cm thick	0,79	0,69	0,61	0,75	0,63	0,55
Round 11–14 cm thick	0,81	0,72	0,66	0,80	0,70	0,62
Chipped up to 15 cm thick	0,77	0,70	0,63	0,76	0,68	0,60
A mixture of round and chipped	0,77	0,70	0,65	0,76	0,68	0,63

Table 2.
Coefficients of full wood content [1].

The full wood ratio determines the volume fraction of wood in a densely folded woodpile. The values of the fullness coefficient for firewood of different sizes are presented in **Table 2**. Assuming that birch chopped firewood 0.5 m long is used for coal production, we will use the value of the wood content factor 0.72 in the calculations.

The process of wood pyrolysis essentially depends on the initial moisture content. This value varies in very wide ranges: from 60 to 65% for freshly sawn raw materials to 15% for air-dry wood. The optimal level of relative humidity of wood intended for pyrolysis is in the range of 15–25%. With a higher humidity, in addition to an increase in fuel consumption and a decrease in the productivity of devices, the mechanical strength of coal decreases. In this regard, the raw material must be pre-dried. With natural drying of wood, a 12-month supply of raw materials is technically justified. With artificial drying, the supply of wood must be at least 3 months [1].

Depending on the calcination temperature and its duration, the content of non-volatile carbon in the coal varies. Tables 2.3 and 2.4 show examples of the distribution of elements in the original wood and decomposition products for the calcination depths, as a result of which the content of non-volatile carbon in coal is 84 and 94%. The data are given in terms of absolutely dry wood. As follows from these tables, an increase in the content of non-volatile carbon leads to a decrease in the yield of coal and an increase in the duration of the process. Thus, the calcination of the coal should only be carried out to the extent required by the customer. For domestic purposes, coal with a non-volatile carbon content of 77–82% is optimal.

Traditionally, charcoal is produced in the process of slow pyrolysis at heating rates of 5–7 K/min and temperatures of 623–873 K. The duration of the pyrolysis process is 1–14 days, depending on the unit productivity. The mass yield of coal is 20–33%. In the world, such equipment produces from 26 to 100 million tons of charcoal per year with a growth trend of about 3% per year [1]. A consequence of the high duration and low efficiency of the process is a significantly higher cost of charcoal in comparison with fossil coal.

Usually, when charring, either direct heating with a hot gas stream flowing through the bed and washing the surface of individual pieces is used, or indirect heating through the wall of the apparatus.

Table 3 shows the classification of charcoal burning equipment according to various criteria affecting the thermal regimes and the speed of the pyrolysis process [2].

Figure 2 shows various types of equipment for the production of charcoal: (a) Combustion furnace VMR (USA) (Duty cycle: 48 hours. Chamber volume: 14 m^3. Loading: 10 m^3. Temperature ~ 450°C. Coal output: 1000 kg (≈30%)); (b) experimental reactor Antala (University of Hawaii, USA) (max. Coal yield ≈ 40%. cycle no more than 1.5 hours. Heating of raw materials in a sealed reactor leads to volatility and pressure rise to 7–10 bar at 350°C); (c) UVP-5US mobile incinerator (CIS) (Working cycle: 8–12 hours. Retort volume: 4.5 m^3. Coal output: 30–32%.); (d) charcoal kiln unitary enterprise "ECO-CARBON" (Ukraine) (Working cycle: 18–34 hours. Consumption of firewood for the furnace up to 0.2 m^3/day).

The main stages of the charcoal production process are: drying, endothermic pyrolysis, exothermic pyrolysis, calcination and cooling.

During the drying process, the wood is heated to a temperature of 100–120°C and the water evaporates. Due to the fact that with excessively intensive drying, vapors tear wood, the temperature of the heat carrier in known technologies does not exceed 200–220°C (average temperature in the chamber). At the same time, while the wood is not dry, its temperature (at normal pressure) remains about 100–120°C. The composition of the wood is practically unchanged. Drying is usually divided into two stages. At the first stage (heating and actual drying), the relative humidity is brought to 5%. The second stage – drying is carried out already in the pyrolysis process.

The stage of endothermic pyrolysis includes heating wood before the decomposition of hemicelluloses, removal of bound moisture, decomposition of hemicelluloses and individual lignin fragments, heating to the temperature of exothermic pyrolysis.

Criterion	Modifications
By orientation	Horizontal
	Inclined
	Vertical
Capturing liquid products	Without capturing
	Partial capture
	With catching
	With combustion of ASG
By material	Steel (retorts)
	Concrete, etc.
	Brick (oven)
By mobility	Stationary
	Collapsible
	Mobile
By the way of heat supply	With internal heat carrier
	Through the wall
Combined with other operations	Drying, pyrolysis, cooling
	Pyrolysis only
	Drying and pyrolysis
	Pyrolysis and cooling
	Combination with activation
By the method of moving raw materials	Due to the movement of the working area or pushing device
	Under own weight
	In withdrawable devices
	Stationary
	In trolleys
By type of raw material	Unseparated logs
	Chopped firewood
	Sawdust briquettes
	Waste
By mode of operation	Periodic
	Semi-continuous
	Continuous

Table 3.
Systematization of charcoal burning equipment [2].

Exothermic decomposition of wood occurs in the temperature range from 275 to 450°C. At this stage, there is an intensive decomposition of cellulose and lignin, secondary polymerization reactions occur, and resins are formed. The free heat of exothermic pyrolysis is 1000–1150 kJ per 1 kg of wood.

The beginning of intensive decomposition of cellulose and wood corresponds to a temperature of 280–290°C, lignin decomposes at a temperature of about 350–450°C.

Figure 2.
Equipment for the production of charcoal: (a) combustion furnace VMR (USA); (b) experimental reactor Antala (University of Hawaii, USA); (c) UVP-5US mobile incinerator (CIS); (d) charcoal kiln unitary enterprise "EKO-CARBON" (Ukraine).

At this stage, it is necessary to provide intensive heat removal to prevent the possibility of an uncontrolled process. The gases generated by pyrolysis can be used for combustion in the combustion chamber. In the event of an excessive temperature rise, intensive removal of the temperature is required. The completion of wood pyrolysis is most often determined by the absence of active combustion of pyrolysis gas in the combustion chamber.

Calcination is necessary to remove from the carbon skeleton the residues of volatile substances retained by adsorption. At the same time, the content of non-volatile carbon in coal increases and the mass yield of charcoal decreases. In the course of calcination, the structure of coal also changes – functional groups containing oxygen and hydrogen are split off from it, and the bonds between carbon atoms change.

In the course of calcination, endothermic and exothermic reactions proceed in parallel, the total balance of the stage is endothermic. However, with a high degree of accuracy, it can be assumed that the heat consumption during the calcination process is necessary only for heating the charcoal.

During the cooling process, the temperature of the charcoal is lowered to 120°C. This process is most efficiently carried out using steam. Lower temperatures can be achieved using cold, non-condensable pyrolysis gases. After coal is unloaded from the reactor, it cools completely without oxygen.

Of the many factors that determine the yield and composition of biomass pyrolysis products, the main ones are the final process temperature and pressure in the apparatus, the heating rate of the material, the type of heat carrier used, the method of heat exchange and the presence of chemical additives, as well as the type and quality of raw materials.

It is known that the process of pyrolysis of plant biomass is influenced by various parameters, such as temperature, composition and structure of raw materials, moisture content of raw materials, environmental pressure, etc.

The influence of pressure on the process of wood pyrolysis has been studied for quite a long time [4–8].

Klason et al. [4, 5] more than 100 years ago, studying the pyrolysis of cellulose and woody biomass at a temperature of 400°C, found that the formation of charcoal is due to the occurrence of primary and secondary chemical processes. According to these authors, as a result of the course of primary processes, the formation of charcoal and a large amount of various hydrocarbons occurs, which then decompose with the formation of charcoal and volatile components CO_2, CO, H_2, etc. pyrolysis, the yield of charcoal is increased.

The results of Klason et al. [4, 5] were later confirmed in studies of pyrolysis of various types of biomass and various conditions of this process [6–8].

Despite extensive studies of the influence of pressure and other factors on the formation of charcoal in the process of pyrolysis of biomass, the chemical mechanism of the influence of pressure has not yet been clarified, which necessitates a study of this effect when developing an effective technology for the production of charcoal from biomass.

A laboratory reactor for the thermochemical conversion of wood was developed and constructed to work out the modes of the process of obtaining charcoal. The diagram of the reactor is shown in **Figure 3**.

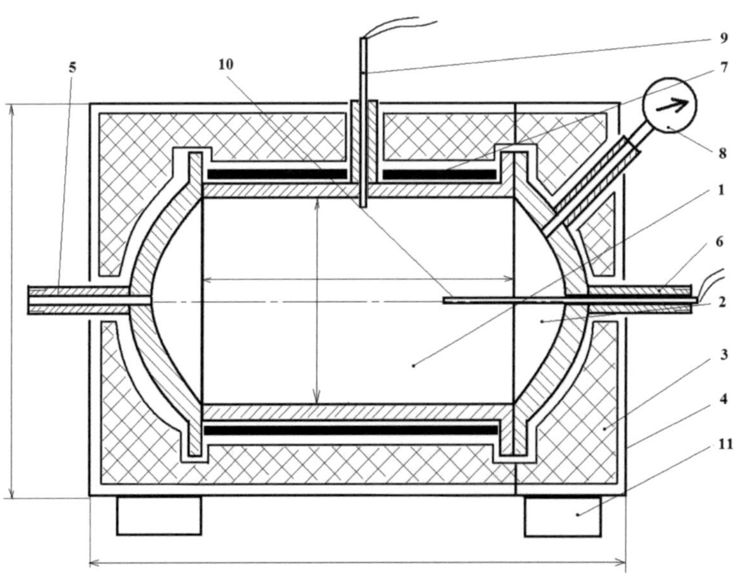

Figure 3.
Schematic of a laboratory reactor.

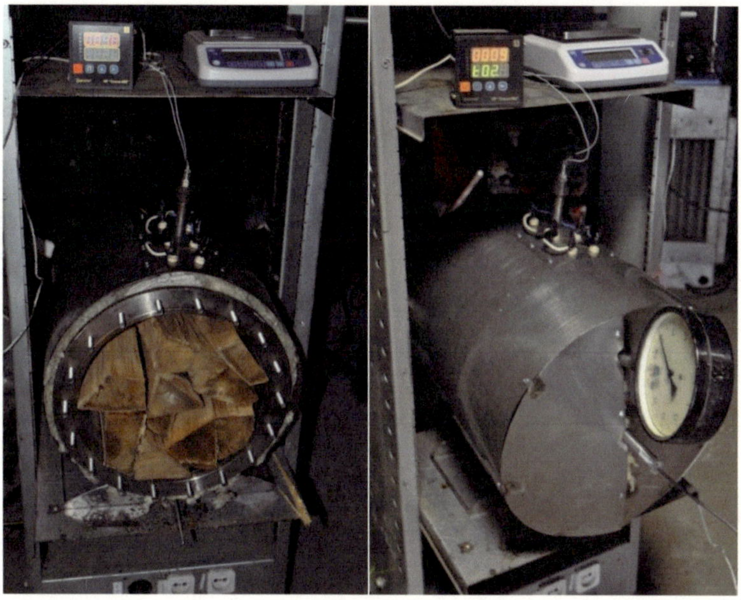

Figure 4.
Laboratory bench for the production of charcoal.

The main elements of the laboratory reactor are housing 1, cover 2, thermal insulation 3, casing 4, fittings 5–6, electric heater 7, monometer 8, thermocouples 9–10, support legs 11.

The general view of the reactor is shown in **Figure 4**.

The raw material used was birch wood in the form of firewood and chips.

The design of the reactor allows pyrolysis of biomass both at atmospheric and overpressure. For this, fitting 5 was plugged, and an adjustable safety valve was attached to fitting 6 instead of thermocouple 10, which was triggered at a pressure of 0.3–0.7 MPa.

2. Description of experiments

To reveal the effect of excess pressure on the yield of solid pyrolysis products (charcoal), experimental studies of the pyrolysis of wood chips with a particle size of 17x8x6 mm were carried out at atmospheric and excess pressure, at various temperatures and pyrolysis times. The moisture content of the wood is 14.2%. The density of the wood is 506.4 kg/m^3, the ash content is 0.23%. The moisture content was determined using a FAUNA-M moisture meter (manufactured by OOO Lenta, RF). **Table 4** shows the elemental composition of the wood raw material used. Measurement error of elemental composition 0.1%.

During pyrolysis without overpressure, raw wood was placed in a reactor. Then thermocouple 10 was placed in the wood layer (**Figure 3**). After that, the electric heater 7 was turned on. The temperature inside the wood was determined using a thermocouple 10. At the moment when the temperature exceeds 110°C, it can be concluded that the drying process is complete. Thus, the drying time of the chips was determined.

After that, the temperature in the reactor increased to 400°C. The thermocouple 10 was used to measure the temperature inside the reactor.

Component	C	O	Si	Al
Content, %	54,9	44,8	0,2	0,1

Table 4.
Elemental composition of used wood raw materials.

The process of cooling the obtained charcoal was carried out in a natural way (with the heater turned off) and with the supply of water through the choke 5. In this case, the change in temperature inside the coal layer was recorded and the time of their cooling was determined.

When pyrolysis of wood with excess pressure increases with temperature inside the reactor, the release of gaseous products begins, which accumulate in the reactor, creating excess pressure. When the required pressure is reached, the valve is activated, and the excess of gaseous pyrolysis products leaves the reactor, which allows maintaining a constant operating pressure in the reactor.

Experiments were carried out at a pressure inside the reactor of 0.1, 0.3, 0.5, 0.7 MPa. The amount of energy consumed for the coal production process was the same for all experiments and amounted to 34.89 MJ.

Figure 5 shows a general view of the resulting charcoal.

There was no excess pressure in the reactor (pyrolysis gases were freely leaving the reactor). The total pyrolysis time was 600 minutes (10 hours). The mass yield of coal was 25.1% (29.2% on dry weight of wood). The calorific value of coal is 34.86 MJ/kg. The calorific value was determined on a V-08MA “K” calorimeter.

In the course of the experiments, it was revealed that with the same consumed energy, the temperature inside the reactor during pyrolysis with excess pressure is higher than at atmospheric pressure. This is due to the fact that the thermal energy released during the operation of the heater accumulates inside the reactor until it begins to be consumed in the drying process or endothermic pyrolysis. However, under excess pressure, these reactions proceed at higher temperatures. For example, at a pressure of 0.1 MPa, an active process of moisture evaporation was observed at a temperature of 104 C, and at a pressure of 0.7 MPa - 185 C. This agrees with the reference data on the dependence of the temperature of water evaporation on pressure.

The mass yield of coal was 27.8% (32.4% per dry weight of wood with a thermodynamically equilibrium solid carbon yield of 41–42% in the pressure range

Figure 5.
General view of the resulting charcoal.

Pressure, MPa	0,1	0,3	0,5	0,7
Xc, %	25,11	28,68	31,96	32,40

Table 5.
Dependence of the mass yield of charcoal on the pressure in the reactor.

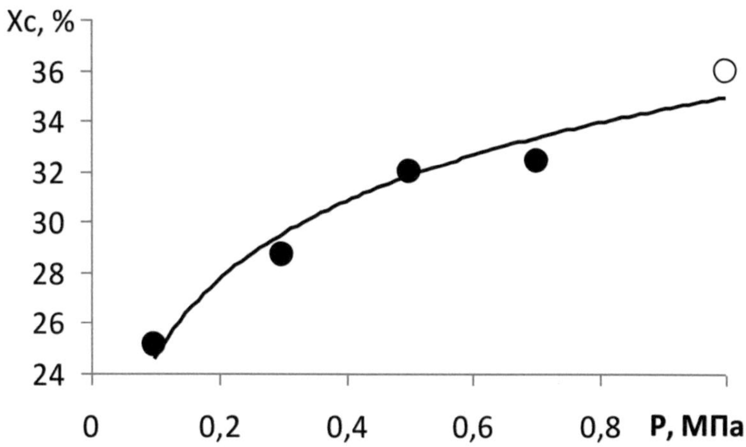

Figure 6.
Graph of the dependence of the mass yield of charcoal on the pressure in the reactor.

of 0.12–0.7 MPa). This is 1.12 times more than in the experiment without excess pressure. The calorific value of coal is 30.23 MJ/kg.

The operating time in all experiments is the same, the voltage on the heater was also constant, which means that the amount of energy expended on coal production is the same (34.9 MJ).

Table 5 and **Figure 6** shows the dependence of the mass yield of charcoal to dry mass on the pressure in the reactor.

The mass yield of charcoal to dry matter was determined by the expression

$$X_c = \frac{M_c \cdot 100}{M_0 \cdot \left(1 - \frac{W}{100}\right)},$$

where M_C – mass of coal, g;

M_0 – mass of the feedstock, g;

W – moisture content of the feedstock, %.

In **Figure 6**, the dark symbols are the data obtained by the authors, and the light symbol is the data obtained at the University of Hawaii (USA) [9] at a pressure of 1 MPa.

An increase in pressure leads to an increase in the residence time of gaseous pyrolysis products in the reactor and their contact with the feedstock and coal, i.e. brings the conditions in the reactor to equilibrium.

Table 6 shows the values of the calorific value of the obtained charcoal at various pressures.

Table 6 shows that with increasing pressure in the reactor, the calorific value of the resulting charcoal decreases. This can be explained by the fact that coal contains tar, the calorific value is higher than that of pure coal. We have determined that the

Pressure, MPa	0,1	0,3	0,5	0,7
Qc, MJ/kg	34,86	33,69	31,04	30,23

Table 6.
Dependence of the calorific value of charcoal on the pressure in the reactor.

The elements	Raw wood	Coal			
		1 atmosphere	3 atmospheres	5 atmospheres	7 atmospheres
C, %	54,9	89,1	95,3	94,02	96,4
O, %	44,8	8,4	4,0	5,66	2,9
Si, %	0,2	0,2	0,1	0,01	—
Al, %	0,1	—	—	0,03	—
Ca, %	—	1,3	0,4	0,1	0,3
K, %	—	0,3	0,1	0,03	0,2
Mg, %	—	0,2	—	0,03	0,1
Mn, %	—	0,2	—	0,05	—
P, %	—	0,1	—	0,02	—
S, %	—	—	—	—	—
Fe, %	—	—	—	0,04	—

Table 7.
Data of elemental composition for charcoal obtained at various pressures. Measurement error of elemental composition 0.1%.

calorific value of pure coal (without tar) is 27.01 MJ/kg. It is close to the value of the calorific value of the coal obtained by us at a pressure of 0.7 MPa. High purity activated carbon was taken as a reference. At the same time, pyrolysis resins have a higher calorific value. Therefore, the presence of resins increases the calorific value of the resulting product, but decreases its quality. Based on this, we can conclude that with increasing pressure, the quality of coal increases.

This is also confirmed by the results of the elemental analysis of the obtained products. **Table 7** shows the elemental composition data for charcoal obtained at various pressures. The elemental composition was determined by X-ray spectral microanalysis using an Oxford Instruments X-MaxN energy dispersive analyzer operating in conjunction with a LEO1455VP scanning electron microscope with a sensitivity of 0.1 at.%.

Figure 7 shows a micrograph of the used wood and a graph of the elemental composition.

Figure 8 shows micrographs (left) and graphs of the elemental composition (right) of charcoal obtained at a pressure of 0.1 MPa (a, b), 0.3 MPa (c, d), 0.7 MPa (e, f).

As can be seen from **Figure 8**, the charcoal composition obtained at a pressure of 0.1 MPa contains amorphous resinous inclusions (**Figure 8a**). At a pressure of 0.3 MPa, inclusions are absent (**Figure 8c**). At a pressure of 0.7 MPa, the general view of coal under a microscope coincides with wood (**Figures 7a** and **8e**). This indicates the complete preservation of the cellular structure of wood during its pyrolysis under a pressure of 0.7 MPa.

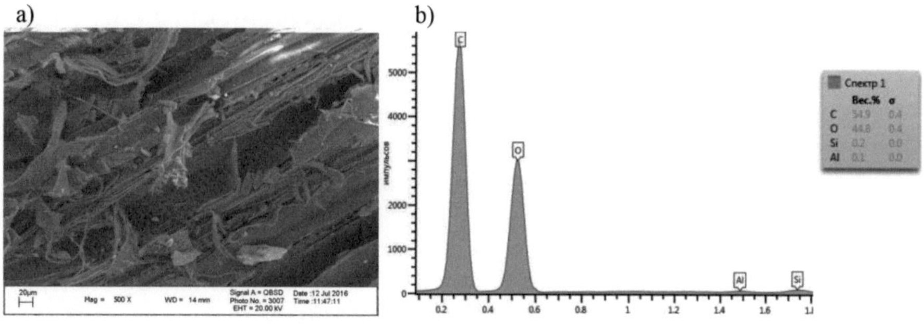

Figure 7.
Micrograph (a) of used wood and graph of elemental composition (b).

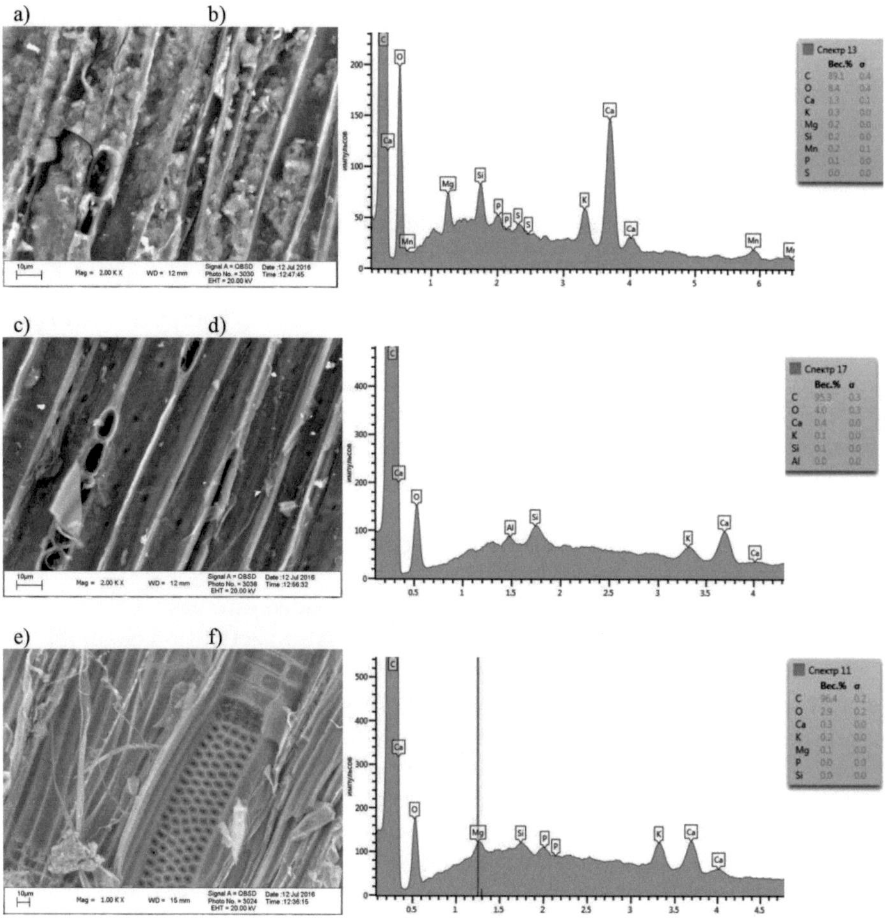

Figure 8.
Micrographs (left) and graphs of the elemental composition (right) of charcoal obtained at a pressure of 0.1 MPa (a, b), 0.3 MPa (c, d), 0.7 MPa (e, f).

As noted above, in the case of pyrolysis of birch chips at higher pressures, the temperature in the chip layer reaches the plateau-like area at higher temperatures. An increase in temperature in the layer of chips promotes the process of desorption of oxygen-containing components from the pores of the resulting charcoal and to an increase in the carbon content in it.

Table 7 shows that with an increase in pressure from 0.1 MPa to 0.7 MPa, the oxygen content in charcoal decreases from 8.4% by weight. up to 2.9% weight. In this case, the carbon content increases from 89.1% by weight. up to 96.4% weight.

According to the authors of [4–8], such a change in the yield of charcoal and the carbon content in it is due to the occurrence of secondary chemical reactions, in which the decomposition of the primary pyrolysis products occurs with the formation of charcoal and volatile components. The positive effect of pressure in this case is due to the difficulty of the diffusion release of the components formed during the decomposition of the wood pulp. This hypothesis was confirmed by us through experimental studies of the influence of the difficulty in the yield of gaseous pyrolysis products on the quantitative yield and composition of the obtained coals.

3. Description of the experimental setup

The authors have developed and manufactured an experimental plant for the pyrolysis of plant biomass under pressure, which makes it possible to obtain a highly porous carbon material, the parameters of which (sorption capacity, total pore volume, bulk density, elemental composition) meet the requirements for activated carbon.

During the development of the installation, the technological scheme of thermal steam activation was adopted. In this case, the installation consists of three blocks shown in **Figure 9**.

The arrows show the movement of the steam. Each unit is equipped with a pressure relief valve. On block 1 - for 8 atmospheres, on blocks 2 and 3 - for 10 atmospheres (to ensure, first of all, the movement of steam into the pyrolysis chamber and activation). Each block is made in the form of cylinders with an inner diameter of 0.1 m. The blocks (chambers) are interconnected by adapters in the form of tubes with an inner diameter of 0.02 m. The thickness of the walls of the chambers and adapters is 0.005 m.

The installation diagram is shown in **Figure 10**.

The experimental setup works as follows.

At the beginning of work, wood raw material is loaded into basket 15 (see **Figure 10**) and placed into the cavity of the pyrolysis and activation chamber.

After loading the basket, the pyrolysis and activation chamber is hermetically closed by the chamber lid 14, having previously installed a brass O-ring. Water with a volume of 2 liters is poured into the steam generator and hermetically closed with a lid. After that, the serviceability of the devices and the reliability of the fastening of the installation elements are checked. Then the fan 17, the temperature controller "Sosna-002" and the heaters 2, 9 and 13 on the steam generator, superheater and pyrolysis and activation chamber are turned on.

The required pressure in the installation is maintained by valves 6 and 20 installed on the steam generator and cooler.

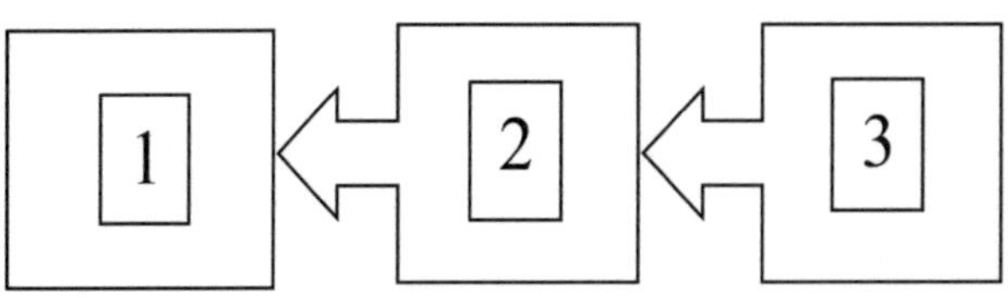

Figure 9.
Block diagram of the experimental setup. (1) pyrolysis and activation chamber, (2) superheater, and (3) steam generator.

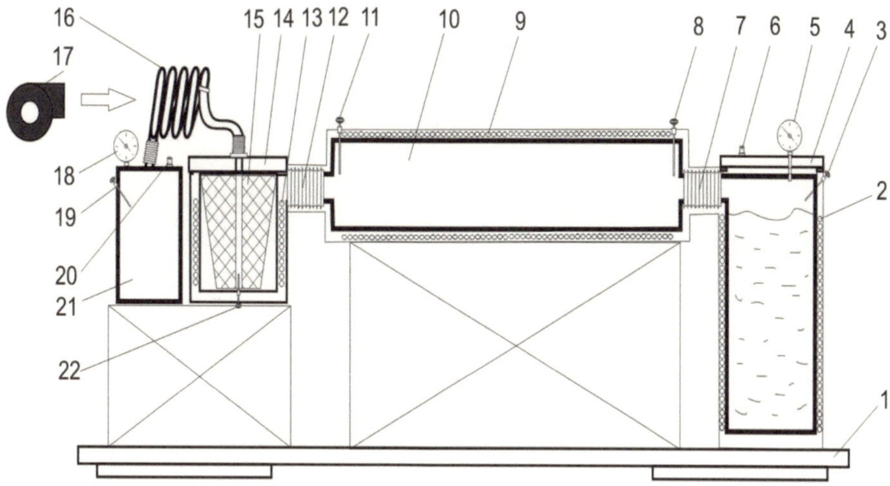

Figure 10.
Diagram of the experimental setup.

The general view of the installation is shown in **Figure 11**.

A general view of the loaded basket is shown in **Figure 12**.

To control the pressure level, pressure gauges are installed on the steam generator and cooler.

The temperature regime in the installation is maintained through the use of electric heaters on the steam generator, superheater and pyrolysis and activation chamber. The temperatures in the chambers are controlled using thermocouples 3, 8, 11 and 19 (see **Figure 10**) installed in the steam generator, superheater, in the pyrolysis chamber and in the cooler.

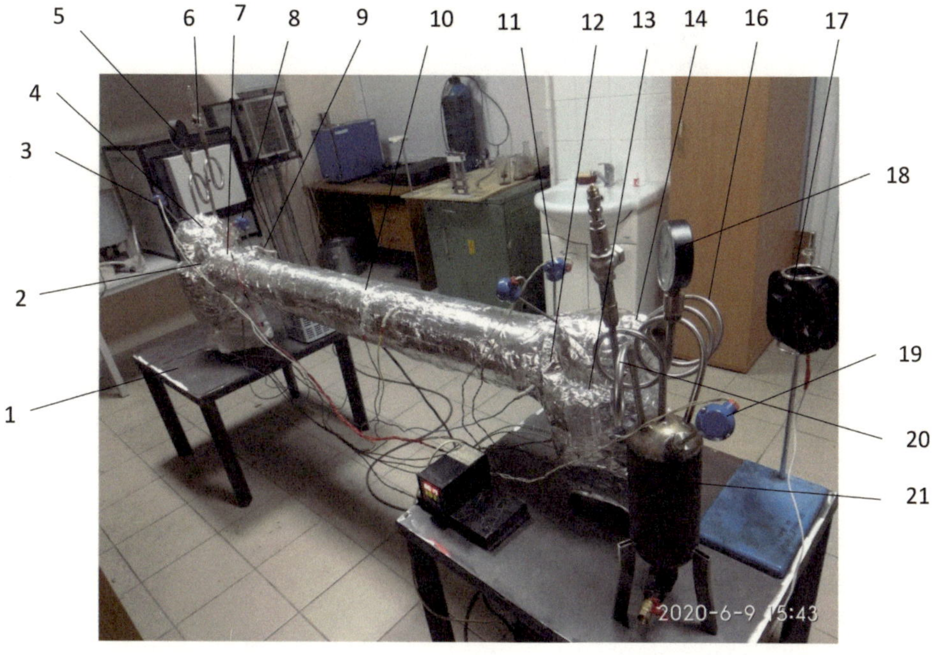

Figure 11.
General view of the experimental setup.

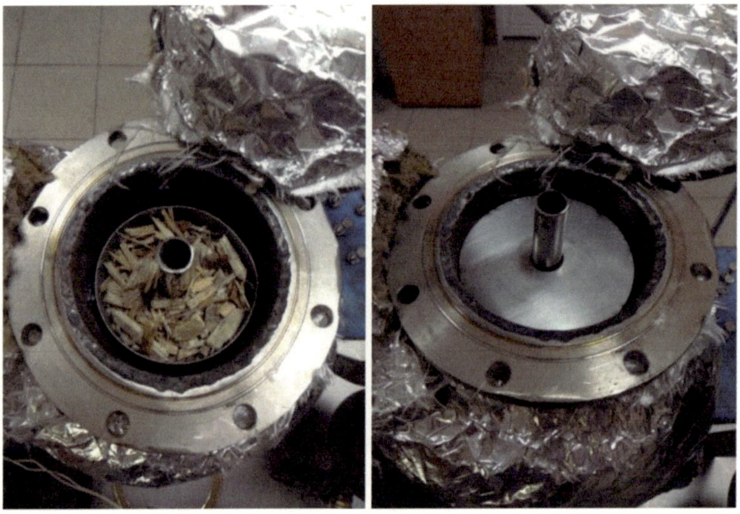

Figure 12.
Loading the basket into the pyrolysis chamber and activation.

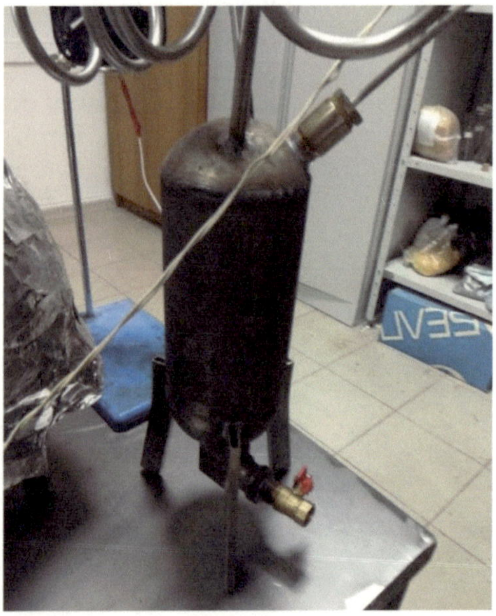

Figure 13.
General view of the cooler.

The gaseous products of the conversion, together with the superheated steam during thermal expansion, create an overpressure in the installation. In the case when the pressure exceeds the operating pressure (8 atm), the surplus of gaseous products leaves the unit through the safety valves.

Heavy hydrocarbons released in the pyrolysis and activation chamber during the operation of the unit enter the cooler (**Figure 13**) through the heat exchanger. In the cooler, heavy hydrocarbons and steam are condensed. To remove them, the cooler is equipped with an outlet pipe and a valve.

To intensify the process of cooling the exhaust gaseous products and ensure their condensation, a fan is installed that creates an air flow to enhance heat transfer.

4. Description of the experiment

During the experiment, birch wood chips with a fraction size of 35x8x4 mm were used as raw materials. The humidity was 9.6% (determined by the EVLAS-2 M device). Ash content 0.11%. The density of the wood is 380.95 kg/m^3. The bulk density of the chips was 187.27 kg/m^3. The calorific value of wood raw material is 17269.79 kJ/kg. The mass of wood in the basket was 200.2 g.

After loading the raw material and turning on the heater, the pyrolysis and activation chamber was brought to a thermal regime corresponding to the onset of pyrolysis (250 C). After that (55 minutes after the start of the experiment), the heaters on the steam generator and the superheater were turned on. At the same time, at the end of the pyrolysis of plant biomass (wood chips), superheated steam began to flow in the pyrolysis and activation chamber. Pyrolysis started 55 minutes after the heaters were turned on and lasted 75 minutes. The maximum pyrolysis temperature was 650 C. At the end of the pyrolysis, calcination and simultaneous activation with superheated steam for 2 hours and 30 minutes took place. The superheated steam temperature was 760 C. The total process time was 4 hours and 40 minutes. At the end of the work, the heaters were turned off. After cooling all the elements of the installation and, as a consequence, reducing the pressure to atmospheric, the lid of the pyrolysis and activation chamber was opened and the basket with samples of the finished solid product was removed.

The general view of the obtained solid pyrolysis products is shown in **Figure 14**. The mass yield of solid products was 15.5%.

After that, the basket was again loaded with wood chips and installed in the pyrolysis chamber. At the same time, the valves were removed so that during the operation of the installation in the chambers the pressure was close to atmospheric.

Then the experiment was repeated and solid pyrolysis products were obtained, the mass yield of which was 9.4%.

It can be seen that at a pressure of 8 atm, the mass yield of solid products is higher than at atmospheric pressure. This confirms the results previously obtained by the authors and described in [10, 11].

Figure 14.
General view of solid products of wood chips conversion.

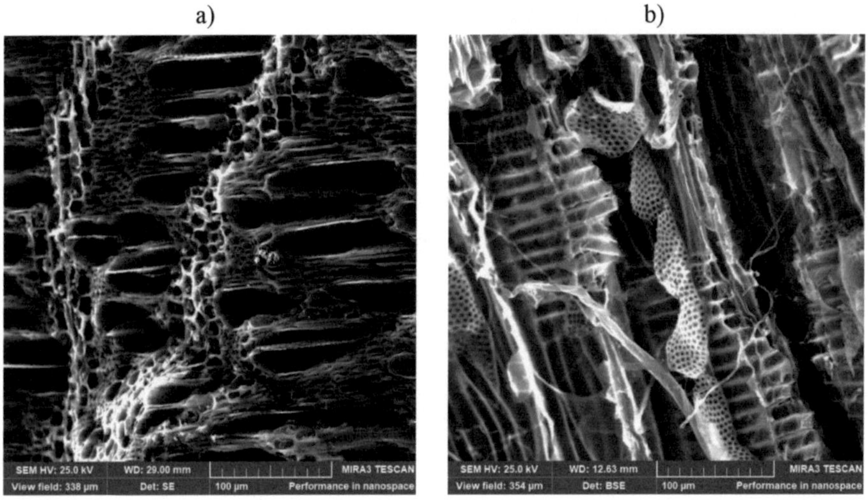

Figure 15.
Micrographs of samples of solid pyrolysis products obtained at a pressure of 1 atm. (a) and 8 atm. (b).

The elements	Raw wood	Coal	
		1 atmosphere	8 atmospheres
C, %	55,18	88,3	93,7
O, %	44,52	8,1	3,1
Si, %	0,15	0,7	2,0
Al, %	0,08	0,5	0,3
Ca, %	0,01	1,1	—
K, %	0,05	0,3	0,4
Mg, %	0,01	0,4	0,3
Mn, %	—	0,1	—
P, %	—	0,2	—
S, %	—	0,1	0,2
Fe, %	0,1	0,2	—

Table 8.
Elemental composition of raw materials and charcoal obtained at different pressures.

Investigations of the composition and properties of solid conversion products were carried out. **Figure 15** shows micrographs of samples of solid pyrolysis products obtained at a pressure of 1 atm. (a) and 8 atm. (b).

The figure shows that in the sample obtained at a pressure of 8 atm. More pores are observed on the surface than in the sample obtained at a pressure of 1 atm.

The elemental composition of the samples is shown in **Table 8**. Measurement error of elemental composition 0.1%.

The table shows that the carbon content in the sample obtained at a pressure of 8 atm. Higher than in the sample obtained at atmospheric pressure. This is consistent with the results of the studies described in [10, 11].

The bulk density of the obtained solid product was 70.85 kg/m^3.

The total pore volume is 3.985 cm^3/g.

The values of the specific surface of the samples were determined. For solid pyrolysis products obtained at 1 atm., These indicators were: 50.31 m^2/g. For the products obtained at 8 atm., They are equal to: 664.99 m^2/g. The iodine content in the samples of solid pyrolysis products was not observed.

All characteristics of the obtained solid products of thermochemical conversion of wood meet the requirements of the standards for activated carbon.

The results of testing an experimental installation for thermochemical conversion of plant biomass under pressure showed the effectiveness of using such equipment to obtain highly porous carbon materials, the parameters of which meet the requirements for activated carbon. The biomass pyrolysis temperature was 650 °C, the superheated steam temperature was 760°C. The total time of the activated carbon production process was 4 hours and 40 minutes. Activation time 2 hours 30 minutes. The mass yield of the obtained solid product corresponds to 15.5%. According to the indicators revealed, the resulting product meets the requirements of the standards, which makes it possible to draw a conclusion about the possibility of using the selected technological scheme for creating industrial samples of pyrolysis equipment for producing activated carbons from woody biomass.

5. Conclusion

The paper describes the methods of obtaining wood (highly porous carbon materials) based on plant (wood) raw materials, and the equipment used to implement these processes, the use of charcoal. The conclusion is made about the advantage of obtaining charcoal under excess pressure.

An experimental study of pyrolysis of birch chips on a laboratory bench in the pressure range from 0.1 to 0.7 MPa showed that an increase in pressure at which pyrolysis is carried out promotes an increase in the mass yield of charcoal and an increase in the carbon content in it. At the same time, there is a decrease in the calorific value of the resulting charcoal, which is due to the release of oxygen-containing components from the porous structure of coal, which have a higher calorific value in comparison with pure carbon.

A description of the developed and manufactured experimental setup is presented, which consists of a steam generator, a superheater, a pyrolysis and activation chamber, a cooler with a heat exchanger with forced convection, which are interconnected sections. Experiments were carried out confirming that excessive pressure increases the mass yield of solid products of thermochemical conversion of plant biomass. Were determined the physical parameters of the obtained solid products of pyrolysis and physical activation at an excess pressure of 0.8 MPa.

Author details

Sergey Vasilevich[1] and Dmitry Degterov[2*]

1 Belarusian State Academy of Aviation, Belarus

2 Institute of Power Engineering of the National Academy of Sciences of Belarus, Belarus

*Address all correspondence to: ddegterov@mail.ru

IntechOpen

© 2021 The Author(s). Licensee IntechOpen. This chapter is distributed under the terms of the Creative Commons Attribution License (http://creativecommons.org/licenses/by/3.0), which permits unrestricted use, distribution, and reproduction in any medium, provided the original work is properly cited.

References

[1] New Handbook of Chemist and Technologist. Raw materials and products of the industry of organic and inorganic substances. Part II, – SPb.: ANO NPO Professional. 2007;**2005**:1144

[2] Market overview of activated (active) carbon in the CIS. An association of independent experts in the field of mineral resources, metallurgy and chemical industry. M. In: 193 p. 2019

[3] Nikolas Hagemann, Kurt Spokas, Hans-Peter Schmidt, Ralf Kägi, Marc Anton Böhler and Thomas D. Bucheli. Activated Carbon, Biochar and Charcoal. Linkages and synergies across pyrogenic Carbon's ABCs. Water. 2018;**10**:182. DOI: 10.3390/w10020182

[4] Klason P, Heidenstam G, Norlin E. Untersuchungen zur Holzverkohlung. I. Die trockene Distillation der Cellulose. Z. Angew. Chem. 1909;**25**:1205-1214

[5] Klason P, Heidenstam G, von Kiefer NEU z HIID t D d H, Fichte B u BZAC. Bd. 1910;**26**:1252-1254

[6] Mok WS-L. Antal M.J. Jr. Effecs of pressure on biomass pyrolysis. I cellulose pyrolysis products. Thermochimica Acta. V. 1983;**68**: 155-164

[7] Wang L, Trninic M, Skreiberg O, et al. Is elevated pressure required to achieve a high fixed-carbon yield of charcoal from biomass. Part 1: Round-Robin results for three different corncob materials. Energy Fuels. 2011;**25**: 3251-3265

[8] Basile L, Tugnoli A, Cozzani V. The role of pressure in the heat of pyrolysis of a Lignocellulosic biomass. Chem. Eng. Trans. 2015;**43**:451-456

[9] Mok WS-L. Antal M.J. Jr. Effecs of pressure on biomass pyrolysis. I cellulose pyrolysis products. Thermochimica Acta. V. 1983;**68**:155-164

[10] Vasilevich SV, Malko MV. Bogach. N., Degterev D.V., Asadchiy A. N. Study of the process of obtaining charcoal by pyrolysis under pressure. Vesti NAS of Belarus. Ser. fiz. tehn. navuk. 2017;**3**: 64-71

[11] Vasilevich SB, Malko MV, Degterov DV, Asadchiy AN. Computational study of the yield of solid products of wood pyrolysis at elevated pressure. Energy. Proceedings of higher educational institutions and energy associations of the CIS. 2020;**63**(3):253-263

Chapter 2

Catalytic Fast Pyrolysis of Lignocellulosic Biomass to Benzene, Toluene, and Xylenes

William H. Gong

Abstract

Catalytic Fast Pyrolysis is a rapid method to depolymerize lignocellulose to its constituent components of hemicellulose, cellulose, and lignin. The pyrolysis reaction in absence of oxygen occurs at a very high heating rate to a targeted temperature of 400 to 600 °C for very short residence time. Vapors which are not condensed and are then contacted with a catalyst that is efficient to deoxygenate and aromatize the pyrolyzed biomass. One class of highly valuable material that is produced is a mixture of benzene, toluene, and xylenes. From this mixture, *para*-xylene is extracted for further upgrading to polyethylene terephthalate, a commodity polyester which has a demand in excess of 80 million tonnes/year. Addressed within this review is the catalytic fast pyrolysis, catalysts examined, process chemistry, challenges, and investigation of solutions.

Keywords: CFP, Catalytic Fast Pyrolysis, BTX, Benzene, Toluene, Xylene, Lignocellulose, Cellulose, Hemicellulose, Lignin, Minerals

1. Introduction

para-Xylene (pX) is a key raw material in the production polyethylene terephthalate (PTA) before it is combined with mono ethylene glycol (MEG) to produce polyethylene terephthalate (PET), one of the largest and fastest globally growing polymers. Demand for pX is driven by the global polyester growth. Global polyester consumption is currently around 80 million tonnes/year. Using application of PET in fiber, this polyester is produced in 57.7 million metric tonne in 2019. Since 2008, the average growth in production has an annual increase of 5.9% (see **Figure 1**) [1]. Industry forecasts are that polyester will continue to grow at approximately 4% annually. While there is also increased focus on recycling polyester (a.k.a. rPET), most industry experts also see continued demand growth in virgin material due to limits on the availability of recycled polyester and need for virgin material to meet quality and specifications of many applications. Over the span between 2008 and 2019, rPET produced has ranged from 2 to 9% of total global PET for fiber applications. Currently, PTA is produced from pX where 0.58 tonnes of pX are required to produce one tonne of polyester, resulting in global demand for pX of about 50 mm tonnes (**Figure 2**).

Figure 3 displays the supply chain to produce PET. Virtually all PET is produced from fossil fuel resources. pX is produced from mixed xylenes that are derived from crude oil, and MEG is produced from ethylene primarily derived from natural gas. There is a commercial bio-based route to MEG that starts from ethanol produced

IntechOpen

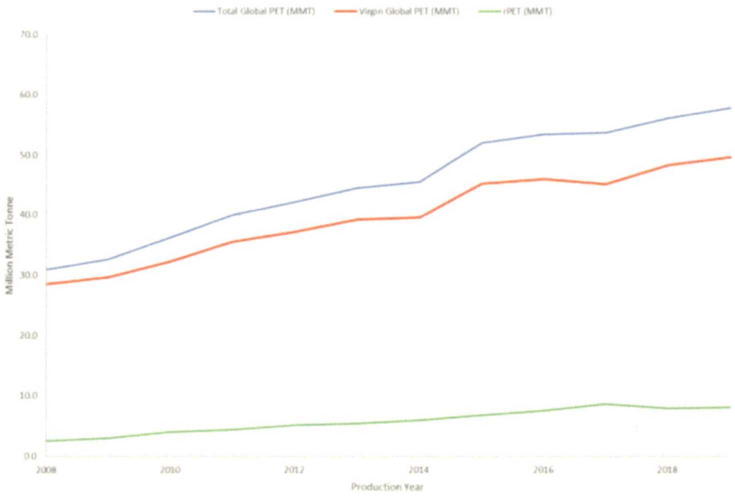

Figure 1.
Annual PET Fiber produced from 2008 to 2019.

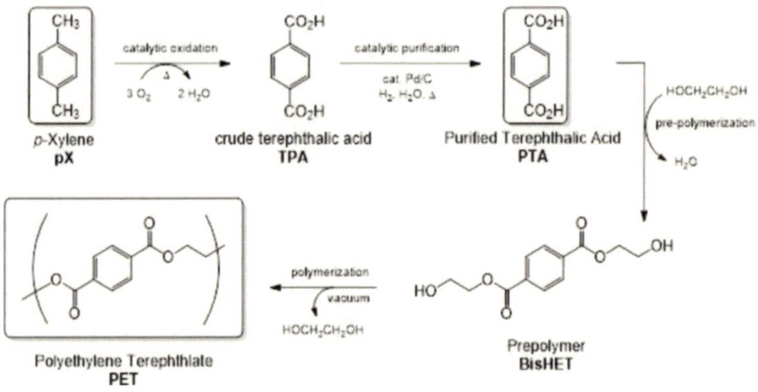

Figure 2.
Pathway from p-xylene to polyethylene terephthalate (PET).

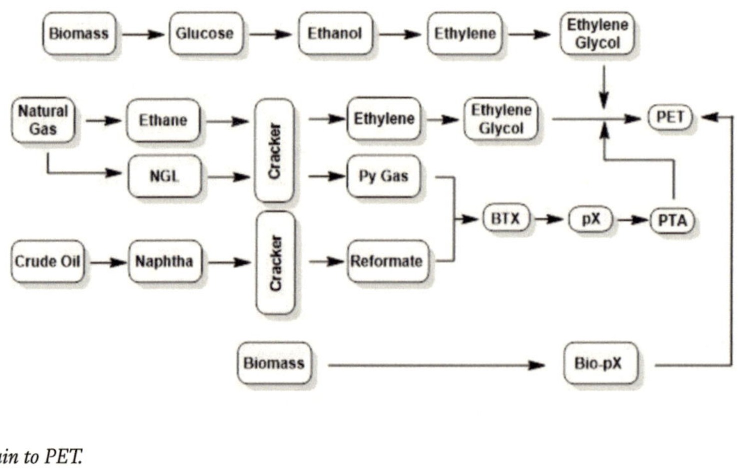

Figure 3.
Supply chain to PET.

from biomass by dehydration to ethylene, oxygenation to ethylene oxide, and hydrolysis to MEG. Bio-based MEG is used to produce a partial bio (~30%) PET. At the current time, there are no commercial routes to bio-pX.

Conventional pX technology is based on the isomerization of mixed xylenes which has been practiced commercially for decades. The mixed xylenes are produced by catalytic reforming of naphtha, which produces a C_8 stream that includes the three xylene isomers and ethylbenzene. Other routes to mixed xylenes are from the pyrolysis gasoline stream produced in a naphtha steam cracker. The main commercial routes used to separate the pX isomer are crystallization, which takes advantage of the different freeze points of the ortho-, meta- and para- isomers, or via selective adsorption separation processes.

As a result of growing concerns about climate change, companies, governments, and consumers are increasing their focus on ways to reduce the carbon footprint of materials and offer more sustainable products. Consequently, companies are looking at new and innovative technologies to produce pX from renewable biomass resources rather than from crude oil, to enable production of a 100% bio polyester product. This report focuses catalytic fast pyrolysis (CFP) of lignocellulose to bio-oils which are refined to bio-pX with the objective to reduce carbon footprint in comparison to petro-pX. There are other processes currently being developed for bio-pX, but a discussion of these processes are beyond the scope of this paper.

As the world is emerging from the global pandemic of 2020–2021, increasingly, big companies have turned their focus on sustainability and establishing aggressive targets to reduce the carbon footprint of their products and over time move toward a net-zero carbon profile. It was earlier in 2009 Coca-Cola announced their rollout of the PlantBottle™ [2]. Coca-Cola's mission is to reduce reliance on non-renewable resources, and reduce CO_2 emission in polyethylene terephthalate (PET). Currently, only 30% of the PET in PlantBottle™ is made from bio-MEG. A commercialized process for bio-PTA process would enable Coca-Cola (and others) to produce 100% bio-PET PlantBottle™. Advantageously, bio-PET production would be a "drop-in" into existing PET assets. By 2012, Coca-Cola, Ford, Heinz, Nike, and Procter & Gamble founded the "Plant PET Technology Collaborative" and announced their mission [3]:

"PTC members are committed to supporting and championing research, expanding knowledge and accelerating technology development to enable commercially viable, more sustainably sourced 100% plant-based PET plastic while reducing the use of fossil fuels". Similarly, Danone, Nestlé Waters, Origin Materials and PepsiCo formed the NaturALL Bottle Alliance in 2018. Since Coca-Cola's 2009 announcement, many have gravitated to the challenge to develop bio-PTA.

The reduced carbon footprint of bio-pX, and consequently bio-polyester, can contribute to meeting these carbon reduction goals for companies that participate in the polyester value chain, and this is the key business driver behind the increasing interest in routes to bio pX. The bottom line: once bio-pX is available, oxidation of bio-pX to bio-PTA would produce 100% bio-PET.

2. Biomass catalytic pyrolysis and upgrading to aromatics

Anellotech, a Pearl, NY start-up chemical company founded in 2008, has developed a feedstock flexible, in situ catalytic fast pyrolysis process to convert low-cost lignocellulosic biomass solids to a range of hydrocarbons containing valuable benzene, toluene, and xylenes (BTX). These monoaromatics are further refined to produce bio-pX which are valuable feedstocks for bio-PTA [4–13]. In situ CFP technology was originally developed by Dr. George W. Huber, Harvey Spangler Professor of Chemical Engineering at the University of Wisconsin-Madison.

Anellotech has partnered with Suntory, a global consumer beverage company. The technology has been jointly developed by IFPEN, Axen (a subsidiary of

IFPEN), and Johnson Matthey. IFPEN is tasked with scale-up, Axen for technology marketing, and Johnson Matthey for catalyst development. On May 6, 2019, Anellotech announced successful six months of operations of their Bio-TCatTM continuous process in their Silsbee, TX pilot plant. The bio-pX process yield ranged from 22 to 24%, 98% C_6+ aromatics. Currently, Anellotech is seeking partners and funding to construct their first commercial CFP based bio-pX plant (500 bone-dry tonnes/day of loblolly pine and target production of 40,000 tonnes BTX and C_9+ aromatics per year).

Pyrolysis is an extraordinarily complex, thermochemical reaction which can be operated in slow or fast mode to yield products ranging from solids to liquids and gas. Pyrolysis is challenged by high feedstock complexity and the multiphasic nature of this process. More than 300 compounds are found in an acidic, aqueous, unstable, and viscous heterogeneous mixture (a.k.a. pyrolysis oil) of esters, alcohols, ketones, aldehydes, and carboxylic acids [14–16]. By themselves, these oxygenates have little fuel value. However, they can be catalytically upgraded to aromatic hydrocarbons, particularly bio-pX. The upgrading catalyst must maintain stability and tolerate wide range of oxygenates [17]. CFP can convert the entire plant, not just sugars, which makes this thermochemical process advantageous to fermentation [18].

Zeolites, particularly HZMS-5, have proven to be effective catalysts for pyrolysis oil upgrading to benzene, toluene, and xylenes (BTX). Crystalline zeolites have well-defined microporous structures with pore sizes ranging from 5 to 12 Å. Zeolites are acidic materials with controllable acid site densities. CFP must operate under O_2-free, optionally ambient pressure, and does not require H_2 (a.k.a. hydroprocessing), but operating under H_2 atmosphere might offer advantage of decreased coking (discussed later in this chapter).

Temperature is an important parameter. In CFP, a moderate temperature of approximately 450 to 550 °C is optimum to produce mainly liquids, but the heating and subsequent cooling rates must be extremely high. The biomass must be ground to a finely divided material with averaged particle sizes of less than 3 mm. The cost of the process is inversely proportional to the averaged particle size. The pyrolysis vapor must condense quickly to avoid unwanted secondary reactions such as cracking. Alkali and alkaline earth metal ions in biomass feed pose substantial challenges to the process (addressed in Pretreatment) [17, 19–33].

3. In situ vs. ex situ CFP

There are two process schemes that can be employed in CFP: in situ and ex situ [32, 34]. Pyrolysis and catalytic upgrading that occur in the same zone is in situ. This "one-pot" process is lower in cost than ex situ. However, in situ CFP is sensitive biomass feed content of inorganics, and minimum catalyst/biomass ratio where 10:1 gives highest yield of aromatics [35].

Pyrolysis and catalytic vapor upgrade that occur in separate zones (two-stage) is ex situ CFP. This mode of operation affords the advantage to exclude biochar and metallic ionic impurities from the catalyst which can extend catalyst lifetime. Additionally, the temperature of pyrolysis and vapor upgrade can be independently controlled, an option not available for in situ CFP. BioBTX operates an ex situ pyrolysis process, and has developed an integrated cascading pyrolysis reactor. Their two-stage process excludes non-volatile impurities from entering catalyst pores and this provides the benefit of improved catalyst vapor contact [26].

A comparison of in situ and ex situ CFP of hybrid poplar at 700 °C was conducted in a microreactor. In situ CFP produced 5.4% olefins and 26.1% aromatics.

While operating in an ex situ mode, up to 17.4% olefins and 18.9% aromatics were made. The carbonaceous material yields for in situ and ex situ were 31.3% and 18.6%, respectively. Temperature strongly influenced the product distribution in ex situ. A comparison of both modes of operations in a mini-plant for CFP of pinewood was conducted [36]. An analysis of ex situ economics has also been published [37].

4. Mechanism

Fast pyrolysis rapidly depolymerizes lignocellulose to its constituents: hemicellulose, cellulose, and lignin. Under these conditions, cellulose is depolymerized to glucose, hemicellulose to C_5 sugars, and lignin to phenolic compounds [22, 38, 39]. Once glucose is made, it is quickly dehydrated to anhydrosugars which are then converted to furanics. Furanics are diffused into zeolite pores where it is combined with in situ olefins and undergoe Diels Alder cyclization and dehydration to aromatics. Zeolite imparts a "confinement effect" for chemicals undergoing dehydrocyclization within the catalyst interior and once formed, monoaromatics exit the catalyst [40]. At a molecular level, at 450–600 °C, the furanic oxygen coordinates to Al of the zeolite. Deoxygenation of organics result in elimination of CO_2, CO, and H_2O. As a test of this hypothesis, furan and olefins were used as models in HZSM-5 catalyzed upgrading. HZSM-5 has a three-dimensional pore structure with intersecting systems of 10-membered ring pores. The dimension of ring pores is nearly ideal for production of pX: 5.1 x 5.5 Å (sinusoidal pores), and 5.3 x 5.6 Å (straight pores) [41]. HZM-5's utility in conversion of oxygenates to aromatics was demonstrated by Mobil's commercial methanol-to-gasoline process. Within HZSM-5, ethylene and furan did not yield benzene, but replacement of ethylene by propylene led to toluene [42]. Likewise, 2-methylfuran and propylene produced xylenes. In the absence of olefins, Diels Alder dimerization and dehydration of furan results in benzofuran (Eq. (1)). Graphitic coke was also deposited onto the catalyst, which contributed to catalyst deactivation in about 30 minutes. When the upgrading temperature was increased to 650 °C, the selectivity was shifted to olefins.

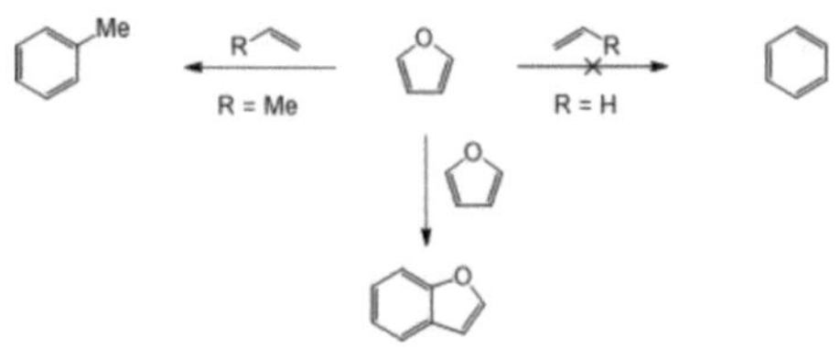

(1)

Other investigators have examined conversion of methylated furan over Ga/HZSM-5 [43]. Even though model compounds were used, the reaction network was extremely complex due to numerous secondary reactions at 300–500 °C. Coking became a dominant reaction at temperature above 500 °C.

A "technical catalyst" such as extruded Al_2O_3 -HZSM-5 was explored in CFP of pine wood and cellulose at 500 °C [44]. The focus of the study was on effect of pyrolysis temperature. The type of coke made was highly dependent on the feedstock. Yet, even by 2020, the mechanism of CFP remained elusive. Gaps still exist between theoretical models and applied pyrolysis. Thus far, there is still insufficient information generated from a single biomass source [45]. High aromatics selectivity and strong resistance to catalyst deactivation are keys to the success of CFP for bio-pX. Therefore, extensive efforts have been directed toward zeolite modifications.

5. Zeolites for CFP

Zeolites are microporous crystalline materials of SiO_2 and Al_2O_3 that have Brønsted and Lewis acidic sites [17, 46]. HZSM-5 also have ordered structure microporous structures. In the refining process, zeolites are integral to crude oil conversions in the gas phase. The feed for these refining processes are bereft of minerals, oxygenates, and water. Under these conditions, zeolite lifetimes are quite long. However, conversion of biomass occurs in a condensed, aqueous phase, which are highly challenging conditions for zeolites. Oxygenate conversions are difficult to achieve because of their tendency to coke. In addition to BTX, CFP produces a very large volume of wide variety of hydrocarbons, valuable for fuel (fuel gas, gasoline and diesel blendstock). Therefore, the literature on CFP catalyst development is quite voluminous due to interests in bio-fuels.

A comparison of HZSM-5, HY, H-BEA, and H-mordenite for pyrolysis oil upgrading has been made [37, 47]. A method to evaluate catalysts for ex situ vapor upgrading is the pyroprobe analytical pyrolyzer. This important tool provides high heating rates and means to analyze complex mixtures of products [48]. An examination of feeds ranging from glucose, xylitol, cellulose, to cellobiose was explored with catalyst candidates of HZSM-5, silicalite, H-BEA, HY, and SiO_2/Al_2O_3. Based on this study, HZSM-5 has been found to be the highest performing catalyst. Pore characteristics and shape selectivity for monoaromatics of HZSM-5 are superior to other zeolites. Another comparison study was conducted on HZSM-5, MCM-41, and HY [49]. HZSM-5 was again the most effective zeolite among this group to produce up to 33.1% BTX and 86.4% aromatics selectivity.

The silica-alumina ratio (SAR) of HZSM-5 is also an important parameter for deoxygenation reactions [50]. As the SAR is reduced, the density of Brønsted acids increases. Other zeolites of similar SAR were inferior to HZSM-5 because they were not microporous. Mesoporous zeolites were not selective for aromatics. For example, sulfated ZrO_2 is a superacid, a property important for aromatics formation, but this metal oxide was ineffective because of the absence of shape-selectivity. Coupling ZrO_2 to MCM-41 resulted in higher levels of pyrolysis vapor deoxygenation and was found to be stable against coking.

6. Zeolite modifications

Because HZSM-5 plays such a vital role in CFP, numerous investigations have been launched to determine whether modifications of the physical structure can improve aromatics selectivity and minimize coking. Modification of zeolite's pore mouth sizes can improve the catalyst for improving of BTX selectivities [51, 52]. Specifically, narrowing the pore mouth openings by chemical deposition techniques was explored [53]. The deposited chemicals also covered external acid sites, which can improve pX selectivity [5]. Phosphorous-modified HZSM-5 promoted greater space confinement and increased pX selectivity. A comparison of various deposited silicon materials was conducted by KH550, TEOS, and methyl silicone oil [54, 55]. KH550 modification of HZSM-5 increased furan conversion to 24.5% aromatics vs. 18.8% unmodified HZSM-5 (see reference [5]). Coking of the modified catalyst was also reduced from 44.1 to 26.7%. A decrease in benzene and increase in toluene and naphthalenes selectivities were observed. Chemical liquid depositions of TiO_2 to HZSM-5 also resulted in improvements to BTX yields from 17.0 to 23.5%, and reduced levels of coking [56].

The other direction to zeolite modifications is to create mesopores. Unilamellar mesoporous MFI nanosheets (UMN) were tested against another mesoporous material, Al-SBA-15 [57]. UMN has stronger acid sites which result in greater cracking and deoxygenation activities. Ketones and alcohols were converted to furanics and aromatics. Another method to produce mesopores is NaOH desilication of HZSM-5 [58]. This method was successful to create intraparticle mesoporous materials which then improved diffusion. By coupling ZrO_2 with desilicated zeolites, a more active zeolite was accompanied with longer catalyst lifetimes [19]. The resulting coke was "softer" (i.e. less aromatic) and was easier to remove. A two-stage microporous and mesoporous vapor upgrading system was examined [59]. In this case, HZSM-5 and MCM-41 (3:1) were evaluated. By increasing the MCM-41 loading, more xylenes were made than toluene. A comparison of chemical liquid deposition (CLD) and acid dealumination (AD) of HZSM-5 was conducted [60]. CLD decreased pore size while AD increased pore size. In both cases, decrease of strong acid sites occurred. CLD boosted BTX yield to 37.2% while AD produced 30.4%. In both cases there was a decrease in indene, naphthalenes, and coke precursors.

Most recently, Li et al. reported development of a core-shell catalyst of HZSM-5@MCM-41 made from an external recrystallization method [61]. HZSM-5@MCM-41 required a lower temperature of 500 °C to produce hydrocarbons, versus 550 °C for unmodified HZSM-5. Monoaromatic hydrocarbons produced from rape straw pyrolysis was as high as 11.43 wt%, which is 2.5 times higher than obtained from HZSM-5. Furthermore, HZSM-5@MCM-41 produced less graphitic coke.

7. Metal doped HZSM-5

The literature is replete with studies of metals doped HZSM-5 to investigate their effects of CFP catalyst. Zeolites metalated even at low concentrations beneficially resulted in accelerated deoxygenation by decarbonylation, decarboxylation, dehydration, and olefin aromatization of pyrolysis products. In some cases, there were reduced coke formation by decreasing selectivities to polycyclic aromatic hydrocarbons. **Table 1** summarizes the metals dopant and their references. Among these metals, gallium, iron, zinc, molybdenum, nickel, and copper increased for deoxygenations, dehydrocyclization, and aromatization.

Investigators also examined combinations of metals to determine whether there were accumulated benefits in CFP. For example, Ga and Ni are perhaps the most investigated metal additives, and they have been combined for CFP [63]. Ga-Ni reduced acidity, and reduced formation of polyaromatic hydrocarbons which could explain the reduced coking. Another combination is Cu-Ni which was examined in hydroprocessing (CFP under H_2 atmosphere) [108]. Ni-Fe produced an additive effect. Individually, Fe increased yields of monoaromatics, while Ni increased yields of naphthalenes [82]. When combined, this catalyst generated more monoaromatics and naphthalenes. The doping of these metals onto HZSM-5 increased the concentration of strong acid sites, but also resulted in more coke formation.

Molybdenum is already a highly active metal when doped in HZSM-5 in zeolites [111]. When Mg was combined with Mo in HZSM-5, higher selectivities to monoaromatic hydrocarbons and decreased polyaromatics selectivities were achieved. Phosphorous modified HZSM-5 and doped with Zn decreased external acidity of catalyst surface while the internal acid sites were left unchanged. This modification resulted in improvements in this CFP catalyst [107]. HF modification of HZSM-5 and later doped by Ni produced a catalyst that increased yields of BTX [110]. The acid modification created mesopores and reduced Brønsted acidity via dealumination.

Certain combinations of metals and nonmetals behaved as "noble-metal-like" catalysts [106]. Therefore Ni-Mo_2N was prepared and confirmed that Ni further extended performance of Mo_2N [112].

Entry	Metals/Additives	References
1	Ga	[24, 26, 62–79]
2	Fe	[26, 63, 66, 75, 80–88]
3	Pt	[72, 89–92]
4	Zn	[26, 63, 77–80, 85, 86, 93–96]
5	Mo	[24, 63, 76, 77, 90, 91, 97–100]
6	Ni	[24, 26, 63, 67, 73, 82, 87, 90, 92, 96, 100–102]
7	Co	[26, 63, 73, 80, 85, 90, 92, 96, 102]
8	Cu	[63, 92, 96, 97, 101, 103]
9	Mn	[63]
10	La	[86, 93, 94]
11	Mg	[93, 94, 96]
12	Ce	[93, 94]
13	Ca	[86, 96, 104]
14	Pd	[73]
15	Sn	[105]
16	WO_3	[91]
17	Mo_2N	[106]
18	W_2P	[106]
19	MoP	[106]
20	WP	[106]
21	Zn/P	[107]
22	Ga-Ni	[63]
23	Ni-Cu	[108, 109]
24	Ni-HF	[110]
25	Fe-Ni	[82]
26	Mo-Mg	[111]
27	Ni-Mo_2N	[112]
28	Ce-Zr	[113]
29	Ga-Mo	[114]
30	Fe-Ca	[115]

Table 1.
Metals and additives to HZSM-5 for CFP.

8. Catalyst deactivation

In almost every publication concerning CFP catalyst or process development, investigators have identified catalyst deactivation to be the single greatest challenge to this process. The catalyst lifetime can be as short as minutes. There are three types of deactivations and of which, two are irreversible deactivations. The most common form of deactivation results from coke buildup on acid sites within the zeolite pores and renders the catalyst inactive [116]. Coking is most rapid when the catalyst/biomass ratio is less than one and particularly at low SAR. The second type of deactivation results from metal oxides deposited in zeolite which can result in irreversible deactivation if these salts are not removed [117]. Thirdly, zeolite dealumination can occur and result in irreversible deactivation.

Coke deactivation most rapidly occurs among highly acidic zeolites of low silica/alumina ratio (SAR) [118]. Ironically, low zeolite SAR are most active for aromatics [82, 119]. Individual lignocellulosic components caused deactivation differently [120]. Depolymerized lignin mainly deactivates by coking on zeolite surfaces because lignin monomers (mostly phenolics) cannot diffuse into the microchannels of HZSM-5 [121]. Cellulose pyrolysis products are responsible for coking within the interior of the zeolite and this leads to occlusion [121]. A study into deactivation mechanism was conducted from in situ CFP pine wood [121]. Fresh catalyst produced mainly aromatics and olefins with no detectable levels of oxygenates. As the campaign progresses, the catalyst/biomass ratio began to decrease, and oxygenates began to breakthrough, including phenolics. When the catalyst/biomass ratio reached 1:3, complete deactivation will have occurred.

Not all oxygenates deactivate HZSM-5 at the same rate [122]. One team examined the effects of upgrading ethanol-only, ethanol+acetic acid, ethanol+ethyl acetate, and ethanol+acetaldehyde. Acetic acid in the vapor strongly adsorbed onto zeolite surfaces and promoted formation of polyaromatic hydrocarbon precursors to coke. Acetaldehyde deactivated 10X more than ethanol. While ethyl acetate did not directly adsorb onto catalyst surfaces, the reaction conditions rapidly hydrolyzed ethyl acetate to acetic acid which then immediately deactivated the catalyst.

An extruded HZSM-5 (a.k.a. "technical catalyst") containing an alumina binder was also examined in its deactivation in ex situ CFP of cellulose and pinewood [44]. CFP of cellulose generates smaller oxygenates, which diffused into the zeolite to produce a "catalytic coke", made via a ring-growth mechanism involving H-transfer at temperatures above 200 °C [123]. Initial formation of monoaromatics led to formation of polyaromatic hydrocarbons, coke precursors. It is not surprising to learn that the coking mechanism of oxygenates is similar in pathways to coke formation in petroleum refining [118]. Lignin forms "thermal coke" from condensation of lignin monomers, and easier to remove because they accumulate on the surfaces of the zeolite [118]. High temperature combustion readily regenerates coke-deactivated HZSM-5.

The presence of water liberated from the pyrolysis can accumulate in the catalyst [118, 124]. Later, at high regeneration temperature, steaming can result in dealumination of the zeolite via leaching. In addition to removal of strong acid-site aluminum, there can be a loss of surface area even after one regeneration cycle [44, 51]. As a further complication, during the regeneration, hot spots can appear and irreversibly damage the catalyst. Successive regeneration increases the risk of further dealumination and reduction in micropore volumes, particularly at high regeneration temperature of 650–670 °C [118]. Therefore, these investigators recommended limiting the O_2 concentration to 15% and include 5% steam to preserve catalyst lifetime [125]. Use of this method was found to preserve catalyst stability for up to 30 regeneration cycles. The formation of aromatics and olefins was stable at about 31.3% [125].

Alkaline and alkaline earth metal ions (AAEMs) are part of the plant's nutrient to enable the plant to grow [15, 126]. Lignocellulose is comprised of four different types of materials: lignin, cellulose, hemicellulose, and inorganics. AAEMs and other metal salts are represented by the inorganics fraction. A measurement of ash content provides a direct correlation of mineral concentration in the feed. Mechanistically, alkali metal salts also ion-exchanged with the proton of Brønsted acid sites and this reduces zeolite activity [118]. This was demonstrated from an on-purpose potassium ion exchange with HZSM-5 (up to 2.85 wt%) which resulted in drastic reduction of catalyst acidity, leading to lower yields of carbons [127]. Within the pyrolysis process itself, AAEMs can catalyze unwanted secondary reactions, including cracking, decarboxylation, and dehydration, resulting in a loss of liquid yields [34]. Carbon oxides and water were increased by 10% due to these inorganic components, changing the product distribution [118, 128]. Among the metal ions that are highly detrimental to the process are K > Na > Ca, and the concentration of potassium in the feed is inversely proportional to aromatics [129]. Furthermore, Xiao et al. studied the effect of potassium salts on CFP of a model feed, 2-methylfuran [130]. The experiments were conducted by deposition of potassium salts into HZSM-5. Neutral KCl was less harmful than basic K_2CO_3 which produced lower yields of aromatic hydrocarbons and gases. The negative effects were due to loss of acidity and erosion of physical structure of HZSM-5, particularly when the K_2CO_3 concentration was as high as 1.0 wt%.

Magnesium is also present, but it is not as harmful as the other metal ions. Other inorganic ions that need to be monitored include Cl, S, P, and N [131]. Particularly, CFP of guayule, a perennial woody shrub, can produce upgraded products containing 80 to 700 ppm sulfur which can be detrimental to any sulfur sensitive downstream processes [132].

Because of the nature of the process, it is in situ CFP that is most sensitive to AAMEs in the feed. Even at levels as low as 0.1 wt% in the feed, AAEMs can greatly impact the pyrolysis process. Aggravating this situation is the variation of feedstock which also varies the inorganic component concentrations. Switchgrass as a feed results in high ash content [118, 131]. Versus pinewood (0.49 wt%), Switchgrass can contain around 2.6 wt% ash comprised of metal oxides. Crop selection is crucial to preserve performance of CFP [15, 133]. Feed pretreatment can be an option to minimize the deleterious effects of AAEMs.

9. CFP feed pretreatment

Two main strategies have been published to combat the challenges of pyrolysis: torrefaction and acid-washing of lignocellulose. Torrefaction is low temperature, slow pyrolysis of less than 60 minutes, and installed upstream to CFP. During torrefaction, removal of undesirable acetic acid and guaiacol from the feed improves the downstream CFP. Typical temperature of torrefaction ranges from 250 to 300 °C [112]. When torrefaction was tested on pinewood conversion at 250 °C, the resulting CFP step produced 30% more aromatics. Other feedstocks such as corn cobs produced a lower O/C ratio in the CFP product [134].

Acid-washing of finely divided lignocellulose can extract AAEMs from the feed. The wash step is conducted under very mild conditions, but it is necessary that thorough drying is conducted prior to CFP [51]. For example, rice husks are washed at 30 °C by acetic acid [84, 135]. Use of acids such as acetic acid or mineral acids can add significant variable costs to the process. One investigator recommended using acetic acid recovered from the pyrolysis process for the wash. Extreme aqueous washing must be avoided as the extraction can result in undesirable changes in hemicellulose and cellulose. Formation of sticky materials can coat catalyst bed particles and risk defluidization [136]. Aqueous HCl washing of the spent catalyst to remove deposited minerals is also necessary to maintain catalyst lifetime [118].

A comparison of both processes was studied to determine their effects on aromatics yield. Acid-leaching promoted the formation of levoglucosan. Torrefaction promoted the formation of catechols and guaicols. It was determined that acid leaching + CFP made less aromatics than torrefaction + CFP [137]. Both processes were conducted in series in the CFP of rice husk [135]. Lower levels of volatiles were made in the pyrolysis result in lower yields of oxygenates and higher yields of aromatics. However, severe torrefaction conditions due to longer residence time can result in cross-linking of lignocellulose, resulting in more charring and reduced formation of aromatics. A combination of both processes did result in higher yields of products [136].

One team of investigators described development of ultrasonic pretreatment [138]. This mechanical method excludes use of toxic chemicals. Ultrasound promotes cleaving of α-O-4 and β-O-4 linkages and overall decreased crystallinity of lignocellulose and increased yield of pyrolysis oils. CFP can accommodate a wide

Feedstock	References
Beatle killed trees	[139]
Pine wood	[36, 60, 67, 97, 101, 106, 111, 121, 128, 140–146]
Pine bark	[58]
Pinecones	[145]
Pinyon juniper	[58]
Beech wood	[58]
Maple wood	[148]
Hybrid poplar	[58, 147]
Yellow poplar	[76]
Red oak	[147]
Oakwood	[149]
Quercus Mongolia	[74]
Eucalyptus	[63, 68, 87]
Willow wood	[66]
Guayule Bagasse	[150]
Corn cobs	[141, 151]
Corn stover residue	[58]
Corn stalk	[88]
Straw	[141, 152]
Sunflower Stalk	[153]
Switch grass	[58, 127, 131]
Mandarin residue	[72]
Citrus unshiu	[119]
Rice Husk	[84, 109, 135, 154–156]
Bamboo sawdust	[56, 157]
Giant cane	[158]
Sugarcane bagasse	[147]
Wheat straw	[91, 98, 159]
Rape straw	[61]
Xylitol	[160]
Cotton stalk	[80]
Jatroha residues	[73]
Miscanthus	[57, 161]
Peanut shells	[162]

Feedstock	References
Microalgae	[70, 140]
Seaweed	[140]
Fish discards	[140]
Guaiacol	[52]
Acetic Acid	[52]
Furanics	[5, 38, 42, 43, 52, 54, 78, 130, 148, 163–168]
Glucose	[148, 160, 169, 170]
Lignin	[121, 141, 171]
Hemicellulose	[141, 171]
Cellulose	[44, 57, 80, 113, 120, 121, 141, 160, 161, 171–174]
Cellbios	[160]
Glycerol	[175, 176]

Table 2.
Feedstocks used in CFP.

variety of feedstocks. **Table 2** summarizes the feedstocks examined, which include woody plants of varying levels of lignin, cellulose, and hemicellulose. The most frequently studied woody biomass for pyrolysis was pinewood. Also included in this table are references to studies of model feeds.

Glucose is one of the intermediate products made in pyrolysis [169]. This monosaccharide is rapidly converted to anhydrosugars before they are dehydrated to furanics and subsequently to aromatics. Sophisticated isotopically labeled glucose experiments permitted an elucidation of the mechanism of glucose conversion [170]. Single ring aromatics contained ^{13}C labels and later these labels began to appear in naphthalene rings, evidence for a ring growth mechanism from monoaromatics to polyaromatics. Other studies included a study of the conversion of other model feeds to aromatics under identical conditions [80, 121, 140–142, 147, 160, 161, 171]. Glycerol CFP is of interest because of its availability from the production of fatty acid methyl esters for bio-diesel [175, 176].

10. Catalytic Co-pyrolysis (CCP)

Investigators recognized that pyrolysis produced improved product quality when a second feed was added to the process [22, 24]. The co-feed can include coal, plastics, tire, and sludge. N.Y. Chen et al. first reported in 1986 that the hydrogen to carbon effective ratio (H/Ceff) can predict the outcome of pyrolysis [177]. This ratio is calculated according to the following formula (Eq. (2)).

$$\frac{H/C_{eff} = H - 2(O)}{C} \tag{2}$$

When this ratio is less than one, coke is expected to be made in high yields. For example, glucose has a ratio of zero, sorbitol has 0.333, glycerol has 0.667, xylitol has 0.400 and petroleum is greater than two [160]. As a feed, lignocellulose is highly deficient in hydrogen. It might be plausible to increase this ratio by co-feeding a material rich in hydrogen [19, 27]. By increasing this ratio, not only is coke reduced, but higher yields of aromatics and olefins result. Other means to increase this ratio is to pretreat the feed by hydrogenolysis [178]. An ex situ co-pyrolysis strategy to increase BTX yield was cleverly conducted by removing BTX from the product stream to leave behind polyaromatic hydrocarbons. This high molecular weight material was then hydrogenated to produce polycyclic aliphatic hydrocarbons. This hydrogenated fraction was then co-pyrolyzed with glycerol to produce higher yields of BTX [176].

The upgrading of furanics to aromatics was benefitted by the co-addition of olefins [42, 119, 163, 177, 179, 180]. While co-addition of ethylene and propylene was effective to produce more aromatics, investigators recognized that the process must now incorporate flammable gases. An alternative to olefins utilization was to co-add methanol to the pyrolysis. Under the upgrading conditions, methanol to in situ olefins can occur to produce the dienophiles required for furanic Diels Alder reactions. Therefore, 2-methylfuran (2-MF) and methanol co-addition resulted in boosting aromatics yield while coke was reduced [168]. When the MeOH/2-MF ratio was increased from zero to 3:1, 2-MF conversion was increased from 39.8% to as high as 96.5% at an optimal ratio of 2:1. Of course, not all furanics were equally converted. A comparison of 2,5-dimethylfuran (DMF), furan, and furfural revealed that electron-rich DMF quickly reacted while the electron-deficient furfural lagged behind: DMF > furan > furfural [94]. In addition to aromatics, CO_2, CO, and other olefins were also made at 400–600 °C [38]. Unavoidably, formation of graphitic coke on catalyst surfaces resulted in its deactivation within 30 minutes. Methane was examined as a co-feed in the catalytic upgrading [164]. This co-feed was beneficial, an indication that methane could be activated by this catalyst to yield more aromatics, but coke formation also increased.

CCP was applied to a mixture of sawdust and methanol. This combination reduced coke and char and improved aromatics yields [77, 93, 181]. However, the researchers could not exclude the possibility that a background methanol to aromatics process was also occurring. Another team examined CCP of other co-feeds such as methanol, polyethylene terephthalate (PET), and calcium formate with pinewood [182]. Of these three, calcium formate was found to be most effective at calcium formate/pinewood of 3:1. In fact, co-feeding calcium formate did not produce a large increase in aromatics vs. control (10.65% vs. 9.23%), but this co-pyrolysis reduced the polyaromatic hydrocarbons yield (1.94% vs. 2.49%). The authors speculated that the calcium salt helped to maintain catalyst activity within the hydrocarbon pool of the zeolite.

A patent was granted to Anellotech for co-feeding oxygenates such as acetone into pyrolysis to enhanced pX yield [183]. The oxygenates utilized are low value byproducts from other industrial processes. Such oxygenates include acetone from the phenol process, furfural, hexanol, and hexanoic acid from other sources. Furfural was less effective, but the formation of xylenes was enhanced by hexanol and hexanoic acid co-additions. Other oxygenates include fermentation products [184]. Co-pyrolysis of citrus unshiu peel and alcohols, ketones, and furanics has been reported [119]. In this case, two different zeolites were compared: HZSM-5 (23) and HBEA (25). The key difference between both zeolites was that HBEA produced more coke.

Tail Gas Reactive or Recycle Pyrolysis (TGRP) is an interesting process that recycles non-condensable pyrolysis gases to the pyrolysis step [150, 166, 185, 186]. Recycling this gas to pyrolysis provides a reductive, low acidic atmosphere which benefits CFP by increasing the H/Ceff. The condensable gas fraction was separated, including unwanted carboxylic acids and furfural which are detrimental to vapor upgrading. Up to 10X more BTX was produced when TGRP was incorporated in comparison to CFP-only.

Other non-conventional processes could also improve CFP yields. One example is addition of ball bearings to the catalyst bed to increase residence time which provided higher hot surface areas [187]. This modification increased cracking reaction and produced 3X more BTX. The downside is formation of more polyaromatic hydrocarbons. An alternative to thermal heating, microwave energy can also be used. Microwave assisted pyrolysis (MOP) does not require agitation and fluidization [56, 188].

11. Outlook for CFP

CFP is a rapid process to produce a high volume of hydrocarbons of which BTX is a small fraction. CFP must be co-located with a petroleum refining complex to take in bio-naphtha which could be directed to bio-gasoline production. The refining complex must also include a pX extraction process to separate pX from mX and oX as well as a process to isomerize the pX lean stream. Every means to improve CFP should be taken, including pretreatment, torrefaction, co-pyrolysis with hydrogen-rich feed, and tail gas recycled pyrolysis to increase BTX yields. Investigations into development of an even more active metalated HZSM-5 catalyst to increase monoaromatics selectivity and resistance to coking will remain subjects of intense interests.

Author details

William H. Gong
Independent Technical Consultant, Lombard, Illinois, USA

*Address all correspondence to: whgong@comcast.com

IntechOpen

© 2021 The Author(s). Licensee IntechOpen. This chapter is distributed under the terms of the Creative Commons Attribution License (http://creativecommons.org/licenses/by/3.0), which permits unrestricted use, distribution, and reproduction in any medium, provided the original work is properly cited.

References

[1] "Preferred Fiber & Materials Market Report 2020", *Textile Exchange*, https://textileexchange.org/wp-content/uploads/2020/06/Textile-Exchange_Preferred-Fiber-Material-Market-Report_2020.pdf;

[2] Coca-Cola Press Release. Coca-Cola Introduces Plant-Based Plastic Bottles, May 18, 2009.

[3] P.B. Smith, Bio-Based Sources for Terephthalic Acid. In: Green Polymer Chemistry: Biobased Materials and Biocatalysis, ed. H.N. Cheng, American Chemical Society, Washington, DC, 2015, pp. 453-469.

[4] Tullo A. Anellotech Advancing Bio-Aromatics. Chem. Eng. News. 2016, 94.

[5] Cheng YT, Wang Z, Gilbert CJ, Fan W, Huber GW. Production of p-Xylene from Biomass by Catalytic Fast Pyrolysis using ZSM-5 Catalysts with Reduced Pore Openings. Angew Chem Int Ed. 2012; 51: 11097-11100.

[6] https://www.ifpenergiesnouvelles.com/article/bio-paraxylene-production-path-100-bio-based-bottles

[7] https://anellotech.com/sites/default/files/Anellotech_Technology%20Update_2019.05.06.pdf

[8] Scott A. Investors Chase Bioaromatics. C&EN. February 24, 2020, 18-19.

[9] Huber GW, Cheng YT, Carlson T, Vispute T, Jae J, Tompsett G. Catalytic Pyrolysis of Solid Biomass and Related Biofuels, Aromatic, and Olefin Compounds. US 8,277,643 B2, October 2, 2012.

[10] Rahman MM, Liu R, Cai J. Selective Aerobic Oxidation of Furfural to Maleic Anhydride with Heterogeneous Mo-V-O Catalysts. Fuel Process Technol. 2018; 180: 32-46.

[11] Dickerson T, Soria J. Dehydro-Aromatization of Cyclohexene-carboxylic Acids by Sulfuric Acid: Critical Route for Bio-Based Terephthalic Acid Synthesis. Energies. 2013; 6: 514-538.

[12] Balasundram V, Ibraham N, Kasmani RM, Isha R, Hamid MKA, Hasbullah H. Catalytic Upgrading of Biomass-Derived Pyrolysis Vapour over Metal-Modified HZSM-5 into BTX: A Comprehensive Review. Biomass Conv. Bioref. 2020; doi.org/10.1007/s13399-020-00909-5.

[13] A.V. Bridgwater. Review of Fast Pyrolysis of Biomass and Product Upgrading. Biomass Bioenergy. 38 (2012), 68-94.

[14] Bridgewater AV, Peacocke GVC. Fast Pyrolysis Processes for Biomass. Renew Sustain Energy Rev. 2000; 4: 1-73.

[15] Bridgwater T. Challenges and Opportunities in Fast Pyrolysis of Biomass: Part I. Johnson Matthey Technol Rev. 2018; 62: 118-130.

[16] Bridgewater AV. Production of High Grade Fuels and Chemicals from Catalytic Pyrolysis of Biomass. Catal Today. 1996; 28: 285-295.

[17] Taarnning E, Osmundsen CM, Yang X, Voss B, Andersen SI, Christensen CH. Zeolite-Catalyzed Biomass Conversion to Fuels and Chemicals. Energy Environ Sci 2011; 4: 793-804.

[18] Mettler MS, Vlachos DG, Dauenhauer PJ. Top Ten Fundamental Challenges of Biomass Pyrolysis for Biofuels. Energy Environ Sci 2012; 5: 7797-7809.

[19] Liu J, Hou Q, Ju M, Ji P, Sun Q, Li W. Biomass Pyrolysis Technology by Catalytic Fast Pyrolysis, Catalytic Co-Pyrolysis and Microwave-Assisted

Pyrolysis: A Review. Catalysts. 2020; 10: 742.

[20] Yan K, Li H. State of the Art and Perspectives in Catalytic Conversion Mechanism of Biomass to Bio-Aromatics. Energy Fuel. 2021; 35: 45-62.

[21] Cai R, Pei X, Pan H, Wan K, Chen H, Zhang Z, et al. Biomass Catalytic Pyrolysis over Zeolite Catalysts with an Emphasis on Porosity and Acidity: A State-of-the-Art Review. Energy Fuel. 2020; 34: 11771-11790.

[22] Wang G, Dai Y, Yang H, Xiong Q, Wang K, Zhou J, et al. A Review of Recent Advances in Biomass Pyrolysis. Energy Fuel. 2020; 34: 15557-15578.

[23] Liu R, Rahman MM, Sarker M, Chai M, Li C, Cai J. A Review on the Catalytic Pyrolysis of Biomass for the Bio-Oil Production with ZSM-5: Focus on Structure. Fuel Process Technol. 2020; 199: 106301.

[24] Rahman MM, Liu R, Cai J. Catalytic Fast Pyrolysis of Biomass over Zeolites for High Quality Bio-Oil - A Review. Fuel Process Technol. 2018; 180: 32-46.

[25] Shi Y, Xing E, Wu K, Wang J, Yang M, Wu Y. Recent Progress on Upgrading of Bio-Oil to Hydrocarbons over Metal/Zeolite Bifunctional Catalysts. Catal Sci Technol 2017; 7: 2385-2415.

[26] Lok CM van Doorne J, Almansa GA. Promoted ZSM-5 Catalysts for the Production of bio-Aromatics, a Review. Renew Sustain Energy Rev. 2019 113: 109248.

[27] Chen X, Che Q, Li S, Liu Z, Yang H, Chen Y, et al. Recent Developments in Lignocellulosic Biomass Catalytic Fast Pyrolysis: Strategies for the Optimization of Bio-Oil Quality and Yield. Fuel Process Technol. 2019; 196: 106180.

[28] Huber GW, Iborra S, Corma A. Synthesis of Transportation Fuels from Biomass: Chemistry, Catalysts, and Engineering. Chem Rev. 2006; 106: 4044-4098.

[29] Tan S, Zhang Z, Sun J, Wang Q. Recent Progress of Catalytic Pyrolysis of Biomass by HZSM-5. Chin J Catal. 2013; 34: 641-650.

[30] Rezaei PS, Shafaghat H, Daud WMSW. Production of Green Aromatics and Olefins by Catalytic Cracking of Oxygenate Compounds Derived from Biomass Pyrolysis: A Review. Appl Catal A: General. 2014; 469: 490-511.

[31] Grams J, Ruppert AM. Development of Heterogeneous Catalysts for Thermo-Chemical Conversion of Lignocellulosic Biomass. Energies. 2017; 10: 545.

[32] Galadima A, Muraza O. In situ Pyrolysis of Biomass with Zeolite Catalysts for Bioaromatics/Gasoline Production: A Review. Energy Con. Manage. 2015; 105: 338-354.

[33] Mukarakate C, Evans RJ, Deutch S, Evans T, Starace AK, ten Dam J, et al. Reforming Biomass Derived Pyrolysis Bio-Oil Aqueous Phase to Fuels. Energy Fuels. 2017; 31: 1600-1607.

[34] Mullen CA. Upgrading of Biomass via Catalytic Fast Pyrolysis (CFP). In: Chemical Catalysts for Biomass Upgrading, ed. M. Crocker and E. Santillan-Jimenez, Wiley-VCH Verlag GmbH & Co. KGaA, Weinheim, Germany, 2020, pp. 1-33.

[35] Wang Y, Liu Q. Manganese-Catalyzed Dehydrogenative/Deoxygenative Coupling of Alcohols. Synlett. 2020; 31: 1464-1473.

[36] Yildiz G, Pronk M, Djokic M, van Geem KM, Ronsse F, van Duren R, et al. Validation of a New Set-Up for Continuous Catalytic Fast Pyrolysis of Biomass Coupled with Vapour Phase Upgrading. J Anal Appl Pyrolysis 2013; 103: 343-351.

[37] Jeong J, Lee HW, Jang SH, Ryu S, Kim YM, Park R, et al. In-situ Catalytic Fast Pyrolysis of Pinecone over HY Catalysts. Catalysts. 2019; 9: 1034.

[38] Cheng YT, Huber GW. Chemistry of Furan Conversion into Aromatics and Olefins over HZSM-5: A Model Biomass Conversion Reaction. ACS Catal. 2011; 1: 611-628.

[39] Ding SM, Fan MH, Wang TJ, Li QX. Transformation of Biomass into Aromatics with Zeolite Catalysts. Chin J Chem Phys. 2014; 27: 361-367.

[40] Galadima A, Muraza O. In situ Fast Pyrolysis of Biomass with Zeolite Catalysts for Bioaromatics/Gasoline Production: A Review. Energy Conv Manage. 2015; 105: 338-354.

[41] Ramasamu KK, Wang Y. Ethanol Conversion to Hydrocarbons on HZSM-5: Effect of Reaction Conditions and Si/Al Ratio on the Product Distributions. Catal Today. 2014; 237: 89-99.

[42] Cheng YT, Huber GW. Production of Targeted Aromatics by using Diels-Alders Classes of Reactions with Furans and Olefins over ZSM-5. Green Chem. 2012; 14: 3114-3125.

[43] Uslamin EA, Kosinov NA, Pidko EA, Hensen EJM. Catalytic Conversion of Furanic Compounds over Ga-Modified ZSM-5 Zeolites as a Route to Biomass-Derived Aromatics. Green Chem. 2018; 20: 3818-3827.

[44] Luna-Murillo B, Pala M, Paioni AL, Baldus M, Ronsse F, Prins W, et al. Catalytic Fast Pyrolysis of Biomass: Catalyst Characterization Reveals the Feed-Dependent Deactivation of a Technical ZSM-5-Based Catalyst. ACS Sustain Chem Eng. 2021; 9: 291-304.

[45] Hu B, Zhang B, Xie WL, Jiang XY, Liu J, Lu Q. Biomass Catalytic Pyrolysis over Zeolite Catalysts with an Emphasis on Porosity and Acidity: A State-of-the-Art Review. Energy Fuel. 2020; 34: 10384-10440.

[46] Ennaert T,van Aelst J, Dijkmans J, de Clerq R, Schutyser W, Dusselier M, et al. Potential and Challenges of Zeolite Chemistry in the Catalytic Conversion of Biomass. Chem Soc Rev. 2016; 45: 584-611.

[47] Adjaye JD, Bakhshi NN. Production of Hydrocarbons by Catalytic Upgrading of a Fast Pyrolysis Bio-Oil. Part I: Conversion over Various Catalysts. Fuel Process Technol. 1995; 45: 161-183.

[48] Carlson TR, Tompsett GA, Conner WC, Huber GW. Aromatic Production from Catalytic Fast Pyrolysis of Biomass-Derived Feedstocks. Top Catal. 2009; 52: 241-252.

[49] Zhu JF, Wang JC, Li QX. Transformation of Bio-Oil into BTX by Bio-Oil Catalytic Cracking. Chin J Chem Phys. 2013; 26: 477-483.

[50] Kelkar S, Saffron CM, Andreassi K, Li Z, Murkute A, Miller DJ, et al. A Survey of Catalysts for Aromatics from Fast Pyrolysis of Biomass. Appl Catal B: Environ. 2015; 174-175: 85-95.

[51] Deischter J, Schute K, Neves DS, Ebert BE, Blank LM, Palkovits R. Aromatisation of Bio-Derivable Isobutyraldehyde over HZSM-5 Zeolite Catalysts. Green Chem. 2019; 21: 1710-1717.

[52] Che Q, Yi W, Liu Y, Wang X, Yang H, Chen H. Effect of Mesopores in ZSM-5 on the Catalytic Conversion of Acetic Acid, Furfural, and Guaiacol. Energy Fuels. 2021; 35: 6022-6029.

[53] Huber GW, Cheng YT, Wang Z, Fan W. Method for Converting a Hydrocarbonaceous Material to a Fluid Hydrocarbon Product Comprising p-Xylene. US 2013/0324772 A1, December 5, 2013.

[54] Zhang H, Luo M, Xiao R, Shao S, Jin B, Xiao G, et al. Catalytic Conversion of Biomass Pyrolysis-Derived Compounds with Chemical Liquid Deposition (CLD) Modified ZSM-5. Bioresour Technol 2014; 155: 57-62.

[55] Zhang B, Zhong Z, Chen P, Ruan R. Microwave-Assisted Catalytic Fast Pyrolysis of Biomass Bio-Oil Production using Chemical Vapor Deposition Modified HZSM-5 Catalyst. Bioresour Technol. 2015; 197: 79-84.

[56] Du H, Zhong Z, Zhang B, Shi K, Li Z. Ex-situ Catalytic Upgrading of Vapors from Microwave-Assisted Pyrolysis of Bamboo with Chemical Liquid Deposition Modified H-ZSM-5 to Enhance Aromatics Production. J Anal Appl Pyrolysis. 2020; 149: 104857.

[57] Lee LW, Park SH, Jeon JK, Ryoo R, Kim W, Suh DJ, et al. Upgrading of Bio-Oil Derived from Biomass Constituents over Hierarchical Unilamellar Mesoporous MFI Nanosheets. Catal Today. 2014; 232: 119-126.

[58] Li J, Li X, Zhou G, Wang, W, Wang C, Komarneni S, et al. Catalytic Fast Pyrolysis of Biomass with Mesoporous ZSM-5 Zeolites Prepared by Desilication with NaOH Solutions. Appl Catal A: General. 2014; 470: 115-122.

[59] Ratnasari DK, Yang W, Jonsson PG. Kinetic Study of an H-ZSM-5/Al-MCM-41 Catalyst Mixture and its Application in Lignocellulose Biomass Pyrolysis. Energy Fuels. 2019; 33: 5630-5637.

[60] Zhang H, Shao S, Luo M, Xiao R. The Comparison of Chemical Liquid Deposition and Acid Dealumination Modified ZSM-5 for Catalytic Pyrolysis of Pinewood using Pyrolysis-Gas Chromatography/Mass Spectrometry. Bioresour Technol. 2017; 244: 726-732.

[61] Li X, Ge S, Shao S, Lv Z, Xiang X, Cai Y. Promoted Production of Hydrocarbons in the Catalytic Pyrolysis of Rape Straw over Composite HZSM-5@MCM-41 Catalysts. J Anal Appl Pyrolysis. 2021; 157: 105067.

[62] Cheng YT, Jae J, Shi J, Fan W, Huber GW. Production of Renewable Aromatic Compounds by Catalytic Fast Pyrolysis of Lignocellulosic Biomass with Bifunctional Ga/ZSM-5 Catalysts. Angew Chem Int Ed. 2012; 51: 1387-1390.

[63] Schultz EL, Mullen CA, Boateng AA. Aromatic Hydrocarbon Production from Eucalyptus Urophylla Pyrolysis over Several Metal-Modified ZSM-5 Catalysts. Energy Technol. 2017; 5: 196-204.

[64] Park HJ, Heo HS, Jeon JK, Kim J, Ryoo R, Jeong KE, et al. Highly Valuable Chemicals Production from Catalytic Upgrading of Radiata Pine Sawdust-Derived Pyrolytic Vapors over Mesoporous MFI Zeolites. Appl Catal B: Environ. 2010; 95: 365-373.

[65] Uslamin EA, Luna-Murillo B, Kosinov N, Bruijnincx, Pidko EA, Weckhuysen BM, et al. Gallium-Promoted HZSM-5 Zeolites as Efficient Catalysts for the Aromatization of Biomass-Derived Furan. Chem Eng Sci. 2019; 198: 305-316.

[66] Zhang H, Zheng J, Xiao R. Catalytic Pyrolysis of Willow Wood with Me/ZSM-5 (Me = Mg, K, Fe, Ga, Ni) to Produce Aromatics and Olefins. Bioresour Technol. 2013; 8: 5612-5621.

[67] Yung MM, Stanton AR, Iiisa K, French RJ, Orton KA, Magrini KA. Multiscale Evaluation of Catalytic Upgrading of Biomass Pyrolysis Vapors on Ni- and Ga-Modified ZSM-5. Energy Fuel. 2016; 30: 9471-9479.

[68] Mullen CA, Tarves PC, Raymundo LM, Schultz EL, Boateng AA, Trierweiler JO. Fluidized Bed Catalytic Pyrolysis of Eucalyptus over HZSM-5: Effect of Acid Density and Gallium Modification on Catalyst Deactivation. Energy Fuel. 2018; 32: 1771-1778.

[69] Kelkar S, Saffron CM, Li Z, Kim SS, Pinnavaia TJ, Miller DJ, Kriegel R. Aromatics from Biomass Pyrolysis Vapour using a Bifunctional Mesoporous Catalyst. Green Chem. 2014; 16: 803-812.

[70] Du Z, Ma X, Li Y, Chen P, Liu Y, Lin X, et al. Production of Aromatic Hydrocarbons by Catalytic Pyrolysis of Microalgae with Zeolites: Catalyst Screening in a Pyroprobe. Bioresour Technol. 2013; 139: 397-401.

[71] Schutlz EL, Mullen CA, Boateng AA. Aromatic Hydrocarbon Production from Eucalyptus urophylla Pyrolysis over Several Metal-Modified ZSM-5 Catalysts. Energy Technol. 2017: 5: 196-204.

[72] Kim JW, Park SH, Jung J, Jeon JK, Ko CH, Jeong KE, et al. Catalytic Pyrolysis of Mandarin Residue from the Mandarin Juice Processing Industry. Bioresour. Technol. 2013; 136: 431-436.

[73] Vichaphund S, Aht-ong D, Sricharoenchaikul V, Atong D. Production of Aromatic Compounds from Catalytic Fast Pyrolysis of Jatropha Residues using Metal/HZSM-5 Prepared by Ion-Exchange and Impregnation Methods. Renew Energy. 2015; 79: 28-37.

[74] Farooq A, Moogi S, Kwon EE, Lee J, Kim YM, Jae J, et al. Catalytic Upgrading of Quercus Mongolica under Methane Environment to Obtain High Yield of Bioaromatics. Environ Pollut. 2021; 272: 116016.

[75] Dai G, Wang S, Zou Q, Huang S. Improvement of Aromatics Production from Catalytic Pyrolysis of Cellulose over Metal-Modified Hierarchical HZSM-5. Fuel Process. Technol. 2018; 179: 319-323.

[76] Moogi S, Jae J, Kannapu HPR, Ahmed A, Park ED, Park YK. Enhancement of Aromatics from Catalytic Pyrolysis of Yellow Poplar: Role of Hydrogen and Methane Decomposition. Bioresour Technol. 2020; 315: 123835.

[77] Cai Q, Yu T, Meng X, Zhang S. Selective Generation of Aromatic Hydrocarbons from Hydrotreating-Cracking of Bio-Oil Light Fraction with MOx Modified HZSM-5 (M = Ga, Mo and Zn). Fuel Process Technol. 2020; 204: 106424.

[78] Espindola JS, Gilbert CJ, Perez-Lopez OW, Trierweiler JO, Huber GW. Conversion of Furan over Galliumm and Zinc Promoted ZSM-5: The Effect of Metal and Acid Sites. Fuel Process Technol. 2020; 201: 106319.

[79] He Y, Zhu L, Luo Y, Fan M, Yang M, Zhang Y, et al. Selective Upgrading of Biomass Pyrolysis Oil into Renewable p-Xylene with Multifunctional M/SiO_2/HZSM-5 Catalyst. Fuel Process. Technol. 2021; 213: 106674.

[80] Li P, Chen X, Wang X, Shao J, Lin G, Yang H, et al. Catalytic Upgrading of Fast Pyrolysis Products with Fe-, Zr-, and Co-Modified Zeolites Based on Pyrolyzer-GC/MS Analysis. Energy Fuels. 2017; 31: 3979-3986.

[81] Mullen CA, Boateng AA. Production of Aromatic Hydrocarbons via Catalytic Pyrolysis of Biomass over Fe-Modified HZSM-5 Zeolites. ACS Sustain Chem Eng. 2015; 3: 1623-1631.

[82] Persson H, Duman I, Wang S, Pettersson LJ, Yang W. Catalytic Pyrolysis over Transition Metal-Modified Zeolites: A Comparative Study between Catalyst Activity and Deactivation. J Anal Appl Pyrolysis. 2019; 138: 54-61.

[83] Sun L, Zhang X, Chen L, Zhao B, Yang S, Xie X. Comparison of Catalytic Fast Pyrolysis of Biomass to Aromatic Hydrocarbons over ZSM-5 and Fe/ZSM-5 Catalysts. J Anal Appl Pyrolysis. 2016; 121: 342-346.

[84] Zhang S, Zhang H, Liu X, Zhu S, Hu L, Zhang Q. Upgrading of Bio-Oil from Catalytic Pyrolysis of Pretreated Rice Husk over Fe-Modified ZSM-5

Zeolite Catalyst. Fuel Process Technol. 2018; 175: 17-25.

[85] Li P, Li D, Yang H, Wang X, Chen H. Effects of Fe-, Zr-, and Co-Modified Zeolites and Pretreatments on Catalytic Upgrading of Biomass Fast Pyrolysis. Energy Fuels. 2016; 30: 3004-3013.

[86] Che Q, Yang M, Wang X, Yang Q, Williams LR, Yang H. et al. Influence of Physicochemical Properties of Metal Modified ZSM-5 Catalyst on Benzene, Toluene and Xylene Production of Biomass Catalytic Pyrolysis. Bioresour. Technol. 2019; 278: 248-254.

[87] Arteaga-Perez LE, Capiro OG, Romero R, Degado A, Olivera P, Ronsse F, et al. In situ Catalytic Fast Pyrolysis of Crude and Torrefied Eucaplyptus Globulus using Carbon Aerogel-Supported Catalysts. Energy. 2017; 128: 701-712.

[88] Cai M, Liu R, He Y. Effects of SiO_2/Al_2O_3 Ratio and Fe Loading Rate on Selection of Aromatics and Kinetics of Corn Stalk Catalytic Pyrolysis. Fuel Process Technol. 2020; 206: 106458.

[89] Wang JX, Cao JP, Zhao XY, Liu SN, Ren XY, Zhang LY, et al. In Situ Upgrading of Cellulose Pyrolysis Volatiles using Hydrofluorinated and Platinum-Loaded HZSM-5 for High Selectivity Production of Light Aromatics. Ind Eng Chem Res 2019; 58: 22193-22201.

[90] Thangalazhy-Gopakumar S, Adhikari S, Gupta RB. Catalytic Pyrolysis of Biomass over H+ZSM-5 under Hydrogen Pressure. J. Am. Chem. Soc. 2012; 26: 5300-5306.

[91] Eschenbacher A, Saraeian A, Shanks RH, Mentzel UV, Jensen PA, Henriksen UB, et al. Micro-Pyrolyzer Screening of Hydrodeoxygenation Catalysts for Efficient Conversion of Straw-Derived Pyrolysis Vapors. J Anal Appl Pyrolysis. 2020; 150: 104868.

[92] Stanton AR, Iisa K, Yung MM, Magrini KA. Optimizing Catalytic Pyrolysis of Rubber Seed Oil for Light Aromatics and Anti-Deactivation of ZSM-5. J Anal Appl Pyrolysis. 2018; 135: 199-208.

[93] Chang R, Zhu L, Jin F, Fan M, Liu J, Jia Q, et al. Production of Bio-Based *p*-Xylene via Catalytic Pyrolysis of Biomass over Metal Oxide-Modified HZSM-5 Zeolites. J Chem Technol Biotechnol. 2018; 93: 3292-3301.

[94] Zhu L, Fan M, Wang Y, Wang S, He Y, Li Q. Selective Conversion of Furans to *p*-Xylene with Surface-Modified Zeolites. J Chem Technol Biotechnol. 2019; 94: 2876-2887.

[95] Su X, Zan W, Bai X, Wang G, Wu W. Synthesis of Microscale and Nanoscale ZSM-5 Zeolites: Effect of Particle Size and Acidity of Zn Modified ZSM-5 Zeolites on Aromatization. Catal Sci Technol. 2017; 7; 1943-1952.

[96] Zheng Y, Wang F, Yang X, Huang Y, Liu C, Zheng Z, et al. Study on Aromatics Production via the Catalytic Pyrolysis Vapor Upgrading of Biomass using Metal-Loaded Modified H-ZSM-5. J Anal Appl Pyrolysis. 2017; 126: 169-179.

[97] Huang Y, Wei L, Crandall Z, Julson J, Gu Z. Combining Mo-Cu/HZSM-5 with a Two-Stage Catalytic Pyrolysis System for Pine Sawdust Thermal Conversion. Fuel. 2015; 150: 656-663.

[98] Eschenbacher A, Saraeian A, Shanks BH, Mentzel UV, Jensen PA, Henriksen UB, et al. Performance-Screening of Metal-Impregnated Industrial HZSM-5/γ-Al_2O_2 Extrudates for Deoxygenation and Hydro-deoxygenation of Fast Pyrolysis Vapors. J Anal Appl Pyrolysis. 2020; 150: 104892.

[99] Sun L, Wang Z, Chen L, Yang S, Xie X, Gao M, et al. Catalytic Fast

Pyrolysis of Biomass into Aromatic Hydrocarbons over Mo-Modified ZSM-5 Catalysts. Catalysts. 2020; 10: 1051.

[100] Ren YX, Cao JP, Zhao SX, Zhao XY, Liu TL, Feng XB, et al. Encapsulation Ni in HZSM-5 for Catalytic Hydropyrolysis of Biomass to Light Aromatics. Fuel Process Technol 2021; 218: 106854.

[101] Kumar R, Strezov V, Kan T, Weldekidan H, He J, Jahan S. Investigating the Effect of Mono- and Bimetallic/Zeolite Catalysts on Hydrocarbon Production during Bio-Oil Upgrading from Ex Situ Pyrolysis of Biomass. Energy Fuel. 2020; 34: 389-400.

[102] Illiopoulou EF, Stefanidis SD, Kalogiannis KG, Delimitis A, Lappas AA, Triantafyllidis KS. Catalytic Upgrading of Biomass Pyrolysis Vapors using Transition Metal-Modified ZSM-5 Zeolite. Appl Catal B: Environ. 2012; 127: 281-290.

[103] Chaihad N, Anniwaer A, Karnjanakom S, Kasai Y, Kongparakul S, Samart C, et al. In-situ Catalytic Upgrading of Bio-Oil Derived from Fast Pyrolysis of Sunflower Stalk to Aromatic Hydrocarbons over Bifunctional Cu-Loaded HZSM-5. J Anal Appl Pyrolysis. 2021; 155: 105079.

[104] Che Q, Yang M, Wang X, Chen X, Chen W, Yang Q, et al. Aromatics Production with Metal Oxides and ZSM-5 as Catalysts in Catalytic Pyrolysis of Wood Sawdust. Fuel Process Technol. 2019; 188: 146-152.

[105] Shang J, Fu G, Cai Z, Feng X, Tuo Y, Zhou X, et al. Regulating Light Olefins or Aromatics Production in ex-situ Catalytic Pyrolysis of Biomass by Engineering the Structure of Tin Modified ZSM-5 Catalyst. Bioresour Technol. 2021; 330: 124975.

[106] Lu Q, Guo H, Zhou M, Cui M, Dong C, Yang UP. Selective Preparation of Monocyclic Aromatic Hydrocarbons from Catalytic Cracking of Biomass Fast Pyrolysis Vapors over Mo_2N/HZSM-5 Catalyst. Fuel Process Technol. 2018; 173: 134-142.

[107] Zhang J, Qian W, Kong C, Wei F. Increasing *para*-Xylene Selectivity in Making Aromatics from Methanol with Surface-Modified Zn/P/ZSM-5 Catalyst. ACS Catal. 2015; 5: 2982-2988.

[108] Yin W, Koekhorst A, Venderbosch RH, Bykova MV, Khromova SA, Yakovlev VA, et al. Catalytic Hydrotreatment of Fast Pyrolysis Liquids in Batch and Continuous Set-Ups using a Bimetallic Ni-Cu Catalyst with a High Metal Content. Catal Sci Technol. 2016; 6: 5899-5915.

[109] Chen H, Cheng H, Zhou F, Chen K, Qiao K, Lu X, et al. Catalytic Fast Pyrolysis of Rice Straw to Aromatic Compounds over Hierarchical HZSM-5 Produced by Alkali Treatment and Metal-Modification. J Anal Appl Pyrolysis. 2018; 131: 76-84.

[110] Wang JX, Cao JP, Zhao XY, Liu SN, Ren XY, Zhao M, et al. Enhancement of Light Aromatics from Catalytic Fast Pyrolysis of Cellulose over Bifunctional Hierarchical HZSM-5 Modified by Hydrogen Fluoride and Nickel/Hydrogen Fluoride. Bioresour Technol. 2019; 278: 116-123.

[111] Li K, Zhang G, Wang Z, Hu B, Lu Q. Calcium Formate Assisted Catalytic Pyrolysis of Pine for Enhanced Production of Monocyclic Aromatic Hydrocarbons over Bimetal-Modified HZSM-5. Bioresour Technol. 2020; 315: 123805

[112] Liu Q, Wang J, Zhou J, Yu Z, Kwang K. Promotion of Monocyclic Aromatics by Catalytic Fast Pyrolysis of Biomass with Modified HZSM-5. J Anal Appl Pyrolysis. 2021; 153: 104964.

[113] Li W, Zhu Y, Li S, Lu Y, Wang J, Zhu K, et al. Catalytic Fast Pyrolysis of

Cellulose over $Ce_{0.8}Zr_{0.2}$-xAlxO_2 Catalysts to Produce Aromatic Hydrocarbons: Analytical Py-GCxGC/MS. Fuel Process Technol. 2020; 205: 106438.

[114] Cai Q, Yu T, Zhang S. Enhanced Aromatic Hydrocarbon Production from Bio-Oil Hydrotreating-Cracking by Mo-Ga Modified HZSM-5. Fuel. 2020; 269: 117386.

[115] Lu Q, Yuan S, Liu C, Zhang T, Xie X, Deng X, et al. A Fe-Ca/SiO_2 Catalyst for Efficient Production of Light Aromatics from Catalytic Pyrolysis of Biomass. Fuel. 2020; 279: 118500.

[116] Illiopoulou EF, Stefanidis S, Kalogiannis K, Psarras AC, Delimitis A, Triantafyllidis KS, et al. Pilot-Scale Validation of Co-ZSM-5 Catalyst Performance in the Catalytic Upgrading of Biomass Pyrolysis Vapours. Green Chem. 2014; 16: 662-674.

[117] Wang D, Zhu Y, Chen J, Li W, Luo F, Li S, et al. Catalytic Upgrading of Lignocellulosic Biomass Pyrolysis Vapors: Insights into Physicochemical Changes in ZSM-5. J Anal Appl Pyrolysis. 2021; 156: 105123.

[118] Lijun PKW, Zhang H, Vitidsant T, Reubroycharoen P, Xiao R. Influence of Inorganic Matter in Biomass on the Catalytic Production of Aromatics and Olefins in a Fluidized-Bed Reactor. Energy Fuels. 2017; 31: 6120-6131.

[119] Kim BS, Kim YM, Jae J, Watanabe C, Kim S, Jung SC, et al. Pyrolysis and Catalytic Upgrading of Citrus unshiu Peel. Bioresour Technol. 2015; 194: 312-319.

[120] Stanton AR, Iisa K, Mukarakate C, Nimlos MR. Role of Biopolymers in the Deactivation of ZSM-5 During Catalytic Fast Pyrolysis of Biomass. ACS Sustain Chem Eng. 2018; 6: 10030-10038.

[121] Kukarakate C, Zhang X, Stanton AR, Robichaud DJ, Ciesielski PN, Malhotra K, et al. Real-Time Monitoring of the Deacivation of HZSM-5 during Upgrading of Pine Pyrolysis Vapors. Green Chem. 2014; 16: 1444-14461.

[122] Ramasamy KK, Gerber MA, Flake M, Zhang H, Wang Y. Conversion of Biomass-Derived Small Oxygenates over HZSM-5 and its Deactivation Mechanism. Green Chem. 2014; 16: 748-760.

[123] Zhang H, Shao S, Xiao R, Shen D, Zeng J. Characterization of Coke Deposition in the Catalytic Fast Pyrolysis of Biomass Derivatives. Energy Fuels. 2014; 28; 52-57.

[124] Liu C, Wang H, Karim AM, Sun J, Wang Y. Catalytic Fast Pyrolysis of Lignocellulosic Biomass. Chem Soc Rev. 2014; 43: 7594-7623.

[125] Shao S, Zhang H, Xiao R, Li X, Cai Y. Controlled Regeneration of ZSM-5 Catalysts in the Combined Oxygen and Steam Atmosphere used for Catalytic Pyrolysis of Biomass-Derivates. Energy Conv Manage. 2018; 155: 175-181.

[126] Guidicianni P, Gariulo v, Grottola CM, Alfe M, Ferreiro AI, Mendus MAA, et al. Inherent Metal Elements in Biomass Pyrolysis: A Review. Energy Fuels. 2021; 35: 5407-5478.

[127] Mullen CA, Botaeng AA, Dadson RB, Hashem FM. Biological Mineral Range Effects on Biomass Conversion to Aromatic Hydrocarbons via Catalytic Fast Pyrolysis over HZSM-5. Energy Fuel. 2014; 28: 7014-7024.

[128] Yildiz G, Ronsse F, Venderbosch R, van Duren R, Kersten SRA, Prins W. Effect of Biomass Ash in Catalytic Fast Pyrolysis of Pine Wood. Appl Catal B: Environ. 2015; 168-169: 203-211.

[129] Mahadevan R, Adhikari S, Shakya R, Wang K, Dayton D, Lehrich M, et al. Effect of Alkali and

Alkaline Earth Metals on in-situ Catalytic Fast Pyrolysis of Lignocellulosic Biomass: A Microreacftor Study. Energy Fuels. 2016; 30: 3045-3056.

[130] Xiao J, Yang M, Che Q, Chen Y, Chen X, Yang H, et al. Effect of Potassium on Catalytic Characteristics of ZSM-5 Zeolite in Fast Pyrolysis of Biomass-Based Furan. J Anal Appl Pyrolysis. 2021; 157: 105230.

[131] Mullen CA, Boateng AA. Accumulation of Inorganic Impurities on HZSM-5 Zeolites during Catalytic Fast Pyrolysis of Switchgrass. Ind Eng Chem Res. 2013; 52: 17156-17161

[132] Ma N, Song Y, Han f, Waterhouse GIN, Li Y, Ai S. Highly Selective Hydrogenation of 5-Hydroxymethylfurfural to 2,5-Dimethylfuran at Low Temperature over a Co-N-C/NiAl-MMO Catalyst. Catal Sci Technol. 2020; 10: 4010-4018.

[133] Wang K, Zhang J, Shanks BH, Brown RC. The Deleterious Effect of Inorganic Salts on Hydrocarbon Yields from Catalytic Pyrolysis of Lignocellulosic Biomass and its Mitigation. Appl. Energy. 148 (2015), 115-120.

[134] Zheng A, Zhao Z, Huang Z, Zhao K, Wei G, Wang X, et al. Catalytic Fast Pyrolysis of Biomass Pretreated by Torrefaction with Varying Severity. Energy Fuel. 2014; 2014; 28: 5804-5811.

[135] Zhang S, Zhu S, Zhang H, Chen T, Xiong Y. Catalytic Fast Pyrolysis of Rice Husk: Effect of Coupling Leaching with Torrefaction Pretreatment. J Anal Appl Pyrolysis. 2018; 133: 91-96.

[136] Xin X, Torr KM, Pang S, van de Pas D, Cooke-Willis M, Mercader FDM. Catalytic Fast Pyrolysis of Demineralized Biomass in a Fluidized Bed Reactor: Effects of Acid-Leaching and Torrefaction Pretreatments. Energy Fuel. 2020; 34: 568-578.

[137] Xin X, Pang S, de Miguel Mercader F, Torr KM. The Effect of Biomass Pretreatment on Catalytic Pyrolysis Products of Pine Wood by Py-GC/MS and Principal Component Analysis. J Anal Appl Pyrolysis. 2019; 138: 145-153.

[138] Karnjanakom S, Guan G, Asep B, Hao X, Kongparakul S, Samart C, et al. Catalytic Upgrading of Bio-Oil over Cu/MCM-41 and Cu/KIT-6 Prepared by β-Cyclodextrin-Assisted Coimpregnation Method. J. Phys. Chem. C. 2016; 120: 3396-3407.

[139] Luo G, Resende FLP. In-Situ and Ex-Situ Upgrading of Pyrolysis Vapors from Beetle-Killed Trees. Fuel. 2016; 166: 367-375.

[140] Lorenzetti C, Conti R, Fabbri D, Yanik J. A Comparative Study on the Catalytic Effect of H-ZSM5 on Upgrading of Pyrolysis Vapors Derived from Lignocellulosic and Proteinaceous Biomass. Fuel. 2016; 166: 446-452.

[141] Zheng A, Zhao Z, Chang S, Huang Z, Wu H, Wang X, et al. Effect of Crystal Size of ZSM-5 on the Aromatic Yield and Selectivity from Catalytic Fast Pyrolysis of Biomass. J Mol Catal A: Chemical. 2014; 383-384: 23-30.

[142] Mante OD, Agblevor FA. Catalytic Pyrolysis for the Production of Refinery-Ready Biocrude Oils from Six Different Biomass Sources. Green Chem. 2014; 16: 3364-3377.

[143] Muragappan K, Mukarakate C, Budhi S, Shetty M, Nimlos MR, Roman-Leshkov Y. Supported Molydenum Oxides as Effective Catalysts for the Catalytic Fast Pyrolysis of Lignocellulosic Biomass. Green Chem. 2016; 16: 5548-5557.

[144] Rahman MdM, Chai M, Sarker M, Liu R. Catalytic Pyrolysis of Pinewood over ZSM-5 and CaO for Aromatic Hydrocarbon: Analytical Py-GC/MS Study. J Energy Inst. 2020; 93: 425-435.

[145] Neupane S, Adhikari S, Wang Z, Ragauskas AJ, Pu Y. Effect of Torrefaction on Biomass Structure and Hydrocarbon Production from Fast Pyrolysis. Green Chem. 2015; 17: 2406-2417.

[146] Srinivasan V, Adhikari S, Chattanathan SA, Park S. Catalytic Pyrolysis of Torrefied Biomass for Hydrocarbons Production. Energy Fuels. 2012; 26: 7347-7353.

[147] Ghorbannezhad P, Firouzabadi MD, Ghasemian A, de Wild PJ, Heeres HJ. Sugarcane Bagasse *ex-situ* Catalytic Fast Pyrolysis for the Production of Benzene, Toluene, and Xylenes (BTX). J Anal Appl Pyrolysis. 2018; 131: 1-8.

[148] Foster AJ, Jae J, Cheng YT, Huber GW, Lobo RF. Optimizing the Aromatic Yield and Distribution from Catalytic Fast Pyrolysis of Biomass over ZSM-5. Appl Catal A: General. 2012; 423-424; 154-161.

[149] Mihalcik DJ, Boateng AA, Mullen CA, Goldberg NM. Packed-Bed Catalytic Cracking of Oak-Derived Pyrolysis Vapors. Ind Chem Eng Res. 2011; 50: 13304-13312.

[150] Elkasabi Y, Mullen CA, Boateng AA. Distillation and Isolation of Commodity Chemicals from Bio-Oil Made by Tail-Gas Reactive Pyrolysis. ACS Sustain Chem Eng. 2014; 2: 2042-2052.

[151] Zhang H, Xiao R, Huang H, Xiao G. Comparison of Non-Catalytic and Catalytic Fast Pyrolysis of Corncob in a Fluidized Bed Reactor. Bioresour Technol. 2009; 100: 1428-1434.

[152] Zhang H, Xiao R, Jin B, Shen D, Chen r, Xiao G. Catalytic Fast Pyrolysis of Straw Biomass in an Internally Interconnected Fluidized Bed to Produce Aromatics and Olefins: Effect of Different Catalysts. Bioresourc Technol. 2013; 137: 82-87.

[153] Karnajankom S, Siriya-Umporn T, Bayu A, Kongparakul S, Samart C, Fushimi C, et al. High Selectivity and Stability of Mg-Doped Al-MCM-41 for in situ Catalytic Upgrading Fast Pyrolysis Bio-Oil. Energy Conv Manage. 2017; 142: 272-285.

[154] Li Z, Zhong Z, Zhang B, Wang W, Seufitelli GVS, Resende FLP. Effect of Alkali-Treated HZSM-5 Zeolite on the Production of Aromatic Hydrocarbons from Microwave Assisted Catalytic Fast Pyrolysis (MACFP) of Rice Husk. Sci Total Environ. 2020; 703: 134605.

[155] Li Z, Zhong Z, Zhang B, Wang W, Wu W. Catalytic Fast Pyrolysis of Rice Husk over Hierarchical Micro-Mesoporous Composite Molecular Sieve: Analytical Py-GC/MS Study. J Anal Appl Pyrolysis. 2019; 138: 103-113.

[156] Zhang Z, Cheng H, Chen H, Chen K, Lu X, Ouyang P, et al. Enhancement in the Aromatic Yield from the Catalytic Fast Pyrolysis of Rice Straw over hexadecyl Trimethyl Ammonium Bromide Modified Hierarchical HZSM-5. Bioresour Technol. 2018; 256: 241-246.

[157] Du H, Zhang Z, Zhang B, Shi K, Li Z. Ex-situ Catalytic Upgrading of Vapors from Microwave-Assisted Pyrolysis of Bamboo with Chemical Liquid Deposition Modified HZSM-5 to Enhance Aromatics Production. J Anal Appl Pyrolysis. 2020; 149: 104857.

[158] Chandler DS, Resende FLP. Comparison between Catalytic Fast Pyrolysis and Cataytic Fast Hydropyrolysis for the Production of Liquid Fuels in a Fluidized Bed Reactor. Energy Fuel. 2019; 33: 3199-3209.

[159] Hu C, Zhang H, Xiao R. Effects of Nascent Char on ex-situ Catalytic Fast Pyrolysis of Wheat Straw. Energy Conv. Management. 2018; 177: 765-772.

[160] Carlson TR, Vispute TP, Huber GW. Green Gasoline by Catalytic Fast Pyrolysis of Solid Biomass Derived

Compounds. ChemSusChem. 2008; 1: 397-400.

[161] Gamliel DP, Cho HJ, Fan W, Valla JA. On the Effectiveness of Tailored Mesoporous MFI Zeolites for Biomass Catalytic Fast Pyrolysis. Appl Catal A: General. 2016; 522: 109-119.

[162] Fang S, Shi C, Li P, Zhang Y, Chang C. Effects of Metal-Modified ZSM-5 Catalysts on Product Characteristics Based on the Py-GC/MS of Peanut Shells. Ind Chem Eng Res. 2020; 59: 17307-17314.

[163] Zheng A, Zhao Z, Chang S, Huang Z, Zhao K, Wu H, et al. Maximum Synergistic Effect in the Coupling Conversion of Bio-Derived Furans and Methanol over ZSM-5 for Enhancing Aromatic Production. Green Chem. 2014; 16: 2580-2586.

[164] Gunawardena DA, Fernando DS. Screening of Transition Metal/Oxide-Impregnated ZSM-5 Catalysts for Deoxygenation of Biomass Oxygenates via Direct Methane Intervention. Biofuels. 2018; 9: 113-120.

[165] Gou J, Wang Z, Li C, Qi X, Vattipalli V, Cheng YT, et al. The Effects of ZSM-5 Mesoporosity and Morphology on the Catalytic Fast Pyrolysis. Green Chem. 2017; 19; 3549-3557.

[166] Gilbert CJ, Espindola JS, Conner WC, Trierweiler JO, Huber GW. The Effect of Water on Furan Conversion over ZSM-5. ChemCatChem. 2014; 6: 2497-2500.

[167] Zhao Y, Pan T, Zuo Y, Guo QX, Fu Y. Production of Aromatic Hydrocarbons through Catalytic Pyrolysis of 5-Hydroxymethylfurfural from Biomass. Bioresourc Technol. 2013; 147: 37-42.

[168] Wang C, Si Z, Wu X, Lv W, Bi K, Zhang X, Chen L, et al. Mechanism Study of Aromatics Production from Furans with Methanol over Zeolite Catalysts. J Anal Appl Pyrolysis. 2019; 139: 87-95.

[169] Jae J, Tomsett GA, Foster AJ, Hammon KD, Auerbach SM, Lobo RF, et al. Investigation into the Shape Selectivity of Zeolite Catalysts for Biomass Conversion. J Catal. 2011; 279: 257-268.

[170] Carlson TR, Jae J, Huber GW. Mechanistic Insights from Isotopic Studies of Glucose Conversion to Aromatics over ZSM-5. ChemCatChem. 2009; 1: 107-110.

[171] Wang K, Kim KH, Brown RC. Catalytic Pyrolysis of Individual Components of Lignocellulosic Biomass. Green Chem. 2014; 16: 727-735.

[172] Hoff TC, Gardner DW, Thilakaratne R, Wang K, Hansen TW, Brown RC, et al. Tailoring ZSM-5 Zeolites for Fast Pyrolysis of Biomass to Aromatic Hydrocarbons. ChemSusChem. 2016; 9: 1473-1482.

[173] Karanjkar PU, Coolman RJ, Huber GW, Blatnik MT, Almalkie S, de Bruyn Kops SM, et al. Production of Aromatics by Catalytic Fast Pyrolysis of Cellulose in a Bubbling Fluidized Bed Reactor. AIChE J. 2014; 60: 1320-1335.

[174] Hassan NS, Jalil AA, Hitam AA, Vo DVN, Nabgan W. Biofuels and Renewable Chemicals Production by Cataytic Pyrolysis of Cellulose: A Review. Environ Chem Lett. 2020; 18: 1625-1648.

[175] He S, Muizebelt I, Heeres A, Schenk NJ, Blees R, Heeres HJ. Catalytic Pyrolysis of Crude Glycerol over Shaped ZSM-5/Bentonite Catalysts for Bio-BTX Synthesis. Appl Catal B: Environ. 2018; 235: 45-55.

[176] Genuino HC, Muizebelt I, Heeres A, Schenk NJ, Winkelman JGM, Heeres HJ. An Improved Catalytic Pyrolysis Concept for Renewable

Aromatics from Biomass Involving a Recycling Strategy for Co-Produced Polycyclic Aromatic Hydrocarbons. Green Chem. 2019; 21: 3802-3806.

[177] Chen NY, Degnan TF, Koenig LR. Liquid Fuel from Carbohydrates. Chemtech. 1986; 16: 506-511.

[178] Zhang H, Cheng YT, Vispute TP, Xiao R, Huber GW. Catalytic Conversion of Biomass-Derived Feedstocks into Olefins and Aromatics with ZSM-5: The Hydrogen to Carbon Effective Ratio. Energy Environ Sci. 2011; 4: 2297-2307.

[179] Carlson TR, Cheng YT, Jae J, Huber GW. Production of Green Aromatics and Olefins by Catalyzing Fast Pyrolysis of Wood Sawdust. Energy Environ Sci. 2011; 4: 145-161.

[180] Huber GW, Zhang H, Carlson T. Method for Producing Fluid Hydrocarbons. US 2013/0060070 A1, March 7, 2013.

[181] Zhang H, Carlson TR, Xiao R, Huber GW. Catalytic Fast Pyrolysis of Wood and Alcohol Mixtures in a Fluidized Bed Reactor. Green Chem. 2012; 14: 98-110.

[182] Li K, Wang ZX, Zhang G, Cui MS, Lu Q, Yang YP. Selective Production of Monocyclic Aromatic Hydrocarbons from ex situ Catalytic Fast Pyrolysis of Pine over the HZSM-5 Catalyst with Calcium Formate as a Hydrogen Source. Sustain Energy Fuels. 2020; 4: 538-548.

[183] Mazanec T, Schmelzer E, Pesa F. Chemical Intermediates by Catalytic Fast Pyrolysis Process. US 9,249,080 B2, February 2, 2016.

[184] Sudarsanam P, Peeters E, Makshina EV, Parvulescu EV, Sels BF. Advances in Porous and Nanoscale Catalysts for Viable Biomass Conversion. Chem Soc Rev. 2019; 48: 2366-2421.

[185] Elkasabi Y, Mullen CA, Boateng AA. Aqueous Extractive Upgrading of Bio-Oils Created by Tail-Gas Reactive Pyrolysis to Produce Pure Hydrocarbons and Phenols. ACS Sustain Chem Eng. 2015; 3: 2809-2816.

[186] Elkasabi Y, Wyatt V, Jones K, Strahan GD, Mullen CA, Boateng AA. Hydrocarbons Extracted from Advanced Pyrolysis Bio-Oils: Characterization and Refining. Energy Fuel. 2020; 34: 483-490.

[187] Horne PA, Williams PT. Upgrading of Biomass-Derived Pyrolytic Vapours over Zeolite ZSM-5 Catalyst: Effect of Catalyst Dilution on Product Yields. Fuel. 1996; 75: 1043-1050.

[188] Zhang X, Rajagopalan K, Lei H, Ruan R, Sharma BK. An Overview of Novel Concept in Biomass Pyrolysis: Microwave Irradiation. Sustain Energy Fuels. 2017; 1: 1664-1699.

Chapter 3

On the Feasibility of the Closed Cycle Local Economy Based on Pyrolysis

Brice H. Leeds

Abstract

Modern technologies allow much higher rate of prosperity than normally seen around the globe. The development can be achieved based on the local resources. Pyrolysis should play the central role on the initial stage of the processing of the biomass. The practical experience made by the development engineers allow to make solid conclusion about the current problems faced during the introduction of the technology. The introduction of pyrolysis technologies, especially in developing countries, faces non-engineering problems. The assessment of the economic viability of the introduced technological process can be performed based on the coverage of the global deficit, primarily in terms of resources substituted as inputs. To motivate the local population, it is proposed to introduce a large number of small production facilities with a unified system of coordination of producers and partial transfer of ownership rights. The regional economy should undergo the balanced growth and be protected by cheap quasi real regional currency and expensive accumulative currency. The reginal saving rate can be regulated, allowing for the accelerated rate of regional investments. New green technologies based on pyrolysis should allow for the new regional economic policy named "Green economic defensive initiative".

Keywords: pyrolysis, solar energy, biogas, bioplastic, mineral fertilizers, green technologies, regional economy, utilization of local resources, processing of biomass, sustainable product, economic viability, coverage of the global deficit, market economy, protection by currency, balanced growth, industrial farming, regional saving rate, green economic defensive initiative, civil society

1. Introduction

One of the modern ways to test a new technology is to implement it in the developing world. Pyrolysis can be viewed as a lowtech technology suitable for these purposes. It was possible over the last 10 years to try to make many attempts to use the technology. It became very fast quite clear that the prime problem is not with the technology, but rather with the economic factors. A universal economic solution is needed. There should be no conceptual difference between the developing and the developed world especially now in the age of globalization. People tend to act economically similarly in any society with a certain variety that has to be incorporated in the general solution. There is no any fundamental advantage in a developed country over the developing country in the age of globalization. In a stagnating region in a developed country the situation is even worse as in a developing country where people normally are not deprived of the hope of improvement. This text is based not as much on the academic studies as on 10 years of practical experience of the author of this discourse working as a development engineer in the field of the sustainable technologies. There is a strong believe that or there will be very much attention shown towards the complex economic solutions or there is no way the pyrolysis-based technologies will find the broad popular support.

2. Green economic defensive initiative

If an investment in a technology is not profitable, any investments in this technology will be just a dream regardless of how much this technology is beneficial for the humanity. Of course, the state can get involved and finance the implementation of the technology, but for a limited period of time. The development of the technology is way more stable process when the technology grows naturally during its industrial implementation. Time works for the technology and the technology grows with time.

Newfoundland could have been a rich and economically self-sustainable territory, but nowadays if not the recent discovery of the shelf oil the province would have been just poor. This is the direct contradiction to the modern state of development of technologies and the availability of natural resources on the island. There are winds, waves and solar power for energy. There are ores on the island and the neighboring coast. There is carbon in the form of the very rich on vegetation sea shelf.

As always there is the discussion between professionals of different trades about the measures that could have been taken by the state to improve the situation, and where would have been the border between the state involvement and the well-organized popular movement. How far the state must go in its endeavors to support and control is a very controversial topic. Internet grew out of the USA military technologies, but the leading USA corporations advocate for the free market development. Sweden is trying to exercise the tight political control over its development and is a success, but North Korea with its totalitarian control over everything is rather an economic failure.

The main problem with the governmental funding is that it is limited and quite often poorly selective. It relies on already existing sources of the national income and non-profit investments what excludes the possibility of self-propulsion. A solution without the idea of self-propulsion is the recipe for a failure in the long run.

In plain words, the task is to build a self-sufficient economy if not the closed cycle economy within the conditions of the global free trade economy with the orientation on the global markets. The contradictory in theory task is not a contradictory task in reality. Historically there are examples that rather support the claim [1]. The concentration of production in one spot in the global economy will automatically lead to the spill over of the production to other regions and to the migration of the labour force to this region thus rendering the idea of the closed economy unsustainable.

Possible solutions with the macroeconomic instruments for the local economy in context of the global sustainable development would be protection by currency, protection by import taxes and redistribution based on various economic features. All these solutions are secondary after the civil society is already in place. Without the civil society any investments in any production beyond the tradesperson level are impossible. These are macroeconomics topics.

The discussion on this level of analysis comes every time the economic problem is out of scale of the microeconomic discussions.

The objective of this paper is to discuss the possibility of any solutions on the local level. An engineer is the key figure for the local solution. A cash generating product should be invented, costs of the production should be minimized. Additional problems should be considered, such as sustainability of production in the long run, suitable industrial organization, and many others.

The recent history knows a number of spectacular failures of companies and governments making investments in new technologies. There are a few examples to mention. There were some talks circulating in the professional environment concerning the German investments in the solar energy projects in the northern Africa. There were some articles in the media concerning Italian attempts to produce components of TV sets in Russia. Both projects failed because of the lack of interest of the local population. Definitely, the companies were bringing wealth to the local communities, but the population of the local communities felt that they were turning into the colonial deprived population. Direct resistance or very negligent fulfillment of duties are normally the outcome. In both cases the tremendous financial losses were the result.

What is absolutely necessary to do from the earliest point of the foundation of a new enterprise is to establish the proper relations with the local labour force and all active parties involved with the project. Here comes the engineer to the forefront as the leader of the working and the conveyer of the technical fulfillment of the project.

A possible solution for the involved participation of the local labour force could be industrial farming. This term is coined by the author of this discourse. The idea is straightforward and simple, though the technical and the organizational side of the project is a challenge. It was noticed that classical cottage economy was a possible economic solution in terms of the involvement of the local population in an economic activity. Small proprietors and at the same time producers develop positive attitude towards the means and the process of production as far as there is a demand for their product and the flow of cash allowing for the accumulation of wealth [2].

Economic studies based on the German reality clearly show that small productional units are no less efficient than big corporations. Studies based on the American reality show that small economic units have lower income per worker than big corporations. Putting all together it is possible to conclude that local productional units in junction with the centralized system of coordination, procurement, and sales is possible and necessary in order to compete successfully on the international markets. The centralized organization must take over such questions as labour education, sustainability of production and certification of production processes and produced goods, overcoming the barrier of the initial capital when required.

Let us take a hypothetical country with low industrial development for the ease of analysis. Let us say, we want to organize the solar panels production of energy in the form of the industrial farming, let say somewhere on the African soils with the idea of sales to the European Union. Clearly this idea with the modern development of energy transmission technologies makes little sense. The resistance in electric cables consumes most of the energy over the long distances. And actually, it is the plague of Canada, where the waterpower plants in the North of the country with all their potential find no consumers as all the consumers are far away in the South. Nevertheless, for the purposes of the economic analysis, this could be a proper example.

There is no question that local peoples of Africa are well organized in terms of their local cultures for the purposes of solving of the local problems arising out of the state of the local development. But the existing patterns are not suitable for any investments in any complex production. Any investments are not secure, no appropriate technical stuff, and there is no mechanism to secure the return on the investments made. All these missing parts are still to invent in order to secure any growth of the local wealth and stabilization of the local society as the whole.

There is a need for a sustainable product, a product that can be sold repetitiously for profit in the long run without the disruption of the local investment patterns. In case of an African scenario everything has to be done pretty much from the scratch. Sale markets have to be found and a product for these markets invented, local labour markets invented and the payments for the labour invented and better in the manner of the appropriate solution in the long run. The technological process suitable for the local reality should be put together and suppliers found. In other words, no sustainable development without the sustainable economic and industrial organizations. No sustainable economic and industrial organizations without the sustainable product.

In the developed countries the situation is not so much different. Though the exchange markets are in place it does not mean that the local economy effectively protected. It can be overwhelmed by businesses from much stronger or cheaper economic regions and suffer from flight of labour and businesses. It can suffer from the local social immobility being unable to initiate a new production up to the standards of the global economic and technical development. Even if all the necessary economic institutions are in place a new sustainable product has to be found and the sustainable production has to be organized.

Any commercial production is dependable on the cash inflow or the revenue created through the sale of a commercial product. The product that can be sold in the long run is central to the process. Existence in the long run means that markets accept the product, its revenue covers all the expenses, and its production does not disrupt the social and natural environment.

There are always concerns about the saturation of the existing markets. There is always a fear that to introduce a new product to the existing markets or to increase the already existing volumes of goods for sale is almost impossible. There is the perception that the modern markets are contracting and not expanding. The counterargument would be the assessment of the global markets by the coverage of the global markets with the existing supplies. From this prospective the global markets exhibit an enormous potential for growth. The growing population is outstripping the globally available resources. The majority of the global population is in poverty. Definitely, everything that can close the gap is feasible in theory. In praxis this means that or a new product finds its markets without contributing to the global problem of shortage like jewelry or the global shortage coverage analysis can validate the future existence of the product. The proposed sequence of analysis from the sustainable product, to the organization of production is: labour involvement and remuneration, labour education and social impacts. Technical aspects—the organization of the production itself, the system of provision, storage and delivery. Business solution—financing, marketing, and sales.

Stepping aside from the proposed sequence of analysis the first question to discuss is the protection of the profit pro unit of the product sold of the production in question. The latest trend in the economic development is the protection of the local producers by means of the local currency. This method has very much potential und fully recognized by certain members of the European Union. The main exchange currency of the European Union is euro. Some countries like Check Republic keep their own currencies for the local circulation. The floating rate of exchange of currencies allows bringing in balance the imports and exports of a nation. The buildup of a stack of currency of one country by another country will immediately trigger at least in theory the adjustment of the exchange rates thus rendering the deficit or surplus in the balance of trade impossible. As the result the national industries as a whole are protected from the foreign competition and the imports match the exports.

At this point many professional people may wonder what kind of protection and from what it would be. It is quite clear that the material wealth is determined mainly through the availability of material resources and productivity of labour. But, first of all, we are living less and less in the world with the dominant material culture in terms of the material production. Alone the video games with the whole army of the support personnel working in the virtual reality prove the point. And besides this, it is about the protection of the sustainable balanced development. The developed industrial centers may produce cheaper and better. This leads to the depletion of the local financial power, regional stagnation and the flight of labour and businesses. The introduction of the local currencies leads to the immediately less efficient production but gives the opportunity for the local economic powers to grow. Maybe it is not so evident when one looks at the USA, but when someone looks at Africa next door to Europe one sees an absolutely different picture. In the States there are very strong agglomerations of the national industries and the rest are most of the time thinly populated areas. In Europe, it is reach western European states and just over the border—poverty.

The national economic strategy based on the idea of the global technological leadership is pretty much impossible for the majority of the nations. What is absolutely possible is to be not worth than the rest of the world and sometimes even better. There are ways to protect the local industries, pay for the trade deficit, learn from the rest of the world and even go forward and contribute to the world's development as the provider of the new ideas. Let us call the strategy the economic defensive initiative. If these strategy is based on the local renewable resources it has the right to be called the "Green economic defensive initiative".

In other words, with the local currency as the mechanism of redistribution that makes local prices cheaper than the foreign goods at the expense of the local economic efficiency, there is a chance that there will be no flight of the local industries and labour out of the region. The protection of the local industries by means of other economic institutions is not necessary. The facilitation by such means as well-developed infrastructure is not a bad idea. An alternative to the protection by currency would be direct transfers paid by the federal government to regions to sustain their local economic development.

It is still quite unclear how to introduce the scheme imitating the national local currency in a community or even a territory. So far with only a few experiments on the go it is too early to say to what extent the scheme may work. In an African remote local community, it is possible to introduce something like a local currency with the clearing center somewhere in a Western bank. Actually, RBC bank of Canada was ready to help out to run the scheme for any community, for an African, Latin American or East European. The support was to be provided for the financial operations of an enterprise and extended for the individual accounts of the labour involved.

3. Balanced reginal economy

Another a very powerful tool of the economic analysis that has to be mentioned here is the analysis of the national economy by the structure of the national employment. The proportionate number of people employed by different sectors of an economy is based on the productivity of labour in these sectors and the gross output by each of the sectors. Next step after the employment structure analysis would be the input–output matrices and planned economy, but this is not the initial purpose of the analysis. The employment structure analysis is a very powerful tool for finding bottle necks by productivity in the economy and justification of the investments for the improvement of productivity by sectors of the economy. This tool also is very handy when the economy is not based on the free market money for goods transactions, but goods for goods exchange. For example, if a local community owns a query and this a community in a developing country, then a financial payment will not increase the well-being of the community as much as for example building a school, a hospital or alike.

The employment structure analysis allows to estimate the number of people needed by sectors before the investments in the subsidiary industries are done. Proper investments can be vital for the successful development of the enterprise. Possible fields of investments: education of labour and procurement of the means of production, professional organizations and certification, protection of the industries and provision of supplies of the input materials. It is about trades people in the service of the local community and the enterprise. One of the possible ways to support the local tradesmen is the provision of facilities where the small industries can be successfully located, supported and function. It is very important that financial adjudication services are provided in addition. The European history knows plenty of successful examples of the organization of this type.

From the start it is necessary to decide if unions should take place in the organizational and social life of the future enterprise, should unions be a part of the local regional economy or not. Unions can be a very important factor of labour organization for the purposes of the self-identification and solution of the local problems, problems associated with the labour itself as well as technical problems on the production site [3]. Unions can be the cementing factor for the local community. Unions can keep the moral of the working force high. On the other hand, unions can be a tyrant that will drive the production costs high and force the acceptance by the enterprise of the inefficient solutions in all of the aspects of the industrial organization.

Should there be unions or not is an open question. The modern trend in the global economy and especially in the USA is to avoid unions in the economic life. Such organizations as Walmart succeeded at keeping unions at bay. The managers in the company are trained to counteract any intentions of the creation of the unionized labour. On the other hand, such countries as Germany are very successful at cooperation with the well self-organized labour.

Regardless if this is a unionized or not unionized enterprise, the same problems will arise, and someone has to solve them. If there is no union, bound to the enterprise officials must be appointed. If there is a union, there should be a form of control over its activities that is often even prescribed by the law.

Labour has to be paid and the payments have to be secured. There should be a place where savings can be made. In case of the retirement the retirement payments must be secured and carried out. All these operations often are impossible in the developing countries. In the developed countries the private financial organizations can take over the functions of the state allowing for more financial freedom and allowing for more just redistribution systems. For the developing countries it is the security that is the prime issue. It could be as simple as robbery or complicated as corruption or ill devised financial or taxation policies. Financial institutions located abroad with the accounts for each of the involved in the labour force could be a solution. Of course, the situation is not as simple as this, and many more questions will arise as, for example, what exchange rates and for what currencies should be applied and when exactly and where transfers should be made, or how to ensure that exactly the right recipients are getting the payments at the pay-off.

It is necessary to remember that one of the prime goals here is the sustainable economic development of the region. An artificial introduction of a local currency can really help a region to accelerate its sustainable growth. It is possible to reduce unnecessary imports substituting them with local products and use currency to buy the stuff that promotes growth. The economic model that helps in this situation is designed to view big agglomerations of population as self-supplying systems with extra import to substitute for the lack of the local resources or production.

A good example would be the idea of steam engine transportation based on the solar energy. For a developing country to buy foreign goods is prohibitively expensive, as these purchases deplete the national stack of currency at the adverse rate of exchange. Thus, imported fuel is too expensive in comparison to a locally produced fuel. The currency spent on the foreign fuel could have been spent on goods promoting sustainable growth. A possible local substitute could be steam produced with the solar power and supplemented with biofuel to keep up the temperature in the kettle on the go.

Naturally, additional questions arise, it is still unclear to what degree it is possible to control the quality of the product produced in a developing country. In the example of the solar and pyrolysis oil run locomotive, who can guarantee the cleanness of the product and who will pay and with what means for a travel with this locomotive. One may wonder, if it is enough just to organize a profitable business to generate the wealth for the local development or a much tighter control will be needed.

An economic policy can be evaluated based on the stack of cash in the possession of different groups in a society. It does not matter how the volume of money is defined. The discourse on the conversion between M1, M2, and M3 money is not a goal of this paper and will be avoided here. Based on the stacks of currency is possible to assess what consumption pattern will exhibit a nation in the short run or what economic policies will bring the nation out of the depression balance. For a local economy this means the predictability of the social impact with the emergency of a new profitable industry.

The same way the rates of the national or reginal inflation can be predicted based on the available stacks of currency. It is possible to imagine that with the high rate of inflation money stacks are being depleted faster than normally and the national speed of currency circulation slows down as everyone is cautious to spend the rest waiting on the events. Thus, inflation dies out much faster, as it is normally predicted.

Back to the development of the local production. The international borrowing definitely can be the source of the necessary capital, but it also will be the source of the local organization. The model under which the local community finds itself facing the foreign financial institutions helps to answer two questions, under the necessary rate of payment on the borrowed capital what salary levels can be achieved by the members of the local labour force. Definitely, according to this pattern of reasoning the wages achieved will be much higher than a competitive wage. Even if the source of capital comes from the local sources, as the local pension fund for example, it still can be regarded as the foreign borrowing for the ease of analysis.

Through the higher wages local investment funds can be accumulated. Saving in the pension funds, local or national, should also be viewed as the local investment funds. Questions concerning local funds and local investments are open questions for discussions, regulations, and choice of the business organization. It is still to decide who has the right and under what conditions to use the funds for the investment purposes.

A similar situation arises if a union takes control over an enterprise or at least plays a very significant part. The questions to be answered how to decide on the establishment or abolishment of jobs and the rate if payments. It is necessary to lay down what role can play the union in the choice of external partners and the conditions of the contracts. Such contracts can be very beneficial for third parties and can lead to the conflict of interests. It has to be established if the employees have the right to receive a premium paid by the external partners and what would be a bribery. Probably the union must have the right to review these contracts and have the power of veto.

The feasibility of wages can be estimated based on the international rate of financial borrowing. The idea is that the foreign financial institutions find themselves in the free competition and the international interest rate is established [4]. Thus, the flight of the local capital does not make much sense and the local producer is facing the international rate of interest as the result of the international financial situation.

4. The necessity of the commercialization of the process

The borrowed capital has to be paid for and paid in the long run. Labour and suppliers have to be paid in the long run. The suitable product has to be found and this product has to be sold for profit in the long run to pay for the contributing factors. Thus, the invention of the sustainable product is the prime aspect of the local economic development.

A sustainable product can be invented based on the regional economic patterns and the outer economic situation. The survival of the local economy will be based on the ability to defend its general industries and to promote the growth of its specialized export industries. The best scenario is when the sectors of the local economy exhibit sustainable proportionate growth and trade deficit is paid by the surplus of the local production.

Investment fonds should not necessarily be of the international nature. Profit from the operations can be redirected towards savings thus creating the investment funds. Savings are generated by both sides of the economy by the businesses and the labour. The created investment funds can be very important for the regional economy as the source for the establishment of the local enterprises. Unions can play a very significant part in this process that should at least in theory promote the business-oriented thinking of the union members. The offspring enterprises can benefit the former labour force and generate even more wealth for the local economy.

Here comes very handy the structure of employment analysis. Investments should manifest themselves in the changing structure of employment of the regional economy especially if the enterprise is dependent on the local resources and the local social climate. The only possibility when investments do not reveal themselves in the changing structure of the local employment is when the investments are done in the capital and the productivity in the sectors is growing equally fast.

The national rate of savings is one of the macroeconomic parameters that describes the performance of the national economy [5]. This parameter can be applied towards the regional economy. This parameter allows to estimate the portion of the national or regional income that goes towards investments. For the analysis of the regional economy this parameter may become especially important if the material or intellectual capital is in big demand and of the imported nature. By readjusting this parameter with the macroeconomic measures, it is possible to meet the financial needs to pay for the necessary imports.

The possible solutions for the readjustment rate of regional savings can be the royalties paid at the exchange of the local currency, local taxation that goes strictly into the investment funds, and all kind of other possible type of royalties. The regional pension funds with their accumulation of funds can play a significant part in this process.

It seemed that the production of electricity deep in Africa is not a sustainable in the long run idea. The empirical results rather were confirming this assertion. This is the experience of the engineers without borders in their efforts to improve the living conditions of peoples of Africa. The local productions for example of solar panels died out first of all due to the poorly understood economic factors. First of all, Engineers Without Borders failed to commercialize the process thus rendering the payments to the providers, labour, and professional stuff not sustainable.

Engineers without borders were not a commercial organization and was not trying to organize a profitable entity in the long run in the first place. One of the reasons it failed to create the sustainable development in the countries of its operations is that it was lacking the will and intelligence to bring the sustainable changes to the local economies.

It is very possible that even when there is the will to commercialize the process the local markets still have to be invented. The situation is straight forward: local resources are not available as no one is selling anything. There are no means of buying something as there are no means of payment in circulation. The same is with the local and international labour, people have to be remunerated for their efforts. Work must provide people with means to sustain their living increasing its standards. With no local markets available the two classical solutions are available: the forced labour camp or the military camp. The first solution means people are forced to work with the bare minimum of supplies, just to sustain their living. In the second case the supplies are shipped from the well-established centers of production and redistribution. The first solution is unacceptable, the second is prohibitively expensive.

The historical solution in the absence of the well-established markets for goods of general consumption in volumes comparable with the national consumption is the process of taxation in real products and assignment. There are plenty of historical examples when whole villages of the peasant population were assigned to certain productions. A classic example would be the early Middle Ages. It does not mean that the peasants have to be the labour force in the production, they can be just the suppliers in goods to provide for the labour, first of all food. For a developing country this solution still can make much sense as the back payments to the population can be made in real goods as medical equipment, building materials or anything alike. To implement this solution in a developed country is rather problematic first of all due to the elaboration of the consumer basket.

Another solution can be borrowed from the Soviet Union. At a certain point the government allowed the sale of the foreign goods to the population for people with the special currency. The system was called Bereska, probably another instance of the national humiliation. To what degree this historical solution can be implemented in the development economics is not quite clear, but the idea is quite promising. It is possible to pay to labour and small suppliers with the nominal records money that can be traded for real high-quality products in the specialized stores. The secondary markets better be tightly controlled or organized or there will be high rate of corruption and criminality. For more details on the criminality spurred by the Western economic aid please read about the recent Somalian history.

The idea of production of electricity in Africa for sales in the European union directly evidently is not a sustainable one. Big production numbers will turn into a trickle by the time they will reach the markets in Europe or elsewhere. Nevertheless, this is the question of sustainability to invent a product suitable for consumption over big distances produced locally. And electricity can still be the one as far as it goes into the product as direct or indirect input. A good example would be coffee that an item of export to Europe. Through coffee electricity can be sold in the European union indirectly.

Another possibility to sell electricity indirectly is to save energy in the chemical bonds. The chemical production requires input of energy. Solar panels and wind or tidal turbines deliver big quantities of energy, but of coordination with demand. The production process on the contrary can be schedule in accordance. Thus, solar energy can be saved in chemical bonds and sold over big distances. A good example could be the idea to sell clean heating oil to the big agglomerations of population.

If the organization of the local market economy is not possible at least in the short run, the organization of a profitable enterprise and the positive influence on the regional development are still possible. What matters is the agglomeration of physical consumers as the end point of financial transactions. Financial capability of the local population is not so necessary as there could be other means of exchange and redistribution. The main income at least on the initial stages of the new local development will come from abroad. It is still the availability of the local labour and availability of certain local resources that play the crucial role.

Selling to the local population quite often can be a problem as the local markets still have to be organized. Even if there is quite dense local population it still can have zero buying power. Here comes handy the money stack analysis to model the creation of the local markets based on the successful local production. Alternative to the creation of the local markets would be the direct redistribution of the consumer products. The problem is that it pretty much entails the creation of the local planned economy with coverage of the whole consumer basket.

The local markets in the typical African context cannot be defined as sufficient in terms of promotion of local industrial development. It is not so much about the financial unattractiveness of these markets. The financial situation finds its manifestation in the material exchange that the local markets can suggest in return for the produced industrial goods. Historically this is exactly how it went on, the parallel development of the agricultural und the industrial sectors of the economy. The main problem is time that was necessary to put together the accidental factors. It took literally centuries, if not longer.

On the other hand, the autonomous development of the local markets is still possible. Again, the example of Newfoundland, it a province in a well-developed western country. Local markets are small, but financially sound. By means of the real exchange economic scenario it is possible to conclude that nothing prohibits the high level of economic/industrial local development. In theory the availability of the local resources should be sufficient for any level of wealth limited only by the level of development of modern technologies.

While the Newfoundland economy is very low on performance the same of Sweeden is quite sound. Both countries are in the northern climate and cannot compete with the rest of the world in the traditional sectors of the economy, like agriculture. Sweeden is flourishing because it can manage its internal sectors of economy and trade with the rest of the world with specialized goods.

A good theoretical question if any local product in surplus can be used as a specialized product for the international trade. The best approach would be to consider is the saturation of the global markets with this product and the possibility of the stagnation of prices for the product by the immediate neighbors in case of the high volumes being supplied [6].

Sweden retained its internal markets even under the new socialist system. This allows for the local development. Investments in the local economy pay off. In case of the developing countries the absence of the well-developed local markets is a big problem. A newly established enterprise should orient itself towards the external markets and something should be improvised for the local situation.

From the historical perspective Sweden was undergoing its pro-socialist transformation at the times when the global competition was not as acute as it is now. By the time of its great changes Sweden was already one of the world most developed countries. Sweden could preserve its economic institutions and develop further. In case of the developing countries the situation can go as bad as no markets for goods for money at all are in existence. What comes very handy is the structure of employment analysis and general planned economy. Good question if it is still possible over a short period of time to nurture the more efficient free market and monetary circulation by introduction of financial payments for labour and suppliers at the newly established enterprises. For example, buying food from the local population, and the local population can get high quality import goods in the specialized stores.

This is where the Engineers without Borders failed by not being able to create long lasting impact on the societies in question. Everything was falling apart after they were leaving. No mechanisms of perpetual maintenance and improvement were introduced.

The USA government followed the same line of reasoning trying to bring industries to the less advanced countries. The well-known to the public cases so far were not very successful. One of them to mention is some operations in the T-shirts production. One enterprise was placed in Vietnam. The main problem with this placement was that the only operation left to the enterprise in Vietnam was earning literally pennies leaving the workers with the wages way below any international standards.

A sustainable product must have the ability to be sold and produced in the long run. In the production process the producer and the working force must feel themselves motivated. The product must be environmentally friendly, face the continuous supply of raw materials, and the enterprise be able to find and educate professional working force. What else must be packed in the definition of the sustainable product is open for further discussions.

The Engineers Without Borders the same way as many other economic powers like international corporations did not manage to create the sustainable product first that would contribute to the continuation of the local production. Like solar panel projects that led nowhere. No one even cared to establish the mechanisms of the economic exchange keeping the production cycle running.

Do not say never ever will happen to transport energy from Africa to Europe. How about producing biofuels, energy can be saved in the carbon chemical ties and brought to the consumer. How about an even more devilish plan? The locally produced in Africa energy will come to Europe as coffee. No one suggests burning coffee beams in furnaces like fuel but drinking it. The idea is the substitution of products, stop bringing low quality fossil fuels from abroad, produce local electric power, produce local heating oil for power plants and for engines, locomotives, heat up with solar, maintain the temperature with the heating oil.

And again, about the USA industrial history. In the second half of the XXs century pretty much all the innovations in the agricultural sector were invented and tested by the research institutes run by the USA government. What makes it to believe that the lack of the local technological development is not justified on the grounds of the lack of the economic exchange markets and the protection of property. In the States the level of the economic development is much higher than in the developing countries, especially in the countries to the south of the European Union. Normally countries do not have the luxury to enjoy the privilege of continuous funding for the research and development purposes. To sustain the sustainable growth the sustainable investments in the research and development is necessary. The best way to achieve it is to build a sustainable industry that can pay for this development and test of the new technical solutions in praxis.

Through the creation of the sustainable product the local markets are getting the chance of incorporation into the global economy. At the same time the local markets must have the protection mechanisms that will ensure the existence of the local industries. The state of affairs of the local peasants being forced to the natural economy is not an example of the well protected local economy, but rather of its defeat.

Without making investments in a technology all the dreams about the implementation of the technology will remain just a dream. For any investments to take place the technology has to be financially attractive. The implementation financed by the state is not a great solution as it exists only as far as there is funding. The task of an engineer is to improve the technology reducing costs up to the point where it is cheap enough to compete with alternatives. If it does not work, it does not mean that this is the end of the application of the technology. There are plenty of neighboring factors that should be considered. Quite often the whole technology should be revised. Production with the technology requires inputs, has side effects, and produces products. The implementation of the technology should be reviewed on all the stages of the production process.

Speaking about inputs it is necessary to keep in mind that there are all kinds of them. There are inputs during the construction and maintenance of the production process, and inputs that are used to create the product itself. When it comes to practice, the difference in approach is evident. The initial inputs are the ones that often are taken out of consideration during the readjustment process.

On the financial side of the analysis inputs are the costs that will be put against the cash flow generated by the sales of the products created with the inputs. It is possible to minimize the expenditures on the inputs, on the waste handling and to increase the revenue per unit of sales. With the growing costs of inputs and the price of disposal of waste the avoidance of waste is picking up in importance. It happens that the byproduct of production can be the most revenue generating seller.

Smart industrial organization can play the vital role in the development of the relations with the labour involved and the local population. The proposed idea is the idea of the industrial farming. In the case of the local energy production this idea would be called the energy harvesting farming. The idea is to try to avoid using mega plants and rely more on small production units operated half independently by the local population at the site. This organization creates the positive attitude towards the production in question as it makes people feel more like the proprietors than the exploited colonial labour. The not interested population as isolated islands can be simply circumvented without much damage to the production as a whole.

Historically the cottage economy did not survive the competition against the fast-growing agglomerating companies. It does not mean though that with the modern industrial organization the idea has no future. Many big corporations build their operations on franchising. The organizing authority will deal with such problems as securing the suppliers, solving internal disputes, controlling prices and quality of the product, dealing with the side effects and securing the ways of distribution of the product. Other important issues would be optimization of the production and the education of labour.

5. Invention of a sustainable product

Speaking about the optimization of production in the rural areas, wide implementation of pyrolysis can be the breaking through technology. Following the same logic of optimization, the pyrolysis production cannot be just the source of the local cheap fuel. A much mor sophisticated sustainable products have to be invented.

If it is energy in form of the biofuel it has to come together with the invention of the local means of transportation like the solar-biofuel powered locomotives. If it is biogas, it should be produced with the hinder thought of using biogas to produce bioplastic to use as the input for solar lens for the local heating installations. The latter can be used as the source of energy for the local production as drying of coffee beans or alike, for cooking, or for heating of the local housing. Imports should be reduced, and exports created. The import of plastics and fuels should be diminished, and the foreign currency used for something more useful. In this case a sustainable product like coffee will pay towards the balance of trade.

Not only coffee, there is a vast field for product inventions. While a new product for sales over the long distances can be invented the production waste can be reduced. The mineral residue from the pyrolysis won from the biomass can be used as mineral fertilizers [7]. Historically people were burning forests to use ashes as mineral fertilizers. Nowadays the mineral residue can be packed and sold, even in small packages as pot flower fertilizers.

It is not yet quite clear if there is a place for a much more ambitious projects than mineral fertilizers from the pyrolysis process mineral residue. Bioplastics may be created and used for solar energy cooking and heating thus reducing the use of wood for heating purposes and deforestation in general. Mineral residue can be used if produced in very big quantities for the purposes of reversing the process of desertification.

Here comes very handy another idea—the idea of the processing of the seaweed. Normally the natural circulation of water washes off the landmass very much of the useful substances, like the mineral components of the soils. Water dissolves and the sea water contains all the possible elements and substances. The next question would be of how much of this stuff can be won back from the seaweed. Seaweed normally grows on the shelf from the stuff accumulated from the landmass. The most trivial idea is that the whole palette of products can be designed around one single industrial process more complex as turning biomass into biofuel and pyrolysis gas.

Depletion of the global resources probably will be the prime driving force behind the further development of the pyrolysis-based technologies. Bioplastic based substitutes can be won from the biomass [8]. Of special interest here would be the production from the seaweed. As the world is facing the problem of the global dispersion of the valuable materials researchers are looking for the technologies to reverse the process. The natural circulation of water slowly moves everything to the world ocean. An open question is how much of this stuff is accumulated in the seaweed and can be won back from the pyrolysis residue.

Energy saved in the chemical bonds can be transferred over big distances and used again. Thus, there is at least one good argument to use local cheap energy. In the south it can be in the form of a solar power plant, in the north—the tidal or wind energy, and everywhere the energy of the solar panels can be generated and used. At least, the accumulation of the energy in the chemical bonds lifts the problem of the time inconsistency between the production of the energy and the demand for the energy. And definitely the alternative energy should not be only used for the conversion of one type of energy into another. Any industrial process needs energy and this energy can be generated cheaply and locally.

Another sustainable biochemical process that may be successfully combined with pyrolysis is biogas. The positive side of the process is that it can be a very labour non-intensive passive process. Any problem with the thermal processing of any biomass is the high content of moisture. The biogas production involves the slow consumption of biomass by bacteria that results in the production of biogas and slush. The water molecules are being freed in the process and thus can be drained. If the resulting residue can be a good feeding material for the pyrolysis process is still to be researched.

Recycling performed in the sustainable way based on the pyrolysis or gasification still can be considered for an industry to be placed in the region suffering under underdevelopment. The idea behind is the same, the carbon-based part of the waste can be separated from the rest of the waste material. Metals can be extracted based on their magnetic qualities. The rest is the rest. The whole process make sense though only if there is an abundance of very cheap energy to start with.

Processing of biowaste can be an absolutely different story from the financial perspective. Biowaste can be concentrated at the site of its generation and be treated locally. A good example here would be an animal farm. The animal waste is already quite energy saturated what makes the production of biofuels quite lucrative. Another aspect is that the mineral component in the animal waste is very high thus making the animal waste a potential stock for the production of mineral fertilizers.

Fresh water reservoirs can become a potent source for the production of biofuels and bioplastics. Water reservoirs allow for quite fast and very intensive growth of all kind of organisms capable of photosynthesis. Also, these reservoirs accumulate the run-offs of the fields. Water organisms use these substances during the life cycle. Processing of these organisms means the further utilization of fertilizers, not to mention the production of high-level energy substances based on carbon. The high density of the biomass and the ability to float and be carried down the stream can considerably low the costs of the further processing. The project can be combined with the extensive water treatment facilities and the fish and poultry farming.

Minimization of costs of production may still involve the invention of the special technical solutions and choice of local or alternative materials. Speaking about the European perspective such solutions as cheap metals from Siberia can be implemented. Speaking about the rural environment of Africa such local solutions as the local production of bricks can be considered. In every environment where the preliminary extensive process in the form of biogas takes place the mobile inflatable systems can be transported to the site of the production located at the site of the biomass.

The pyrolysis-based technology can become the technological foundation for the emerging high-tech power cell technologies. In case the world decides to convert its technologies and rely on the renewables the pyrolysis-based technologies can be used to produce fuel for power cells. These are the newly emerging technologies and to elaborate on the topic is not yet possible.

6. Conclusion

The modern technologies allow to achieve much higher standards of living, as it is quite often seen even in the developed countries, and these high standards of living normally can be achieved through the utilization of the local resources. Pyrolysis based technologies can be successfully used to win substitutes for many non-renewable resources from biomass. Closed cycle local technological development based on pyrolysis is possible but remains rather a dream. The international progress makes the intensive trade with the surrounding world unavoidable, and the absence of the suitable local economic institutions make the implementation of the technologies by the regional producers impossible. New economic solutions and introduction of the new green technologies are necessary to sustain the balanced local development.

Author details

Brice H. Leeds[1,2,3]

1 Studies of Economics: University of Mannheim, Mannheim, Germany

2 Bachelor of Engineering: Dalhousie University, Halifax, Canada

3 Currently: Technische Universität Berlin, Berlin, Germany

*Address all correspondence to: briceleeds@web.de

IntechOpen

© 2021 The Author(s). Licensee IntechOpen. This chapter is distributed under the terms of the Creative Commons Attribution License (http://creativecommons.org/licenses/by/3.0), which permits unrestricted use, distribution, and reproduction in any medium, provided the original work is properly cited.

References

[1] Pruitt B. Self-sufficiency and the agricultural economy of eighteenth-century Massachusetts. The William and Mary Quarterly 1984;41(3):334-364. DOI:10.2307/1922729

[2] Handy J. "The enchantment of property". Arthur young, enclosure, and the cottage economy in England, 1770-1840. J Agrar Change 2019;19: 711-728. DOI:10.1111/joac.12334

[3] Spencer B. Unions and Learning in a Global Economy: International and Comparative Perspectives. Toronto: Thompson Educational Publishing; 2002.

[4] Edwards S. The pricing of bonds and bank loans in international markets: An empirical analysis of developing countries' foreign borrowing. European Economic Review 1986;30(3):565-589. DOI:10.1016/0014-2921(86)90009-7

[5] Summers LH. Issues in National Savings Policy. National Bureau of Economic Research Working Paper Series. No. 1710. September 1985. DOI:10.3386/w1710. http://www.nber.org/papers/w1710

[6] Poole R, Clarke GP, Clarke DB. Competition and saturation in West European grocery retailing. Environment and Planning A: Economy and Space 2006;38(11):2129-2156. DOI:10.1068/a3816

[7] Naveed AQ, Md. Mukhlesur R, Seunggun W, Soomin S, Changsix R. Biochar properties and eco-friendly applications for climate change mitigation, waste management, and wastewater treatment: A review. Renewable and Sustainable Energy Reviews 2017;79:255-273. DOI:10.1016/j.rser.2017.05.057

[8] Snell KD, Peoples OP. PHA bioplastic: A value-added coproduct for biomass biorefineries. Biofuels Bioprod Bioref 2009;3:456-467. DOI:10.1002/bbb.161

Chapter 4

Advances in the Pyrolysis Process and the Generation of Bioenergy

Vittor Rodrigues Santos Alves

Abstract

The reduction of environmental impacts caused by emissions of greenhouse gases has become an internationalized goal. In this context the development of technologies capable of producing energy from clean or renewable sources has gained broad prominence, among them the fast pyrolysis is a type of thermochemical process capable of converting biomass and agroindustrial waste into a liquid product called bio-oil that has a wide range of applications in the bioenergy scenario. For this type of technology to be consolidated as an alternative source of renewable energy, economic, political and environmental incentives are necessary, as well as research development to improve the conversion processes, such as reactor types, logistics in obtaining and pre-treating potential biomass, improvement and conversion routes for bio-oil obtained in renewable biofuels or chemicals with higher added value. This chapter covers the fundamentals of thermal conversion of biomass into bio-oil and the most studied processes to convert bio-oil into a product with better properties, such as deoxygenation and energy densification.

Keywords: biomass, fast pyrolysis, bio-oil, bioenergy, environmental

1. Introduction

The demand for clean energy sources and the accompanying reduction in greenhouse gas emissions have gained increasing emphasis on environmental policy agendas around the world. One of the current portraits of these discussions was in 2021, when the Climate Summit was promoted, in which the political leaders of each nation set social, technological and political goals for the coming years in order to mimic the agents causing natural and climate changes. Among the most discussed measures, mainly by countries in the northern hemisphere, is the attack on the consumption of fossil fuels such as diesel, gasoline and mineral coal.

Biomass, urban and agricultural waste have great potential to be converted into products with added energy value. Although there are several routes for converting biomass into fuels (biological, thermal and physical–chemical), most of them, as biological, are highly selective, offering little variability of the final product. In thermal processing there is the possibility of producing a range of complex products such as chemicals, advanced biofuels and integrated electrical energy. Although the production of charcoal, through the carbonization process, is an old and well-known process, it was during the oil crisis in the 1970s that research was directed towards obtaining renewable fuels from lignocellulosic materials, where fast pyrolysis stood out. In this process, the carbonaceous material is subjected to moderate temperatures (in the range of 500°C) at high heating rates and in the absence of oxygen, thus, the release of organic vapors and gases occurs at low resistance times in the reactor, in addition to the formation of a char fraction [1, 2].

The biomass pyrolysis process in general can be understood as a thermal fractionation of the lignocellulosic complex into three predominant fractions: char, gases and liquids (condensable vapors). The greater or lesser yield of each of these fractions is adjusted according to the input parameters and adjustment of process variables such as temperature, reactor type, biomass composition, biomass physical properties and so on. In the fast pyrolysis process, all these parameters are conditioned in such a way as to provide low residence times in the vapor phase reactor. When being condensed, these vapors form a liquid phase composed of a variety of organic compounds called bio-oil.

Bio-oil has a wide range of applications. Due to its varied composition (alcohols, organic acids, sugars, phenolic aromatics, etc.) it can be used to obtain chemical species through refining and extraction, raw material for obtaining advanced biofuels, in addition to liquid biofuels for processes of direct burning of biomass is not possible. In addition, as it is a liquid product, its handling and transportation are facilitated in relation to solid biomass. Finally, bio-oil can have its quality improved by modifications directly in the pyrolysis process such as downstream upgrading processes.

Thus, this work presents an overview of the fast pyrolysis of biomass for bio-oil production, the properties of bio-oil and its general application, showing the main processes (reactors) developed and new trends for the improvement of the pyrolysis process aiming to obtain more efficient bioenergy sources.

2. Fast pyrolysis: general aspects

Thermochemical processes have by their nature the conversion of raw materials into products that have greater added energy value. These processes occur having heat as the agent of transformation, either by release or demand of the process. **Figure 1** shows a summary of the main endothermic processes [3–9].

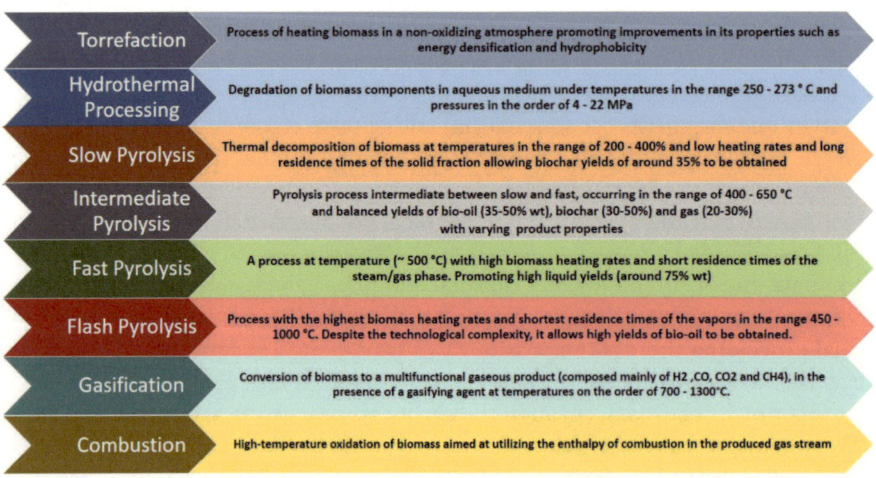

Figure 1.
Diagram showing the main characteristics of the most conventional thermochemical processes.

Property		Heavy oil fuel	Biomass feedstock			
			Mahogany waste wood	Oak/ maple	Birch	Pine wood
Elemental analysis (dry basis), % wt	C	83–86	50.2	—	44.0	45.7
	H	11–14	6.6	—	6.9	7.0
	N	<1	<0.4	0.2	<0.1	<0.1
	O	<0.1	42.6	—	49.0	47.0
	S	<4	<0.06	<0.01	0.00	0.02
	Ash	0.03	0.15	0.01	0.004	0.03
Moisture, % wt		0.1	21.4	22.0	18.9	17.0
pH		—	5.64	2.5	2.5	2.4
Density, kg/L		0.94	0.95	1.18	1.25	1.24
HHV, MJ/kg		44	29.52	17.0	16.5[a]	17.2[a]
Viscosity (40°C), cSt		5.5–24	4.6	50[b]	28[b]	28[b]
Solids (char), % wt		—	—	0.045	0.06	0.03
Distillation residue, % wt		<1	—	not distillable	—	—
Flash point, °C			68.0	55	62	95
Pour point, °C		−18	13.5	−25	−24	−19
Ref.		[10–13]	[14]	[15]	[16, 17]	[16, 17]

[a]*as LHV.*
[b]*at 50°C.*

Table 1.
Main properties of bio-oil from different feedstock compared to heavy fuel oil.

During the fast pyrolysis process, the lignocellulosic structure of the biomass is degraded and converted mostly under specific reaction conditions, generating a main product called bio-oil. **Table 1** shows an overview summary of the main physical–chemical properties of a typical bio-oil compared to a heavy fuel oil.

In order to quantitatively obtain organic products in the form of bio-oil in an expressive way, it is necessary to maintain some parameters and variables in an essential way during fast pyrolysis.

- High heating rates of the particle, for this it is necessary that the feed is in reduced particle size and moisture [18, 19].

- Good temperature control, allowing the process to occur in temperature ranges close to 500°C [18, 19].

- Low residence times of the vapor phase (below 2 s) as well as a fast cooling of these [18, 20].

By carefully following these parameters, it is possible to obtain bio-oil yields of up to 75% on a dry basis depending on the technology (reactor type) employed and the nature of the material. Moreover, depending on its composition and the way it is collected, the bio-oil can form two phases, a lighter one with a higher presence of water and a heavier one with a higher concentration of organic fraction.

3. Bio oil properties and its applications

In general, biomass is composed essentially of three organic components: cellulose, hemicellulose and lignin. Besides these there is also an inorganic fraction (ash) and the water associated with the structure of the material in the form of moisture. The composition of these fractions is an important reference in the quality of bio-oil to be obtained. **Table 2** shows a general summary of the lignocellulosic composition of various biomasses.

During the pyrolysis process these three polymers decompose at virtually the same time, however each has different thermal decomposition kinetics [21], which can lead not only to different yields of bio-oil and its sub products (biochar and gas) but also generate a variation in the distribution of organic functions that make up the bio-oil [22].

Due to the different physicochemical properties of cellulose, hemicellulose and lignin, they tend to produce different products. While cellulose tends to contribute to higher bio-oil yields, higher lignin concentration contributes more significantly to biochar formation [20, 23, 24]. The organic composition of bio-oil is extremely variable and presents several classes of organic composts. **Figure 2** shows a deviation map of these organic functions that typically compose the pyrolysis bio-oil in relation to the thermal reactivity of cellulose, hemicellulose and lignin [25].

In addition, the inorganic fraction of biomass (ash) can also influence the characteristics of bio-oil. Some studies show that biomasses with alkali rich ash, such as potassium (K), and phosphorus can exhibit catalytic properties and alter the distribution of organic compounds present in bio-oil [26]. A number of other characteristics of bio-oil can be described to assess its quality, such as appearance, typically characterized by a dark brown or reddish color; odor, which is characteristic of acidic flavors and can even cause airway and visual irritation upon prolonged exposure; and miscibility, which is a complex factor as it is not miscible with petroleum fuels due to the large presence of polar and oxygenated compounds,

Biomass	**Composition, % wt**				**Ref.**
	Cellulose	**Hemicellulose**	**Lignin**	**Other**	
Softwood	41	24	28	7	[9, 20]
Oakwood	34.5	18.6	28	—	
Birchwood	35.7	25.1	19.3	—	
Hardwood	39	35	20	7	
Leaves	15–20	80–85	0		
Olive husk	24	23–24	48–49		
Miscanthus	24	44	17		
Sugarcane bagasse	19–24	32–48	23–32	—	
Willow plant	50	19	25	—	
Rice straw	30	25	12	33	
Wheat Straw	40	28	17	15	
Nut shell	25–30	25–35	30–40		
Walnut shell	25–26	22–23	52–53		
Corncob	50.5	31	15		

Table 2.
Lignocellulosic composition of various biomasses.

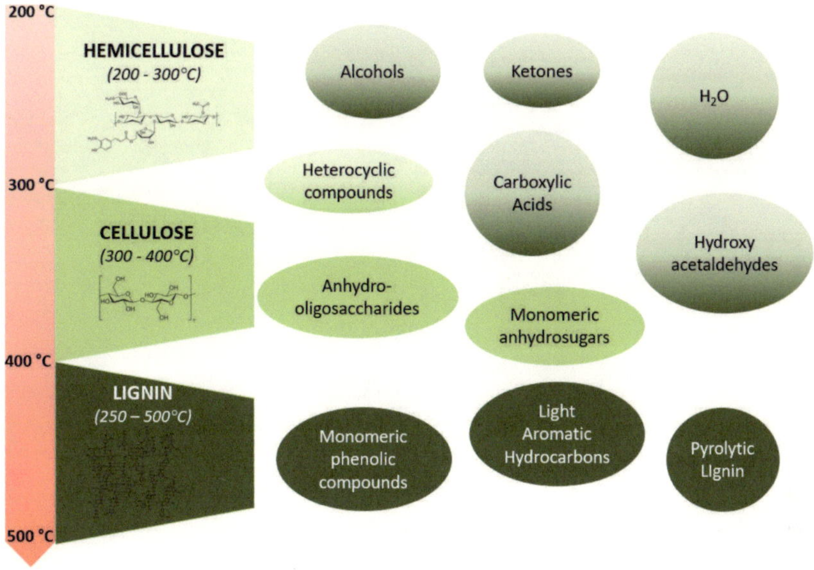

Figure 2.
Thermal decomposition of lignocellulosic compounds and their products (adapted from [25]).

and is not water soluble as excessive addition can lead to the formation of two liquid phases [19, 20].

Due to its varied composition, a number of applications have been attributed to bio-oil. **Figure 3** illustrates some of the main applications.

The pyrolysis bio-oil presents compounds in high concentrations such as phenols, guaiacs, acetic acid among others and with the advancement of separation techniques, such as liquid–liquid extraction, with the addition of solvents (hexane, chloroform etc.) has allowed the recovery of fractions with high concentrations of these compounds that have high added value for the industry, production phenolic resins, organic acids among others [27–30]. In addition, the greatest potential for use of bio-oil is in its use for the production of biofuels, being necessary, firstly, the application of upgrading steps to improve its properties.

Bio-oil can also be used in combustion or co-combustion systems with other fuel oils to obtain heat in industrial processes. However, the applications of bio-oil in these combustion systems are traditional for combustion of heavy or

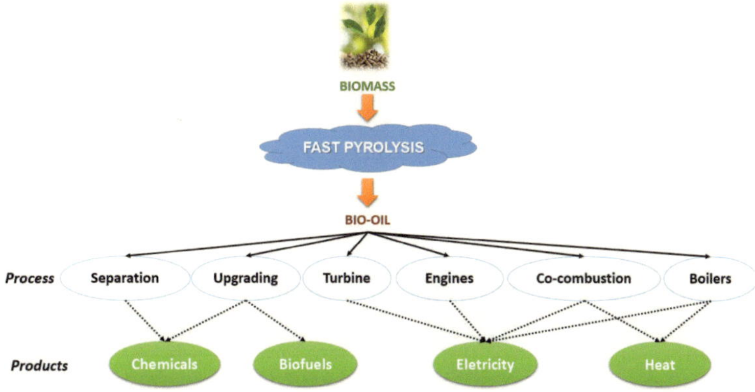

Figure 3.
Bio-oil applications.

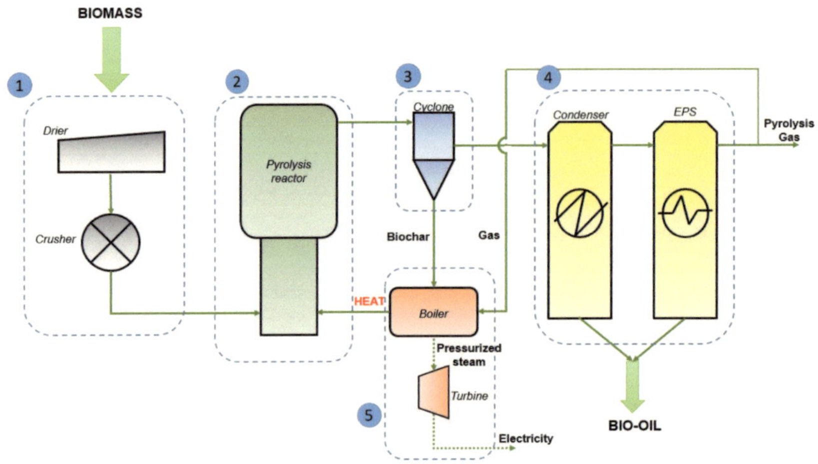

Figure 4.
General system to biomass fast pyrolysis.

mineral oils due to their physical–chemical characteristics completely different and their instability that can impact on essential parameters for a quality combustion of oil as the quality of atomization, ignition, tendency to coke formation, vaporization rate of drops, clogging among others. However, having a good understanding of the properties of the bio-oil used and the ratio some benefits are reported in co-combustion operations such as the reduction of SO_2 and thermal NOx emissions [27–30].

4. Reactors to fast pyrolysis

The design of a fast pyrolysis process has at its core the reactor where the appropriate heat and mass transfer conditions will be provided for the thermochemical processing of the material. In general, a fast pyrolysis process consists of a biomass fitting unit (1) to fit the required particle size and moisture conditions, the pyrolysis reactor (2) where the bio-oil (vapors) and by-products (biochar and non-condensable gases) are formed, cyclone(s) and/or particulate removal systems (3), condensation unit and bio-oil collection (4) and finally a thermal energy conversion and integration system in the system (5) as illustrated in **Figure 4**.

The pyrolysis is an endothermic process, as the goal of fast pyrolysis is to obtain higher yields in liquids, the other by-products (biochar and gas) can be used as fuels in the thermal energy generation unit providing the necessary entapia, to perform the pyrolysis reactions. Fast pyrolysis reactors in general can be separated into those that use a gaseous agent for heat transfer in the reactor and those that do not use gas as a fluid, and most of them are already in commercial scale. A **Table 3** mostra um resumo dos principais reatores utilizados nos processos de pirólise dus prncipais características e seu status comercialização.

5. Pyrolysis and bio-oil improvement strategies

Bio-oil obtained from lignocellulosic feedstock with fast pyrolysis has many disadvantages in front of fossil fuels such as high water content, presence of small particles of coal and alkaline ash, acidity, low calorific value and thermal stability that make it difficult to application of bio-oil in the power system such as

Reactor type	Main features	Limitations	References
Circulating fluidized bed	Easy scale-up. Efficient heat and mass transfer. Flexibility in system design.	Low solid phase residence time (similar to gases) leading to higher coal yields. Complex hydrodynamics.	[1, 31–39]
Bubbling fluidized bed	Good heat and mass transfer, easy control, operational simplicity. High bio-oil yields.	High pre-treatment accuracy of biomass, requiring low particle sizes	
Rotating cone	It uses centrifugal force to move the particles. Uses little carrier gas. Yields in bio-oil in the range of 60–70%.	A more complex in-grid system is required with cone pyrolysis reactor, riser for transport and recycle of the sand particles, fluidized bed combustion chamber.	
Auger or screw	The process takes place during mechanical conveying. The heat carrier can be the walls or ceramic/sand particles. Suitable for heterogeneous biomass/waste or waste that is difficult to transport and handle.	Longer residence time of the vapors. Lower bio-oil yield compared to fluid bed reactors.	
Ablative	Heat transfer is via reactor part contact, allowing the use of larger diameter parts and without the need for carrier gas.	Limitation in the scale-up mainly concerning the reactor surface.	
Spouted bed	Very high heat and mass transfer coefficients. Short residence times. High liquid yields (up to 70%).	Complex scale-up accompanied by high pressure drop.	

Table 3.
Description of the main fast pyrolysis reactors and their characteristics.

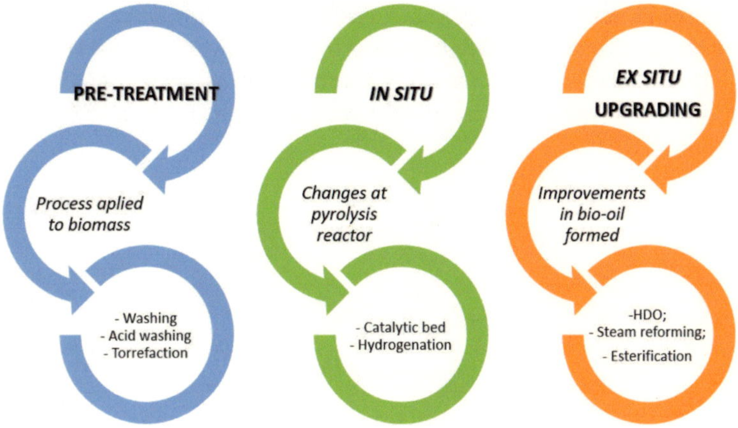

Figure 5.
Different processes and strategies to improve the fast pyrolysis of bio-oil quality.

engines [40]. The currently technological research shows that there are basically three ways to improve the performance of the fast pyrolysis and the quality of the obtained bio-oil. The **Figure 5** shows these deferments strategies.

5.1 Pre-treatment

Several authors have already reported the effect of physical treatments (i.e. treatments that do not alter the chemical composition or chemical structure of biomass) of biomass on pyrolysis processes. The most relevant one reported is the biomass grinding aiming at its particle size reduction, which has a positive impact on all fast pyrolysis reactors, providing an increase in the bio-oil yield [41].

The washing (or leaching) it's a process used to remove or modified inorganic compounds of the biomass or modified the lignocellulosic complex. The H_2O washing it's a simple and low cost process able to remove mainly alkalis (K, Ca), Cl and S [42, 43]. The Acid washing it's a more complete but expensive kind of washing with (organic or inorganic) acid diluted, this process its able to remove near all components (alkalis, halogens, heavy metals etc) found in biomass ashes promoting the complete demineralization [42, 44]. In addition to these effects, the acid treatment of biomass (washing and acid infusion) can lead to the formation of bio-oil rich in sugars [45, 46]. Studies have pointed out [46] showed that there is considerable enrichment in sugars, especially levoglusan (up to 55% of cellulose) during the acid infusion of corn stover. These results were mainly attributed to the catalysis of the decomposition of the cellulose structure.

Another process considered a pre-treatment of biomass for fast pyrolysis is Dry Torrefaction. This process is like mild pyrolysis; occurring in the range of 200–280°C and in an inert atmosphere, it promotes a thermal pre-degradation of lignocellulosic polymers, generating a biomass with the potential to produce a bio-oil with better quality [47, 48]. In parallel, the Wet Torrefaction (WT) process can also qualitatively improve the biomass properties for the fast pyrolysis process. It occurs in aqueous phase and ambient to moderate pressures, with water at high temperatures it removes alkalis, carbonates and halogens. Research shows that bio-oil produced from biomass pre-treated with WT produced bio-oil with less phenols, ketones, furans and richer in sugar, especially levoglucosan. However, in addition to energy consumption, this process generates a biomass with high water content [46, 49–52].

The process of acid treatment and torrefaction of biomass can be combined to improve the performance of fast pyrolysis both quantitatively (higher bio-oil yield) and qualitatively (bio-oil with better properties) as shown in **Figure 6**. Studies showed that the coupling of acid washed with acetic acid (1% wt) and

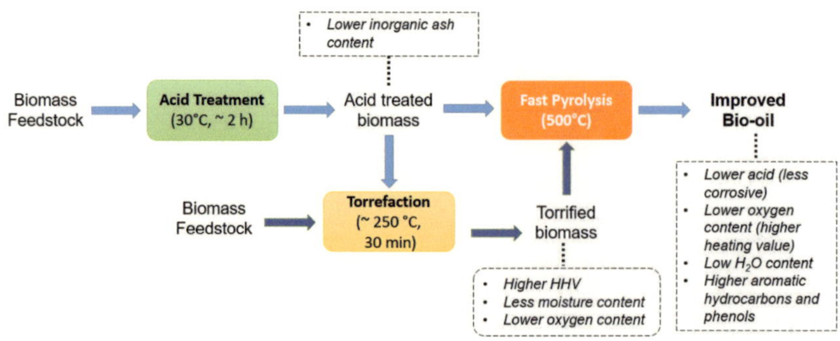

Figure 6.
A combination of acid treatment and dry torrefaction to improve the fast pyrolysis process performance (adapted from [53]).

dry torrefied (270°C) in the pretreatment of *P. radiata* allowed achieving 57.8% of bio-oil yield and improving the quality of bio-oil with low H_2O and organic acids content [54].

5.2 In situ upgrading

Other strategy to improve the properties of the bio-oil obtained by pyrolysis process is the modification in situ, i.e., inside the own reactor by change of the variables and the parameters. Two strategies that have gained prominence and show good results in obtaining a bio-oil with good qualities are hydrogenation and catalytic pyrolysis, and the hybrid process with both.

The addition of H2 as a pyrolysis gas agent in the absence of catalysts promotes hydrogenation reactions mainly with the volatiles released and may have various effects such as increased formation of less oxygenated species, greater selectivity in phenolic compounds among others. For example, the effect of adding hydrogen in the carrier gas (up to 15% vol) in the pyrolysis of food waste in a dawndraft reactor promoted the formation of a bio-oil more selective in aromatic, phenolic compounds while reducing the amount of heterocycline nitrogenous species such as (quinoline) [55]. Other biomasses, such as poplar wood showed the same behavior upon pyrolysis in H2 atmosphere (6% vol) compared to pure N2 atmosphere obtaining higher yield and selectivity in phenolic species [56]. Another important effect of the addition of H2 in the pyrolysis process is that, regardless of the presence of catalysts, it can promote the production of bio-oil with higher H/C ratio which is highly beneficial for the use of bio-oil as liquid fuel [57].

Recycling the pyrolysis gas to the reactor as a reactive atmosphere is a potential alternative, although its effect is not yet clear. When fast pyrolysis of biomass takes place under a N_2 atmosphere, the main gaseous products of pyrolysis are CO, CO_2, H_2, CH_4 and low levels of light hydrocarbons, therefore, when this gas is recycled to the reactor, it can promote a reactive atmosphere and not more inert. Studies in fluidized bed reactors, in the temperature range of 430–500°C, show that the increase in gas recycling rates can lead to a decrease in the production of organic liquids, but some biomasses, such as oak and switch grass, show good results. Deoxygenation effect and HHV increase in your bio-oil with gas recycling rates of up to 80%. However, other biomasses, such as *Pennycress Presscake* and pine wood, had no effect on the yield and composition of bio-oil with the pyrolysis gas recycled in the reactor [58, 59].

Among all the in situ alternatives to improve the obtainment of higher quality bio-oil, catalytic pyrolysis stands out as the most promising. The process basically consists of the insertion of a solid catalyst in the pyrolysis reactor capable of acting in the set of pyrolysis reactions, promoting the formation of bio-oil with better properties and selectivity in the components of interest of interest. Several authors have already published works testing different catalysts, in different process configurations, attesting to diverse improvements in obtaining bio-oil. Among these improvements in the use of catalysts in the fast pyrolysis process, it is worth mentioning: [60].

- The use of some catalysts can provide lower temperature operation. Considering that the pyrolysis process is endothermic, this would lead to a reduction in energy costs [61, 62].

- Reduction of bio-oil acidity and possible corrosion problems due to less formation of acidic compounds.

- High degree of deoxygenation, increased H/C ratio and associated energy densification [63].

- Considerable increase in the selectivity of bio-oil products, reducing the molecular weight range of the hydrocarbons obtained (C_5–C_{12}) [64].

- Depending on the catalyst(s) selected, there may be a significant increase in aromatic compounds (up to 50%), giving greater stability to the bio-oil and providing a potential use as a fuel [60, 65].

Despite promising experimental results, the fast pyrolysis process using catalysts still needs to overcome some obstacles to reach the industrial level. The basis for these challenges is the development of catalytic reactors capable of supporting the pyrolysis atmosphere without expressive rates of catalyst deactivation, expanding the selection of desired products in bio-oil, in addition to efficient developments in the recovery and regeneration of catalysts [41, 60]. **Table 4** below shows a set of

Feedstock	Process parameters	Catalyst	Yield	Deoxygenation	Ref.
Loblolly pine	Circulating fluidized bed 36 kg_{biom}/h 520°C	γ-Al_2O_3	—	13–6–24.2% wt O	[66]
Forest pine woodchips	Auger reactor 450°C 2 kg_{biom}/h	CaO CaO.MgO	48–50%	24.2–25.4% wt O	[67]
Pine wood	Auger reactor 500°C 0,2 kg_{biom}//h	ZSM-5 based acidic catalys	50,3%	—	[68]
Pine wood	Bubbling fluidized bed reactor	Sspray-dried 40% ZSM-5	45.5–50.0%	—	[69]
Microalgae (Enteromorpha prolifera) + HDPE plastics	Fixed bed reator 550°C EP:plastic:catalyst ratio 1:1:2	HZSM-5	71.4%	13.5% wt O	[70]
Miscanthus	Conical spouted reactor 400–600°C 1–5 catalyst/biomass	ZSM-5	37.5% [500°C]	1,7% of phenols and 5,5% wt other oxygenates [500°C]	[71]
Waste tires + pistachio seeds	Fixed bed 500°C 5% wt catalyst	Ni-Mo/ Al_2O_3	63%	5.3% wt (oxygen)	[72]
Lingnin	Fixed bed reactor 450°C	K_2CO_3	29%	O/C=1,40	[73]
L. japonica + polypropylene	Fixed bed reactor 500°C	Al-SBA-15	50,32%	~ 10% of oxygenate compounds	[74]
Pine wood	Fluidized bed reactor	H-MOR-20	17,6% (free water)	—	[75]

Table 4.
Some research about catalytic pyrolysis.

works on catalytic pyrolysis of biomass at different scales, the characteristics of the catalysts used as well as the yield in bio-oil.

5.3 *Ex situ* upgrading

Ex situ processes are carried out outside the pyrolysis reactor, ie outside the main reaction zone. Esses processos podem ser de natureza física ou físico-química, promovendo alguma melhoria na qualidade do bio-óleo bruto gerado na pirólise. Dentre esses, pode se citar: [76–78].

- Physical methods
    - Solvend adition
    - Emulsification
    - Hot vapor filtration
- Chemical
    - Catalytic hydrodeoxygenation (HDO)
    - Steam reforming
    - Esterification
    - Supercritical fluids

Although there are several strategies to improve the quality of bio-oil, the most studied processes, aiming at the industrial potential, are hot vapor filtration (HVF), catalytic hydrodeoxygenation (HDO) and steam reforming.

HVF is one of the most common and simple able to improve some properties of bio-oil. It consists of passing the pyrolysis vapors through a filtering medium, even at higher temperatures. This method, in addition to being more efficient than traditional cyclones for removing small particles of coal and ash, can add better properties to the condensed bio-oil later on [48, 79].

Some types of filter (fixed bed glass wool, ceramic candles) and biomass raw material (sugarcane, rice and cassava waste) for pyrolysis were tested in the configuration: reactor + hot filter. In short, the addition of the filter introduces a longer residence time of the bio-oil organic vapor at high temperatures, promotes cracking reactions of organic molecules and causing a loss in bio-oil yield (around to 5%), increase in yield of the gas in addition to the higher content of H_2O in the bio-oil. But the condensed bio-oil tends to present a series of improvements such as: practically free of char and ash, less viscous. There may also be a certain deoxygenation of the bio-oil (decrease in the O/C molar ratio) but this effect is not fully understood, depending on the type of hot filter, biomass and operating conditions of the pyrolysis [79–81].

Another upgrading process, of the great highlighted is the hydrodeoxygenation (HDO) of the bio-oil. The HDO it's a particular case of hydrotreatment process, where the bio-oil is its reacted with H_2 under specific conditions of temperature, pressure, catalystic, and fluidynamis (**Figure 7**). In this way, the organic compounds of the bio-oil are submitted to a set of reactions, mainly hydrogenation and hydrodeoxygenation, providing an improvement in properties, forming less oxygenated and more stable compounds.

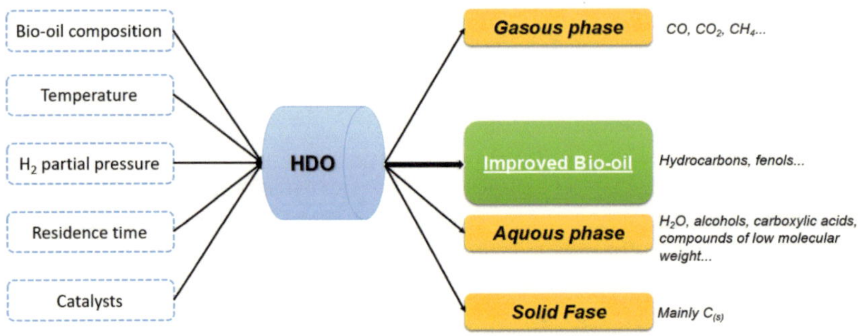

Figure 7.
Schematic of the catalytic hydrodeoxygenation process, its variables and products.

Good control of the variables and parameters is essential to obtain a bio-oil with a high degree of deoxygenation and with lower H2 consumption. The process occurs in the range 200–450°C and higher temperatures, in general, increase the degree of deoxygenation but at the expense of a decrease in the yield of improved bio-oil due to higher gas production. The partial pressure of H2 is a variable of great relevance in the process and a minimum value of around 80 bar is required for good solubilization of the bio-oil, increased catalytic activity and minimization of adverse effects such as repolymerization of the bio-oil forming unwanted solid products [82, 83].

Finally the selection of the catalyst, jutanly with the support, is fundamental. The first catalysts tested in this process were CoMo/γ-Al2O3 and NiMo/γ-Al2O3 due to their use in oil refineries for nitrogen and sulfur removal processes. About these a number of sulfide, noble metal, and transition catalysts have been studied, each presenting different advantages with respect to higher catalytic activity, selectivity, and deactivation rates upon the presence of inorganic elements such as sulfur and ash present in the bio-oil [83]. The catalytic hydrodeoxygenation becomes a promising alternative for obtaining a bio-oil of higher quality and opening the possibility for the production of biofuels from it.

The steam reforming process emerges as another alternative of great potential for the valorization of lignocellulosic currents and hydrogen production. H2 is a gaseous fuel with high added value due to its energy density and its combustion is free of carbon emissions. In this process, bio-oil, as well as fossil fuels, react with H2O vapor at temperatures in the range of 700–1000°C, in the presence of a catalyst (usually nickel-based) offering as main product the H_2-rich syngas, along with CO_2 [76–78]. The main advantage of this process is the simultaneous production of high value-added fuel (H2), and it allows the assimilation of CCSU (Carbon Capture Storage and Utilization) technologies. But one of the major disadvantages is the high energy demand to carry out the process.

The overall balance of this process can be given by Eq (1): [77, 78].

$$C_nH_mO_k + (2n-k)H2O \rightarrow nCO2 + \left(2n + \frac{m}{2k}\right)H_2 \tag{1}$$

Where the coefficient expresses the maximum possible H_2 yield per mole of carbon fed. The steam reforming of bio-oil can be performed mainly in fixed bed, fluidized bed or staged bed reactors. The process can be carried out using a wide range of catalysts (Dolomite, Ru, Ni, Co, Rh) and supports (Al_2O_3, ZrO_2, $MgAl_2O_4$, etc.) but Ni based catalysts are the most active but also have the highest deactivation rates due to coke formation. However, changes in the processes, mainly concerning the Vapor/Carbon ratio, besides the temperature and the catalytic support, can minimize the impact of the decay of the catalytic activity by coke formation in the steam reforming process [84, 85].

6. Conclusions

The process of fast pyrolysis is an alternative of great potential for the conversion of biomass and waste into renewable fuels and chemicals of high added value contributing to the reduction of emissions of greenhouse gases, with potential for application and integration in biorefinery projects. Technically some technologies such as fluid bed and rotating cone reactors are emerging for commercial applications and the main product generated, bio-oil, has a variable and complex composition requiring stages of improvement and stabilization. In this sense, several studies point out routes to improve the quality of bio-oil obtained from the pre-treatment of biomass (such as washing and torrefaction), in situ improvements and downstream processes such as HDO and steam reforming capable of converting bio-oil into a stream with higher added value.

Author details

Vittor Rodrigues Santos Alves
Foundation to Support the Technological Research Institute of São Paulo (FIPT) – Bioenergy and Energy Efficiency Laboratory (LBE - IPT), São Paulo, Brazil

*Address all correspondence to: vittoralves@ipt.br

IntechOpen

© 2021 The Author(s). Licensee IntechOpen. This chapter is distributed under the terms of the Creative Commons Attribution License (http://creativecommons.org/licenses/by/3.0), which permits unrestricted use, distribution, and reproduction in any medium, provided the original work is properly cited.

References

[1] Bridgwater, A. V. (2012). Review of fast pyrolysis of biomass and product upgrading. Biomass and Bioenergy, 38, 68-94. DOI:10.1016/j.biombioe.2011.01.048

[2] Czernik, S., and Bridgwater, A. V. (2004). Overview of applications of biomass fast pyrolysis oil. Energy and Fuels, 18(2), 590-598. DOI:10.1021/ef034067u

[3] Das Shreya, Mohanty Samanyita, Sahu Gayatri, Rana Mausami, and Pilli Kiran (2021). Biochar: A sustainable approach for improving soil health and environment. DOI:10.5772/intechopen.97136.

[4] Tekin, K., Karagöz, S., and Bektaş, S. (2014). A review of hydrothermal biomass processing. Renewable and Sustainable Energy Reviews, 40, 673-687. DOI:10.1016/j.rser.2014.07.216

[5] Ribeiro, J.M.C., Godina, R., Matias, J.C.D.O., Nunes, L.J.R. Future perspectives of biomass torrefaction: Review of the current state-of-the-art and research development. Sustainability 2018, 10, 2323. DOI:10.3390/su10072323

[6] Verma Anil, Shankar Ravi, and Mondal Prasenjit (2018). A review on pyrolysis of biomass and the impacts of operating conditions on product yield, quality, and upgradation. DOI:10.1007/978-981-13-1307-3_10

[7] Kazawadi Deodatus, Ntalikwa Justin, and Kombe Godlisten (2021). A review of intermediate pyrolysis as a technology of biomass conversion for coproduction of biooil and adsorption biochar. Journal of Renewable Energy. 2021. DOI:10.1155/2021/5533780

[8] Jahirul, M.I., Rasul Mohammad, Chowdhury Ashfaque, and Ashwath Nanjappa (2012). Biofuels production through biomass pyrolysis—A technological review. Energies. 5. 4952-5001.DOI:10.3390/en5124952.

[9] Molino, A., Larocca, V., Chianese, S., Musmarra, D. Biofuels production by biomass gasification: A review. Energies 2018, 11, 811. DOI:10.3390/en11040811

[10] Uzun, B., Apaydın Varol, E., and Pütün, E. (2016). Pyrolysis: A sustainable way from biomass to biofuels and biochar. In: V. Bruckman, E. Apaydın Varol, B. Uzun, and J. Liu (Eds.), Biochar: A Regional Supply Chain Approach in View of Climate Change Mitigation (pp. 239-265). Cambridge: Cambridge University Press. DOI:10.1017/9781316337974.013

[11] Oasmaa Anja, Elliott Douglas, and Müller, S. (2009). Quality control in fast pyrolysis bio-oil production and use. Environmental Progress and Sustainable Energy. 28. 404-409. DOI:10.1002/ep.10382.

[12] Zhou Xiaowei, Broadbelt, L.J., and Vinu Ravikrishnan (2016). Mechanistic understanding of thermochemical conversion of polymers and lignocellulosic biomass. DOI:10.1016/bs.ache.2016.09.002.

[13] Cheng Shouyun, Wei Lin, Zhao Xianhui, and Julson James (2016). Application, deactivation, and regeneration of heterogeneous catalysts in bio-oil upgrading. Catalysts. 6. 195. DOI:10.3390/catal6120195.

[14] Chukwuneke, J. L., Ewulonu, M. C., Chukwujike, I. C., and Okolie, P. C. (2019). Physico-chemical analysis of pyrolyzed bio-oil from swietenia macrophylla (mahogany) wood. Heliyon, 5(6), e01790. DOI:10.1016/j.heliyon.2019.e01790

[15] Oasmaa, A., Peacocke, C., Gust, S., Meier, D., and McLellan, R. (2005).

Norms and standards for pyrolysis liquids. End-user requirements and specifications. Energy and Fuels, 19(5), 2155-2163. DOI:10.1021/ef040094o

[16] Ringer, M, Putsche, V, and Scahill, J. Large-scale pyrolysis oil production: A technology assessment and economic analysis. United States: N. p., 2006. Web. DOI:10.2172/894989.

[17] Elliott, D C. Analysis and comparison of biomass pyrolysis/gasification condensates: Final report. United States: N. p., 1986. Web. DOI:10.2172/7049810.

[18] Bridgewater Anthony. (2004). Biomass fast pyrolysis. Thermal Science. 8. 21-50. DOI:10.2298/TSCI0402021B.

[19] Bridgwater, A. (1999). Fast Pyrolysis of Biomass: A Handbook.

[20] Jahirul, M.I., Rasul Mohammad, Chowdhury Ashfaque, and Ashwath Nanjappa (2012). Biofuels production through biomass pyrolysis—A technological review. Energies. 5. 4952-5001. DOI:10.3390/en5124952.

[21] Yang, W., Ponzio, A., Lucas, C., Blasiak, W. Performance analysis of a fixed-bed biomass gasifier using high-temperature air. Fuel Proc. Technol. 2006, 87, 235-245. DOI:10.1016/j.fuproc.2005.08.004

[22] Garcia-Nunez, J. A., Pelaez-Samaniego, M. R., Garcia-Perez, M. E., Fonts, I., Abrego, J., Westerhof, R. J. M., and Garcia-Perez, M. (2017). Historical developments of pyrolysis reactors: A review. Energy and Fuels, 31(6), 5751-5775. DOI:10.1021/acs.energyfuels.7b00641

[23] Venderbosch Robbie, and Prins, W. (2010). Fast pyrolysis technology development. Biofuels, Bioproducts and Biorefining. 4. 178-208. DOI:10.1002/bbb.205.

[24] Chouhan Ashish, and Sarma Anil (2013). Critical analysis of process parameters for bio-oil production via pyrolysis of biomass: A review. Recent Patents on Engineering. 7. 98-114. DOI:10.2174/1872212111307999000 5.

[25] Arnold, S., Moss, K., Henkel, M., and Hausmann, R. (2017). Biotechnological perspectives of pyrolysis oil for a bio-based economy. Trends in Biotechnology, 35(10), 925-936. DOI:10.1016/j.tibtech.2017.06.003

[26] Gómez, N., Banks, S. W., Nowakowski, D. J., Rosas, J. G., Cara, J., Sánchez, M. E., and Bridgwater, A. V. (2018). Effect of temperature on product performance of a high ash biomass during fast pyrolysis and its bio-oil storage evaluation. Fuel Processing Technology, 172, 97-105. DOI:10.1016/j.fuproc.2017.11.021

[27] Shah, Z., CV, R., AC, M., and DS, R. (2017). Separation of phenol from bio-oil produced from pyrolysis of agricultural wastes. Modern Chemistry and Applications, 05(01). DOI:10.4172/2329-6798.1000199

[28] Panwar, N.L., and Paul Arjun. (2020). An overview of recent development in bio-oil upgrading and separation techniques. Environmental Engineering Research. 26. DOI:10.4491/eer.2020.382.

[29] Lehto Jani, Oasmaa Anja, Solantausta Yrjö, Kytö Matti, and Chiaramonti David. (2018). Fuel oil quality and combustion of fast pyrolysis bio-oils. DOI:10.13140/RG.2.2.15925.99042.

[30] Hou Shuhn-Shyurng, Huang Wei-Cheng, Rizal Mokhammad, and Lin Ta-Hui. (2016). Co-firing of fast pyrolysis bio-oil and heavy fuel oil in a 300-kWth furnace. Applied Sciences. 6. 326. DOI:10.3390/app6110326.

[31] Freire José Teixeira, and Dermeval José Mazzini Sartori. 1992. Tópicos Especiais Em Secagem. São Carlos, SP: UFSCar

[32] Alvarez, J., Lopez, G., Amutio, M., Bilbao, J., and Olazar, M. (2014). Bio-oil production from rice husk fast pyrolysis in a conical spouted bed reactor. Fuel, 128, 162-169. DOI:10.1016/j.fuel.2014.02.074

[33] Fernandez-Akarregi, A. R., Makibar, J., Lopez, G., Amutio, M., and Olazar, M. (2013). Design and operation of a conical spouted bed reactor pilot plant (25kg/h) for biomass fast pyrolysis. Fuel Processing Technology, 112, 48-56. DOI:10.1016/j.fuproc.2013.02.022

[34] Alvarez, J., Hooshdaran, B., Cortazar, M., Amutio, M., Lopez, G., Freire, F. B., Olazar, M. (2018). Valorization of citrus wastes by fast pyrolysis in a conical spouted bed reactor. Fuel, 224, 111-120. DOI:10.1016/j.fuel.2018.03.028

[35] Orozco, S., Alvarez, J., Lopez, G., Artetxe, M., Bilbao, J., Olazar, M. Pyrolysis of plastic wastes in a fountain confined conical spouted bed reactor: Determination of stable operating conditions. Energy Convers. Manag. 2021, 229, 113768. DOI:10.1016/j.enconman.2020.113768

[36] Sharifzadeh, M., Sadeqzadeh, M., Guo, M., Borhani, T. N., Murthy Konda, N. V. S. N., Garcia, M. C., Shah, N. (2019). The multi-scale challenges of biomass fast pyrolysis and bio-oil upgrading: Review of the state of art and future research directions. Progress in Energy and Combustion Science, 71, 1-80. DOI:10.1016/j.pecs.2018.10.006

[37] Vamvuka, D. (2011). Bio-oil, solid and gaseous biofuels from biomass pyrolysis processes—An overview. International Journal of Energy Research, 35(10), 835-862. DOI:10.1002/er.1804

[38] Yu X, Makkawi Y, Ocone R, Huard M, Briens C, Berruti F. A CFD study of biomass pyrolysis in a downer reactor equipped with a novel gas–solid separator—I: Hydrodynamic performance. Fuel Process Technol 2014; 126:366-382. DOI:10.1016/j.fuproc.2014.05.020.

[39] Perkins, G., Bhaskar, T., and Konarova, M. (2018). Process development status of fast pyrolysis technologies for the manufacture of renewable transport fuels from biomass. Renewable and Sustainable Energy Reviews, 90, 292-315. DOI:10.1016/j.rser.2018.03.048

[40] Agblevor, F.A., Besler, S., 1996. Inorganic compounds in biomass feedstocks. 1. Effect on the quality of fast pyrolysis oils. Energy Fuels 10, 293-298. DOI:10.1021/ef950202u

[41] Choi, H.S., Choi, Y.S., Park, H.C. Fast pyrolysis characteristics of lignocellulosic biomass with varying reaction conditions. Renew. Energy 2012, 42, 131-135. DOI:10.1016/j.renene.2011.08.049

[42] Chang, S. H. (2019). Rice husk and its pretreatments for bio-oil production via fast pyrolysis: A review. BioEnergy Research. DOI:10.1007/s12155-019-10059-w

[43] Deng L, Zhang T, Che D. (2013) Effect of water washing on fuel properties, pyrolysis and combustion characteristics, and ash fusibility of biomass. Fuel Process Technol 106: 712-720. DOI:10.1016/J.FUPROC.2012.10.006

[44] Ang TN, Ngoh GC, Chua ASM (2013) Comparative study of various pretreatment reagents on rice husk and structural changes assessment of the optimized pretreated rice husk.

Bioresour Technol 135:116-119. DOI:10.1016/J.BIORTECH.2012.09.045

[45] S. Oudenhoven, R.J.M. Westerhof, N. Aldenkamp, D.W.F. Brilman, S.R. Kersten, Demineralization of wood using wood-derived acid: Towards a selective pyrolysis process for fuel and chemicals production, J. Anal. Appl. Pyrol. 103 (2013) 112-118. DOI:10.1016/j.jaap.2012.10.002

[46] Zhou Shuyu, Xue Yuan, Cai Junmeng, Cui Cunhao, Ni Ziang, and Zhou Zhongyue. (2021). An understanding for improved biomass pyrolysis: Toward a systematic comparison of different acid pretreatments. Chemical Engineering Journal. 411. 128513. 10.1016/j.cej.2021.128513.

[47] Mamvura, T.A., Danha, G. Biomass torrefaction as an emerging technology to aid in energy production. Heliyon 2020, 6, e03531. DOI:10.1016/j.heliyon.2020.e03531

[48] Echresh Zadeh Zahra, Abdulkhani Ali, Aboelazayem Omar, Saha Basudeb (2020). Recent insights into lignocellulosic biomass pyrolysis: A critical review on pretreatment, characterization, and products upgrading. Processes. 8. 799. DOI:10.3390/pr8070799.

[49] S.R.G. Oudenhoven, C. Lievens, R.J.M. Westerhof, S.R.A. Kersten, Effect of temperature on the fast pyrolysis of organic-acid leached pinewood: The potential of low temperature pyrolysis, Biomass Bioenergy 89 (2016) 78-90. DOI:10.1016/j.biombioe.2015.12.019

[50] S.R.G. Oudenhoven, R.J.M. Westerhof, S.R.A. Kersten, Fast pyrolysis of organic acid leached wood, straw, hay and bagasse: Improved oil and sugar yields, J. Anal. Appl. Pyrol. 116 (2015) 253-262. DOI:10.1016/j.jaap.2015.09.003

[51] A. Zheng, Z. Zhao, S. Chang, Z. Huang, K. Zhao, G. Wei, F. He, H. Li, Comparison of the effect of wet and dry torrefaction on chemical structure and pyrolysis behavior of corncobs, Bioresour. Technol. 176 (2015) 15-22. DOI:10.1016/j.biortech.2014.10.157

[52] S. Zhang, T. Chen, Y. Xiong, Q. Dong, Effects of wet torrefaction on the physicochemical properties and pyrolysis product properties of rice husk, Energy Convers. Manage. 141 (2017) 403-409. DOI:10.1016/j.enconman.2016.10.002

[53] Srifa, A., Chaiwat, W., Pitakjakpipop, P., Anutrasakda, W., and Faungnawakij, K. (2019). Advances in bio-oil production and upgrading technologies. Sustainable Bioenergy, 167-198. DOI:10.1016/b978-0-12-817654-2.00006-x

[54] Wigley, T., Yip, A. C. K., and Pang, S. (2017). A detailed product analysis of bio-oil from fast pyrolysis of demineralised and torrefied biomass. Journal of Analytical and Applied Pyrolysis, 123, 194-203. DOI:10.1016/j.jaap.2016.12.006

[55] Xu, F., Ming, X., Jia, R., Zhao, M., Wang, B., Qiao, Y., and Tian, Y. (2020). Effects of operating parameters on products yield and volatiles composition during fast pyrolysis of food waste in the presence of hydrogen. Fuel Processing Technology, 210, 106558. DOI:10.1016/j.fuproc.2020.106558

[56] Zhang, Z., Li, K., Ma, S., Cui, M., Lu, Q., and Yang, Y. (2019). Fast pyrolysis of biomass catalyzed by magnetic solid base catalyst in a hydrogen atmosphere for selective production of phenol. Industrial Crops and Products, 137, 495-500. DOI:10.1016/j.indcrop.2019.05.066

[57] Wang, S., Persson, H., Yang, W., and Jönsson, P. G. (2018). Effect of H2 as pyrolytic agent on the product

distribution during catalytic fast pyrolysis of biomass using zeolites. Energy and Fuels, 32(8), 8530-8536. DOI:10.1021/acs.energyfuels.8b01779

[58] Mullen, C. A., Boateng, A. A., and Goldberg, N. M. (2013). Production of deoxygenated biomass fast pyrolysis oils via product gas recycling. Energy and Fuels, 27(7), 3867-3874. DOI:10.1021/ef400739u

[59] Pala, M., Marathe, P. S., Hu, X., Ronsse, F., Prins, W., Kersten, S. R. A., Westerhof, R. J. M. (2020). Recycling of product gas does not affect fast pyrolysis oil yield and composition. Journal of Analytical and Applied Pyrolysis, 104794. DOI:10.1016/j.jaap.2020.104794

[60] Fadillah, G., Fatimah, I., Sahroni, I., Musawwa, M.M., Mahlia, T.M.I., Muraza, O. Recent progress in low-cost catalysts for pyrolysis of plastic waste to fuels. Catalysts 2021, 11, 837. DOI:10.3390/catal11070837

[61] Mertinkat, J., et al., 1999. Cracking catalysts used as fluidized bed material in the Hamburg pyrolysis process. Journal of Analytical and Applied Pyrolysis 49 (1-2), 87-95. DOI:10.1016/S0165-2370(98)00103-X

[62] Onu, P., et al., 1999. Thermal and catalytic decomposition of polyethylene and polypropylene. Journal of Analytical and Applied Pyrolysis 49 (1), 145e153. DOI:10.1016/S0165-2370 (98)00109-0

[63] French, R., Czernik, S., 2010. Catalytic pyrolysis of biomass for biofuels production. Fuel Processing Technology 91 (1), 25e32. DOI:10.1016/j.fuproc.2009.08.011

[64] Muhammad, C., Onwudili, J.A., Williams, P.T., 2015b. Thermal degradation of real-world waste plastics and simulated mixed plastics in a two-stage pyrolysisecatalysis reactor for fuel production. Energy and Fuels 29 (4), 2601e2609. DOI:10.1021/ef502749h

[65] Carlson, T.R., Vispute, T.P., Huber, G.W., 2008. Green gasoline by catalytic fast pyrolysis of solid biomass derived compounds. ChemSusChem 1 (5), 397e400. DOI:10.1002/cssc.200800018

[66] Mante, O. D., Dayton, D. C., Carpenter, J. R., Wang, K., and Peters, J. E. (2018). Pilot-scale catalytic fast pyrolysis of loblolly pine over γ-Al2O3 catalyst. Fuel, 214, 569-579. DOI:10.1016/j.fuel.2017.11.073

[67] Veses, A., Aznar, M., Martínez, I., Martínez, J. D., López, J. M., Navarro, M. V., García, T. (2014). Catalytic pyrolysis of wood biomass in an auger reactor using calcium-based catalysts. Bioresource Technology, 162, 250 258. DOI:10.1016/j.biortech.2014.03.146

[68] Yildiz, G., Pronk, M., Djokic, M., van Geem, K. M., Ronsse, F., van Duren, R., and Prins, W. (2013). Validation of a new set-up for continuous catalytic fast pyrolysis of biomass coupled with vapour phase upgrading. Journal of Analytical and Applied Pyrolysis, 103, 343-351. DOI:10.1016/j.jaap.2013.02.001

[69] Jae, J., Coolman, R., Mountziaris, T. J., and Huber, G. W. (2014). Catalytic fast pyrolysis of lignocellulosic biomass in a process development unit with continual catalyst addition and removal. Chemical Engineering Science, 108, 33-46. DOI:10.1016/j.ces.2013.12.023

[70] Xu, S., Cao, B., Uzoejinwa, B. B., Odey, E. A., Wang, S., Shang, H., Nwakaire, J. N. (2020). Synergistic effects of catalytic co-pyrolysis of macroalgae with waste plastics. Process Safety and Environmental Protection. DOI:10.1016/j.psep.2020.02.001

[71] Du, S., Sun, Y., Gamliel, D. P., Valla, J. A., and Bollas, G. M. (2014). Catalytic pyrolysis of miscanthus × giganteus in a spouted bed reactor. Bioresource Technology, 169, 188-197. DOI:10.1016/j.biortech.2014.06.104

[72] Onay, Ö. (2014). The catalytic co-pyrolysis of waste tires and pistachio seeds. Energy Sources, Part A: Recovery, Utilization, and Environmental Effects, 36(18), 2070-2077. DOI:10.1080/15567036.2013.791900

[73] Rutkowski, P. (2011). Pyrolysis of cellulose, xylan and lignin with the K2CO3 and ZnCl2 addition for bio-oil production. Fuel Processing Technology, 92(3), 517-522. DOI:10.1016/j.fuproc.2010.11.006

[74] Lee, H.W., Choi, S.J., Park, S.H. et al. Pyrolysis and co-pyrolysis of Laminaria japonica and polypropylene over mesoporous Al-SBA-15 catalyst. Nanoscale Res Lett 9, 376 (2014). doi:10.1186/1556-276X-9-376

[75] Aho, A., Kumar, N., Eränen, K., Salmi, T., Hupa, M., and Murzin, D. Y. (2008). Catalytic pyrolysis of woody biomass in a fluidized bed reactor: Influence of the zeolite structure. Fuel, 87(12), 2493-2501. DOI:10.1016/j.fuel.2008.02.015

[76] Attia, M., Farag, S., Chaouki, J. Upgrading of oils from biomass and waste: Catalytic hydrodeoxygenation. Catalysts 2020, 10, 1381. DOI:10.3390/catal10121381

[77] Shan Ahamed, T., Anto, S., Mathimani, T., Brindhadevi, K., and Pugazhendhi, A. (2020). Upgrading of bio-oil from thermochemical conversion of various biomass – Mechanism, challenges and opportunities. Fuel, 119329. DOI:10.1016/j.fuel.2020.119329

[78] Gollakota, A. R. K., Reddy, M., Subramanyam, M. D., and Kishore, N. (2016). A review on the upgradation techniques of pyrolysis oil. Renewable and Sustainable Energy Reviews, 58, 1543-1568. DOI:10.1016/j.rser.2015.12.180

[79] Pattiya, A., and Suttibak, S. (2017). Fast pyrolysis of sugarcane residues in a fluidised bed reactor with a hot vapour filter. Journal of the Energy Institute, 90(1), 110-119. DOI:10.1016/j.joei.2015.10.001

[80] Chen, T., Wu, C., Liu, R., Fei, W., and Liu, S. (2011). Effect of hot vapor filtration on the characterization of bio-oil from rice husks with fast pyrolysis in a fluidized-bed reactor. Bioresource Technology, 102(10), 6178-6185. DOI:10.1016/j.biortech.2011.02.023

[81] Pattiya, A., and Suttibak, S. (2012). Production of bio-oil via fast pyrolysis of agricultural residues from cassava plantations in a fluidised-bed reactor with a hot vapour filtration unit. Journal of Analytical and Applied Pyrolysis, 95, 227-235. DOI:10.1016/j.jaap.2012.02.010

[82] Tran N, Uemura Y, Chowdhury S, Ramli A. A review of bio-oil upgrading by catalytic hydrodeoxygenation. AMM 2014;625:255-258. DOI:10.4028/www.scientific.net/amm.625.255.

[83] Oyedun, A. O., Patel, M., Kumar, M., and Kumar, A. (2019). The upgrading of bio-oil via hydrodeoxygenation. Chemical Catalysts for Biomass Upgrading, 35-60. DOI:10.1002/9783527814794.ch2

[84] Zhao, Z., Situmorang, Y. A., An, P., Chaihad, N., Wang, J., Hao, X., Guan, G. (2020). Hydrogen production from catalytic steam reforming of bio-oils: A critical review. Chemical Engineering and Technology. DOI:10.1002/ceat.201900487

[85] Zhang, Y., Brown, T. R., Hu, G., and Brown, R. C. (2013). Techno-economic analysis of two bio-oil upgrading pathways. Chemical Engineering Journal, 225, 895-904. DOI:10.1016/j.cej.2013.01.030

Chapter 5

Cheminformatics Applied to Analytical Pyrolysis of Lignocellulosic Materials

Jorge Reyes-Rivera

Abstract

Pyrolysis-Gas Chromatography/Mass Spectrometry has been used to characterize a wide variety of polymers. The main objective is to infer the attributes of materials in relation to their chemical composition. Applications of this technique include the development of new improved materials in the industry. Furthermore, due to the growing interest in biorefinery, it has been used to study plant biomass (lignocellulose) as a renewable energy source. This chapter describes a procedure for characterization and classification of polymeric materials using analytical pyrolysis and cheminformatics. Application of omics tools for spectral deconvolution/alignment and compound identification/annotation on the Py-GC/MS chromatograms is also described. Statistical noise is generated by production of numerous small uninformative compounds during pyrolysis. Such noise is reduced by cheminformatics here detailed and this facilitate the interpretation of results. Furthermore, some inferences made by comparison of the identified compounds to those annotated with a biological role in specialized databases are exemplified. This cheminformatic procedure has allowed to characterize in detail, and classify congruently, different lignocellulosic samples, even using different Py-GC/MS equipment. This method can also be applied to characterize other polymers, as well as to make inferences about their structure, function, resistance and health risk based on their chemical composition.

Keywords: Biomass pyrolysis, polymeric materials characterization, cheminformatics, multivariate comparative analysis, Py-GC/MS

1. Introduction

The largest repository of lignocellulosic biomass is generated by the cell walls of plants [1]. Its main chemical components are cellulose, hemicelluloses and lignin. The proportions are variable but close to 4:3:3, respectively, and the element content is 50% C, 6% H, 44% O y ≤ 0.4% N, for resources such as wood [1]. Because biomass is a renewable resource, its study for the production of energy and value-added aromatic compounds has gained importance in recent decades [2, 3]. It has been considered that lignocellulosic biomass as a renewable energy source would satisfy around 25% of energy requirements [4]. Thus, CO_2 sequestered by plants during photosynthesis would balance the CO_2 generated by biofuels and their use would not contribute to global warming [5, 6]. On the other hand, after cellulose, lignin is the most abundant polymer in nature and the main natural source of aromatic compounds [1, 7]. For this reason, lignin is important in the chemical industry and it has been projected as a replacement for aromatic polymers derived from fossil fuels [8].

Lignocellulosic biomass, like other non-volatile complex materials, cannot be directly analyzed in its original state by gas chromatography. Therefore, one of the most common methods for its analysis is the Pyrolysis-Gas Chromatography/Mass Spectrometry (Py-GC/MS). This method consists of the rapid heating of the materials under analysis (close 300°C), to break the covalent bonds and produce individual fragments. The compounds derived from pyrolysis pass through a capillary column of fused silica in a Gas Chromatograph using an inert gas as carrier (e.g., He). Then the fragments are separated based on their retention times. The selective fragmentation pattern caused by Electron Impact and the m/z ratio for each pyrolysis product are registered by a detector on a Mass Spectrometer. Finally, each compound is identified by comparing its mass spectrum to those in the reference electronic libraries (NIST, MONA, etc.) or to the mass spectra produced by analytical standards [9–12]. The sequential combination of these three processes in Py-GC/MS makes it a versatile and powerful tool for the analysis of lignocellulosic materials and other complex mixtures, such as polymers and copolymers [3, 13].

Analytical pyrolysis is currently implemented as a standard method for determining the ratio of H/G/S subunits in plant biomass, agricultural and industrial waste, soil samples and organic matter [6]. This technique has also been useful to elucidate the series of reactions and products derived from the pyrolysis of carbohydrates [14, 15] and lignins [16, 17]. It has been applied for monitoring changes during the delignification and bleaching process as well as for the characterization of different lignocellulosic materials [12]. In addition, it has been used to determine the S/G ratio in lignin of drought-resistant succulent species with results highly comparable to other characterization techniques [18]. On the other hand, its high sensitivity has enabled the detection of hundreds of chemical compounds, including less abundant monomers in lignin, such as acetylated subunits (i.e., sinapyl and coniferyl acetates [19]) and 5-hydroxyguaiacyl units [20]. Recently, Py-GC/MS applied to the analysis of cacti spines, with the use of cheminformatics, allowed a detailed characterization of lignocellulosic matrix, as well as the classification of the samples from a chemotaxonomic approach [21].

1.1 Advantages of Py-GC/MS

Different advantages confer great versatility of application to Py-GC/MS. Firstly, its efficiency, precision and relatively low operating costs [6] make it a suitable routine technique. In addition, it is a fast technique that requires a very small sample size [22, 23]. Volatilization of samples by pyrolysis minimizes the need for pre-isolation, even when analyzing macromolecules in complex mixtures [24]. Therefore, it can be used to analyze a wide variety of materials: e.g., fibers and textiles, wood, bark and paper, artistic materials, synthetic polymers and heteropolymers [12, 13]. Likewise, comparable and reproducible results can be obtained when the conditions of the analysis are kept constant: i.e., carrier gas, heating rate, maximum temperature, homogeneous particle size and removal of non-structural compounds [18, 21]. Therefore, samples with the same composition will produce the same derivatives of pyrolysis [13, 21]. On the other hand, the advantages of the coupled GC/MS system are associated with a high speed, specificity and sensitivity, in both the separation of the pyrolysis products and in their identification [9, 12]. In addition, Py-GC/MS allows the identification of compounds without the necessity of standards.

It enables the comparison to commercial or open access libraries, including some already curated for different classes of chemical compounds [21, 25–28]. Finally, the raw data generated can be exported for quantitative or qualitative analysis [29, 30].

1.2 Issues related to Py-GC/MS

Although the many advantages and applications of Py-GC/MS are evident, different authors consider some problematic aspects. The main ones are: 1) pyrolysis produces a large amount of compounds, therefore, is necessary to deal with the vast amount of information registered by the Mass Spectrometer. 2) Only one part of the compounds produced can be unambiguously identified. 3) Low availability of mass spectra in databases and reference libraries. 4) Altogether, this makes the interpretation of the results from analytical pyrolysis difficult. However, most of these problems can be solved if cheminformatics is applied to the data resulting from Py-GC/MS.

The following sections will describe the use of omics tools for the deconvolution of mass spectra, as well as the alignment and annotation of the compounds identified in the chromatograms (**Figure 1**). This process is useful to compare different samples obtained by Py-GC/MS, under the same operating conditions, even using different equipment. In addition, different multivariate methods will be described to minimize the statistical noise generated by numerous uninformative compounds (i.e., those derived from carbohydrates). Together, the use of omics tools and multivariate methods facilitate the interpretation of the results of analytical pyrolysis. The processes detailed here may also be applicable to Py-GC/MS analysis of materials other than lignocellulosics (i.e., polymers, copolymers, soil samples and organic matter). In addition, they can be applied to raw data generated by other chromatography systems coupled to mass spectrometry (i.e., GC/MS/MS, LC/MS, and LC/MS/MS), including different equipment and output formats.

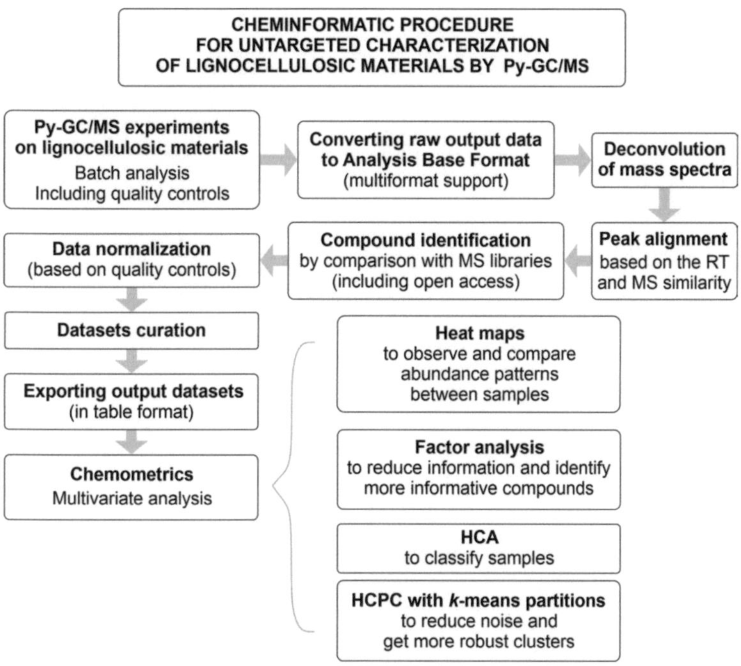

Figure 1.
Untargeted cheminformatics workflow for analysis of lignocellulosic materials by Py-GC/MS.

1.3 Common problems in Py-GC/MS and contribution of cheminformatics for their solution

Some apparent methodological problems attributed to pyrolysis are associated with the conditions necessary for the analysis of specific materials. Lourenço *et al.* [12], point out that care must be taken with the pyrolysis temperature when analyzing materials rich in suberin, such as barks. The main problem is that suberin decomposes at temperatures in the range of 550–600°C [31]. Therefore, this is an aspect to take into account if it is required to know the composition of this polymer within lignocellulosic samples [12]. Another problem referred in various works is that Py-GC/MS cannot guarantee an entirely quantitative determination. However, some authors have successfully carried out quantitative analyses in the optimization of aromatic hydrocarbon production from biomass [29]. Also for the quantification of only small amounts of aromatic hydrocarbons, applying the external calibration method [3, 30].

The amount of information that is generated as a result of the entire process can be challenging aspect. One analysis of 45 minutes by Py-GC/MS on lignocellulosic samples can generate up to 2,729 mass spectra [21]. However, after cheminformatics and manual curation of the datasets, the authors were able to unambiguously recognize 451 compounds, including some putative isomers. Another common problem is the displacement of the peaks in the chromatograms for samples with different chemical composition. For example, the displacement of the peak corresponding to levoglucosan in Py-GC/MS chromatograms for syringil-rich wood [18]. The displacement is due to the absence of acetovanillone in the samples. Therefore, the peak of levoglucosan appears at a Retention Time (RT) of 22.72 min, while in species that produce acetovanillone it is observed at 23.55 min (**Figure 2**). The above effect is problematic when it is required to directly process a batch of several samples with differential compositions. There are two reasons: 1) the process would be very time consuming if several species are analyzed and all the peaks identified by Py-GC/MS are compared one by one (about 40 compounds per sample, using

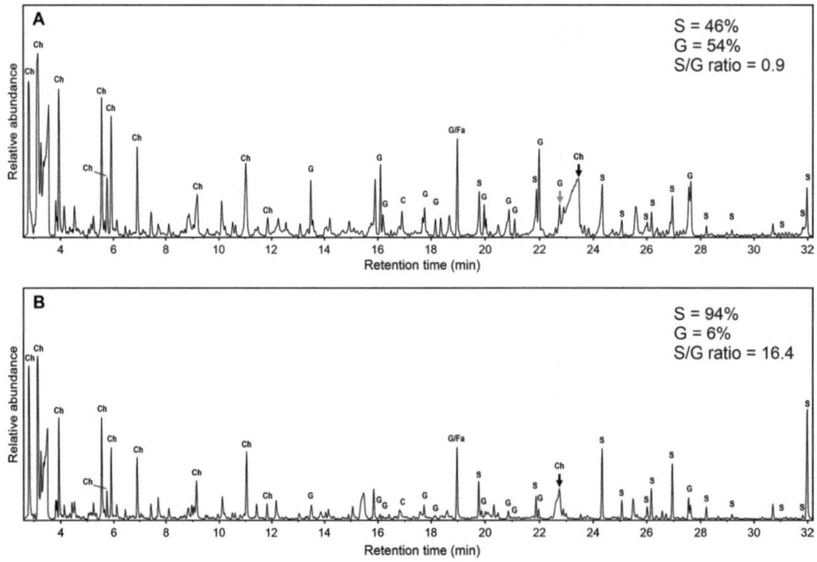

Figure 2.
Displacement of the peaks. Py-GC/MS chromatograms from extractives-free wood in cacti: A) Pilosocereus chrysacanthus and B) Ferocactus hamatacanthus. Displacement of levoglucosan (black arrows) is due to the absence of acetovanillone (gray arrows) in samples with 94% of syringil units [18]. The origin of the compounds is marked with letters: Ch, carbohydrates; G, guaiacyl subunits; S, syringil subunits, Fa, ferulates.

the native GC/MS software). This implies that the analysis has to be limited only to differences in the relative abundance, or the presence/absence, of only certain compounds. 2) If the raw datasets from the chromatograms are compared directly, using any multivariate method, the peak displacement would cause methodological bias because equivalent compounds are not being compared. Cheminformatics analysis solves this problem by automating the alignment of mass spectra and the identification of compounds for a batch of samples.

On the other hand, the high degree of degradation caused by the high temperatures used in pyrolysis represents, by far, the main problem of this technique. Therefore, this technique is considered to be of little use to characterize molecules larger than monomers or dimers in biopolymers such as lignin [6]. In addition, it is considered that the large number of derivatives makes the description of the chemical composition of sample difficult. Therefore, the detailed interpretation of the results is difficult and probably not necessary [3]. For example, when analyzing carbohydrate samples, low molecular weight derivatives can originate from hexoses or pentoses [12, 32]. The reason is that cellulose and hemicelluloses involve similar thermal degradation pathways, therefore a large part of the derivatives produced are the same [33, 34]. The reason is that cellulose pyrolysis causes the heterolytic cleavage of the glycosidic C—O bonds. In addition, it involves complex reactions and different pathways to give rise to anhydro sugars and numerous compounds with low molecular weight: i.e., acetic acid, 1-hydroxybutan-2-one, hydroxyacetaldehyde, 1-hydroxypropan-2-one and 2-furaldehyde [15, 35, 36]. A large part of these small compounds can also be originated from the decomposition of hemicelluloses. For example, 2-furaldehyde and acetic acid can be produced from the degradation of xylan [12, 37, 38]. On the other hand, there are contrary cases, but they also

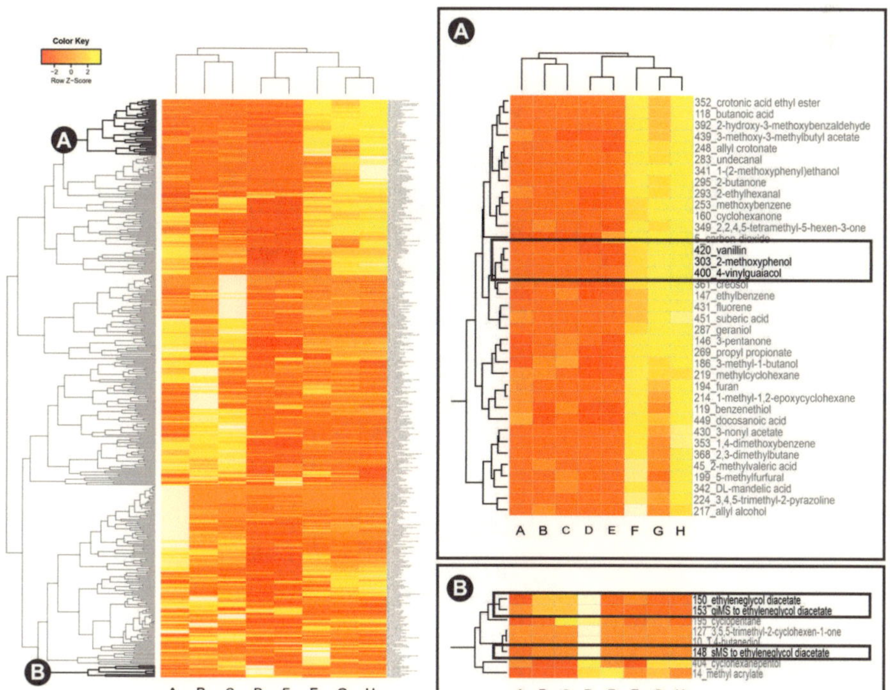

Figure 3.
Complete profile of the compounds identified for eight samples of lignocellulosic materials. A) Cluster corresponding to Guaiacyl lignin derivatives. B) Abundance patterns for carbohydrates derivatives. Similar (sMS) or quasi identical (qiMS) mass spectra.

contribute to the ambiguity in the identification of the compounds and their origin. Particularly when different ions are produced by the same class of compounds. The case of pyrans and furans is an example of compounds with ambiguous origin; both, with different molecular ions, can derive from the degradation of cellulose or hemicelluloses [12]. In this sense, the use of cheminformatics makes it possible to identify the abundance patterns of the compounds in a batch of samples. Based on this, it can be inferred if there are coincidences in the behavior of the pyrolysis products (**Figure 3**). In this way, it is possible to infer whether different compounds have the same origin, or rule out differences due to the operating conditions of the method or the characteristics of the samples [21].

For example, 2,5-dimetylfuran and 4-methyl-2*H*-pyran correspond to different molecular ions, but have the same average mass (96.13 Da) with similar RT, 4.64 min and 4.74 min, respectively (*see* Supplementary Materials of [21]). Based on the observed abundance patterns, it can be deduced that they are related to

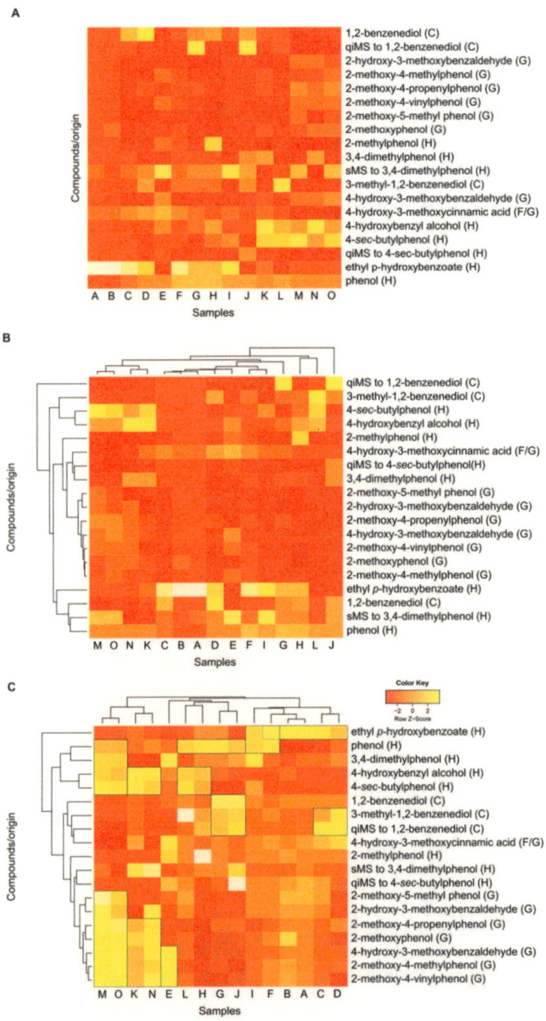

Figure 4.
Representation of the importance of using standardized data for the interpretation of the results. Non-standardized data: A) just ordered alphabetically; it is not possible to identify abundance patterns. B) Data arranged based on the HCA; trace compounds are overshadowed by the most abundant ones. C) Standardized data; compounds with the same origin share patterns of abundance and high similarity.

two different groups of compounds. Another example includes guaiacols, which are derived from guaiacyl (G) units. Under the same conditions of pyrolysis and composition of the samples, their abundance patterns should be the same. In the clustering analysis (CA) of **Figure 3A**, the guaiacols appear together forming a single group. For carbohydrate derivatives, abundance patterns with high similarity can also be identified for related compounds or putative isomers. **Figure 3B** shows the abundance patterns for ethyleneglycol diacetate and compounds with *quasi* identical (qiMS) or similar (sMS) mass spectrum. Another similar example is the independent origin of catechols and guaiacols in some lignocellulosic samples [21]. Catechols can be produced from guaiacols by secondary reactions at high temperatures [12, 21, 36]. However, as seen in **Figure 4**, the catechol abundance patterns across the samples, under the same experimental conditions, are clearly different from those samples with a predominance of G lignin. Therefore, catechols can be considered independently derived from those derivatives from G lignin.

2. Cheminformatics applied to Py-GC/MS

Increased computational capacity, development of powerful deconvolution algorithms and technological advances in analysis equipment have allowed the design of specialized software for chemical analysis. Areas such as omics sciences have particularly benefited from the rise of cheminformatics [26]. However, the application of untargeted analysis is becoming broader and is no longer restricted to the discovery and characterization of compounds in metabolomics. In this sense, it is possible to use the spectral deconvolution software for the processing of the data resulting from Py-GC/MS [21]. Open source software follows the same principle as native GC/MS software for spectra deconvolution and compound identification. However, it allows the use of different input formats for the raw datasets, regardless of the type, resolution and brand of the GC/MS equipment [26, 28]. In addition, different parameters can be adjusted to improve the informative quality of the results; e.g., the parameters used for deconvolution, the use of quality controls and normalization of the relative abundances for a batch of samples, alignment parameters and identification of compounds, use of different reference libraries for mass spectra, retention indices and times of retention. Because Py-GC/MS produces a large number of derived compounds, a lot of information is generated (i.e., mass spectra recorded by the detector in the MS). Omics tools allow deconvolution of all acquired mass spectra for a batch of samples in independent experiments. Basically, the peaks are detected by deconvolution of the mass spectra, smoothing the data points by the least squares method or by linear weighted smoothing average [28, 39]. Afterwards, both the first and second derivatives are considered together with the amplitude of the ions to identify the noise threshold. Based on the noise levels, the initial retention times are calculated for each peak. For the final detection of the peaks, the unsmoothed raw chromatogram is used as a control [28]. The deconvoluted spectra for the batch of samples are aligned based on the similarity of their mass spectra and their RTs. Finally, they are compared with those spectra in the reference MS libraries and the compounds can be identified based on the maximum fit of their RT, RI and mass spectra [26]. Additionally, the deconvoluted datasets for a batch of samples can be normalized and exported in table format. The information contained in the output file is important for comparative analysis: i.e., EI fragmentation pattern, quant mass (m/z of the main ion), averaged RT, InChIKey, total similarity with the reference spectrum and relative abundance of each compound normalized for the entire batch of samples [28]. This information can be used for comparative analysis by multivariate methods. Alternatively, it can be compared with databases such as the Chemical Entities of Biological Interest (ChEBI) ontology [25], to infer biological characteristics of the original samples based on their derivatives from pyrolysis [21].

The comparative analysis of lignocellulosic samples is highly favored by the normalization process on the data obtained for a batch of samples [21]. The normalization of the deviations of the MS signal intensities is carried out including a series of quality control (QC) samples. The QC samples are one or more samples obtained by combining all samples in the batch. For lignocellulosic materials it is suitable to use alternately one QC sample for every five samples analyzed [21]. The data obtained from the measurement of the QC samples are smoothed by the Lowess of the single-degree least-squares. The coefficients generated on the QC samples are interpolated using the cubic spline and finally all the datasets are aligned based on the spline interpolation result [28].

Additionally, the unknown compounds can be annotated using their elemental formulas and *in silico* mass spectra fragmentation based on public spectral databases, such as MassBank, LipidBlast and GNPS [27, 28]. Currently most open access MS reference libraries are focused on the compounds of interest; i.e., metabolomics and lipidomics. Several of them include precursors or derivatives of lignocellulosic biomass, such as anhydro sugars, furans, pyrans and phenols and their derivatives. Actually, as the areas of application of omics tools diversify (for spectral deconvolution and compound annotation) it can be expected that the diversity and number of compounds incorporated in open access databases will increase.

2.1 Multivariate analysis on exported Py-GC/MS data

Interpretation of the results obtained by Py-GC/MS is a complex process. This is due to the large number of compounds that are generated by pyrolysis and the little information provided by compounds with ambiguous origin, often very numerous (as described above). Multivariate analysis applied to Py-GC/MS data from various materials helps to make data management easier, reduce the information obtained and facilitate the interpretation of the results. It has been used to characterize lignocellulosic samples and other biological samples [40–43].

A common application of Py-GC/MS material analysis is the classification of samples based on the similarities of the compounds they produce. For example, to evaluate different experimental systems [44, 45] or for the optimization of two different methods [46]. It was recently used to characterize and classify lignocellulosic samples applying cheminformatics from a chemotaxonomic approach [21].

Classification of the observations into groups requires the calculation of the distance between each pair of observations. As a result, a distance matrix is obtained, also called a dissimilarity matrix. The distance most commonly used by computational algorithms is the Euclidean distance [47]; i.e., the root sum-of-squares of differences for a set of vectors [48]. As a result, observations with high values of features will be grouped together, likewise, observations with low features values will be grouped together.

Apart from the normalization performed by the spectral deconvolution software on the output datasets, it is highly recommended to standardize the variables before measuring the dissimilarities between observations [49]. This step is considered necessary as it can have a great impact on the results of the analysis on biological data [49, 50]. **Figure 4** represents the differences between non-standardized data and standardized data. In standardization, the values of each variable are weighed by a scale factor in order to give more weight to the small but potentially significant changes in signal intensity [51]. Thus, the standard deviation and the mean usually take values of one and zero, respectively. On the other hand, standardization will help to obtain equivalent similarities regardless of the distance method used (e.g.,

Euclidian, Manhattan, Correlation or Eisen). For example, when using standardized data, there is a functional relationship between Pearson's correlation coefficient and the standardized Euclidean distance, so that both results are comparable [48].

2.2 Groupings by *k*-means partition

The *k*-means algorithm is commonly used in the partition of *N*-dimensional population into *k* series based on a sample [52, 53]. Where *k*-series corresponds to the number of clusters to be calculated, arbitrarily specified by the researcher. The algorithm consists of classifying objects forming *k* clusters, so that for each group the intra-class similarity is minimized, but in turn, each group is as different as possible from the rest [54, 55]. Since the members of each cluster are the most similar to each other, the centre (centroid) of each group is represented by the respective mean. Briefly, the standard procedure for the computational algorithm is as follows: 1) the researcher specifies an arbitrary number of *k* clusters to be calculate. Alternatively, centroids can also be specified; 2) if the centroids are not specified, they are obtained randomly for each group; 3) by calculating the Euclidean distance, each object is assigned to its closest centroid; 4) the centroids are updated considering the recently incorporated objects; 5) each observation is reviewed with respect to the other clusters to confirm their membership to the respective group. The assignment and update steps are repeated until convergence or the total number of iterations are reached [53]. This method implies advantages when the author has prior knowledge of the analyzed data. For example, in taxonomy, the number of *k* clusters can refer to the number of data classes to classify [56, 57] or to the taxa that are known or those that want to be tested [21]. In the validation or optimization analysis of methods, it could correspond to the number of systems or criteria that are being considered [58]. An optimal number of *k* clusters can be more efficient when combined with other multivariate analysis techniques; e.g., in analysis of hierarchical clustering on principal components with partition of *k*-means (HCPC), which will be explained in the subsequent sections. If there is not enough information to select a specific number of *k* clusters, the optimal number of *k* partitions can be inferred using the "elbow" method [49, 59, 60]. The method consists of applying the *k*-means algorithm to the data, adopting different numbers of *k* clusters. Then graphically represent the internal variance of the groups, using the number of groups and their respective total within-cluster sum-of-squares (WCSS). The optimal number of *k* clusters will be indicated by the point where the slope of the WCSS tends to flatten, that is, where the variance is minimized [59, 61, 62]. Due to the randomness with which the initial centroids are selected, it is possible to observe variation in the clusters obtained when replicating the analysis. A suggested solution is to calculate the *k*-means algorithm several times and select the number of *k* clusters that generates the lowest WCSS [49]. Furthermore, it is suggested to compare different indices and select an optimal number of *k* clusters based on the majority rule (**Figure 5**).

2.3 Principal component analysis

Among multivariate analyses, Principal Component Analysis (PCA) is the most common method for extracting information from large datasets generated by analytical pyrolysis [3, 12]. The PCA has different objectives, it is mainly used to reduce the dimensions of the datasets by extracting the most important information. In addition, it is useful to simplify the description of the data series and to analyze the structure of the observations and variables [63–65]. The PCA generates principal components (PC) that result from linear combinations of the original variables (e.g., the identified compounds). The number of these new variables can

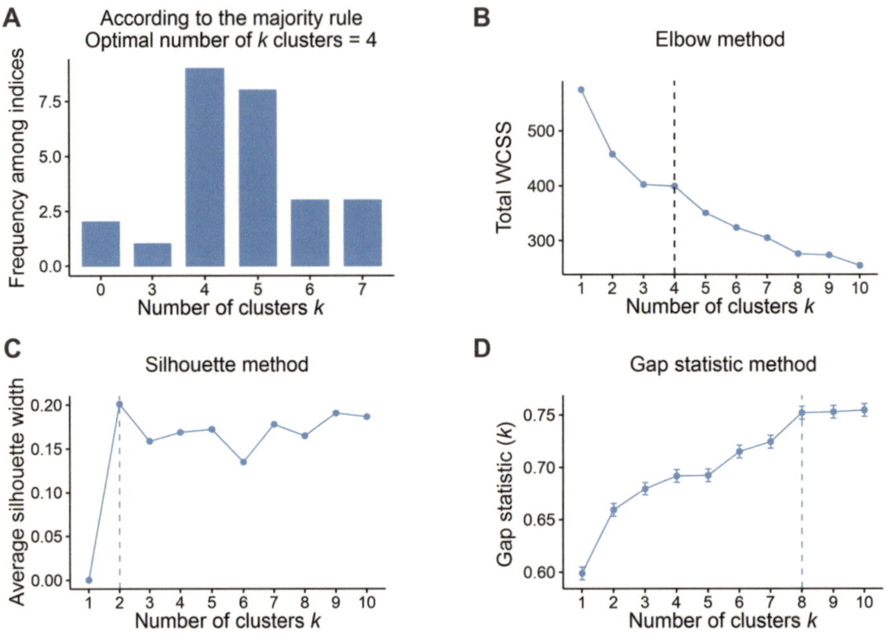

Figure 5.
Comparison of different methods for calculating the optimal number of k clusters. A) Optimal number of k clusters suggested by the majority rule by analysing all indexes. B) Elbow method. C) Silhouette method. D) Gap Statistic method.

be arbitrarily defined. Commonly, the first component explains the largest possible variance of the dataset and the second, being orthogonal to the first, will be calculated to represent the largest possible variance. The factor scores correspond to the values of these new variables for the original observations (e.g., relative abundances of the compounds). The eigenvalues associated with each component correspond to the sum of the squared factor scores for each component. Thus, the contribution of each observation to a component (i.e., importance of the observation) is represented by the ratio of the square factor score of the observation by the eigenvalue associated with that component. Contributions for a given component can take values from zero to one, so the sum of all contributions for that component is equal to one [65]. Alternatively, the correlation of the two new variables generated by the PCA can be represented by a biplot [66]. Thus, it is possible to know the compounds that contribute the most to the sets obtained in the PCA (**Figure 6**). As stated, the first two components extracted by the PCA represent the largest variances for the data series. However, to determine the optimal number of components to consider, it is suggested to perform the "scree" test, plotting the eigenvalues as a decreasing function of their size [64]. In the graph, an "elbow" will be observed after the point where the slope of the curve decreases (flattens), therefore the optimal number of components must include all the components before that point (**Figure 7A**).

2.4 Classification of samples using only the most informative compounds

Multivariate analyses are very useful when working with a large number of data. If lignocellulosic samples are analyzed by Py-GC/MS and the deconvolution method is applied, hundreds of derived compounds can be expected for each sample [21]. The PCA and clustering analysis allows separately to reduce the dimensionality of the datasets, identify relationships between the variables, and quantify the significance of the variables that can explain the resulting clusters [67].

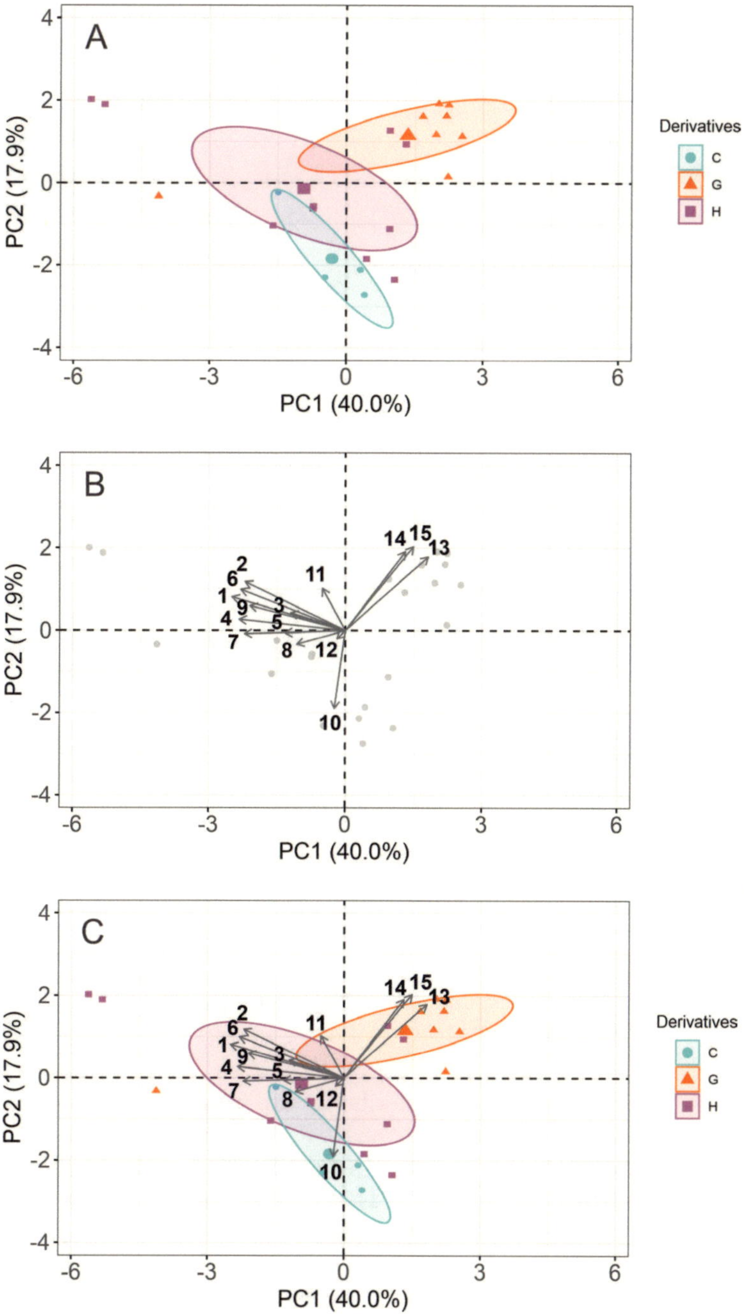

Figure 6.
PCA results: the correlation between the variables generated by the PCA for lignin derivatives is shown. A) Compounds clustered according to their origin: C, catechols; H, phenols; G, guaiacols. B) Biplot that represents the correlation between variables. C) Confidence intervals for the correlation between variables; ellipses represent a significance level of 99%.

The dimensionality of the data directly influences the results; the higher the dimensionality the classifications obtained will be more reliable [68, 69]. For the analysis of chemical compounds in materials the optimal relation of data points to variables is 6:1 or higher, with an absolute minimum of 3:1 [69–71]. However, to achieve these high proportions in the optimal ratio it is necessary to increase the number of experiments. An alternative to achieve the optimal relationship

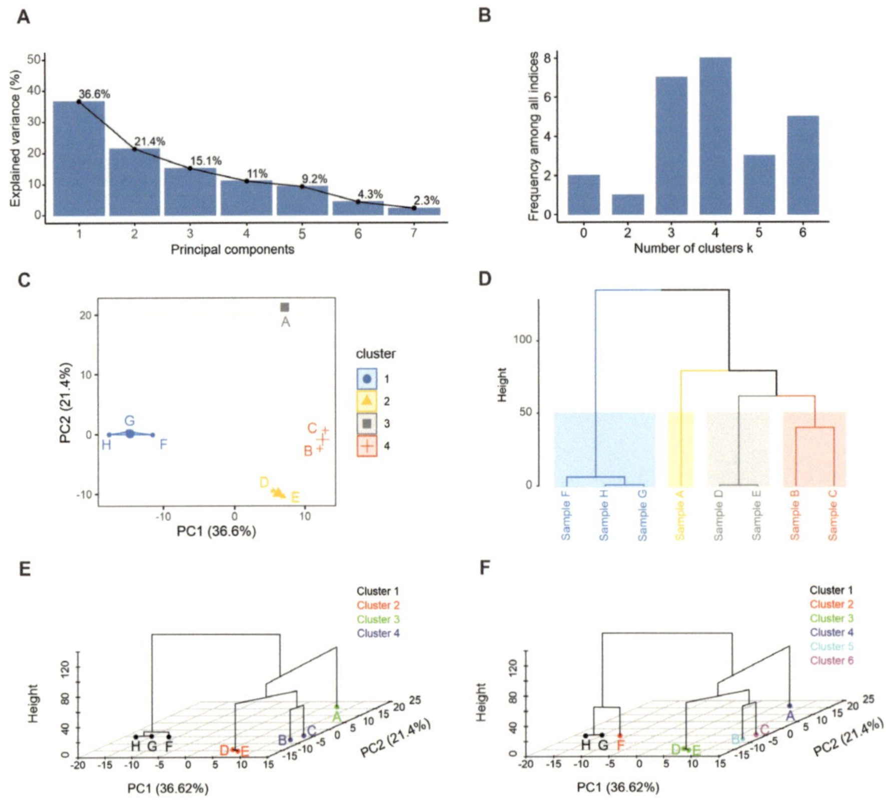

Figure 7.
HCPC analysis for minimizing noise resulting in Py-CG/MS analysis. A) Scree plot, to determine the number of components that explain most of the variance. Number of components used = 5. B) Optimal number of k clusters. Optimal k clusters suggested by the majority rule = 4. C) Factorization of the data series using the PCA. D) Initial hierarchical clustering on the reduced matrix generated by the PCA. E) Clustering obtained using the number of k clusters suggested by the majority rule (the same suggested by the "elbow" method). F) Clusters obtained using a non-optimal number of k clusters.

when it is not possible to increase the number of experiments is by reducing the number of variables [68]. In that sense, the HCPC analysis is a very powerful tool (**Figure 7A–C**). Compared with PCA and CA, the HCPC analysis increases the objectivity and robustness of the results. That is, the classifications are restricted only to the dimensions that contain the most significant information [67, 72]. In this way, the statistical noise caused by the many uninformative derivatives of pyrolysis is minimized [21]. In addition, it improves the visualization of the data and provides information on the variables (i.e., compounds) that contribute predominantly to the resulting clusters [21, 67]. The HCPC is an exploratory statistical analysis whose computational algorithm can be summarized in three steps: first, the reduction of dimensions can be by any factorial method. PCA for quantitative variables, multiple correspondence analysis for categorical data, or multiple factor analysis to jointly integrate different data blocks [72, 73]. This step allows the determination of the relationships between the concentrations of most abundant compounds and the trace compounds. In addition, it simplifies the dataset by reducing the number of variables to only two principal components that explain most of the variance [74] (**Figure 7C**). Second, the hierarchical cluster analysis (HCA), by using the Euclidean distance, form clusters of samples according to the similarities in their chemical composition [73, 74] (**Figure 7D**). Each object is treated as a single cluster and pairs of groups are successively merged until all clusters merge into one large group [48]. The algorithm uses Ward's method to minimize the total intragroup variance [47, 72, 75]. Finally, the partition with *k*-means allows to stabilize the groupings obtained by the HCA [67, 73] (**Figure 7E**). In this way, the HCPC applied to the data resulting from Py-GC/MS of lignocellulosic materials allows the samples to be classified based on the abundance patterns of the most informative compounds. That is, statistical noise generated by uninformative, ambiguous, or noisy compounds is suppressed [21].

2.5 Simplified visualization of abundance and similarity patterns from Py-GC/MS data

The heat map method is a simple but highly efficient tool for the graphical representation of large datasets (**Figure 3**). This method is very useful in studies where it is necessary to interpret a large amount of quantitative data; e.g., metabolomics, proteomics, lipidomics, and genomics [76–78]. The quantitative data (i.e., relative abundances of the ions detected by the MS) are represented in different color scales in the format of a two-dimensional matrix [79, 80]. The basic structure of the matrix is given by columns and rows; each column represents a sample and each row represents a compound [76]. The quantitative values correspond to the relative abundance for each compound in each sample. For a certain range of values a particular color is assigned. The highest relative abundances are represented by one end of the color scale and the lowest abundances are represented by the opposite end of that color scale [77]. Additionally, the columns and rows of the matrix are rearranged to recognize significant patterns in the heat map. To do this, rows and columns with similar profiles are arranged so that they are closer to each other, making these profiles easily visible to the eye [79, 80]. The permutation of rows and columns is made based on the result of the CA on the correlation matrix of the variables for each set of variables [77]. Alternatively, the dendrograms resulting from the CA can be represented at the edges of the matrix, both for the samples and for the compounds [77, 79, 81, 82]. This form of representation of the relative abundances is so efficient that after rearranging the rows and columns of the matrices the abundance patterns of the compounds become obvious [76, 83].

The standardization (e.g., Z-transformation) of the variables from each series of variables highly influences the correct representation of the similarity patterns obtained [77, 80]. If raw, non-standardized data are used, the low relative abundances will be obscured by the higher relative abundances (**Figure 4A–C**). When using transformed data it is possible to infer that those compounds with similar abundance patterns imply equal origins [21, 79].

An interactive variant of the heat map method has been referred by several authors in the field of metabolomics [76, 84, 85]. Of course, this can also be applied to Py-GC/MS data. This online variant allows the visualization of important information from the mass spectra on the matrix. Metadata such as mass spectrum, retention time, extracted ion chromatograms (EICs), box and whisker plots as well as matches for each compound can be displayed in real time for each observation [76, 86].

On the other hand, alternative methods for interpreting the data resulting from Py-GC/MS have emerged recently. The Van Krevelen (VK) diagrams have been successfully applied for interpretation of high resolution GC/MS data [3, 87]. These diagrams allows to visualize the chemical composition of complex chemical mixtures by plotting the H:C ratio against the O:C ratio for every compound in the mixture [6]. Thus, the VK diagrams provide information about the classes of compounds present and allow accurately evaluate the number of compounds in a sample [88]. Furthermore, van Krevelen diagrams play an important role in the deconvolution of high resolution MS spectra for complex lignin samples [6].

3. Potential areas of cheminformatics applied to Py-GC/MS

Due to its versatility, Py-GC/MS has been successfully applied to different areas of knowledge. Among these areas, cheminformatics reviewed in this chapter also has important application opportunities. Environmental, chemical and materials sciences, engineering, energy and biorefinery, biology, biotechnology, and conservation and restoration of cultural heritage are among the most cited in the literature. The fields of application are also varied; for example, in the development and optimization of the properties of new materials and resources, such as synthetic polymers, resins and biofuels [3, 10, 11, 13]. On the other hand, several samples of environmental materials have been characterized by analytical pyrolysis; e.g., organic matter, soil and pollutants in different natural substrates [89–91]. In addition, Kush [13] list a series of applications for polymers, in which the following can be highlighted: 1) identification of polymers through the use of reference libraries, 2) qualitative analysis of copolymers, 3) investigation of thermal stability and kinetics degradation of polymers and copolymers and 4) determination of monomers in polymers and volatile organic compounds.

4. Conclusions

Cheminformatics detailed in this chapter can be applied to the analysis of any type of polymeric materials by Py-GC/MS. The use of open access software to deconvolution of mass spectra streamlines the processing of the resulting data series for a large number of samples. The computational processing capacity of current equipment makes this technique suitable for any laboratory with a Py-GC/MS equipment. In a few minutes a large number of samples can be processed: e.g., deconvolution, alignment and identification of compounds for 30 samples can take about 30 min. On the other hand, the interpretation of the results is greatly aided by the use of the chemometric techniques exemplified here. In addition, cheminformatics makes it possible to compare the mass spectra of the studied compounds, not only with commercial databases, but with other open access databases. Some of the open access databases contain relevant biological information about the compounds (e.g., the ontology of CheBI, MassBank, LipidBlast and GNPS). This is important in studies of materials (e.g., in the case of elements with carcinogenic potential), or of biological interest (e.g., in samples with antibacterial, antibiotic, or medicinal properties). There are currently a significant number of open access MS libraries. Actually, with the diversification of the application field for deconvolution software it is expected that the number of mass spectra in open access libraries will increase. Finally, studies like this leave open the possibility of knowing most of the chemical compounds that take part in the decomposition and secondary reactions during pyrolysis of polymeric materials.

Acknowledgements

JR-R thanks to DGAPA, UNAM by the postdoctoral fellowship [Programa de becas posdoctorales en la UNAM; communiqué 113/2017] and to Facultad de Estudios Superiores Zaragoza, UNAM for supporting this work.

Author details

Jorge Reyes-Rivera
Facultad de Estudios Superiores Zaragoza, Universidad Nacional Autónoma de México, Mexico City, Mexico

*Address all correspondence to: jorge.reyes@zaragoza.unam.mx

IntechOpen

© 2021 The Author(s). Licensee IntechOpen. This chapter is distributed under the terms of the Creative Commons Attribution License (http://creativecommons.org/licenses/by/3.0), which permits unrestricted use, distribution, and reproduction in any medium, provided the original work is properly cited.

References

[1] Chen, H. Chemical composition and structure of natural lignocellulose. In Biotechnology of lignocellulose. Dordrecht: Springer; 2014. p. 25-71

[2] Bulushev DA, Ross JR. Catalysis for conversion of biomass to fuels via pyrolysis and gasification: a review. Catal Today. 2011;171:1-13

[3] Grams, J. Chromatographic analysis of bio-oil formed in fast pyrolysis of lignocellulosic biomass. Reviews in Analytical Chemistry. 2020;39(1), 65-77

[4] Briens, C., Piskorz, J., & Berruti, F. Biomass valorization for fuel and chemicals production--A review. International Journal of Chemical Reactor Engineering. 2008;6(1)

[5] Agblevor, F. A., Evans, R. J., & Johnson, K. D. Molecular-beam mass-spectrometric analysis of lignocellulosic materials: I. Herbaceous biomass. Journal of Analytical and Applied Pyrolysis. 1994;30(2), 125-144

[6] Letourneau, D. R., & Volmer, D. A. Mass spectrometry-based methods for the advanced characterization and structural analysis of lignin: A review. Mass Spectrometry Reviews. 2021

[7] Sun, Z., Fridrich, B., de Santi, A., Elangovan, S., & Barta, K. Bright side of lignin depolymerization: toward new platform chemicals. Chemical reviews, 2018;118(2), 614-678

[8] Prothmann, J., Li, K., Hulteberg, C., Spégel, P., Sandahl, M., & Turner, C. Nontargeted Analysis Strategy for the Identification of Phenolic Compounds in Complex Technical Lignin Samples. ChemSusChem. 2020;13(17), 4605

[9] Bridgwater, A. V., & Peacocke, G. V. C. Fast pyrolysis processes for biomass. Renewable and Sustainable Energy Reviews. 2000;4(1), 1-73. doi:10.1016/s1364-0321(99)00007-6

[10] French, R., & Czernik, S. Catalytic pyrolysis of biomass for biofuels production. Fuel Processing Technology. 2010;91(1), 25-32

[11] Isahak, W. N. R. W., Hisham, M. W., Yarmo, M. A., & Hin, T. Y. Y. A review on bio-oil production from biomass by using pyrolysis method. Renewable and sustainable energy reviews. 2012;16(8), 5910-5923

[12] Lourenço, A., Gominho, J., & Pereira, H. Chemical characterization of lignocellulosic materials by analytical pyrolysis. In Analytical Pyrolysis. IntechOpen; 2018

[13] Kusch P. Pyrolysis-Gas Chromatography/Mass Spectrometry of Polymeric Materials, Advanced Gas Chromatography - Progress in Agricultural, Biomedical and Industrial Applications. Dr. Mustafa Ali Mohd (Ed.); 2012. ISBN: 978-953-51-0298-4

[14] Yang H, Yan R, Chen H, Lee DH, Zheng C. Characteristics of hemicellulose, cellulose and lignin pyrolysis. Fuel. 2007;86:1781-1788

[15] Lu, Q., Yang, X. C., Dong, C. Q., Zhang, Z. F., Zhang, X. M., & Zhu, X. F. Influence of pyrolysis temperature and time on the cellulose fast pyrolysis products: Analytical Py-GC/MS study. Journal of Analytical and Applied Pyrolysis. 2011;92(2), 430-438

[16] Faix, O., Meier, D., & Fortmann, I. Thermal degradation products of wood. A collection of electron-impact (EI) mass spectra of monomeric lignin derived products. Holz als Roh-und Werkstoff. 1990;48(9), 351-354

[17] Ralph, J., & Hatfield, R. D. Pyrolysis-GC-MS characterization of forage materials. Journal of Agricultural and Food Chemistry. 1991;39(8), 1426-1437

[18] Reyes-Rivera, J., Soto-Hernández, M., Canché-Escamilla, G., & Terrazas, T. Structural characterization of lignin in four cacti wood: implications of lignification in the growth form and succulence. Frontiers in plant science. 2018;9, 1518

[19] del Río, J. C., Gutiérrez, A., & Martínez, Á. T. Identifying acetylated lignin units in non-wood fibers using pyrolysis-gas chromatography/mass spectrometry. Rapid communications in mass spectrometry. 2004;18(11), 1181-1185

[20] del Río, J. C., Martínez, Á. T., & Gutiérrez, A. Presence of 5-hydroxyguaiacyl units as native lignin constituents in plants as seen by Py-GC/MS. Journal of analytical and applied pyrolysis. 2007;79(1-2), 33-38

[21] Reyes-Rivera, J., Solano, E., Terrazas, T., Soto-Hernández, M., Arias, S., Almanza-Arjona, Y. C., & Polindara-García, L. A. Classification of lignocellulosic matrix of spines in Cactaceae by Py-GC/MS combined with omic tools and multivariate analysis: A chemotaxonomic approach. Journal of Analytical and Applied Pyrolysis. 2020;148, 104796

[22] Meier, D., & Faix, O. Pyrolysis-gas chromatography-mass spectrometry. In Methods in lignin chemistry. Berlin, Heidelberg. Springer; 1992. p. 177-199

[23] Brunow, G., Lundquist, K., & Gellerstedt, G. Lignin. In Analytical methods in wood chemistry, pulping, and papermaking. Berlin, Heidelberg: Springer; 1999. p. 77-124

[24] Wampler, T. P. Analytical pyrolysis: An overview. In: Wampler T.P., editor. Applied Pyrolysis Handbook. 2nd ed. New York: Taylor Francis Group; 2007. p. 288

[25] Degtyarenko, K., De Matos, P., Ennis, M., Hastings, J., Zbinden, M., McNaught, A., ... & Ashburner, M. ChEBI: a database and ontology for chemical entities of biological interest. Nucleic acids research. 2007;36(suppl_1), D344-D350

[26] Tsugawa, H., Cajka, T., Kind, T., Ma, Y., Higgins, B., Ikeda, K., ... & Arita, M. MS-DIAL: data-independent MS/MS deconvolution for comprehensive metabolome analysis. Nature methods. 2015;12(6), 523-526

[27] Tsugawa, H., Kind, T., Nakabayashi, R., Yukihira, D., Tanaka, W., Cajka, T., ... & Arita, M. Hydrogen rearrangement rules: computational MS/MS fragmentation and structure elucidation using MS-FINDER software. Analytical chemistry. 2016;88(16), 7946-7958

[28] Lai, Z., Tsugawa, H., Wohlgemuth, G., Mehta, S., Mueller, M., Zheng, Y., ... & Fiehn, O. Identifying metabolites by integrating metabolome databases with mass spectrometry cheminformatics. Nature methods. 2018;15(1), 53-56

[29] Zhang, B., Zhong, Z., Ding, K., & Song, Z. Production of aromatic hydrocarbons from catalytic co-pyrolysis of biomass and high density polyethylene: analytical Py–GC/MS study. Fuel. 2015;139, 622-628

[30] Lu, Q., Zhou, M. X., Li, W. T., Wang, X., Cui, M. S., & Yang, Y. P. Catalytic fast pyrolysis of biomass with noble metal-like catalysts to produce high-grade bio-oil: analytical Py-GC/MS study. Catalysis today. 2018;302, 169-179

[31] Marques, A. V., & Pereira, H. Aliphatic bio-oils from corks: A Py–GC/MS study. Journal of Analytical and Applied Pyrolysis. 2014;109, 29-40

[32] Faix O, Fortman I, Bremer J, Meier D. Thermal degradation products of wood. Gas chromatographic separation and mass spectrometric characterization of polysaccharide derived products. Holz Roh Werkst. 1991;49:213-219

[33] Luo, Z., Wang, S., Liao, Y., & Cen, K. Mechanism study of cellulose rapid pyrolysis. Industrial & engineering chemistry research. 2004;43(18), 5605-5610

[34] Zhu X, Lu Q. Production of chemicals from selective fast pyrolysis of biomass. In: Momba M, Bux F, editors. Croatia: Biomass. Sciyo; 2010. p. 147-16

[35] Demirbas A. Pyrolysis mechanisms of biomass materials. Energy Sources, Part A: Recovery, Utilization, and Environmental Effects. 2009;31(13):1186-1193

[36] Kawamoto, H. Lignin pyrolysis reactions. Journal of Wood Science. 2017;63(2), 117-132

[37] Ponder, G. R., & Richards, G. N. Thermal synthesis and pyrolysis of a xylan. Carbohydrate Research. 1991;218, 143-155

[38] Dobele, G., Rossinskaja, G., Telysheva, G., Meier, D., & Faix, O. Cellulose dehydration and depolymerization reactions during pyrolysis in the presence of phosphoric acid. Journal of Analytical and Applied Pyrolysis. 1999;49(1-2), 307-317

[39] Savitzky, A., & Golay, M. J. Smoothing and differentiation of data by simplified least squares procedures. Analytical chemistry. 1964;36(8), 1627-1639

[40] Zodrow, E. L., & Mastalerz, M. Chemotaxonomy for naturally macerated tree-fern cuticles (Medullosales and Marattiales), Carboniferous Sydney and Mabou sub-basins, Nova Scotia, Canada. International Journal of Coal Geology. 2001;47(3-4), 255-275

[41] Alves, A., Gierlinger, N., Schwanninger, M., & Rodrigues, J. Analytical pyrolysis as a direct method to determine the lignin content in wood: Part 3. Evaluation of species-specific and tissue-specific differences in softwood lignin composition using principal component analysis. Journal of Analytical and Applied Pyrolysis. 2009;85(1-2), 30-37

[42] Mattonai, M., Licursi, D., Antonetti, C., Galletti, A. M. R., & Ribechini, E. Py-GC/MS and HPLC-DAD characterization of hazelnut shell and cuticle: Insights into possible re-evaluation of waste biomass. Journal of Analytical and Applied Pyrolysis. 2017;127, 321-328

[43] Xin, X., Pang, S., de Miguel Mercader, F., & Torr, K. M. The effect of biomass pretreatment on catalytic pyrolysis products of pine wood by Py-GC/MS and principal component analysis. Journal of Analytical and Applied Pyrolysis. 2019;138, 145-153

[44] Gómez, X., Meredith, W., Fernández, C., Sánchez-García, M., Díez-Antolínez, R., Garzón-Santos, J., & Snape, C. E. Evaluating the effect of biochar addition on the anaerobic digestion of swine manure: application of Py-GC/MS. Environmental Science and Pollution Research, 2018;25(25), 25600-25611

[45] Raja Sabaradin, R. Z., & Osman, R. Evaluation of evidence value of car primer using pyrolysis-gas chromatography-mass spectrometry (Py-GC-MS) and chemometrics. Science Letters (ScL). 2021;15(1), 45-37

[46] Maurer, J., Buffaz, K., Massonnet, G., Roussel, C., & Burnier, C. Optimization of a Py-GC/MS method for silicone-based lubricants analysis. Journal of Analytical and Applied Pyrolysis. 2020;149, 104861

[47] Murtagh, F., & Legendre, P. Ward's hierarchical agglomerative clustering method: which algorithms implement

Ward's criterion?. Journal of classification. 2014;31(3), 274-295

[48] Kassambara, A. Practical guide to cluster analysis in R: Unsupervised machine learning. Sthda. 2017;1

[49] van den Berg, R. A., Hoefsloot, H. C., Westerhuis, J. A., Smilde, A. K., & van der Werf, M. J. Centering, scaling, and transformations: improving the biological information content of metabolomics data. BMC genomics. 2006;7(1), 1-15

[50] Smilde, A. K., van der Werf, M. J., Bijlsma, S., van der Werff-van der Vat, B. J., & Jellema, R. H. Fusion of mass spectrometry-based metabolomics data. Analytical chemistry. 2005;77(20), 6729-6736

[51] Bouhlel, J., Bouveresse, D. J. R., Abouelkaram, S., Baéza, E., Jondreville, C., Travel, A., ... & Rutledge, D. N. Comparison of common components analysis with principal components analysis and independent components analysis: Application to SPME-GC-MS volatolomic signatures. Talanta. 2018;178, 854-863

[52] MacQueen, J. Some methods for classification and analysis of multivariate observations. In Proceedings of the fifth Berkeley symposium on mathematical statistics and probability. 1967;1(14), 281-297

[53] Kalra, M., Lal, N., & Qamar, S. K-mean clustering algorithm approach for data mining of heterogeneous data. In Information and Communication Technology for Sustainable Development Singapore: Springer; 2018. p. 61-70

[54] Arthur, D., & Vassilvitskii, S. How slow is the k-means method?. In Proceedings of the twenty-second annual symposium on Computational geometry. 2006; 144-153

[55] Nazeer, K. A., & Sebastian, M. P. Improving the Accuracy and Efficiency of the k-means Clustering Algorithm. In Proceedings of the world congress on engineering: London: Association of Engineers. 2009;1, 1-3

[56] Asosheh, A., & Ramezani, N. A comprehensive taxonomy of DDOS attacks and defense mechanism applying in a smart classification. WSEAS Transactions on Computers. 2008;7(4), 281-290

[57] Pandey, K. K., & Shukla, D. A study of clustering taxonomy for big data mining with optimized clustering MapReduce model. decision making. 2019;1(5), 30

[58] Verma, N., Sharma, V., Kumar, R., Sharma, R., Joshi, M. C., Umapathy, G. R., ... & Chopra, S. On the spectroscopic examination of printed documents by using a field emission scanning electron microscope with energy-dispersive X-ray spectroscopy (FE-SEM-EDS) and chemometric methods: application in forensic science. Analytical and bioanalytical chemistry. 2019;411(16), 3477-3495

[59] Kuraria, A., Jharbade, N., & Soni, M. Centroid Selection Process Using WCSS and Elbow Method for K-Mean Clustering Algorithm in Data Mining. International Journal of Scientific Research in Science, Engineering and Technology. 2018;190-195

[60] Syakur, M. A., Khotimah, B. K., Rochman, E. M. S., & Satoto, B. D. (, April). Integration k-means clustering method and elbow method for identification of the best customer profile cluster. In IOP Conference Series: Materials Science and Engineering. IOP Publishing. 2018;336(1), 012017

[61] Marutho, D., Handaka, S. H., & Wijaya, E. The determination of cluster number at k-mean using elbow method and purity evaluation on headline news.

In 2018 International Seminar on Application for Technology of Information and Communication. 2018; 533-538

[62] Duong, M. Q., Le Hong Lam, B. T. M., Tu, G. Q. H., & Hieu, N. H. Combination of K-Mean clustering and elbow technique in mitigating losses of distribution network. GMSARN International. 2019;153-158

[63] Jolliffe, I. T. Principal Component Analysis, Encyclopedia of Statistics in Behavioral Science. 2002;30 (3), 487

[64] Kassambara, A. Multivariate Analysis II: Practical Guide to Principal Component Methods in R. 2017

[65] Abdi, H., & Williams, L. J. Principal component analysis. Wiley interdisciplinary reviews: computational statistics. 2010;2(4), 433-459

[66] Gower, J. C., Lubbe, S. G., & Le Roux, N. J. Understanding biplots. John Wiley & Sons. 2011

[67] Strbova, K., Ruzickova, J., & Raclavska, H. Application of multivariate statistical analysis using organic compounds: Source identification at a local scale (Napajedla, Czechia). Journal of environmental management. 2019;238, 434-441

[68] Goodner, K. L., Dreher, J. G., & Rouseff, R. L. The dangers of creating false classifications due to noise in electronic nose and similar multivariate analyses. Sensors and Actuators B: Chemical. 2001;80(3), 261-266

[69] Brenet, S., John-Herpin, A., Gallat, F. X., Musnier, B., Buhot, A., Herrier, C., ... & Hou, Y. Highly-selective optoelectronic nose based on surface plasmon resonance imaging for sensing volatile organic compounds. Analytical chemistry. 2018;90(16), 9879-9887

[70] Wang, B., Cancilla, J. C., Torrecilla, J. S., & Haick, H. Artificial sensing intelligence with silicon nanowires for ultraselective detection in the gas phase. Nano letters. 2014;14(2), 933-938

[71] Shehada, N., Cancilla, J. C., Torrecilla, J. S., Pariente, E. S., Brönstrup, G., Christiansen, S., ... & Haick, H. Silicon nanowire sensors enable diagnosis of patients via exhaled breath. ACS nano. 2016;10(7), 7047-7057

[72] Husson, F., Josse, J., & Pages, J. Principal component methods-hierarchical clustering-partitional clustering: why would we need to choose for visualizing data. Applied Mathematics Department. 2010;1-17

[73] Moyne, O., Castelli, F., Bicout, D. J., Boccard, J., Camara, B., Cournoyer, B., & Le Gouëllec, A. Metabotypes of *Pseudomonas aeruginosa* Correlate with Antibiotic Resistance, Virulence and Clinical Outcome in Cystic Fibrosis Chronic Infections. Metabolites. 2021;11(2), 63

[74] Merrot, P., Juillot, F., Noël, V., Lefebvre, P., Brest, J., Menguy, N., ... & Morin, G. Nickel and iron partitioning between clay minerals, Fe-oxides and Fe-sulfides in lagoon sediments from New Caledonia. Science of the Total Environment. 2019;689, 1212-1227

[75] Ward Jr, J. H. Hierarchical grouping to optimize an objective function. Journal of the American statistical association. 1963;58(301), 236-244

[76] Ivanisevic, J., Benton, H. P., Rinehart, D., Epstein, A., Kurczy, M. E., Boska, M. D., ... & Siuzdak, G. An interactive cluster heat map to visualize and explore multidimensional metabolomic data. Metabolomics. 2015;11(4), 1029-1034

[77] Haarman, B. C. B., Riemersma-Van der Lek, R. F., Nolen, W. A., Mendes, R., Drexhage, H. A., & Burger, H. Feature-expression heat maps–A new visual

method to explore complex associations between two variable sets. Journal of biomedical informatics. 2015;53, 156-161

[78] Gu, Z., Eils, R., & Schlesner, M. Complex heatmaps reveal patterns and correlations in multidimensional genomic data. Bioinformatics. 2016;32(18), 2847-2849

[79] Gehlenborg, N., & Wong, B. Heat maps. Nature Methods. 2012;9(3), 213

[80] Key, M. A tutorial in displaying mass spectrometry-based proteomic data using heat maps. BMC bioinformatics. 2012;13(16), 1-13

[81] Sneath, P. H. The application of computers to taxonomy. Microbiology. 1957;17(1), 201-226

[82] Ling, R. L. A computer generated aid for cluster analysis. Communications of the ACM. 1973;16(6), 355-361

[83] Vita, F., Taiti, C., Pompeiano, A., Bazihizina, N., Lucarotti, V., Mancuso, S., & Alpi, A. Volatile organic compounds in truffle (Tuber magnatum Pico): comparison of samples from different regions of Italy and from different seasons. Scientific reports. 2015;5(1), 1-15

[84] Patti, G. J., Tautenhahn, R., Rinehart, D., Cho, K., Shriver, L. P., Manchester, M., & Siuzdak, G. A view from above: cloud plots to visualize global metabolomic data. Analytical chemistry. 2013;85(2), 798-804

[85] Tautenhahn, R., Cho, K., Uritboonthai, W., Zhu, Z., Patti, G. J., & Siuzdak, G. An accelerated workflow for untargeted metabolomics using the METLIN database. Nature biotechnology. 2012;30(9), 826-828

[86] Gowda, H., Ivanisevic, J., Johnson, C. H., Kurczy, M. E., Benton, H. P., Rinehart, D., ... & Siuzdak, G. Interactive XCMS Online: simplifying advanced metabolomic data processing and subsequent statistical analyses. Analytical chemistry. 2014;86(14), 6931-6939

[87] Kim, J. Y., Park, J., Hwang, H., Kim, J. K., Song, I. K., & Choi, J. W. Catalytic depolymerization of lignin macromolecule to alkylated phenols over various metal catalysts in supercritical tert-butanol. Journal of Analytical and Applied Pyrolysis. 2015;113, 99-106

[88] Brockman, S. A., Roden, E. V., & Hegeman, A. D. Van Krevelen diagram visualization of high resolution-mass spectrometry metabolomics data with OpenVanKrevelen. Metabolomics. 2018;14(4), 1-5

[89] Fabbri, D., Trombini, C., & Vassura, I. Analysis of polystyrene in polluted sediments by pyrolysis—gas chromatography—mass spectrometry. Journal of chromatographic science. 1998;36(12), 600-604

[90] White, D. M., Garland, D. S., Beyer, L., & Yoshikawa, K. Pyrolysis-GC/MS fingerprinting of environmental samples. Journal of Analytical and Applied Pyrolysis. 2004;71(1), 107-118

[91] Campo, J., Nierop, K. G., Cammeraat, E., Andreu, V., & Rubio, J. L. Application of pyrolysis-gas chromatography/mass spectrometry to study changes in the organic matter of macro-and microaggregates of a Mediterranean soil upon heating. Journal of Chromatography A. 2011;1218(30), 4817-4827

Chapter 6

Biowastes as a Potential Energy Source in Africa

Deodatus Kazawadi, Justin Ntalikwa and Godlisten Kombe

Abstract

High population and industrialization have brought the need for a reliable and sustainable source of energy and protection of the environment. Although Africa has a low energy consumption capacity (3.4% of the global share in 2019), its high population growth rate and industrialization predict high energy demand in the future. Reliable and available energy resources are required to protect the environment and create energy dependency. Despite Africa's low energy consumption capacity (3.4% of global consumption in 2019), its rapid population growth rate and industrialization indicate future significant energy demand. The current high production of biowastes with high energy content and their low utilization provides an opportunity for energy dependency, crop value addition, creation of jobs, and protection of the environment. The chapter has identified that the African population of 1.203 billion in 2017 consumed 928 Mtoe of energy and this demand is expected to increase in years to come. The energy mix has been identified to depend on fossil fuels with little consideration of biowastes. The biowaste is reported to contain 20.1 TWh in 2025. Biowaste is currently underutilized, and there are few conversion methods available. Government and non-government investments have been reported to be making efforts to improve bioenergy and biowaste usage. The prevailing challenges have been low proven technologies, poor energy policy, low population knowledge, and poor investments. Biowastes use can be increased when environmental laws and legislation are tightened, energy policy strengthened and enforced, cheap and appropriate technologies are introduced, and the population Education is provided. It is expected that when biowastes are well utilized, energy will be available even in disadvantaged (remote) areas at an affordable price for the developing continent of Africa.

Keywords: municipal solid wastes, crop and forest residues, biowaste, renewable energy

1. Introduction

Energy has been a critical issue in most African countries where most of their populations are deprived of it. With the current population growth, the situation is alarming and needs serious intervention to rescue. It is predicted that one-in-two people added to the global population between 2019 and 2040 will be African. In 2025, Africa's population is predicted to exceed that of both India and China [1]. This increase in the population, combined with an increase in purchasing power, will put additional strain on the existing energy supply, resulting in a significant increase in energy demand. With the current trend of industrialization and population growth, the energy demand is expected to rise and put high pressure on the current fossil fuel resources. In 2018, Africa's energy demand was estimated at 700 TWh of which 70% was consumed by south and northern countries. This energy consumption is forecasted to reach 1600-2300 TWh by 2040 [1]. Such a huge increase in energy demand requires African countries to be prepared for a sustainable solution. The globe's energy resources have mainly been dominated by fossil fuels, which cover around 81% of the total energy supply in 2018 [2]. High dependency on fossil fuels not only brings uncertainty but also leads to global warming and environmental pollution. Therefore, the appropriate use of fossil fuels and the introduction of renewable energy technologies are required for sustainable energy and the environment.

The use of some renewables may be associated with high cost and occurrence, but bioenergy sources are expected to be the most promising option for meeting future energy demands [3]. Its conversion cost is expected to be lowered due to the availability of biomass, low cost, and high energy demand. Due to the global goal to departing from fossil fuels, the incorporation of bioenergy in energy generation has gained attention, mostly in developed countries. Although Africa is blessed with biomass, its use in electricity generation is still low, but expectations in the future are high [1]. The biomass potential is expected to continue to increase due to available cultivatable and fertile land, unutilized wastes, and cheap and simple conversion methods. Also, the utilization of biomass will continue to grow due to initiatives of African countries to intensify energy security. The African Union has the 2063 agenda that aims to have modern, efficient, reliable, and cost-effective renewable energy for all households, businesses, industries, and institutions [4]. This agenda has gained support with investments, among them being the African Development bank since 2016 [4]. High investment of USD $43-55 billion per year compared to USD $ 8-9.2 billion that is currently invested, will continue to stimulate incorporation of renewable energy, among them being biomass. To achieve this goal, understanding the type, quality, quantity, and distribution is necessary.

Although utilization of fresh biomass for energy has a great potential in unlocking energy demand challenges in Africa, it tends to bring competition for land with food crop cultivation. Furthermore, it leads to high utilization of fertilizers and high investment costs. To reduce the dependency of fresh biomass, alternative biomass sources should be considered. The possible sources that have gained attention not only for energy generation, but also for environmental concerns, are; crops and forest residue, non-edible biomass, animal wastes, and Municipal Solid Wastes (MSW). The application of biowaste for energy generation will not only solve the forecasted energy demand but also improve waste management in Africa. Africa is faced with ever-increasing waste management and utilization challenges. Among the challenges that hinder biowastes utilization are poor management, weak legislation, lack of enforcement, low public awareness, corruption, conflict and political instability. While the amount of waste produced in Africa is minimal in comparison to developed nations, waste mismanagement in Africa is already having an impact on human and environmental health [5]. It is estimated that an additional USD$8 billion could be injected into the African economy each year by just diverting waste away from dumpsites or landfills for reuse, recycling, and recovery [5].

Converting biomass waste to energy necessitates a thorough understanding of the potential type and availability of waste feedstock, as well as the potential application of the chosen technology. The knowledge of energy potential and possible appropriate conversion technology affects investors and policymakers. Therefore, there is a need to expose the huge potential of biowaste as an energy source for Africa to shift the alignment from fossil fuels to biowaste energy for sustainable energy and the environment. This chapter mainly focused on biowastes as a potential energy source in Africa. It describes the potential biowaste sources, such as crop and forest residue, MSW, and animal waste, and the potential energy present. Different technologies available for converting have also been discussed. Finally, this chapter proposes a way forward towards sustainable biowaste energy generation for Africa.

2. Energy status and challenges in Africa

Energies in Africa change, but not uniformly. Although new oil and gas discoveries are now accompanying the expansion of renewable energy generation, yet Africa as a whole is straightly above Germany as a country in electricity generation and consumption [6, 7]. The mean energy consumption per capita in most African countries is well below the world average and is equivalent to that of India. Biomass energy is currently Africa's most important source of energy, supplying 47.2% of primary energy demand and more than half of total energy consumption [8]. In 2017, Africa had a total primary energy supply (TPES) of 1,148,475 Mtoe in which biomass, oil, natural gas, and coal contributed 37.93, 37.99, 15.3, and 12.7%, respectively [7]. Of all the energy available, Africa imported 184,768 Mtoe (16%). Transportation, household, and industrial consumers, share were about 70.1, 8.9, and 8.5%, respectively [7]. This indicates that transportation consumes a large percentage of energy, with 29,461 Mtoe imported. Furthermore, in 2017, electricity that was available for use was 1.71% of TPES, which amounted to 1,058.74 TWh from thermal conversion of biomass, natural gas, and coal at 78.55%. It was noted also in the same year, the total energy supplied was 928.051 Mtoe, which is equivalent to 123.75% energy dependency. The energy consumed was sourced from biomass, crude oil and natural gas at percentage shares of 47, 11.3, and 13, respectively.

There has been an increase in electrical energy consumption worldwide from 10,897.94 to 24,738.92 TWh from 1990 to 2018, while Africa increased from 286.04 to 722.84TWh in the same period [9]. Global primary consumption for the same period increased from 83,830 to 123,695 TWh, meaning the electricity consumption share changed from 13–20%. This increase has been attributed to the development and an increase in population. This means that the increase in electricity in the energy mix in Africa should be increased. In 2017, the electricity share was only 9% [7], meaning electricity consumption is below the global average. Globally, energy comes from a range of sources, ranging from biomass, nuclear, fossil, and other renewables. The main source of energy has been fossil fuels.

Biomass has emerged as a viable and sustainable renewable energy source, with an 11% global contribution. According to reports, biomass provided 95% of world heat production in 2019 [2] and generated 637 TWh of electricity. Biomass also accounted for 3% of worldwide transportation energy. Domestic consumption accounts for roughly 85% of its contribution [2]. By 2030, it is estimated that 720-920 million people in Sub-Saharan Africa will rely on biomass and 2.7 billion people worldwide. According to UN Environment Programme research, 65% of countries in Sub-Saharan Africa will still rely on biomass in 2050 [10]. This is the indication that biomass is the main energy source and should well be used efficiently and sustainably.

2.1 Energy mix in Africa

Africa has low energy consumption per person when compared to other countries, falling below the global average of 2 toe. For example, in 2018, with except

South Africa, the consumption was below 0.8 toe, with sub-Sahara Africa having lower than 0.4 toe [1]. Energy is derived from a variety of sources, including coal, oil, gas, nuclear power, biomass, and other renewables. **Figure 1** depicts the energy mix by energy source. Biomass is still the main source of primary energy mostly used in cooking. Except north Africa, biomass contributes around 60% by which no any part of the global depends [1]. In Sub-Saharan Africa, for example, biomass consumption is currently above 80% [4]. Oil is the next energy source in Africa after biomass. In 2018, oil consumption was over 4 million barrels per day, with 60% of that consumed in transportation and diesel used as a generator backup. **Figure 2** depicts energy consumption in Sub-Saharan Africa by sector in 2017. Household is the main consumer of energy of about 64% and mostly depend on biomass.

Africa continues to rely on fossil fuels for electrical generation (92%); oil (42%), natural gas (28%), coal (22%), hydropower (6%), nuclear (1%), and other renewable energy (1%). **Figure 3** shows electricity production sources through thermal heat technology. The resources are limited, and the deposits will be depleted as quickly as possible as the population grows. Since fossil sources do not rejuvenate, the sustainability of them is in compromise. To solve dependency of fossil fuel, Africa should focus on other resources. Apart from biomass, she still has other potentials such as; 10 TW of solar, 350 GW of hydro, 110 GW of wind, and 15 GW of geothermal energy. Due to high investment costs and distribution of other renewable resources, biomass is and will continue to be a significant source of energy due to wide distribution, modular and simple conversion technologies, and appropriate production land. The population growth leads to large production of biowastes. Due to high increase in energy demand, there comes a need to develop the utilization process of wastes for energy generation. Therefore, in future,

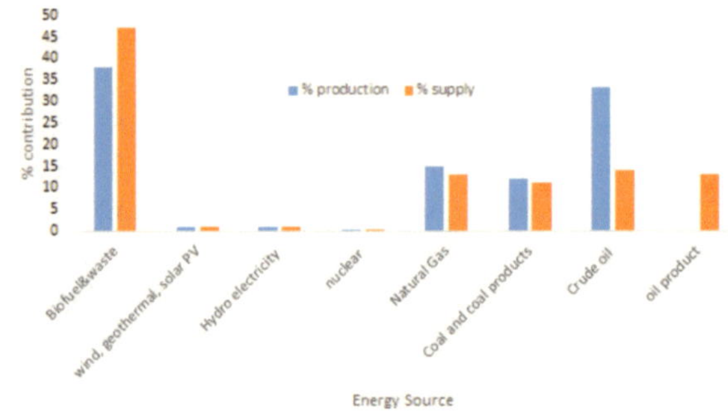

Figure 1.
% share of energy sources 2017 [7].

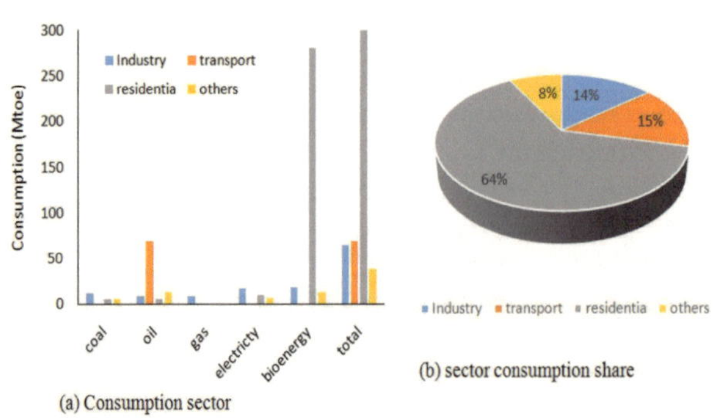

Figure 2.
Energy source consumption by sector in sub-Sahara Africa 2018 [1].

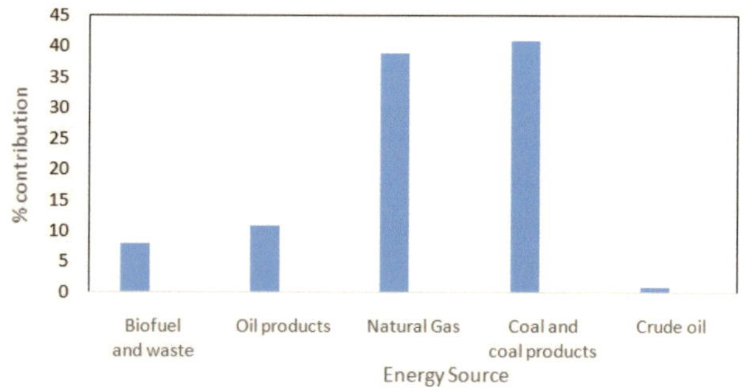

Figure 3.
Contribution of energy sources in electricity production [7].

biowastes are expected to contribute highly to the energy mix and hence lead to energy and environmental sustainability.

2.2 Challenges of energy sources and stability

As reported earlier, Africa energy consumption per capital is the lowest in the world. But also it was stated that the population and development in the future will lead to high energy demand. From earlier analysis, still Africa relies on fossil fuel in electricity generation and biomass in other uses. Transport sector has been identified to consume 64% of total energy consumed. Unfortunately, the production of transportation energy is still low with large importation of oil. The utilization of electricity and natural gas in the transportation sector is still low. This implies that much of transportation energy source is dependent on global oil stability. Lack of local and alternative sources puts uncertainty of both energy production and supply. Unluckily, available renewable sources are not well utilized due to obstacles such as investments. These entire situations put a big challenge of energy stability.

It has been also identified that biomass is the main source of energy. Biomass sources such as forest have challenges on utilization when it comes to deforestation. The study report in 2018 has shown that East and West Africa deforestation since 1900 was 93 and 83%, respectively [11]. Thus, the sustainability of biomass as the source of energy is hindered by deforestation. Furthermore, the competition of land and food crops still hinder the sustainability of biomass. In case of biowaste utilization, high investment, knowledge, and technology make them unfavorable resources. Also the poor governmental policy on sourcing local sources are also a challenge. To avoid instability of energy sources from external situation such as environmental movement and global oil market, Africa should find suitable sources. The dependency of virgin biomass and competition of area for food and energy crops, utilization of less used energy resources should be implemented. The policy and investment should be the priority in utilizing sources that lead to sustainable energy and environment.

Thus, to have energy and environmental stability, the use of biowastes is important. To implement biomass challenges such as variation of composition, large distribution, low energy density and poor energy conversion, appropriate technology should be implemented (**Figure 3**).

3. Status of biowastes potentials and utilization in Africa

The global biowastes production is currently estimated to be over 140 Gt per annum [12]. On global scale, more than 2 Gt of garbage is burned inefficiently,

Africa sub-region	Crop residue (Mt)[1]	Wood residue (m^3)	Animal wastes (Mt)[2]
Eastern Africa	77	114 465	841
Middle Africa	22	16 000	266
Northern Africa	59	121 000	334
Southern Africa	15	55 380	88
Western Africa	101	712 200	565
Total	274	1 019 045	2094
World	3714	231 773 138	8507
% to globe	7.4	0.44	25

[1]*Estimation by FAO for crops Barley, Beans, Maize, Millet, Oats, Potatoes, Rice, Paddy, Rayne, sorghum, Soybeans, Wheat using FAO TIER 1 approximation with factor 0.01.*
[2]*Estimation by FAO using FAO TIER 1 approximation with factor 0.015 for manure excreted for Asses, buffaloes, camels, cattle diary, cattle non-diary, chicken broilers, chicken layers, ducks, goats, horses, llamas, mules, sheep, swine breeding, swine market, and turkey.*

Table 1.
FAO estimation of biowastes in Africa sub-regions compared to globe [13].

accounting for 18% of total global CO_2 emission and significant amount of particulate [12]. Africa produces biowastes, which can be turned into useful energy, from agriculture, forest, and animal residues, as well as MSW. **Table 1** reports the FAO estimate of biowastes. Africa is located in tropical and sub-tropical region and due to good climate conditions; production rate of biomass is 4-5 times the other regions [14] leading to high agricultural and forest wastes. In 2012, 11.519 Gt was produced and 5.098 Gt was collected and in 2025, 28.155 Gt and 18.232 Gt are expected to be produced and collected every year [4]. In 2025, Africa is anticipated to generate 20.1 TWh of biomass waste and collect 12.5 TWh of garbage, compared to 8.4 TWh of biomass waste generation and 3.5 TWh of rubbish collection in 2012 [4]. This still shows low utilization of these biowastes. These wastes can be a good source of energy if they are successfully transformed. For example, in 2019 Barz et al. reported that if agricultural wastes are properly handled, they can contribute 20-40%ofprimaryenergy [14], demonstrating that all biowastes can contribute significantly. According to estimates, processing crop and agro-processing leftovers might provide 4.2 EJ in 2030, forest Residue 1.1 EJ, and animal residue 1.5 EJ. To reduce competing with food in the production of fuel, boost waste economic value, and safeguard the environment (lower GHG), biowastes should be successfully employed as an energy source [14]. The most difficult challenge is to invest in an efficient method of collecting and converting it. As a result, Africa should invest in an efficient method of collecting and converting these wastes to reduce the use of traditional energy sources. Agricultural wastes have the potential for energy production, but they compete with other uses, such as fodder, fertilizer, fiber, and construction, so an effective balance is required.

3.1 Biowastes sources

In Africa, where the main activity is agriculture and forestry with abundant cultivatable land, the biomass wastes are in abundance. The sources of wastes range from crop, forest, and animal residues, and MSW. There is a huge potential with high organic components. The MSW in Africa has high organic components of around 70% and thus making them potential for energy conversion. Unfortunately, these wastes are poorly managed and hence become environmental concern. Effective utilization can significantly offset the use of fossil fuel.

3.1.1 Municipal solid wastes

MSW has emerged worldwide as the source of energy. Report by IEA [9], electricity generated from MSW worldwide increased from 8,313 GWh in 1990 to 38,648 GWh in 2018. Africa is also generating MSW. In 2012 analysis, it was reported that Africa generated 125 million tones with 1,125 PJ but only collected 68 million tones [15]. It is projected that in 2025, 244 million tones with 2,199 PJ will be generated and 168 million tones will be collected [15]. Furthermore, it was reported that if landfill energy recovery technology is used, in 2012, 155 PJ would have been recovered while in 2025, 363 PJ will be recovered. If all wastes are collected and converted to electricity, 62.5TWh could be generated in 2012 and expectation to be 122.2 TWh. With targeted collection, 34.1 TWh and 83.8 TWh could be generated in 2012 and 2025 [4]. The increase in African population and purchasing power will lead to high generation of MSW. For example, in sub-Saharan region, MSW was predicted to increase from 300,000 t/day in 2010 to 3,250,000 t/day in 2100 [16]. This is the huge potential for use.

Apart from high potential, MSW also have high organic contents that become good source of energy. Their poor management leads to environmental pollution through degradation. The average composition of MSW in Africa, for example of sub- Saharan Africa, is about 57-78% organic, 9% paper/cardboard, 13% plastic, 4% glass, 4% metal and 13% other materials. This gives advantages on separation and treatment since large amount is energy part. Fewer amounts of plastics and metals indicate the applicability of low technologies such as anaerobic digestion can be utilized since organic parts are easily degradable. Although of high energy potential from these biodegradable organic waste, MSW remains largely an untapped resource for the continent [5]. Therefore, good policy, investment, and technology are important factor to utilize this energy.

3.1.2 Forest residue

During forest harvesting, up to 60% is left as residue which can be source of energy and soil stabilization. Furthermore, residue continues accumulating due process of forest products, such as production of saw dust. Forest residue can be used as energy source and in production of products such plywood. In 2019, Africa produced 1.02 million cubic meters (0.4% of globe) forest residue [17]. This low globe percentage is due to export of unprocessed forest products. This is the potential source of energy. The 2020 energy potential analysis in forest residue for Africa can produce energy of 10 to 5,254 PJ/yr. [18]. Low investment on technology and facilities, and poor government policy hinders their conversion. Therefore, it is important to invest in collection and conversion technologies for effective energy recovery.

3.1.3 Crop residue

In every crop harvested, the residue obtained is 25-50% (Residue to Product ratio) of the product and sometime is above 100%. These organic parts mostly are left behind to regenerate the soil, as animal feed and source of energy. In 2010, agricultural residues with an energy content of some 90 EJ were reported to have been generated worldwide and is expected to reach 128 EJ in 2050 [19]. These materials

S/No	Country	Year	Dry Wastes (kt)					Energy (PJ)	Energy consumed in 2017 [7]**
			Crops	Forest	Animal	MSW	Total		
1	Cameroon [21]	2018	45 510	2051	5153	971	53685	243	390
2	Zambia [22]	2016	12 193	6559	5	295	19051	281	438
3	Uganda [23]	2013	19 274	5549	14967	7271	47061	260	814
4	Nigeria [24]	2020	215844	4478	17165	36500	273987	1950	9525
5	Ethiopia [25, 26]	2017	53811	18400	48011	19053	72266	750	2033
6	Tanzania [27]	2018	19642	711	5946	624	26924	385	877

**The assumption made is that biowastes are used at small percentages; hence energy contributed by biomass mostly comes from fresh biomass.*

Table 2.
Example of biowaste energy potentials.

are potential for energy although compete with animal feed and soil stabilization. The analysis of 19 crops mostly grown found that crop wastes of production rate of 697.87 million tons per year can generate 10.52 EJ. This potential is enough to support 366 million population (around 33% of 2019 Sub-Saharan Africa population) that had 28.76 GJ per capital [20]. If this energy from crop residue is well utilized and assuming mean biomass energy of 20 GJ/t, the wood biomass that can be saved is around 526 million tons of wood which is equal to 122 million hectare of forest area saved. The controversy of utilizing these wastes is the removal of nutrients from the soil. Currently, uncontrolled burning of these wastes onsite leads to the nutrients loss. A good control of burning and generating energy can generate ashes that can be used as fertilizers.

Table 2 has summarized available data of biowaste from crop, forest and animal residue and MSW of six countries according to reported estimations. Furthermore the primary energy consumption for 2017 has been also recorded. Since the incorporation of biowaste analyzed is almost negligible, it can be said that the biomass contribution comes from fresh biomass. The use of biowastes can reduce this dependency.

3.2 Current management and utilization of biowastes in Africa

Biowaste resources are important for energy, animal feed, and soil stabilization. The level of utilization depends on the investment and technology. In developed countries, the collection, treatment, and conversion have been a priority for incorporation in renewable energies and protection of environment. In Africa, the utilization of biowastes is still at low level. Very small investment and conversion technologies have been implemented. In rural areas, biowastes have been little used as animal feed and sources of energy. Due to poor technology and investment, biowastes use as sources of energy has been poor. The availability of fresh biomass for animal feed, biowastes have also found little use. These wastes have been accumulating continuously. The main method of handling has been unplanned dumping and uncontrolled combustion. Crop residue burning is more regular feature and a source of greenhouse gas emissions. Crop residue burning is driven by factors like labour scarcity, short turn-around time, socio-economic constraints, ignorance of farmers towards public health issues, and low nutritive value of crop residues. Crop residue burning causes loss of valuable soil nutrients, pollution episodes, and public health issues [28]. Unplanned damping may cause pollution especially when decomposition takes place and produces methane. Improper use of biowastes such as animal wastes especially cooking and heating leads to health problems. It has

Country	Biowastes	Capacity	Reference
Mauritius	MSW-landfill	20.4 GWh	[30]
Kenya	Agricultural waste-digestion	2.8 MW	[30]
Burkina Faso	Crop waste-digestion	0.25 MW	[30]
South Africa	MSW and crop residue -digestion	4.6 MW	[30]

Table 3.
Some of biowaste to energy conversion plants.

been reported by WHO that 4.3 million deaths have occurred due to use of these wastes in cooking and heating [29]. The only advantage of poor handling of biowastes in rural area is to maintain soil quality. In rural areas, the generation of MSW is low while forest, animal, and crop wastes are high. Animal waste is commonly used as organic fertilizer and hence maintain soil ecosystem.

In urban area, the production of MSW is high while animal, forest, and crop residue is low. The big problem has been handling of MSW. Some efforts have been made in the collection and dumping of MSW. In Africa, most of collected MSW are unsorted and hence if energy conversion is decided to be done gets the challenge. In Africa, only a small amount of MSW is collected and dumped in landfill with no plan of energy recovery. There has been uncontrolled dumping and combustion which all lead to energy loss and environmental pollution.

Although there has been little effort to effectively utilize biowastes in Africa, some uses such as cooking, electricity generation, and transport fuel have been realized [30]. **Table 3** summarizes some of the plants generating electricity. Energy recovery technologies, such as landfill gas recovery and bio digesters for the organic fraction of MSW and industrial biomass are currently very limited in their implementation in Africa [31]. Gasification, direct combustion, and carbonization technologies have been in low utilization. Only noticeable biowaste waste to energy conversion has been in the use of bagasse in sugar plantation. In Tanzania, Kilombero, Mtibwa, TPC, and Kagera sugar companies have daily generation of 21,914 tons per day of bagasse and electricity installation of 40 MW [32]. The use of forest residue and sisal in Tanzania has led to the production of 21 MW. Currently, African only recycles 4% of waste produced and the African Union plans by 2023, 50% should be recycled. Other ways of managing biowaste have been producing composite manure, and animal feed.

3.3 Challenges of utilizing biowastes in Africa

The use of biowastes in energy mix has not been fully done. Lack of appropriate technologies, investment, knowledge, and policy has hindered its utilization. Poor realization of biowaste potential has led to low investment and hence utilization. The wide large of properties of biowaste due to low level of farming, production of variety of biowastes, still brings challenges in the choice of universal technology and hence need segregation stages. Furthermore, government and financial institution have little interest in funding of innovation, purchasing, and utilization of appropriate technologies [33]. Therefore, efforts are needed to increase investment, education, and introduction of appropriate conversion technology.

To increase the use of biowastes in energy mix, the following can be done. The government should have appropriate policy that promotes use of biowastes. These includes; increase of incentives, removal of fossil fuel subsides, promotion of a net zero emission, and promotion research and development of appropriate technology.

Education should be made to population on the need of using biowastes, separation of components of MSW in the disposal, discourage uncontrolled burning, introduction of simple and cheap conversion technology at household level. The peasants and farmers should be educated on the effective ways of preparation, collection, disposal, and utilization of wastes. To reduce variation obstacle, the co-generation and small processing plants should emphasized. To increase awareness, organization or companies that add value should be created.

4. Current conversion technologies of bio wastes to energy

Conversion of biowaste aims to recover energy, remove residues, and increase material value [3]. Effective conversion technologies should address financial requirements, effective recovery of energy, waste removal, and environmental protection. The development, improvement, and implementation of conversion technologies have been made due to surge in wastes that pollute environment and need for energy sustainability. The need for land has emerged as another challenge. For example, Waste to Energy (WTE) plant that can treat 30 million tons of MSW for 30 years only will use 10 hectare compared to landfill that needs 30 hectare [34].

A number of conversion technologies have been developed and applied in many parts of the world. Advanced waste to energy conversion has led the increase of energy mix from biowastes. The technologies range from direct energy conversion to more upgraded fuel. The technologies have advanced in a way that they produce energy efficiently and meet the requirements for public health while reducing air pollution and obligatory number of dumping locations [3]. These methods include direct combustion/incineration, carbonization, pyrolysis, gasification, anaerobic digestion, and oil pressing for biodiesel production. Some conversion leads to production of feedstock for chemical production. Each method has the advantages and disadvantages and thus limitation in use in African countries. The common and simple technologies in biowaste conversion have been anaerobic digestion and incineration.

4.1 Direct combustion/incineration

This is the oldest and simple, and well utilized conversion technology [34]. It deals with direct heating of fuel in presence of oxygen. It takes place at high temperature (850-1200°C) and release energy in form of heat [3]. It has been used in many biowaste plants ranging from bagasse, MSW to saw dust. Apart from energy recovery, the bottom and fly ashes can be used in other applications such as construction. The challenges of this technology are that feedstock should be of low moisture content and thus limiting many of African MSW which have high moisture content to above 40% [35]. It also may lead to pollution if poorly controlled, since it operates at high temperatures and thus releasing pollutant gases and dust. Its advantages and appropriateness in African biowastes are that it is simple, mature, and low investment cost especially if heat is the final conversion. It is also appropriate in dry biowastes such as forest residue since no pretreatment costs [3] and thus utilizing wood and crop residues produced in Africa.

4.2 Carbonization

It is the process of heating biowaste at low temperature in the absence of oxygen [36]. This process upgrades biowastes to produce usable charcoal. It is mature, cheap, and simple technology that has been used on other biomass for years and years [37, 38]. The technology works efficiently for large size and low moisture contents.

Most of biowastes such as forest and crop residues have been converted into charcoal by this process. Its advantages include the cheapness, simplicity, and ability to handle variety of feedstock including lignocellulosic biomass. This method is appropriate for African biowastes although for fine materials such as sawdust and a wet material such as MSW becomes inappropriate. It is applicable in every part of Africa.

4.3 Gasification

Gasification is a mature technology and a thermochemical conversion of biomaterials in limited oxygen at high temperature (800-1600°C [39, 40]. It produces syngas that can be burnt to produce heat or used as the source of chemicals. The major constituents of syngas are carbon monoxide which poses a lot safety challenges in its handling. It requires high knowledge and thus is not appropriate for household. It is appropriate to Africa when dealing with lignocellulosic and fine biowaste although inappropriate to MSW due to high moisture content. It is appropriate to every part in Africa although due to low investment, sub-Saharan countries may not effectively fit.

4.4 Pyrolysis

Pyrolysis is the thermochemical conversion technology that operates at moderate temperature to produce solid, liquid, and gas at different composition [41, 42]. It can treat waste to produce char, oil, and gas for further energy conversion. The fast pyrolysis is favored in fine feed and operates at high temperature to produce bio oil [43]. Intermediate pyrolysis operates at moderate temperature to produce both char, oil, and gas [44]. It has an ability to handle large size, moderate moisture, and different compositions [45]. It is a technology that fits biowastes including MSW. Its appropriateness in Africa is challenged since it is sophisticated and requires high investment, and thus unfavorable in small scale. Intermediate pyrolysis has a big chance in Africa since it flexible, modular, and operates at moderate temperatures.

4.5 Anaerobic digestion

Anaerobic digestion is process that uses microorganism to breakdown organic biodegradable matter in the absence of oxygen to produce biogas [46]. It is a common technology that has been used in conversion of sugar biomass. It is simple, cheap, and works under low temperature. Its size ranges from small household to large plant. The technology is appropriate in Africa biowaste treatment since it requires small investment, mature technology, and presence of sugar biowastes especially in MSW and animal wastes. Large presence of animal residue in Africa, 25% globe, this technology is the one to be applied in all parts of Africa. Its use in lignocellulosic materials such as forest residue is inappropriate. It also has low kinetics and hence requiring larges reactors and hence investments. The residues after digestion can also be used as farm manure and thus giving it an added advantage.

4.6 Microbial fuel cell (MFC)

Microbial fuel cells use microbes as a biocatalyst and changes directly the chemical energy deposited in chemical compounds into an energy current [3]. This alternative technology reduces sludge formation, refine odor, and remove aeration supplies, all at the rate of waste reaction for electricity creation. Its advantage is the direct conversion to electricity although it has low efficiency. Due to challenges such as more sophisticated ion exchange membranes [47], this method is not appropriate in Africa especially in sub-Saharan African countries.

4.7 Landfilling

This is the technology in which biowastes are deposited and may be connected to biogas recovery and leachate processing system [3]. This method simple and cheap although may lead to environmental pollution and degradation if planning is not well done [3, 48]. This technology is appropriate in Africa since the technology is cheap, mature, and simple.

4.8 Agglomeration

This is the technology of increasing energy density by converting small particle to large particles. This increases energy density and effectiveness in conversion processes such as combustion [49]. This involves the formation of pellets and briquettes from fine biowastes that can be usable household [50]. Its challenge is the use of high energy in compressing the materials. This technology is appropriate in Africa since the technology is simple, presence of binding materials in biowaste, presence of fine materials such as charcoal dust, and cheap. All countries in Africa can utilize it due to presence of fine materials especially during charcoal production.

5. Potential of pyrolysis in biowaste conversion

Pyrolysis is the decomposition of organic materials using heat in the absence of oxygen to produce solid, liquid, and gases [51, 52]. It is said to have lower operating temperatures and emissions of air pollutants compared to combustion and gasification [52]. Pyrolysis has emerged as technology for current and future conversion technology of biomass [53]. The type, quality, and quantity of product depend on how operating parameters are controlled [52]. Its ability to treat almost all types of organic and production of solid, liquid, and gas fuel has made it a prospective technology of biomass conversion. The innovations on reactors, catalysts, and upgrading technologies have increased the importance of pyrolysis. Its level of development has made it a prospective technology for biowastes especially lignocellulosic biowastes. Therefore, pyrolysis is the current and future appropriate technology in the conversion of biowastes in Africa.

5.1 Categories of pyrolysis

Depending on the heating rate, operating temperature, and product distribution, pyrolysis can be classified as **Table 4**. It can be seen that slow pyrolysis targets char, fast and flash pyrolysis target targets bio oil while intermediate targets all. Intermediate pyrolysis has emerged as an alternative pyrolysis that is flexible, modular, economic, and ability to handle different biomass materials. It is appreciated to favor well the treatment of biowastes (**Table 4**).

The quality and distribution of products depends on a well control of parameters. Among important parameters as such as feed composition, temperature, heating rate, catalysts, feed residence time, vapor residence time, particle size, and moisture contents. The composition of feedstock should be made of organic components to lead to effective conversion. For example, cellulose and hemicelluloses produce high bio-oil while lignin can yield up to 40% of its weight as char [57]. The presence of non-organic materials reduces the amount of useful products.

Temperature also dictates the composition and yield of the products [58]. Generally, an increase in temperature increases the amount of bio-oil and gas. Higher pyrolysis temperature favors production of hydrogen, while low temperatures

Property	Slow	Intermediate	Fast	Flash
Heating rate (°C/s)	1	1-10	10-200	>1000
Feed size (mm)	5-50	1-5	<1	<0.5
Reaction temperature (°C)	400-500	400-650	850-1250	>1000
Vapor Residence time (s)	300-550	0.5-20	0.5-10	<1
Feed Water content (%)	Up to 40	Up to 40	<<10	<<10
Bio-oil Yield (%)	20-50	35-50	60-75	60-75
Bio-char Yield (%)	25-35	25-40	10-25	10-25
Gas Yield (%)	20-50	20-30	10-30	10-30

Table 4.
Classification of pyrolysis processes [54–56].

produces more char [59]. The increase in temperature reduces the amount of char but increases its quality due to decrease of the volatile matter in char.

The residence time determines the quality of the products. For example, longer residence time at low temperature favors production of biochar and its quality increases due to favoring the development of micro-and macro-pores of bio-char [60, 61]. Reduced residence time reduces the re-polymerization thus reducing the amount of char [62]. Heating rate affects the quantity, quality, and composition of products. Rapid heating gives higher volatiles and more small reactive char than those produced by a slower heating process [52]. Slower heating rate increases the amount of char due to the secondary char produced from a reaction between the primary char and the volatiles [57].

Water content affects the quality and quantity of products. It promotes the reduction of species in bio-oil and improves the production of light aromatic [63]. Water can also catalyze char formation by acting as steam activator [64, 65]. Large amount of water reduces energy content in the feed; and hence, an optimal content is required. Particle size affects the quality and quantity of products by affecting the heat transfer. The decrease in size increases heating rate and easy escape of condensable products. In addition, small size favors liquid formation, hydrogen and carbon monoxide; while large size favors the formation of char and its quality due to secondary cracking [66–68].

Inert gas carrier (sweeping gas) controls the vapor residence time. Higher flow rates cause rapid removal of products that leads to minimization of secondary reactions such as char formation, thus increasing gas production [67, 69]. Poor properties of bio-oil such as high viscosity, non-volatility, high acidity, corrosiveness, instability upon storage, lower energy density, and immiscibility with fossil fuels caused by the presence of oxygen can be improved using catalysts [70]. The catalyst increases the quality of products through increased cracking, selectivity, and de-oxygenation [71]. This increases the quality of bio-oil and biochar while decreasing their quantities [51, 72]. It has been reported that, the presence of catalyst $HZSM-5$ reduced the amount of oxygenated compounds and thus increased the quality of bio-oil [72, 73]. Studies have also found out that bio-char production increases with an increase of $NaOH$, $NaCl$, CaO, and Na_2SiO_3 [74, 75].

5.2 Pyrolysis in biowaste conversion

Pyrolysis has emerged as the appropriate technology in biowaste conversion although its applicability in household in difficult [52]. It can convert almost all materials; small and large size, lignocellulosic and non-lignocellulosic, wet and dry, and variety of compositions. It has the ability to produce usable product with little or no upgrading. For example, intermediate and fast can produce bio oil which can be used with little or no upgrading. Hydrothermal pyrolysis can convert wet materials to produce usable solid and liquids. Its high reaction rate compared to anaerobic digestion leads to small reactor. Its main challenges to be used biowaste conversion are the investment and operation cost, sensitivity of the process, and difficulty in operation. For Africa, this becomes a promising technology but difficult conversion technology for small and household users.

6. Prospect of biowaste and conversion technologies in Africa

Bioenergy utilization in African is favored by; feedstock availability, availability of proven conversion technologies, large required land for production and plant setting, and increased value chain of bioenergy [33]. Africa aims to incorporate primary sources such as biogas, biodiesel, bioethanol and other biochemicals produced from agricultural, domestic and industrial resources and with the application of simple and indigenous technologies. With increase in population which leads to high demand of energy and pollution of biowaste, the necessity for effective conversion of biowastes is important. An important issue is to have effective conversion technologies that provide required energy at minimum cost. The technologies should be simple and flexible that household can apply. The study and investment of appropriate technologies should be emphasized for sustainable energy and environment.

7. Prospect of biowaste potentials Africa

The biowastes have the opportunity to produce different types of energy resources that can replace fossil fuel. The production of liquid fuels such as bioethanol and biodiesel is important to replace/reduce importation of petrol and diesel, respectively. The following are possible products that can be produced from biowastes.

7.1 Liquid biofuel

From 2017 African energy analysis, 70.1% of 928.051 Mtoe of energy consumed was through transportation. Many of African transportation sectors depend on liquid fuel. This highlights the high demand of liquid fuel. The introduction of biofuel from biowaste can be the solution on importation of liquid fuels. Africa has little use of liquid biofuel such as biodiesel, and ethanol. Up to 2020, no electricity has been produced from liquid biofuel compared to globe 3211 MW. The potential of producing liquid biofuel from waste is high since most of wastes are sugary biomass. The use of technologies such as pyrolysis, especially fast and intermediate pyrolysis, can convert biowaste into liquid fuel. Other technologies such as gasification and anaerobic digestion can produce syngas and biogas, respectively, as feedstock for gas to liquid conversion. The fermentation process can also produce bioethanol. The production of bioethanol from biowaste can reduce the competition with food in bioethanol production. Liquid fuel can also be produced from pressing of waste shells such as coconut and cashew nut shells. Liquid biofuel has a potential of replacing fossil fuel. IRENA 2017 report of five sub-Saharan countries (Ghana, Mozambique, Nigeria, South Africa, and Uganda) [19] has predicted biomass including biowastes can produce all liquid fuel for transportation and twice for industrial electricity and heat by 2050.

7.2 Solid fuel production

African has been depending on solid fuel especially wood charcoal in household needs. Most of charcoal has been produced from forest and thus leading to deforestation. The utilization of biowastes can increase the production of charcoal while reducing deforestation. Forest and crop residues are potential feedstock for charcoal production. The agglomeration such as pelleting has the potential of increase energy quantity in the biowastes. The products can increase the efficiency of other conversion processes such as combustion.

7.3 Economic and environmental sustainability

The use of biowaste in the production of useful energy leads to creation of jobs and increased value of agriculture and forest activities [31]. This can encourage people working in these areas to increase production and hence increase in biowastes and latter economic improvement. The use of biowaste will also increase environmental protection since uncontrolled burning and dumping will be avoided.

8. Prospect of conversion technologies in Africa

There are different technologies that can help to convert biowastes to useful energy. Study done by LTS team in 2017 [76] identified possible suitable conversion

Primary conversion technology	Secondary conversion technology	End use				
		Heat	Electric Power	CHP*	Transport	cooking
Combustion	None	x				
	Steam turbine		x	x		
	Steam engine		x	x		
	Sterling engine		x	x		
	Organic Rankine cycle		x	x		
Gasification	Internal combustion Engine		x	x		
Fast pyrolysis	Combustion	x	x	x	x	
Slow pyrolysis	Combustion	x	x	x		x
Intermediate pyrolysis	Combustion	x	x	x	x	x
Oil pressing	Internal combustion engine		X	x	x	
	Transesterification				x	
Anaerobic digestion	None				x	x
	Internal combustion engine		x	x		
Fermentation	Ethanol fermentation		x	x	x	x
	Butanol fermentation				x	

**CHP means combined heat and electric power.*

Table 5.
List of appropriate biowaste to energy conversion technologies in Africa.

technologies for Africa. In applicability efficiency sequence, anaerobic digestion, gasification, direct combustion, fast pyrolysis, slow pyrolysis, and fermentation was observed [76]. Intermediate pyrolysis can be a good candidate in converting biowastes due to its advantages including; flexibility, ability to handle variety of feedstock, modular, ability to handle high moisture content, and productions of products with good quality [44]. Africa with biowastes that have variety of properties can effectively be handled using intermediate pyrolysis. **Table 5** summarizes possible effective conversion technologies of biowastes in Africa.

Biowaste is now regarded as source of energy for both energy and environment sustainability in Africa. Some polices have been formulated to increase biowaste energy in energy mix. Some funding from both government and private has been provided. Noticeable funding has come from African Development Bank Group. Example of such projects are; USD$ 1 million for Kenya for 10 MW from MSW in 2017, Bioethanol from waste for UDS$ 66 million in 2012, among others [77].

9. Conclusion

Africa is a continent that is undergoing rapid population and economic growth that require sustainable energy sources. Currently, energy production and consumption being below the global average of 2 toe per capital. The energy demand is expected to be 1600-2300 TWh by 2040 compared to 700 TWh in 2018. Increase in population from 1.2 billion in 2017 to 2.07 billion by 2040 and rapid industrialization will increase the energy demand and waste production, leading to the challenge of attaining economic and environment sustainability. Currently, energy comes from biomass at 42% and electricity is mostly from fossil fuels, leading to deforestation and pollution. The introduction of renewables in energy mix has been at low rate due to poor investment, knowledge, policy, location, and technology. Biomass is seen as the appropriate renewable energy due to abundance, affordable conversion technology, and widely distributed. To reduce deforestation and pollution, use of biowaste is appropriate solution. Although biowastes are not widely used, their use is a potential source of energy. The MSW, animal, crop, and forest residues are currently abundant in Africa and their use is a good solution for sustainable energy and environment. It is predicted that these wastes can have 20.1 TWh in 2025. Although biowastes are sources of energy, poor management, technology, and investments have hindered their use. This has led to unplanned dumping and site combustion leading to loss of energy and pollution. The introduction of appropriate technologies such as pyrolysis, gasification, and anaerobic digestion has the potential to produce solid and liquid fuels while increasing value of agriculture. Effective utilization of biowastes with other biomass will lead to energy dependency, for example by 2050; liquid biofuel can be enough for transport needs in countries such as Uganda, Nigeria, and Ghana. It is therefore important for government to formulate policy for biowastes utilization while investing on both private and public utilization plants. Education on the handling of biowastes should also be given to local population. This gives the hope that future energy and environmental sustainability in Africa can be contributed by effective utilization of biowastes.

Acknowledgements

It is our pleasure to give our appreciation to The University of Dodoma through Deodatus Kazawadi PhD funding. This has been a great enforcer in preparing this book chapter.

Conflict of interest

This book chapter has used data from reliable organization and data used are open data. This chapter has not been submitted or published to any publisher.

Author details

Deodatus Kazawadi*, Justin Ntalikwa and Godlisten Kombe
College of Earth Sciences and Engineering, University of Dodoma, Dodoma, Tanzania

*Address all correspondence to: dkazawadi@gmail.com

IntechOpen

© 2021 The Author(s). Licensee IntechOpen. This chapter is distributed under the terms of the Creative Commons Attribution License (http://creativecommons.org/licenses/by/3.0), which permits unrestricted use, distribution, and reproduction in any medium, provided the original work is properly cited.

References

[1] Birol, Fatih, *Africa Energy Outlook 2019*. 2019, IEA.

[2] Wba, *Global Bioenergy Stastics 2020*. 2020, World Bioenergy Association: Sweden.

[3] Awasthi, Mukesh Kumar, Surendra Sarsaiya, Hongyu Chen, Quan Wang, Meijing Wang, Sanjeev Kumar Awasthi, Jiao Li, Tao Liu, Ashok Pandey, and Zengqiang Zhang, *Global Status of Waste-to-Energy Technology*, in *Current developments in biotechnology and bioengineering*. 2019, Elsevier. p. 31-52.

[4] Gyde Lund, Frank Mabirizi, *Atlas of Africa Energy Resources*. 2017, United Nations Environment Programme: Nairobi Kenya.

[5] Godfrey, Linda, Mohamed Tawfic Ahmed, Kidane Giday Gebremedhin, Jamidu Hy Katima, Suzan Oelofse, Oladele Osibanjo, Ulf Henning Richter, and Arsène H Yonli, *Solid waste management in Africa: governance failure or development opportunity.* Regional Development in Africa, 2019: p. 235.

[6] Schiffer, Hans-Wilhelm and Johannes Trüby, *A review of the German energy transition: taking stock, looking ahead, and drawing conclusions for the Middle East and North Africa.* Energy Transitions, 2018. **2**(1): p. 1-14.

[7] Abdalah, Rashid Ali, *Key Africa Energy Statistics Edition 2019*. 2019, African Energy Comission: Algeria.

[8] Iea, *Africa Energy Outlook 2019: World Energy Outlook Special Report*. 2019, IEA: Paris.

[9] Iea. *Electricity generation from biofuels and waste by source, World 1990-2018*. 2018 [cited 2021 14/7]; Available from: https://www.iea.org/data-and-statistics/data-browser/?country=WORLD&fuel=Renewables%20and%20waste&indicator=WasteGenBySource.

[10] Gordon, Philip. *UN study: 'Urgent action' needed to reduce biomass use in Africa*. 2019 [cited 2021 8/7].

[11] Aleman, Julie C, Marta A Jarzyna, and A Carla Staver, *Forest extent and deforestation in tropical Africa since 1900.* Nature ecology & evolution, 2018. **2**(1): p. 26-33.

[12] Tripathi, Nimisha, Colin D Hills, Raj S Singh, and Christopher J Atkinson, *Biomass waste utilisation in low-carbon products: harnessing a major potential resource.* NPJ climate and atmospheric science, 2019. **2**(1): p. 1-10.

[13] Fao. *Livestock manure*. 2021 [cited 2021 17 August]; estimation of livestock manure]. Available from: https://www.fao.org/faostat/en/#data/EMN.

[14] Barz, Mirko, Mitra Kami Delivand, and Konstantin Dinkler, *Agricultural Wastes–a Promising Source for Biogas Production in Developing Countries of the Tropical and Subtropical Regions.* Revista Forestal Mesoamericana Kurú, 2019. **16**(38): p. 2-12.

[15] Scarlat, Nicolae, Vincenzo Motola, Jean François Dallemand, Fabio Monforti-Ferrario, and Linus Mofor, *Evaluation of energy potential of municipal solid waste from African urban areas.* Renewable and Sustainable Energy Reviews, 2015. **50**: p. 1269-1286.

[16] Hoornweg, Daniel, Perinaz Bhada-Tata, and Christopher Kennedy, *Peak waste: When is it likely to occur?* Journal of Industrial Ecology, 2015. **19**(1): p. 117-128.

[17] Fao. *Forestry Production and Trade*. 2019 [cited 2021 15/7]; Available from: http://www.fao.org/faostat/en/#data/FO/visualize.

[18] Thrän, K. Stechera. Brosowski D., *Biomass Potential in Africa*. 2013, IRENA: German.

[19] Jeffrey Skeer, Rodrigo Leme and Yasuko Inoue, *Biofuel Potential in Sub-Saharan Africa: Raising food yields, reducing food waste and utilising residues*. 2017, IRENA: Abu Dhabi.

[20] Duguma, Lalisa, Esther Kamwilu, Peter A Minang, Judith Nzyoka, and Kennedy Muthee, *Ecosystem-Based Approaches to Bioenergy and the Need for Regenerative Supply Options for Africa*. Sustainability, 2020. **12**(20): p. 8588.

[21] Mboumboue, Edouard and Donatien Njomo, *Biomass resources assessment and bioenergy generation for a clean and sustainable development in Cameroon*. Biomass and bioenergy, 2018. **118**: p. 16-23.

[22] Shane, Agabu, Shabbir H Gheewala, Bundit Fungtammasan, Thapat Silalertruksa, Sébastien Bonnet, and Seveliano Phiri, *Bioenergy resource assessment for Zambia*. Renewable and Sustainable Energy Reviews, 2016. **53**: p. 93-104.

[23] Okello, Collins, Stefania Pindozzi, Salvatore Faugno, and Lorenzo Boccia, *Bioenergy potential of agricultural and forest residues in Uganda*. Biomass and bioenergy, 2013. **56**: p. 515-525.

[24] Jekayinfa, Simeon Olatayo, Joseph Ifeolu Orisaleye, and Ralf Pecenka, *An assessment of potential resources for biomass energy in Nigeria*. Resources, 2020. **9**(8): p. 92.

[25] Gabisa, Elias W and Shabbir H Gheewala, *Potential of bio-energy production in Ethiopia based on available biomass residues*. Biomass and bioenergy, 2018. **111**: p. 77-87.

[26] Gelan, Eshetu, *Municipal Solid Waste Management Practices for Achieving Green Architecture Concepts in Addis Ababa, Ethiopia*. Technologies, 2021. **9**(3): p. 48.

[27] Aslam, Zahida, Hu Li, James Hammerton, Gordon Andrews, Andrew Ross, and Jon C Lovett, *Increasing Access to Electricity: An Assessment of the Energy and Power Generation Potential from Biomass Waste Residues in Tanzania*. Energies, 2021. **14**(6): p. 1793.

[28] Venkatramanan, V, Shachi Shah, Shiv Prasad, Anoop Singh, and Ram Prasad, *Assessment of Bioenergy Generation Potential of Agricultural Crop Residues in India*. Circular Economy and Sustainability, 2021: p. 1-14.

[29] Dahunsi, So, At Adesulu-Dahunsi, Co Osueke, Ai Lawal, Tma Olayanju, Jo Ojediran, and Jo Izebere, *Biogas generation from Sorghum bicolor stalk: effect of pretreatment methods and economic feasibility*. Energy Reports, 2019. **5**: p. 584-593.

[30] Surroop, Dinesh, Zumar Ma Bundhoo, and Pravesh Raghoo, *Waste to energy through biogas to improve energy security and to transform Africa's energy landscape*. Current opinion in green and sustainable chemistry, 2019. **18**: p. 79-83.

[31] Godfrey, Linda, Anton Nahman, Arsène H Yonli, Fisseha Gebru Gebremedhin, Jamidu Hy Katima, Kidane Giday Gebremedhin, Mohamed Abdelnaem M Osman, Mohamed Tawfic Ahmed, Mohammed Mohammedbrhan Amin, and Naglaa M Loutfy, *Africa waste management outlook*. 2018.

[32] Bishoge, Obadia Kyetuza, Lingling Zhang, and Witness Gerald Mushi, *The potential renewable energy for sustainable development in Tanzania: A review*. Clean Technologies, 2019. **1**(1): p. 70-88.

[33] Dahunsi, Samuel Olatunde, Omololu Oluwatobi Fagbiele, and Esther Ojima Yusuf, *Bioenergy technologies adoption in Africa: A review of past and current status*. Journal of Cleaner Production, 2020. **264**: p. 121683.

[34] Kumar, Atul and Sukha Ranjan Samadder, *A review on technological options of waste to energy for effective*

management of municipal solid waste. Waste Management, 2017. **69**: p. 407-422.

[35] Fobil, Julius N, Derick Carboo, and Nathaniel A Armah, *Evaluation of municipal solid wastes (MSW) for utilisation in energy production in developing countries.* International journal of environmental technology and management, 2005. **5**(1): p. 76-86.

[36] Ogawa, Makoto, Yasuyuki Okimori, and Fumio Takahashi, *Carbon sequestration by carbonization of biomass and forestation: three case studies.* Mitigation and adaptation strategies for global change, 2006. **11**(2): p. 429-444.

[37] Hosier, Richard H, *Charcoal production and environmental degradation: environmental history, selective harvesting, and post-harvest management.* Energy Policy, 1993. **21**(5): p. 491-509.

[38] Guo, Mingxin, Weiping Song, and Jeremy Buhain, *Bioenergy and biofuels: History, status, and perspective.* Renewable and sustainable energy reviews, 2015. **42**: p. 712-725.

[39] Siedlecki, Marcin, Wiebren De Jong, and Adrian Hm Verkooijen, *Fluidized bed gasification as a mature and reliable technology for the production of bio-syngas and applied in the production of liquid transportation fuels—a review.* Energies, 2011. **4**(3): p. 389-434.

[40] Kirkels, Arjan F and Geert Pj Verbong, *Biomass gasification: Still promising? A 30-year global overview.* Renewable and Sustainable Energy Reviews, 2011. **15**(1): p. 471-481.

[41] Wampler, Thomas P, *Applied pyrolysis handbook*. 2006: CRC press.

[42] Soltes, Ed J and Thomas J Elder, *Pyrolysis*, in *Organic chemicals from biomass*. 2018, CRC Press. p. 63-99.

[43] Venderbosch, Robbie H, *Fast pyrolysis.* Thermochemical processing of biomass: conversion into fuels, chemicals and power, 2019: p. 175-206.

[44] Kazawadi, Deodatus, Justin Ntalikwa, and Godlisten Kombe, *A Review of Intermediate Pyrolysis as a Technology of Biomass Conversion for Coproduction of Biooil and Adsorption Biochar.* Journal of Renewable Energy, 2021. **2021**.

[45] Tinwala, Farha, Pravakar Mohanty, Snehal Parmar, Anant Patel, and Kamal K Pant, *Intermediate pyrolysis of agro-industrial biomasses in bench-scale pyrolyser: product yields and its characterization.* Bioresource Technology, 2015. **188**: p. 258-264.

[46] Khoshand, Afshin, Ali Hasani Bafrani, Mohammad Zahedipour, Seyed Ahmad Mirbagheri, and Majid Ehtehsami, *Prevention of landfill pollution by multicriteria spatial decision support systems (MC-SDSS): development, implementation, and case study.* Environmental Science and Pollution Research, 2018. **25**(9): p. 8415-8431.

[47] Kim, Byung Hong, In Seop Chang, and Geoffrey M Gadd, *Challenges in microbial fuel cell development and operation.* Applied microbiology and biotechnology, 2007. **76**(3): p. 485-494.

[48] El-Fadel, Mutasem, Angelos N Findikakis, and James O Leckie, *Environmental impacts of solid waste landfilling.* Journal of environmental management, 1997. **50**(1): p. 1-25.

[49] Munawar, Sasa Sofyan and Bambang Subiyanto, *Characterization of biomass pellet made from solid waste oil palm industry.* Procedia Environmental Sciences, 2014. **20**: p. 336-341.

[50] Gaqa, Sibongiseni and Paul Watts, *The agglomeration of coal fines using wet microalgae biomass.* Journal of Energy in Southern Africa, 2018. **29**(2): p. 43-50.

[51] Brown, Robert C and Kaige Wang, *Fast Pyrolysis of Biomass: Advances in*

Science and Technology. 2017, Royal Society of Chemistry: Croydon. p. 300.

[52] Czajczyńska, D. , L. Anguilano, H. Ghazal, R Krzyżyńska, A.J. Reynolds, N. Spencer, and H. Jouhara, *Potential of pyrolysis processes in the waste management sector.* Thermal Science and Engineering Progress, 2017. **3**: p. 171-197.

[53] Wang, Guanyu, Yujie Dai, Haiping Yang, Qingang Xiong, Kaige Wang, Jinsong Zhou, Yunchao Li, and Shurong Wang, *A review of recent advances in biomass pyrolysis.* Energy & Fuels, 2020. **34**(12): p. 15557-15578.

[54] Jouhara, Hussam, Darem Ahmad, Inge Van Den Boogaert, Evina Katsou, Stefaan Simons, and Nik Spencer, *Pyrolysis of domestic based feedstock at temperatures up to 300 C.* Thermal Science and Engineering Progress, 2017. **5**(2018): p. 117-143.

[55] Tripathi, Manoj, Jaya Narayan Sahu, and P Ganesan, *Effect of process parameters on production of biochar from biomass waste through pyrolysis: a review.* Renewable and Sustainable Energy Reviews, 2016. **55**: p. 467-481.

[56] Torri, Isadora Dalla Vecchia, Ville Paasikallio, Candice Schmitt Faccini, Rafael Huff, Elina Bastos Caramão, Vera Sacon, Anja Oasmaa, and Claudia Alcaraz Zini, *Bio-oil production of softwood and hardwood forest industry residues through fast and intermediate pyrolysis and its chromatographic characterization.* Bioresource technology, 2016. **200**: p. 680-690.

[57] Basu, Prabir, *Biomass gasification, pyrolysis and torrefaction: practical design and theory*. 2013, Elsevier: India.

[58] Akhtar, Javaid and Noraishah Saidina Amin, *A review on Operating Parameters for Optimum Liquid Oil Yield in Biomass Pyrolysis.* Renewable and Sustainable Energy Reviews, 2012. **16**(7): p. 5101-5109.

[59] Agarwal, Manu, *An Investigation on the Pyrolysis of Municipal Solid Waste*, in *School of Applied Sciences*. 2014, RMIT University: Hyderabad.

[60] Encinar, Jm, Fj Beltran, A Bernalte, A Ramiro, and Jf Gonzalez, *Pyrolysis of two agricultural residues: olive and grape bagasse. Influence of particle size and temperature.* Biomass and Bioenergy, 1996. **11**(5): p. 397-409.

[61] Tsai, Wt, Cy Chang, and Sl Lee, *Preparation and characterization of activated carbons from corn cob.* Carbon, 1997. **35**(8): p. 1198-1200.

[62] Park, Hyun Ju, Young-Kwon Park, and Joo Sik Kim, *Influence of reaction conditions and the char separation system on the production of bio-oil from radiata pine sawdust by fast pyrolysis.* Fuel Processing Technology, 2008. **89**(8): p. 797-802.

[63] Huang, Zhihang, Linbo Qin, Zhe Xu, Wangsheng Chen, Futang Xing, and Jun Han, *The effects of Fe2O3 catalyst on the conversion of organic matter and bio-fuel production during pyrolysis of sewage sludge.* Journal of the Energy Institute, 2018.

[64] Gray, Murray R, William H Corcoran, and George R Gavalas, *Pyrolysis of a wood-derived material. Effects of moisture and ash content.* Industrial & Engineering Chemistry Process Design and Development, 1985. **24**(3): p. 646-651.

[65] Antal, Michael Jerry, Eric Croiset, Xiangfeng Dai, Carlos Dealmeida, William Shu-Lai Mok, Niclas Norberg, Jean-Robert Richard, and Mamoun Al Majthoub, *High-yield biomass charcoal.* Energy & Fuels, 1996. **10**(3): p. 652-658.

[66] Fassinou, Wanignon Ferdinand, Laurent Van De Steene, Siaka Toure, Ghislaine Volle, and Philippe Girard, *Pyrolysis of Pinus pinaster in a two-stage gasifier: Influence of processing parameters*

and thermal cracking of tar. Fuel processing technology, 2009. **90**(1): p. 75-90.

[67] Zhang, Huiyan, Rui Xiao, He Huang, and Gang Xiao, *Comparison of non-catalytic and catalytic fast pyrolysis of corncob in a fluidized bed reactor.* Bioresource Technology, 2009. **100**(3): p. 1428-1434.

[68] Choi, Hang Seok, Yeon Seok Choi, and Hoon Chae Park, *Fast pyrolysis characteristics of lignocellulosic biomass with varying reaction conditions.* Renewable Energy, 2012. **42**: p. 131-135.

[69] Pütün, Ersan, *Catalytic pyrolysis of biomass: Effects of pyrolysis temperature, sweeping gas flow rate and MgO catalyst.* Energy, 2010. **35**(7): p. 2761-2766.

[70] Imran, Ali, Eddy Bramer, Kulathuiyer Seshan, and Gerrit Brem, *Catalytic flash pyrolysis of biomass using different types of zeolite and online vapor fractionation.* Energies, 2016. **9**(3): p. 187.

[71] Al-Salem, Sm, A Antelava, A Constantinou, G Manos, and A Dutta, *A review on Thermal and Catalytic Pyrolysis of Plastic Solid Waste (PSW).* Journal of environmental management, 2017. **197**: p. 177-198.

[72] Miandad, R, Ma Barakat, Asad S Aburiazaiza, M Rehan, and As Nizami, *Catalytic pyrolysis of plastic waste: a review.* Process Safety and Environmental Protection, 2016. **102**: p. 822-838.

[73] Li, Xiangyu, Lu Su, Yujue Wang, Yanqing Yu, Chengwen Wang, Xiaoliang Li, and Zhihua Wang, *Catalytic fast pyrolysis of Kraft lignin with HZSM-5 zeolite for producing aromatic hydrocarbons.* Frontiers of Environmental Science & Engineering, 2012. **6**(3): p. 295-303.

[74] Encinar, José M, Fernando J Beltran, Antonio Ramiro, and Juan F Gonzalez, *Catalyzed pyrolysis of grape and olive bagasse. Influence of catalyst type and chemical treatment.* Industrial & engineering chemistry research, 1997. **36**(10): p. 4176-4183.

[75] Yu, Ying, Junqing Yu, Bing Sun, and Zhiyu Yan, *Influence of catalyst types on the microwave-induced pyrolysis of sewage sludge.* Journal of Analytical and Applied Pyrolysis, 2014. **106**: p. 86-91.

[76] Lts, *Bioenergy for Sustainable Energy Access in Africa*. 2017, LTS International Limited: Edinburgh.

[77] Adbg. *Nigeria - Green Energy & Biofuels (Geb) Bio-Refinery Project - SEFA Project Summary Note*. 2021 [cited 2021 20/7]; Available from: https://www.afdb.org/en/about-us/organisational-structure/complexes/country-regional-programs-policy/country-office-contacts.

Chapter 7

Pyrolysis: A Convenient Route for Production of Eco-Friendly Fuels and Precursors for Chemical and Allied Industries

Jamiu Mosebolatan Jabar

Abstract

Thermochemical decomposition of post harvest agro-wastes (biomass) to solid carbonaceous material called as bio-char, condensable vapors (bio-oils and bio-tars) and non-condensable vapors (bio-gas or syn-gas) is referred as pyrolysis. The yield of these products from biomass pyrolysis depends on temperature and other conditions (such as vapor retention time and heating rate) of thermal decomposition in air or oxygen excluded reactor. Bio-char is often used as adsorbent in treatment of water contaminated with dye effluent from textile industry and/or emerging contaminants from other industries. It is also used in production of supercapacitor for energy storage, fertilizer composite and soil amendment for slow release of nutrients for plants and stabilizing pH, enhances water holding and ion exchange capacity of soil. Bio-oils are used for transportation fuels, soaps and other cosmetics production. Bio-tars are also used for transportation fuels but with high heating values and also as organic solvents in chemical, biological and biochemical laboratories. Non-condensable vapors are mostly used as bio-fuels. Products of biomass pyrolysis are potential alternative eco-friendly precursors for chemical and allied industries.

Keywords: bio-char, bio-fuel, bio-gas, bio-oil, bio-tar

1. Introduction

Post harvest agricultural wastes are promising materials for generating renewable and eco-friendly energy source and raw materials for chemical and allied industries. Agricultural wastes serving as a potential alternative source of energy and starting renewable materials for industrial productions will definitely bring about reduction in demand for non-renewable petrochemical feed stocks often used by most chemical industries today. Other benefits are efficient carbon cycle and reduction in carbon(IV) oxide (CO_2) emissions [1]. Among the agro-wastes, plant-based (lignocellulose) materials are most abundant and disposed indiscriminately, where by constituting nuisance to the environment. Plant-based agricultural wastes are called lignocellulosic materials because their compositions are majorly of lignin and cellulose. Others are hemicellulose, pectin, wax, oil and inorganic matters [2]. These compositions of plant vary from one plant to another. Other factors that

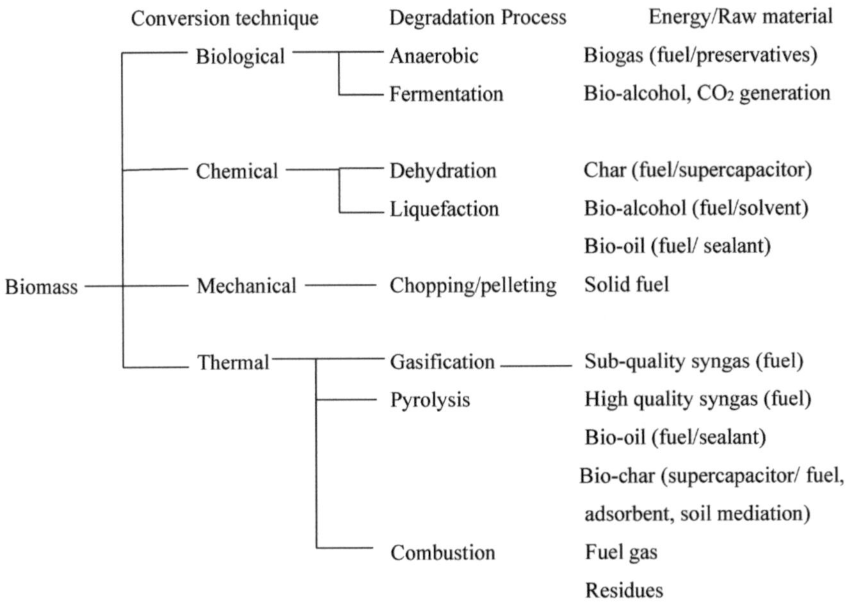

Figure 1.
Various forms of biomass conversion techniques.

contribute to the variation in compositions are age of the plant, processing methods, geological and climatic factor [3, 4]. The plant-based agro-wastes worked on by researchers as energy source and feed stocks for chemical and allied industries are sugarcane bagasse, cola nut pod, cocoa pod, moringa seed pod, oil palm empty fruit bunch, almond leaf, *Bridelia ferruginea* bark, elephant grass etc. [5–9]. Conversion of these agro-wastes to energy source and starting materials for chemical and allied industries will bring a turn around to high dependence on ever depleting and price fluctuating petrochemicals. This will in turn generate wealth from wastes; creating jobs for teaming unemployed youths and keep our environments clean [3].

Several technologies deployed for converting agro-wastes to energy source and precursors for chemical and allied industries are hot acid carbonization by dehydration and anaerobic digestion. Others that are thermally based are torrefaction, pyrolysis, combustion, gasification and hydrothermal process. Among these technologies, pyrolysis is most preferred, because the products which are solid, liquid and gas can easily be stored, transported and handled [10]. **Figure 1** shows summary of various biomass conversion techniques and products obtained. Pyrolysis is a composition of two Greek words, split able into 'pyro' meaning fire and 'lysis' means degradation into many parts. Therefore, pyrolysis can be defined as thermochemical degradation of biomass into biochar (solid carbonaceous material), bio-tar, bio-oil (condensable vapor) and non condensable gases (syngas) in absence or very limited supply of oxidizing agents [10, 11]. Depending on degradation temperature, heating rate and vapor retention time, pyrolysis can be classified as slow, fast or flash pyrolysis. The choice of any of the pyrolysis techniques depends on desired or targeted products [10].

2. Biomass

Biomass waste stream can be categorized into proteineous (animal-based), lignocellulosic (plant-based), municipal solid (biogenic and non biogenic), industrial (sewage sludge) and other wastes. They are referred to as feedstock in bio-refinery. Biomass based on animal and plant residues (briefly described in sub-sections 2.1–2.2) is an area where researchers are currently exploiting for possibility of obtaining sustainable and environmental friendly alternative to fossil fuels/ precursors for chemical and allied industries. The use of biomass for these afore-mentioned purposes reduces demand for fossil fuels, their derivatives and makes our environment greener [1].

2.1 Animal-based biomass

Animal wastes, such as blood, bones, crab shells, hairs, horns, dung, etc. are by-products of livestock rearing. Often seen livestock that generate these wastes are cattle, sheep, chickens, goats, crabs and pigs (**Table 1**). They have been used in raw or pyrolyzed form as source of energy and raw materials for industrial productions by several researchers. Adetuyi and Jabar [9] are among the researchers that have used carbonized animal bone for treatment of wastewater. **Table 1** shows list and applications of some animal-based biomass.

2.2 Lignocellulose-based biomass

Lignocellulosic biomass is carbonaceous plant materials containing lignin and cellulose as major constituents, other are hemicellulose, pectin and other extractives as stated earlier [2, 12].

These compositions of biomass vary from plant species to another. Variation equally exists between/among different parts from the same plant. Factors responsible for this variation are in introduction section [4]. List of some biomasses with their compositions are presented at **Table 2**.

Lignocellulose can be categorized as woody, herbaceous and aquatic biomass [17]. These biomasses have been modified by many researchers for production of biofuels and starting materials for chemical industry.

2.2.1 Woody biomass

These are plant-based resources obtainable from forest; they include trees, leaves, limbs, needles, tops and their residues [17]. Less than two-third of the woody biomasses is converted to timber, others end up as wastes in the forest. The occurrence of huge wastes might be as a result of inappropriate harvesting technique or equipment. Utilization of these wastes called residues as bioenergy for fuel and for generation of raw materials for furniture, pulp and paper industries will be a good

Animal	Residue	Application	Ref.
Cattle	Bone	Char for production shoe sole, breakable plate, soil enhancement	[9]
	Horn	Char for soil enhancement, bio-oil for fuel and cosmetic industry	
Crab	Shell	For wastewater treatment	
Sheep	Hair	Textile fiber production	
	Born	Char for wastewater treatment, supercapacitor production	
Camel	Bone	Char for production shoe sole, breakable plate, soil enhancement	
	Horn	Char for soil enhancement, bio-oil for fuel and cosmetic industry	

Table 1.
Animal-based biomass and applications.

Lignocellulose	Compositions				Ref.
	Hemicellulose	Lignin	Cellulose	Extractives	
				Pectin and others	
Coir fiber	9.0	47.7	40.8	2.5	[12]
Oil palm fiber	13.7	21.6	61.6	3.0	[12]
Palm kernel fiber	14.0	30.0	51.8	4.1	[12]
Sorghum	23.4	17.9	31.4	23.7	[13]
Switchgrass	21.5	21.1	35.8	16.9	[13]
Alfalfa	19.0	13.0	25.0	20.0	[14]
Pine	21.0	26.0	42.0	2.7	[14]
Kenaf fiber	38.0	19.0	49.0	4.7	[15]
Oak	22.0	24.0	46.0	6.0	[16]

Table 2.
Chemical composition of some selected lignocellulose.

waste management approach in the forest. Consequently, this conversion of wastes to wealth will reduce rate of deforestation for fuels and industrial precursors [18].

2.2.2 Herbaceous biomass

Herbaceous is non-woody biomass called annual crops. They include rice, wheat, sorghum, maize, millet, groundnut, soyabean etc. After harvesting, their residues do not constitute much treat to the environment, as they are grown and their residues decay within a year. Alternatively, their residues can be utilized for production of bio-gas, bio-oil and bio-chair as bio-fuels and bio-chemical raw materials for chemical and allied industries [19].

2.2.3 Aquatic biomass

Approximately, one-third of the world is water body. Weeds grow exceptionally on water bodies when compared to terrestrial, whereby enhancing aquatic pollution index. The aquatic biomass includes algae (micro and macro), water hyacinth, kelp, water lily, seaweed etc. These so called aquatic weeds are underutilized; they can be made to serve as raw materials for bio-fuels and bio-chemicals for industrial productions through pyrolysis [20].

3. Pyrolysis

Pyrolysis is the thermochemical decomposition of biomass or other feedstock into biochar, bio-tar, bio-oil, bio-gas and other related products in absence or very limited supply of oxidizing agents [10, 11, 21]. This process can be broadly classified as fast, slow and flash, depending on degradation temperature, heating rate and vapor retention time as stated earlier. Other factors that affect pyrolysis are nature and moisture content of the chosen biomass [22]. **Figure 2** presents possible products of pyrolysis of biomass, while **Figure 3** illustrates pyrolysis mechanism. Incipient pyrolysis products are solid bio-char and condensable vapor. The later breaks down into char, liquid and non-condensable gases through gas-solid and

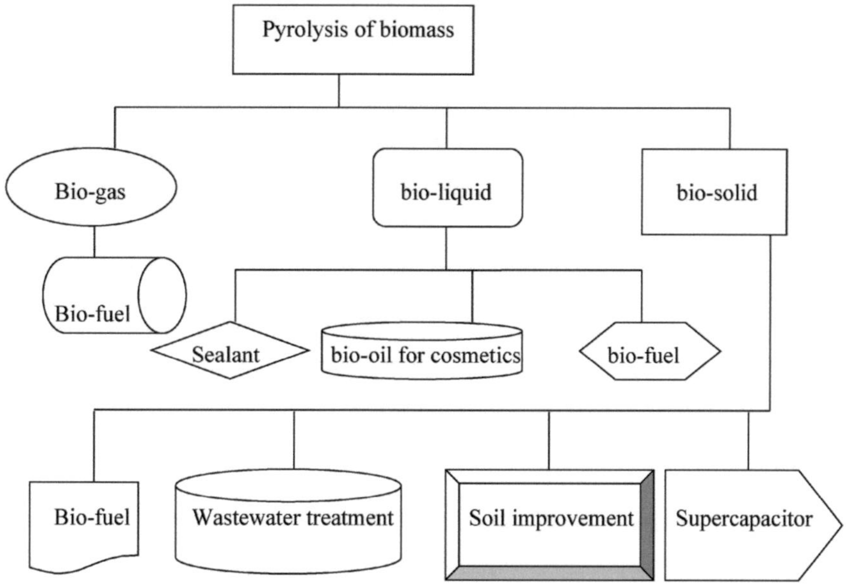

Figure 2.
Possible products of biomass pyrolysis and their applications.

$$C_aH_bO_{d\,(biomass)} \xrightarrow{\text{pyrolysis}} \Sigma_{gas}\, C_eH_fO_g + CO + CO_2 + CH_4 + H_2 + \Sigma_{liquid}\, C_jH_kO_m + C_{char}$$

Figure 3.
Pyrolysis of biomass [11].

gas phase heterogeneous and homogeneous reaction respectively at high temperature. In homogeneous gas phase reaction, condensable vapor cracks into small molecules of non condensable gases (CO, CO_2, H_2O, H_2, N_2 and CH_4). In gas-solid phase heterogeneous reaction, high molecular weight hydrocarbons are broken into low molecular weight hydrocarbons (such as CH_4, C_2H_4, C_2H_6, C_6H_6 etc.). Several reactors called pyrolyzers have been used by many researchers to obtain two or more of these products from biomass. Some of these pyrolyzers are bed, vacuum, microwave and solar reactor [11].

Pyrolysis of lignocellulosic biomass involves several stages of thermal decomposition. Evaporation of moisture occurs between 40 and 115°C, at this stage, biomass dries up as a result of heat absorption that evaporates free moisture and eradicates loosely bonded water [7]. The heat transfers into biomass interior through convention to wipe out all internally bonded moisture. Biomass incipient degradation occurs through decomposition of pectin and other extractives between temperatures of 115 and 180°C, generating syn-gas and organic liquids (such as carbon(II) oxide, carbon(IV) oxide, methane, ethane, ethane, benzene, methanol, pyrogallol and ethanoic acid) [22].

Hemicellulose degrades between temperatures of 180–250°C, giving off considerable amount of carbon(II) oxide, carbon(IV) oxide, water, hydrocarbon of low carbon content, organic liquids and minor bio-oil [23]. Temperature from 250–350°C decomposes cellulose to yield mixture of non-condensable vapor (syn-gas), condensable vapor (bio-oil and tar) and minor bio-char [24]. Proper formation of solid fraction (primary bio-char) commences from thermal decomposition of lignin from 400°C, this is the stage at which primary pyrolysis occurs. As temperature increases from 500°C upward, minor tar and bio-oil contained in the bio-char get separated and high quality bio-char obtained. Although, a low bio-char yield is obtainable from temperature above 700°C, but a high grade bio-char formed [25]. This is the final stage of thermochemical degradation called secondary pyrolysis. It involves cracking of condensable vapor (bio-tar and bio-oil) into non condensable gases (CO, CO_2, CH_4 etc.) and on cooling; part of the products solidifies to form char (secondary bio-char). Summary of the entire pyrolysis process is presented in **Table 3**.

3.1 Fast pyrolysis

This process involves decomposition of biomass at high heating rate (10–250°C/s), high temperature (500–1200°C) and short vapor retention time (<2 s) in the absence of oxygen. Due to high heating rate, biomass is rapidly heated to pyrolysis temperature before it is decomposed. Although, this process yield gaseous, liquid and solid fractions, but exclusively for bio-oil (liquid fuels) production, if pyrolysis temperature is less than 650°C. Quick vapor condensation enhanced formation of bio-oil, usually >50% products fraction is liquid, while gaseous formation is favored if pyrolysis temperature is up to 1000°C (about 30% yield is obtainable) and solid fraction is less 30% of pyrolysis products [26].

3.2 Slow pyrolysis

In contrary to fast pyrolysis, biomass is decomposed at lower heating rate (<1°C/s), lower degradation temperature (500–550°C) and longer vapor retention time (300–1800 s). This process has long been practiced and is otherwise known as conventional pyrolysis. Just like in fast pyrolysis, products of slow pyrolysis of biomass are gases, liquids and solids. Due to longer vapor retention time of condensable and non condensable gaseous fractions, the condensable volatile organic fractions react with one another to form high quantity of solid fraction (bio-char) and low quantity of bio-oil and bio-gas [27]. Therefore, one can say that this process is more desirable for bio-char production.

3.3 Flash pyrolysis

Flash pyrolysis can be called very fast pyrolysis, because it employs very high temperature (800–1200°C), high heating rate (>1000°C/s) and extremely low vapor retention time (below 0.5 s). The condensable and non condensable gases leave the pyrolyzer and cool rapidly to form liquid fuel called bio-oil. Some part of the poor thermally stable solid fraction (bio-char) formed degrades at high operating temperature and dissolves in the liquid fraction (bio-oil), resulting in high yield of bio-oil (>70%). The more the quantity of dissolved bio-char, the more viscose and the poorer the resulting bio-oil [28]. This process is better recommended for syn-gas production.

Composition	Degradation temp. (°C)	Products formed
Moisture	40–115	Dehydration
Extractives	115–180	Syn-gas, organic liquids
Hemicellulose	180–250	Syn-gas, organic liquids and minor bio-oil
Cellulose	250–350	Syn-gas, organic liquids, bio-oil and minor bio-char
Lignin	>400	Bio-oil, bio-char and minor syngas

Table 3.
Degradation temperature of biomass compositions and corresponding products.

4. Pyrolyzer

Pyrolyzer is a reactor where pyrolysis of biomass is done [22]. The design of pyrolyzer depends on desired pyrolysis product(s). Therefore, it is another important factor that determines products obtainable from pyrolysis [10]. Some of designed pyrolyzers are discussed in sub-sections below.

4.1 Bed pyrolyzer

This pyrolyzer can be made of firebrick, steel or concrete and is predesigned for bio-oil production. Its design contains feeding unit, non-condensing vapor collector and vapor condenser, others are filter and ash exit unit [29]. It is operated in such a way that biomass is fed downward in reactor through vertical shaft and counter current gas stream simultaneously fed-in in an upward direction. The biomass is heated by bed of hot solids (such as sand heated by external source or partial combustion of biomass) at high temperature, high heating rate and low vapor retention time in an inert environment to decompose it into non condensable gas, condensable vapor and char. The vapor is rapidly cooled to form high quality bio-oil with yield of about 70% mass of pyrolysis products [30]. Therefore, it is a simple and well designed fast pyrolyzer that gives precedence to production of bio-oil. Merits of this reactor are exemplary heat transfer system, very high relative velocity between the solid and liquid phase, good control of vapor holding time and pyrolysis reaction as a whole. Some of the existing designs are fixed bed, circulating fluidized and bubbling fluidized bed reactor [10].

4.2 Vacuum pyrolyzer

This is a slow pyrolyzer with a moderate biomass degradation temperature, very low heat transfer rate and high vapor retention time. The biomass is feed into the pyrolyzer with the aid of conveyor belt made of metal containing stirrer that agitates the biomass mechanically. The biomass is heated with attached burner for decomposition into gas, liquid and solid. It produces bio-char of yield greater than 50% and bio-oil of less than 30% yield. Therefore, it is designed for bio-char production [31].

4.3 Microwave pyrolyzer

Design of microwave pyrolyzer is done in such a way that biomass is heated to decompose into gaseous, liquid and solid products through heat generated from electrical microwave oven chamber. Inert gas is used to fashion out non oxygen environment. In this pyrolyzer, undesirable output (by-product) is not formed, there is effective heat transfer and heat process can be easily monitored. Among its merits are a very good chemical recovery, efficient control of heat transfer and heating system. This pyrolyzer is exclusively designed for syn-gas production with greater than 30% yield from feedstock biomass [32].

4.4 Solar pyrolyzer

This makes use of clean energy from sun light. It is made of quartz tube containing opaque external wall. The opaque external wall when exposed to high intensity solar radiation is capable of generating temperature higher than 700°C for decomposition of biomass into syn-gas, bio-oil and bio-char [33, 34]. In its design, there is provision for solar energy storage inform of chemical energy. Therefore, at night and during poor sun shine weather, it can still work using already stored energy for generating heat for pyrolysis process. It is a very fast pyrolyzer with bio-oil yield greater than 50%. Although, the yield of gaseous fraction is less than 30%, but is recommended for production of syn-gas because of poor quality of bio-oil formed according to Cornelissen et al. [28].

5. Products of biomass pyrolysis

Thermochemical decomposition of biomass yields mixture of gases, liquid and solid bio-char as major products as stated earlier. The yield of any of the pyrolysis products is largely depends on pyrolysis temperature and biomass' moisture content among others like design of pyrolyzer, vapor retention time and heating rate [22].

5.1 Gas

Mixture of gases formed from primary (non condensable gases) and secondary degradation (tar and volatile organic compounds called condensable vapor) of biomass in the pyrolyzer during pyrolysis after moisture dehydration are called syn-gas. Components of syn-gas vary from one biomass specie to another. Often formed gases from primary degradation of biomass are carbon(II) oxide, carbon(IV) oxide, hydrogen gas, gaseous water, nitrogen and hydrocarbons of lower carbon content (methane, ethane, ethylene). Additional gases are produced through secondary cracking of condensable vapor at high temperature [35]. Very high pyrolysis temperature enhances yield of these mixture of gases and reduces yield of bio-char. The increase in yield of syn-gas might be as a result of thermal decomposition of tar and hydrocarbons (condensable vapor) to produce more oxides of carbon and other gases.

Lower carbon content hydrocarbons (eg methane, ethane, ethane and butane) are often used as domestic cooking fuels. The heating value of syn-gas obtained from secondary degradation of biomass (condensable vapor) is about 82% higher than those from biomass' primary degradation. Hydrogen can be blended with oil to upgrade it to transportation fuel [36]. Hydrogen gas and carbon(II) oxide are mainly used as bio-fuel inform of water gas [11].

Liquefied carbon(IV) oxide is used as major component in production of fire extinguisher. It is also used as cooling agent and preservative by bottling company. Recent research work by Patel [37] claimed possibility of converting CO_2 to starch in the laboratory about nine times more efficient as obtainable from corn plant using combination of chemical catalysts and enzymes. During the conversion process, useful industrial chemical like methanol was obtainable. This will serve as sustainable way of feeding the ever growing world population without need of land, seed, pesticide, fertilizer and water. The approach can equally be said to be a good form of food security and reducing CO_2 emissions [37].

Nitrogen is utilized in production of ammonia from Haber process for preparation of fertilizer by agrochemical industry in order to enhance food production for feeding human population and livestock [38].

5.2 Liquid

Major liquid components of biomass themochemical degradation are bio-oil (bio-crude) and black tar, others are organic solvents. Bio-oil contains mixture of heavy molecular weight hydrocarbons. High pyrolysis temperature coupled with quick vapor condensation result in high bio-oil yield (>70%) [39]. Condensed pyrolysis vapor contains more than three hundred mixture of condensable gaseous compounds. Higher yield of bio-oil is obtainable from agricultural residues when compared with one obtainable from woody biomass. Quality of pyrolysis bio-oil is mostly affected by storage time and temperature. The higher the storage time, the more viscous and poorer quality of the bio-oil due to escape of volatile components. Decline in quality of bio-oil due to aging can be minimized through cool storage temperature. Bio-oil is mainly used as bio-fuel and chemicals for industrial productions. Bio-oil heating value can be maintained by monitoring its metal and water content. It is used as transportation fuel, when mixed in certain ratio with fossil fuel it minimizes greenhouse gas emission that causes major environment pollution, especially in the city. Bio-oil can serve as precursor for production of soap through saponification process in soap and detergent industry. Cosmetic industry uses bio-oil in production of shoe polish, hair and body cream [40]. Organic solvents, such as benzene is a very good precursor in production of important materials, like dyes, pigments, synthetic tyres and textile materials. Others are production of plastics, resins, rubber lubricants, detergents, drugs and pesticides. Benzene is also a good solvent used in chemical, biochemical and biological laboratories, as well as aviation bio-fuel [41]. Other organic solvents like hyroxyketones, hydroxyaldehydes, carboxylic acids, phenolic compounds, sugar and dehydrosugars are important precursors for chemical and allied industry [11].

5.3 Bio-char

The solid component of pyrolysis products is known as bio-char. It is a black amorphous carbonaceous solid matter [42]. Properties and yield of bio-char are dependents of pyrolysis temperature, heating rate, vapor retention time, inert gas flow rate and nature of biomass [43]. A coarse and high yield bio-char is obtained from woody biomass, if pyrolysis temperature between 500 and 550°C, heating rate less than 1°C/s and greater than 450 s vapor retention time are used for its thermo-chemical process under inert environment. In contrary, if a fine bio-char is a desired material, agro-residue would be required as pyrolysis feedstock using degradation temperature greater than 600°C, high heating rate between 10 and 300°C/s and vapor retention time less than 20 s [11]. Bio-char with acceptable properties contains carbon as major component; other fractions are oxygen, hydrogen, nitrogen, sulfur and ash (inorganic metals). Bio-char with a long half life has its carbon to oxygen ratio (C:O) usually greater than 3:2, but one with 1:5 (C:O) has short half life. Inorganic ash component of bio-oil is very small in comparison with fossil fuel and it has low heating value of about 32 MJ/kg, which is about 60% higher than that of lower heating value of the parent material (biomass). Bio-char has many areas of application, ranging from soil enrichment in agro-chemistry, adsorbent in wastewater treatment to energy in electrochemical capacitor (supercapacitor) and bio-fuel [10].

5.3.1 Bio-char for soil amendment

Incessant ability of agricultural activities depends largely on biological, physical and chemical properties of soil. A little change in soil organic carbon has significant impact on plant growth. Farmers mostly use synthetic fertilizers for achieving this goal, but ironically the soil fertility decreases. It is only bio-char that provides soil with organic carbon that boots soil nutrient [22]. Environmental awareness has increased interest of global community for using bio-char as soil amendment agent. Tilling soil with bio-char stabilizes soil pH, enhances water retention capacity and available nutrients for plant growth [44]. Enhancement of soil nutrient by bio-char might be as a result of presence of potassium, magnesium, calcium, iron, zinc, phosphorous, sulfur and nitrogen in the bio-char. Equally, amendment of soil by bio-char reduces leaching out of soil nutrient and aids slow nutrients release to the plant, where by making nutrients available for long period of time. This promotes plant growth and mitigates climate change through consumption of carbon(IV) oxide (greenhouse gas) by growing plants during photosynthesis (smart carbon cycle). Therefore, application of bio-char in soil amendment makes our environment greener and increase production of food for geometrically increasing human population [45].

5.3.2 Bio-char as adsorbent for wastewater treatment

Inadequate water supply globally has given significant attention to the production of sustainable and eco-friendly bio-char based adsorbent for treatment of wastewater containing heavy metals, emerging contaminants and dye effluent from textile industry [7]. Properties of bio-char, such as cation exchange capacity, high surface area, high carbon content, amorphous and hydrophilic nature aid its applicability. Previous research works have established bio-char as suitable material for treatment of wastewater. Jabar and Odusote [3, 6] used bio-char obtained from water lily and oil palm empty bunch fiber for treatment of water contaminated with malachite green and cibacron blue 3G-A (CB) dye respectively. Thue et al. [46] used shell of cashew of Para for treatment of wastewater containing emerging contaminants. Ozsin et al. [47] used chemically activated carbon from agricultural waste of chickpea and its application for heavy metal adsorption. Wood-based bio-char was used as adsorbent for removing toxic elements from wastewater by Shaheen et al. [48], while Gwenzi et al. [49] showed that designed water treatment plant based on biomass derived char was a prospective low cost sustainable technology for producing dirt free water.

5.3.3 Bio-char as energy and energy storage device

Use of bio-char as source of energy for heating was old as human existence. It was popularly used as energy source for blast furnace in production of iron from iron ore called hematite and in petroleum refinery plant for separation of factions using fractional distillation. Researchers are currently focusing on the use of bio-char in production of supercapacitor called electrochemical capacitor [50]. Supercapacitor is an energy storage device that stores and releases electrical energy within a few seconds [51]. It stores energy as electric charge at electrolyte porous carbon electrode interface more than ten times conventional capacitor does [52]. It has to its advantages, high power density and satisfactory energy density [53]. Properties of bio-char, such as excellent electrochemical performance, high surface area, fast electron transport and abundant functional moieties make it suitable for production of supercapacitor [54]. Several researchers have prepared and activated bio-char for production of supercapacitor in electrochemistry. One of these researchers is Vinayagam and his co-researchers [55] who prepared activated carbon from *Sapindus trifoliatus* nut shell for high performance symmetric supercapacitor applications.

6. Conclusion

Pyrolysis is thermochemical decomposition of biomass under inert environment. The products of this process are syn-gas, bio-oil and bio-char. These products are alternative to petrochemicals as they are good sources of renewable and eco-friendly energy and bio-chemical precursors for chemical and allied industry. Several factors including nature of biomass, biomass moisture content and design of pyrolyzer, others are vapor retention time, heating rate and degradation temperature determine quantity and quality of these pyrolysis products.

Conflict of interest

The author declares no conflict of interest.

Author details

Jamiu Mosebolatan Jabar
Textile and Polymer Research Laboratory, Chemistry Department, The Federal University of Technology, Akure, Nigeria

*Address all correspondence to: jmjabar@futa.edu.ng

IntechOpen

© 2021 The Author(s). Licensee IntechOpen. This chapter is distributed under the terms of the Creative Commons Attribution License (http://creativecommons.org/licenses/by/3.0), which permits unrestricted use, distribution, and reproduction in any medium, provided the original work is properly cited.

References

[1] Sanusi S, Isah M, Fatima GM. Synthesis of polyol from sponge gourd (*Luffa aegyptiaca*) seed oil and production of polyurethane foam. United International Journal for Research & Technology. 2020;**02**(02):41-46

[2] Jabar JM. Effect of chemical modification on physicochemical properties of coir, empty fruit bunch and palm kernel fibres. Applied Tropical Agriculture. 2016;**21**(1):153-158

[3] Jabar JM, Odusote YA. Utilization of prepared activated biochar from water lily (Nymphaea lotus) stem for adsorption of malachite green dye from aqueous solution. Biomass Conversion and Biorefinery. 2021;1-12. https://doi.org/10.1007/s13399-021-01399-9

[4] Jabar JM, Adedayo TE, Odusote YA. Green, eco-friendly and sustainable alternative in dyeing cotton fabric using aqueous extract *Mucuna slonaei* F dye: Effects of metal salts pre-mordanting on color strength and fastness properties. Current Research in Green and Sustainable Chemistry. 2021;**4**:100151. DOI: 10.1016/j.crgsc.2021.100151

[5] Jabar JM, Odusote YA, Alabi KA, Ahmed IB. Kinetics and mechanisms of congo-red dye removal from aqueous solution using activated *Moringa oleifera* seed coat as adsorbent. Applied Water Science. 2020;**10**:136. DOI: 10.1007/s13201-020-01221-3

[6] Jabar JM, Odusote YA. Removal of cibacron blue 3G-A (CB) dye from aqueous solution using chemo-physically activated biochar from oil palm empty fruit bunch fiber. Arabian Journal of Chemistry. 2020;**13**:5417-5429. DOI: 10.1016/j.arabjc.2020.03.020

[7] Jabar JM, Owokotomo IA, Ayinde YT, Alafabusuyi AM, Olagunju GO, Mobolaji VO. Characterization of prepared eco-friendly biochar from almond (*Terminalia catappa* L) leaf for sequestration of bromophenol blue (BPB) from aqueous solution. Carbon Letters. 2021;1-14. DOI: 10.1007/s42823-020-00214-1

[8] Jabar JM, Ogunmokun AI, Taleat TAA. Color and fastness properties of mordanted *Bridelia ferruginea* B dyed cellulosic fabric. Fashion and Textiles. 2020;**7**:1. DOI: 10.1186/s4069 1-019-0195-z

[9] Adetuyi AO, Jabar JM. Kinetic and thermodynamic studies of indigo adsorption on some activate bio-solids. Journal of the Chemical Society of Pakistan. 2011;**33**(2):158-165

[10] Zaman CZ, Pal K, Yehye WA, Sagadevan S, Shah ST, Adebisi GA, et al. A Sustainable Way to Generate Energy from Waste. UK: IntechOpen, 2017. DOI: 105772/intechopen69036

[11] Basu P. Pyrolysis. In: Biomass Gasification, Pyrolysis and Torrefaction. USA: CRC Press, 2nd ed. 2013;147-176. DOI: 10.1016/B978-0-12-396488-5.0005-8

[12] Jabar JM. The effect of surface modifications on chemical composition of the oil palm and coir fibres. International Journal of Advanced Research in Chemical Science. 2017;**4**(1):13-21. DOI: 10.20431/2349-0403.0401002

[13] Jung S, Kim S, Chung I. Comparison of lignin cellulose and hemicellulose contents for biofuels utilization among 4 types of lignocellulosic crops. Biomass and Bioenergy. 2015;**83**:322-327. DOI: 10.1016/j.biombioe.2015.10.007

[14] Lupoi JS, Smith EA. Characterization of woody and herbaceous biomasses lignin composition with 1064 nm dispersive multichannel Raman spectroscopy. Applied Spectroscopy. 2012;**120**(1): 810-819. DOI: 10.1366/12-06621

[15] Abdul Kalil HPS, Yusra AFI, Bhat AH, Jawaid M. Cell wall ultrastructure, anatomy, lignin distribution and chemical composition of Malaysia cultivated kenaf fibre. Industrial Crops and Products. 2009;**31**(1):113-121

[16] Petteren RC. The chemical composition of wood. In: Rowell R, editor. Advances in Chemistry Series 207, the Chemistry of Solid Wood. Washington DC, USA: American Chemical Society; 1984; Chap 2. pp. 57-126

[17] Jekayinfa SO, Orisaleye JI, Pecenka R. An assessment of potential resources for biomass energy in Nigeria. Resources. 2020;**9**:92. DOI: 10.3390/resources9080092

[18] Imasuen OI, Oshodi JN, Onyeobi TUS. Protection area for environmental sustainability in Nigeria. Journal of Applied Sciences and Environmental Management. 2013;**17**:53-58

[19] Loi R. World crop residues production and implications of bits uses as bio-fuel. Environment International. 2005;**31**:375-384

[20] Kaura A, Kumar M, Sachdeva S, Puri SK. Aquatic weeds as the next generation feedstock for sustainable bioenergy production. Bioresource Technology. 2018;**251**:390-402

[21] Garrett DE, Mallan GM. Pyrolysis process for solid wastes. US Patent. 1979;**4**:153-514

[22] Panwar NL, Pawar A, Salvi BL. Comprehensive review on production and utilization of biochar. SN Applied Sciences. 2019;**1**:168. DOI: 10.1007/s42452-019-0172-6

[23] Lanzetta M, Blasi DC. Pyrolysis kinetics of wheat and corn straw. Journal of Analytical and Applied Pyrolysis. 1998;**44**:181-192

[24] Yang H, Yan R, Chen H, Lee DH, Liang DT, Zheng C. Pyrolysis of palm oil wastes for enhanced production of hydrogen rich gases. Fuel Processing Technology. 2006;**87**:935-942

[25] Jahirul ML, Rasual MG, Chowdhury AA, Ashwath N. Biofuel production through biomass pyrolysis—a technological review. Energies. 2012;**5**:4952-5001

[26] Venderbosch RH, Prins W. Fast pyrolysis technology development. Biofuels, Bioproducts and Biorefining. 2010;**4**:178-208

[27] Bridgewater AV, Czernik S, Piskorz J. An overview of fast pyrolysis. Process Design of Biomass Thermochemical Conversion. 2001;**2**:977-997

[28] Cornelissen T, Yperman Y, Reggers G, Schreurs S, Carleer R. Flash co-pyrolysis of biomass with polylatic acid. Part 1: Influence on bio-oil yield and heating value. Fuel. 2008;**87**:1031-1041

[29] Altafini CR, Wander PR, Barreto RM. Prediction of working parameter of a waste gasifier through an equilibrium model. Energy Conversion and Management. 2003;**27**:155-171

[30] Lv PM, Xiong ZH, Chang J, Wu CZ, Chen Y, Zhu JX. An experimental study on biomass air-stream gasification in fluidized bed. Bioresource Technology. 2004;**95**:95-101

[31] Roy C, Blanchette D, Korving L, Yang J, De Caumia B. Development of novel vacuum pyrolysis reactor with improved heat transfer potential. In: Bridgewater AV, Boocock DGB, editors. Developments in Thermochemical Biomass Conversion. London, UK: Blackie Academic and Professional; 1997. pp. 351-367

[32] Fernandez Y, Menendez JA. Influence of feed characteristics on the

microwave-assisted pyrolysis used to produce syngas from biomass wastes. Journal of Analytical and Applied Pyrolysis. 2011;**91**:316-322

[33] Hofmann I, Antal MJ. Numerical simulations of the performance of solar fired flash pyrolysis reactor. Solar Energy. 1984;**33**:427-440

[34] Boutin O, Ferrer M, Lede J. Flash pyrolysis of cellulose pellets submitted to a concentrated radiation: Experiments and modeling. Chemical Engineering Science. 2002;**57**:15-25

[35] Wei L, Thomasson JA, Bricka RM, Batcher WD, Columbus EP, Wooten JR. Experimental Study of a Downdraff Gralifier, ASABE Meeting Paper No. 066029. St. Joseph, Michigan: American Society of Agricultural and Biological Engineers; 2006

[36] Ben-Iwo J, Manovic V, Longhurst P. Biomasss resources and biofuels potential for the production of transportation fuel. Renewable and Sustainable Energy Reviews. 2016;**63**:172-192. DOI: 10.1016/j.biombioe.2015.10.007

[37] Patel P. The advance hints at sustainable, efficient factory production of food and industrial chemicals. ACS. 2021;1-3. DOI: 10.1126/science.abh4049

[38] Rouwenhorst KHR, Elishav O, Mosevitzky Lis B, Grader GS, Mounaim-Rousselle C, Roldan A, et al. Future trends. In: Techno-Economic Challenges of Green Ammonia as Energy Vector. USA: Academic Press, 2021;303-319. DOI: 10.1016/B978-0-12-820560-0.00013-18

[39] Gonzalez JF, Roman S, Encinar JM, Martina G. Pyrolysis of various biomass residues and char utililzation for production of activated carbons. Journal of Analytical and Applied Pyrolysis. 2009;**85**:134-141

[40] Fahmi R, Bridgewater AV, Donnison I, Yates I, Jones JM. The effect of lignin and inorganic species in biomass on pyrolysis oil yield, quality and stability. Fuel. 2008;**87**:230-240

[41] Frank H, Stadelhofer JW. Producion and uses of benzene derivatives. Industrial Aromatic Chemistry. 1988:132-235. 10.1007/978-3-642-73432-8_5. Springer Nature. Rutgerwerke AG, Frankfurt/M 11, West Germany

[42] Certin E, Mightadei B, Gupta R, Wall TF. Influence of pyrolysisconditions on the structure and gasification reactivity of biochars. Fuel. 2004;**83**:2139-2150

[43] Bourka J, Manky-Harris M, Fushimi C, Dowaki K, Nunoura T, Antal MJ Jr. Do all carbonized charcoals have the same structure? A model of chemical structure of carbonized charcoal. Industrial & Engineering Chemistry Research. 2007;**46**:5954-5967

[44] Ernsting A, Smolker R. Biochar for climate change mitigation: Fact or friction. 2009. pp. 1-10

[45] Sohi S, Lopez-Capel E, Krull E, Bol R. Biochar, climate change and soil. A review guide future research. In: CSIRO Land and Water Science Report. Canberra, Australia: CSIRO; 2009

[46] Thue PS, Lima DR, Naushad M, Lima EC, de Albuquerque YRT, Dias SLP, et al. High removal of emerging contaminants from wastewater by activated carbons derived from the shell of cashew of Para. Carbon Letters. 2020;1-16. DOI: 10.1007/s4282 3-020-00145-x

[47] Ozsin G, Kılıc M, Apaydın-Varol E, Putun AE. Chemically activated carbon production from agricultural waste of chickpea and its application for heavy metal adsorption: Equilibrium, kinetic, and thermodynamic studies. Applied Water Science. 2019;**9**:56

[48] Shaheen SM, Niazi NK, Hassan NE, Nibi I, Wang H, Tsang DC, et al. Wood-based biochar for the removal of potential ly toxic elements in water and wastewater: A critical review. International Materials Reviews. 2018;1-32. DOI: 10.1080/09506608.2018.1473096

[49] Gwenzi W, Chaukura N, Noubactep C, Mukome FN. Bio-char based water treatment systems as a potential low-cost and sustainable technology for clean water provision. Journal of Environmental Management. 2017;**197**:732-749

[50] Sadeghi Ghazvini AA, Taheri-Nassaj E, Raissi B, Riahifar R, Yaghmaee MS, Shaker M. Co-electrophoretic deposition of Co_3O_4 and grapheme nanoplates for supercapacitor electrode. Materials Letters. 2021;**285**:129195

[51] Fathi AR, Riahofar R, Raissi B, Yaghmaee AS, Ghorbazadeh M. Optimization of cathode material components by means of experimental design for Li-ion batteries. Journal of Electronic Materials. 2020;**49**(11): 6547-6558

[52] Sheng LZ, Zhao YY, Hou BQ, Xiao Z, Fan Z. N-doped layered porous carbon electrodes with high mass loadings for high performance supercapacitors. New Carbon Materials. 2021;**36**(1):179-188

[53] Luo X, Chen Y, Mo Y. A review of charge storage in porous carbon based supercapacitor. New Carbon Materials. 2021;**36**(1):49-68

[54] Shaker M, Ghazvini AS, Cao W, Riahifa R, Ge Q. Biomass-derived porous carbon as electrode—A review. New Carbon Materials. 2021;**36**(3):546-572

[55] Vinayagam M, Babu RS, Sivasamy A, Lucia A, Barros F. Biomass-derived porous activated carbon nanofibres from *Sapindus trifoliatus* nut shells for high-performance symmetric supercapacitor applications. Carbon Letters. 2021;1-11. DOI: 10.1007/s42823-021-00235-4

Chapter 8

Recent Advances in Thermochemical Conversion of Biomass

Meegalla R. Chandraratne and Asfaw G. Daful

Abstract

The chapter focuses on recent trends of biomass conversion into valuable energy, chemicals, gaseous and liquid fuels. Biomass is presently the largest source of renewable energy and the primary bioenergy resource in the world. A comprehensive discussion on different types, sources and compositions of biomass is presented. The most abundant biomass on the earth is lignocellulose and it represents a major carbon source for chemical compounds and biofuels. The chapter presents a thorough review of lignocellulosic biomass and the importance of biomass as a renewable source. It then reviews biomass classification and composition. It introduces the analysis of biomass feedstock. Biomass is converted to energy, chemicals and clean fuels using various conversion techniques such as thermochemical, chemical and biochemical. The chapter provides a thorough examination of thermochemical conversion processes that use high temperatures to break down the bonds of organic matter. It briefly introduces combustion and gasification, followed by a comprehensive review of different pyrolysis techniques.

Keywords: bioenergy, hydrothermal processing, lignocellulosic biomass, biomass composition, pyrolysis, gasification, combustion

1. Introduction

Due to industrialization, population growth, and urbanization, there has been a rapid increase in global energy demand and consumption over the past few decades. Currently, more than 80% of the world energy demand is supplied by fossil fuels. Global attention on alternative resources has increased due to the high prices and limited reserves of fossil fuels, environmental pollution and climate change. As a result, there is renewed interest in the production and use of renewable energy resources such as biomass, solar, wind, geothermal and tidal. However, biomass is the only renewable energy resource to produce transport fuels. Various biomass components, including sugars, starches and lignocellulosic materials can be used to produce transport fuels and potentially limit the use of fossil fuels. On the other hand, biomass residues and wastes (such as agricultural residue, food waste, animal manure, municipal solid waste) as resources for bioenergy production are promising alternatives to reduce environmental issues concerning waste management and disposal. If improperly managed, these residues and wastes could lead to greenhouse gas emissions, pests breeding, insects, foul odor, etc. Converting these residues and wastes, that would otherwise have been left to decompose, into useful bioenergy is environmentally beneficial for waste management as well as to reduce the dependence on fossil fuel [1–3]. Lignocellulosic biomass (LCB) such as agricultural crop residues, wood and forestry residues are readily available, inexpensive and promising resource for biofuels. Biomass can be considered as the best option for meeting future energy demand sustainably. The efficient utilization of biomass resources is essential by having more efficient biomass production and conversion [4].

Biomass is an organic matter of plant or animal origin that is living or was living in the recent past [5]. Biomass is perhaps the oldest source of energy used by humans. People have used biomass as an energy source for cooking, heating, and lighting for thousands of years. The energy contained in biomass originally came from the sun. Plants get their energy from the sun and convert water (H_2O) and carbon dioxide (CO_2) into oxygen (O_2) and sugars (carbohydrates) in the process of photosynthesis. The energy of sunlight is stored in chemical bonds of plant organic matter. When different conversion processes break the bonds between adjacent carbon, hydrogen and oxygen molecules of carbohydrates, these substances release their stored chemical energy [6]. Biomass is a renewable energy source as it can be produced year after year and, therefore, is valuable as a fuel source. Compared to petroleum-based products, the use of biomass represents the potential to reduce CO_2 emissions, resulting in negligible or zero SO_X emissions and lower NO_X emissions [7, 8].

Biomasses include agricultural crop residues, forestry and wood residues, purposely grown bioenergy crops, aquatic biomass, sewage sludge, digestate, animal, industrial, municipal and food waste. Various types of wastes such as cow manure, poultry litter, wastepaper, sewage sludge, municipal and many industrial wastes are treated as biomass because these are a mixture of organic (and non-organic) compounds.

2. Importance of biomass

Non-renewable sources produce a significant portion of current energy supplies globally and their use is associated with the emission of CO_2 into the atmosphere. However, about 10–15% of this demand is covered by renewable resources, making biomass by far the most important renewable energy source used to date. Biomass contributes to 9% - 13% of the total energy supplies in industrialized countries. However, biomass energy is the primary energy source in many developing nations, contributing to about 20–30% of the total energy supplies. In some developing countries, biomass covers 50 to 90% of the total energy demand. In industrialized countries, biomass is used as a modern bioenergy source for industrial (heat, electricity), transportation (biofuels) and domestic (mainly heat) applications [9–11]. A significant part of the traditional use of biomass (firewood, sometimes animal waste) in developing countries is non-commercial and used for cooking and heating, generally by the poorer part of the population. The non-commercial use of biomass is poorly mapped and as a result, the contribution of biomass to the energy supply is not accurately known [12]. Solid biomass is one of the primary energy sources (mainly used for cooking) in many developing countries, especially in rural sub-Saharan Africa and South Asia. Traditional biomass use is not sustainable because it may result in soil quality degradation by depriving soil nutrients and burning it in inefficient cooking stoves can cause high levels of indoor air pollution. Most households in sub-Saharan Africa rely on the traditional use of biomass to meet their daily cooking needs. In countries such as the Democratic Republic of Congo, Ethiopia, Tanzania, Nepal and Nigeria, more than 80% of the total energy demand is met through biomass energy sources. In developing countries, biomass is used to some extent in industries such as brick manufacturing [13, 14].

Crude oil is an excellent source of energy, it is easily transported and energy-rich, it is one of the most energy-dense fuels. The energy density of crude oil and methane are 42 MJ/kg and 55 MJ/kg, respectively. Even coal has a good energy density (27–32 MJ/kg). Compared to these fossil fuels, the energy density of biomass ranges from 8 MJ/kg for greenwood to 20 MJ/kg for dry plant matter. Biomass has low bulk density and causes a major problem during storage, handling and transportation for further processing. The bulk density of biomass varies from around 40 kg/m^3 for loose straw and bagasse, in the range of 80–100 kg/m^3 for agricultural residues and 150–200 kg/m^3 for woody biomass to the highest levels of around 250 kg/m^3 for some wood residues. This translates to needing more biomass to produce for the same amount of heat or power and higher transportation costs, processing costs, etc. [15, 16].

Most of the biomass used today is derived from agricultural and forestry biomass. Agricultural biomass includes the food-based portion and the non-food-based portion of crops. The food-based portion comprises oils and simple carbohydrates of crops such as corn, sugarcane and beet. The non-food-based portion comprises complex carbohydrates of crops such as the leaves, stalks, seed hulls, orchard trimmings, nutshells, rice husk, bagasse, coffee pulp and straw. Forestry biomass includes wood residues associated with the production of timber in the forest, as well as the processing of timber into their final products [17, 18].

The comparatively low energy density of biomass means that issues associated with land use must be taken into consideration. Expansion of land use for biomass production can lead to a high potential for environmental damage due to deforestation, erosion, nutrient runoff, emissions, etc. This reduces any potential benefit of using biomass. Large-scale cultivation of non-food perennial energy crops for bioenergy feedstock is feasible when sufficient land areas are available. The best land for agriculture must be used to grow food crops. To avoid food versus energy conflict, it is important to use infertile/marginal lands for energy crop cultivation with little use of fertilizer or pesticides and potentially needing minimal water. Energy crops should not be grown at the expense of biodiversity [15]. For productive agricultural systems, prospects of biomass production based on the factors provided by nature such as light, soil, water and nutrients with soil and water are considered as the most crucial natural resource constraints. Identifying land areas with minimal disturbance to food production is critical for technically and economically feasible biomass production. To achieve sustainable large-scale biomass production, marginal or abandoned agricultural land has been widely considered as important. Energy crops are adaptive to marginal or abandoned agricultural land. Compared with food crops, energy crops such as switchgrass and miscanthus generally require much less water to grow and are suitable to partially replace the dryland crops [19]. Beyond the vast areas of land needed to grow energy crops, the long-term impact of soil quality due to repeated removal of biomass is a concern. Water usage is another major concern. Biomass may have a moderate carbon footprint, but its water footprint is enormous. Only a small percentage of the biomass produced by photosynthesis is currently being cultivated, harvested and used, but how much can be used sustainably? As with any approach to energy generation, the massive demand for energy stresses the need to be careful in considering the use of biomass for energy generation [15].

Plants grow through photosynthesis by absorbing atmospheric CO_2 and producing carbohydrates that form the building blocks of biomass. Water and sunlight are the other two key ingredients of photosynthesis, which typically convert less than 1% of the energy available in sunlight to chemical energy. When biomass burns, it releases CO_2 back to the atmosphere that the plants had absorbed recently. i.e., the burning of biomass does not add to the total CO_2 inventory of the earth. Therefore, biomass is considered the most important source of green carbon or carbon-neutral fuel. In order to decide the true carbon neutrality, the overall biomass chain needs to be considered, including cultivation (for energy crops), harvesting, drying, storage, transportation and processing. These represent a significant cost, energy needs and CO_2 emissions sources [7]. All these factors must be taken into consideration in life-cycle analysis for sustainability. Biomass plays an integral part in the overall sustainable energy solution, but it is not a panacea. Biomass for biochar production makes land usage more complex. The effect is not only the land usage for biomass supply but also the impacts of adding biochar to soils. The impacts may include increased productivity and, hence, reducing the land area required for food production as well as the potential for biochar to make previously unmanaged or marginal land economically productive, thereby facilitating the conversion of marginal land to agriculture [15, 20].

Besides heat and electricity generation, biomass can be used as a feedstock for biofuel production with technologies already available on the market. Biofuels are liquid or gaseous fuels produced from biomass and can be used as a replacement or blended with fossil-based fuels for different applications. This makes biomass very valuable within future energy systems based on renewable/sustainable sources of energy. The biomass potential for the energy markets needs to be evaluated without affecting the demands for food and fodder as well as for raw materials. The use of biomass has been debated critically on the background of the ongoing environmental and sustainability discussions.

2.1 Biomass classification

Biomass can be classified into different groups depending upon the origin where it is produced, including agricultural biomass, forestry and wood processing residues, dedicated energy crops (crops cultivated solely for energy), aquatic biomass, sewage sludge, digestate (remains of anaerobic digestion), industrial crops, animal, industrial, municipal and food waste. Biomass is also classified based on the chemical composition as carbohydrates, lignin, essential oils, vegetable oils, animal fats and natural resins (gums) [21–24].

Agricultural biomasses are natural products of all agriculture. These include a wide range of agricultural crop residues (the non-food based portion of crops) that are not harvested for commercial use or byproducts from harvesting or processing, such as corn stover (leaves, stalks, husks and corn cobs left in a field after harvest), sugarcane bagasse, straw residues (barley, oats, rice, rye, wheat) from grain production, waste from other food crops, horticulture and food processing. Other plant residues include husks of grains and seeds, coconut shells, fruit stones and nut shells. Straw and other agricultural residues usually have a high ash content and contain chlorides and potassium compounds, which can cause high levels of corrosion in boilers. The problems of corrosion and slagging can be mitigated by burning biomass at lower temperatures [1, 18].

Forestry and wood processing residues include trees not harvested during logging (trees that are not valuable as timber, such as imperfect commercial trees, dead wood and other non-commercial trees), biomass removed during logging (such as crowns and branches from fully-grown trees) in commercial forests, waste from forest and wood processing (such as palm kernel shells, wood pellets, woodchips, leaves, barks, lumps and sawdust) as well as materials removed during forest management operations (such as trunks of smaller trees removed during thinning, dead and dying trees removed during forest control) [1, 25].

Dedicated energy crops are another expanding and potentially larger source of biomass. Energy crops are low maintenance and high yield crop species that give the maximum energy yield. These are grown specifically for their fuel value (energy applications) on marginal land unsuitable for agriculture. Several crops can be readily used as energy sources. There are two types of energy crops, herbaceous and short-rotation woody. Herbaceous energy crops include perennials that are harvested annually after reaching maturity. It takes 2–3 years to reach complete production. These are grasses such as switchgrass, miscanthus, bluestem, elephant grass, bamboo and wheatgrass. They do not require replanting for 15 years or more. The drawback with most non-woody energy crops is that their chemical properties generally make them less suitable for combustion due to the high ash and salt content [18, 26, 27]. The woody crops are grown on short rotations, generally with more intensive management than timber plantations. These fast-growing hardwood trees are harvested within 5–8 years of planting. These crops include poplar, willow, maple, cottonwood, black walnut and sweetgum [25, 27].

Different kinds of algae, plants and microbes found in water form another class of biomass called aquatic biomass. Aquatic biomasses include macroalgae, microalgae, seaweed, kelp, water hyacinth and aquatic plants [24]. Animal, industrial, municipal and food waste and sewage sludge are other important sources of biomass. Animal and human waste biomass includes waste resulting from farm and processing operations, manure of different animals, cooked or uncooked food, fruits, paper and pulps. When these waste materials are treated and converted to useful energy products, not only energy is being produced, but the problem of disposing of these materials is also reduced to a certain extent. Industrial waste involves waste from various manufacturing and industrial processes like paper sludge from paper industry, sugar cane residues from sugar mills, waste from food processing industry, waste oils, textile industry waste and others. Animal and human waste biomass and industrial biomass are categorized differently because industrial biomass may contain different types of toxic chemicals and harmful additives. In contrast, animal and human waste are primarily free of these types of harmful materials [28]. Municipal solid waste (MSW) includes waste from residential, commercial and industrial sectors that contains a significant amount of biomass (such as paper, cardboard, wood, food, leather, textiles and yard trimmings) with energy content. Food waste contains residues from food and drinks manufacture, preparation and processing, post-consumer waste, animal fat, used cooking oil etc. Other biomasses include Industrial crops (crops developed to produce specific chemicals or feedstocks such as kenaf), construction and demolition waste, building material waste, abandoned furniture etc. [18, 25].

2.2 Composition of biomass

The chemical composition of biomass is different from fossil fuels. The most abundant biomass on the earth is LCB, including agricultural biomass, forestry and wood processing residues, dedicated energy crops, industrial crops and food waste, hence, the following sections primarily focus on LCB. Plant biomass is a complex mixture of polymers consisting of three key elements: 42–47% of carbon (C), 40–44% of oxygen (O) and 6% of hydrogen (H), all percentages in dry matter, whose total content generally reaches above 95% [24, 29]. Plant biomass also contains macronutrients such as nitrogen (N), phosphorus (P), potassium (K), magnesium (Mg), sulfur (S), calcium (Ca). These inorganics are required in relatively high amounts (> 0.1% of dry mass) and essential for plant life cycle and biomass production. In addition, plants need a small amount of micronutrients (essential elements required in relatively small amounts, 100 mg/kg of dry mass) such as chlorine (Cl), iron (Fe), boron (B), manganese (Mn), zinc (Zn), copper (Cu), molybdenum (Mo), nickel (Ni). Trace elements like sodium (Na), silicon (Si), selenium (Se), titanium (Ti), vanadium (V), cobalt (Co), aluminum (Al) and other heavy metals may also be present in plant biomass at different levels depending upon the plant species and the environment [24, 30–32].

Plant biomass has a carbon-to-oxygen (C/O) ratio of almost one. Because of this high oxygen level, the energy density of biomass is relatively low relative to fossil fuels. The major component of plant biomass is cellular lignocellulosic material, which is the non-starch fibrous part of the plant materials. Cellulose, hemicellulose and lignin are the three major constituents of LCB constituting the cell wall of plants [5, 15, 21]. The main component of the plant cell wall is cellulose (a linear homogeneous structural polysaccharide composed of D-glucose units with molecular weight (MW) > 100,000), which provides structural support. The second most abundant polymer in LCB is hemicellulose, a ramified heterogeneous structural polysaccharide composed of D-xylose, L-arabinose, D-mannose, D-galactose and D-glucose units. The third most abundant polymer in LCB is lignin, a phenylpropanoid polymer composed of guaiacyl, p-hydroxyphenyl and syringyl units [33, 34]. The compositions (cellulose, hemicellulose and lignin) of common LCB are listed in **Table 1**. Cellulose macromolecules form tough microfibers that function as the skeleton material of the cell wall. The inner space is packed with branched amorphous hemicellulose and lignin linking material. Cellulose connects with hemicellulose or lignin mainly through hydrogen bonds, while hemicellulose connects with lignin via both hydrogen and covalent bonds. Lignins and carbohydrates link tightly together in lignin-carbohydrate complexes, which results in residual carbohydrate or lignin fragments in extracted lignin or hemicellulose samples. Wooden biomasses are usually rich in cellulose, leaves and grasses are rich in hemicellulose and the shells are mostly rich in lignin. Cellulose is thermally more stable than hemicellulose. Knowledge of cellulose, hemicellulose and lignin composition in LCB can be helpful in controlling the product chemistry [32, 37].

In addition to the three major components, some other compounds present in LCB include inorganic compounds and organic extractives. These exist as non-structural components that do not constitute the cell walls or cell layers. Organic extractives can be extracted by nonpolar solvents (such as toluene and hexane) or polar solvents (such as water and alcohol). These include fats, waxes, proteins, terpenes,

Lignocellulosic material	Cellulose %	Hemicellulose %	Lignin %
Hardwood stems	40–55	24–40	18–25
Softwood stems	45–50	25–35	25–35
Almond shells	29.0–31.1	28.0–38.0	27.7–35
Corncobs	45	35	15
Grasses	25–40	35–50	10–30
Wheat straw	30	50	15
Flax straw	36.70	34.40	28.90
Leaves	15–32	80–85	0
Sugar cane bagasse	32–44	27–32	19–24
Coastal Bermuda grass	25	35.7	6.4
Switch grass	45	31.4	12

Table 1.
Composition of common LCB [35, 36].

simple sugars, gums, resins, starches, alkaloids, phenolics, pectins, glycosides, mucilages, saponins and essential oils. Often these compounds are responsible for the smell, color, flavor and natural resistance to rotting of some species. A common classification divides them into aliphatic compounds (mainly fats and waxes), terpenes and terpenoids, and phenolic compounds. The inorganic compounds constitute less than 10% by weight of LCB, forms ash in the pyrolysis process [21, 32].

LCB contains varying amounts of inorganic materials (including alkali and heavy metals, chlorine, phosphorus and sulfur) collectively called ash. The ash contents in LCB depend on feedstock type, the environment in which it was grown, fertilizer use, and contamination with soil particles. Typically, softwood and hardwood have ash contents below 1 wt%, short-rotation woody crops have around 2 wt%. Herbaceous crops have high levels of potassium and silicon and ash contents up to 7 wt%. Ash content is also not uniformly distributed within biomass; bark has higher concentrations of inorganics [38]. Water in wet biomass contains in three phases: bound water (hygroscopic or adsorbed, in cell walls, believed to be hydrogen-bonded to the OH groups of primarily cellulose and hemicelluloses of the biomass); free or unbound water (liquid water in cell cavities or voids of the biomass if the moisture content is higher than the fiber saturation point); and water vapor which fills the cell cavities or voids of the biomass [23]. Depending on the type of LCB, the cellulose, hemicellulose and lignin content fall in the range 40–60%, 15–30% and 10–25%, respectively. Both cellulose and hemicellulose are carbohydrates (polymers of sugars) and can be hydrolyzed into fermentable sugars, which in turn can be converted into fuels and chemicals. Wood biomass contains much higher amounts of the three main components (~90%), while agricultural and herbaceous biomass contains more extractives and ash [21, 32].

2.2.1 Cellulose

Cellulose is the most abundant organic polymer on the planet. It is one of the main structural constituents of the lignocellulose cell wall of green plants and is found in an organized fibrous structure. Cellulose is a polysaccharide consisting of d-glucose (pyranose) units linked by β-1,4 glycosidic bonds. The *β*-linkages in cellulose form linear chains. The degree of polymerization (DP) is about 300–15,000, depending on the plant variety. Cellobiose is the repeating glucose disaccharide of cellulose. Because of its long and linear molecules, cellulose does not dissolve readily in water. The chemical formula of cellulose can be written as $(C_6H_{10}O_5)_n$, where n is the DP. It is the long flexible natural polymer in fibers that predominately gives trees and wood of their strength. Cellulose chains are grouped to form cellulose fibers, which are interlinked by hydrogen bonds and van der Waals forces, resulting in long microfibrils. These microfibrils are arranged as a mesh in the cell wall, giving it strength and shape. Hemicelluloses and lignin cover the cellulose microfibrils. Cellulose is highly stable and resistant to chemical attack because of the high degree of hydrogen bonding between cellulose chains [15, 32, 37, 39, 40].

The hydroxyl (OH) groups present on the inner and outer surfaces of cellulose forms intra- and intermolecular hydrogen bonds, which stiffens the chains and promotes aggregation into a crystalline structure. The crystallinity and stabilization of cellulose mainly originate from the presence of OH groups. As three OH groups are available in each glucose molecule, the inner and outer surfaces of cellulose are covered by OH groups. These OH groups make hydrogen bonds with other OH groups and other groups (such as O, N and S) available in lignocelluloses. The crystalline structure of cellulose leads to chemical stability and provides strength and toughness to the roots and stems of a plant. Cellulose molecules have different orientations throughout the structure resulting in different levels of crystallinity. The energy of hydrogen bonds in water and cellulose is 15 and 28 kJ/mol, respectively and the energy of van der Waals in water is only 0.15 kJ/mol. The strength of cellulose mainly originates from the existence of hydrogen bonds rather than van der Waals forces. The interchain hydrogen bonds introduce order (crystalline) or disorder (amorphous) into cellulose structure, creating two forms of cellulose: crystalline and amorphous. Cellulose requires severe hydrolysis conditions for breaking it into simple glucose units due to its crystalline structure. As suggested by some authors, cellulose consists of three regions: true crystal, subcrystalline (disordered structure in true crystal regions) and or noncrystalline (subscrystalline) regions. The crystallinity index (CrI) usually characterizes the crystallinity of cellulose, increasing CrI leads to decreasing chemical and biological hydrolysis of cellulose [41, 42].

2.2.2 Hemicellulose

Hemicellulose is the second most abundant natural organic polymer after cellulose on the planet. In contrast to the linear or one-dimensional structure of cellulose, hemicellulose is a two-dimensional polymer composed of short chain branched heteropolysaccharides side connections. It is one of the major constituents of plant cell walls and is strongly linked to the surface of cellulose microfibrils. Hemicellulose is a random heterogeneous polysaccharide of pentoses (xylose and arabinose), hexoses (galactose, glucose and galactose) and their acidified derivatives such as glucuronic and galacturonic acids. Because of the branched nature, hemicellulose is amorphous, which is relatively easy to hydrolyze (by dilute acids, bases and hemicellulose enzymes) to its constituent sugars compared to cellulose. The content and structure of hemicellulose differ among LCB. The general nature of the hemicellulose structure depends on the type of plant, with the result that certain types of lignocellulosic materials are easier to hydrolyze than others. The various sugar units are arranged with different substituents and in different proportions. Hemicellulose has a degree of polymerization of 80–200. It is much smaller than cellulose, with a relatively low MW (< 30,000). The general chemical formula of hemicellulose can be written as $(C_6H_8O_4)_m$, where m is the DP. Lateral chains of hemicellulose form the tightly bound network through hydrogen bonds with cellulose microfibrils. It makes a highly rigid matrix of the cellulose-hemicellulose-lignin with interaction of lignin via covalent bonds [15, 32, 37, 39, 43, 44].

The different groups of polysaccharide molecules such as xylans, mannans, galactans and arabinogalactans make up hemicellulose. Xylan is the most common polysaccharide in hemicellulose consists of backbone chains that contain a varying number of D-xylopyranose linked by β –1,4 linkage (70–130 in softwood xylan and 150–200 in hardwood xylan). Mannans are made up of β–1, 4-linked D-mannose backbone mixed with D-glucose and D-galactose residues. These compounds include mannan (made up of mannose monomer), galactomannan (made up of mannose and galactose monomers), glucomannan (made up of mannose and glucose monomers), glucuronic acid (made up of mannose and glucuronic monomers). Galactan is composed of repeating galactose units as a polymer. Arabinogalactans consist of arabinose and galactose monosaccharides. The dominant hemicellulose component in hardwood, agricultural residues and herbaceous crops is xylan, with a small degree of acetylation and arabinose side groups. The main form of hemicellulose in softwood is glucomannan, highly acetylated and containing glucose and mannose [15, 32, 37, 44].

2.2.3 Lignin

After cellulose and hemicellulose, lignin is the third largest heteropolymer that occurs predominantly in the cell walls of woody plants. It is the main non-polysaccharide constituent of plant biomass and the amount of lignin vary widely with plant species. Lignin primarily consists of macromolecules that contain highly branched phenolic compounds. Lignin is composed of three different phenyl propane (three-carbon chain attached to rings of six carbon atoms) monomers, including coniferyl alcohol (guaiacyl propanol), p-coumaryl alcohol (p-hydroxy-phenyl propanol) and syringyl alcohol (sinapyl alcohol). The phenyl propane monomeric units in lignin are linked in different ways (alkyl-aryl, alkyl-alkyl and aryl-aryl ether bonds): through oxygen bridges between two propyl and phenyl groups, between a phenyl and a propyl group or through carbon–carbon bonds between the same groups. Lignin is generally considered as the natural phenolic glue that tightly binds cellulose and hemicellulose of LCB together; thus, leading to a strong cell wall structure and making it insoluble in water. The functions of lignin, an amorphous and highly complex aromatic hydrophobic biopolymer, are (a) to provide mechanical strength to the plants. It plays a cementing role for linkages (van der Waals, hydrogen bond and covalent bond) between cellulose and hemicellulose to form a 3-dimensional structure of lignin-polysaccharide complex in cell wall leading to a strong cell wall structure; (b) to provide sealing for water-conducting system linking roots with leaves. Polysaccharide components of the plant cell wall are hydrophilic and permeable, while lignin is hydrophobic and impermeable. The cross-linking between polysaccharides and lignin is a barrier for water absorption to the cell wall that create vascular tissues for efficient conduction of water in plants. Lignin exists in all vascular plants; and (c) to protect plants against biodegradation. It forms a natural protective shield protecting cellulose and hemicellulose in plants and makes plants resistant to pathogens, oxidative stresses and biodegradation by enzymes and microorganisms [42, 45–47].

The distinctive feature that differentiates lignin from cellulose and hemicelluloses is the presence of aromatic monomers. Lignin is less polar than cellulose or hemicellulose. Physically, cellulose microfibrils encrust hemicellulose whose empty spaces are filled up with lignin. Lignin is embedded within hemicellulose to provide additional rigidity to the plant. These lignin-hemicellulose fibers characterize woody plants, whereas the fibers in herbaceous plants are more loosely bound, indicating a lower amount of lignin [15, 35, 47]. The lignin content of plants varies with species and age. It is originated from not only content but also monomeric units and linkage types. The lignin content of softwood is in the range of 25–40%, which is higher than that of hardwood (18–25%), herbaceous crops (10–20%) and annual plants (10–12%). LCB with lignin percentage up to 40% has been reported. Softwood lignin is primarily made from coniferyl alcohol (>95%), with the rest consisting of coumaryl alcohol derived units and trace amounts of syringyl alcohol derived units. Hardwood lignin is composed of coniferyl alcohol and syringyl alcohol derived units in varying ratios. Lignin in grassy biomass has all three types of monomers. Annual plants lignin is composed of coumaryl alcohol. The elemental composition of lignin is approximately 61–65% carbon, 5–6% hydrogen and the remaining is oxygen. The carbon to oxygen (C/O) atom ratio for lignin is higher than 2:1, which is at least double that of hemicellulose and cellulose, where the C/O ratio is nearly 1:1. Therefore, lignin is a more energy-dense substance than polysaccharides. Lignin structure has many polar and hydroxyl groups allowing the establishment of strong intramolecular and intermolecular hydrogen bonds. These make lignin insoluble in any solvents except alkaline solutions [23, 37, 42, 44, 48].

3. Analysis of biomass

Analysis of biomass feedstock is an integral part of understanding the behavior of biomass in energy use. The proximate analysis, ultimate analysis and higher heating value (HHV) of biomass can provide a clear understanding of its thermochemical conversion characteristics.

3.1 Proximate analysis

The proximate analysis provides information on the biomass in terms of volatile matter (VM), ash content, fixed carbon (FC) and moisture (M). VM of biomass is the condensable and non-condensable gases released from the biomass during heating. The amount of VM depends on the heating rate and the final temperature to which biomass is heated. Ash is the solid residue left after the biomass is completely burned. FC shows the percentage of biomass burned in the solid states, while VM indicates the percentage of biomass burned in the gaseous state. The ash content indicates the quantity of non-combustible ash remaining on the fire grates or ash pit or entrained with flue gases. These are of fundamental importance for biomass energy use and such information for specific LCB are depicted in **Table 2**. Such data provides the furnace designer with essential information for the furnace design, including sizing and location of primary and secondary air supplies, refractory, ash removal and exhaust handling equipment etc. [49, 50].

The composition of ash depends on the type of biomass which includes mostly inorganic residues such as silica, aluminum, iron, calcium and small amounts of magnesium, titanium, sodium and potassium may also be present. Even though ash content of biomass is usually very small, it may play a significant role in biomass combustion or gasification if biomass contains alkali metals (such as potassium) or halides (such as chlorine). Straw, other agricultural residues and grasses generally contain potassium compounds and chlorides are particularly susceptible to this problem and can cause severe agglomeration, fouling and corrosion in boilers or gasifiers. The ash obtained during biomass conversion does not necessarily come from biomass itself but also from other sources such as contamination as well. Biomass can pick up a considerable amount of dirt, soil, rock and other impurities during collection and handling. These also partly contribute to ash content. FC is

Lignocellulosic Material	VM	FC	Ash	Moisture
Hardwood	72.3	25	2.7	7.8
Softwood	70	28.1	1.7	8.8
Almond shells	79.7	4.9	2.3	8.7
Corncobs	80.6	18.2	1.2	9.7
Reed canary grass	82.5	12.1	5.5	4.7
Wheat straw	63	23.5	5.5–13.5	8.5
Flax straw	80.3	8.8	3.0	7.9
Hazelnut seedcoat	71.20	27.0	1.8	6.8
Sugar cane bagasse	84.00	1.64	4.5–9	8.5
Elephant grass	65	14.66	6.0	10

Table 2.
Proximate analysis (wt%) of common LCB [35].

the solid carbon (non-volatile) in the biomass that remains in the char in the pyrolysis process after devolatilization. The amount of FC is related to VM, moisture (M) and ash by the equation: $FC = 1 - M - VM - ASH$ [18, 49].

The relationship between FC and charcoal yield in biomass is positive, while VM and ash relate negatively to charcoal yield. It is expected that the greater biomass VM lead to greater gas production instead of the solid phase. Moisture content will have a significant impact on the biomass conversion process. High moisture content is a major concern in biomass. Biochemical conversion processes can use biomass with high moisture content, while thermochemical conversion processes generally require biomass with low moisture content. However, gasification processes require some moisture to produce hydrogen and the amount of hydrogen produced will increase with moisture content. The moisture content of some biomass, such as water hyacinth, can be very high (> 90%). As the energy used in evaporation is not recovered, moisture drains much of the deliverable energy during conversion [49–51].

3.2 Ultimate analysis

The ultimate analysis provides the composition of biomass, including major elements (C, H, O, S, N and Cl), moisture and ash on a gravimetric basis. The ultimate analysis of common LCB are listed in **Table 3**. The ultimate analysis is generally reported on a dry and ash-free basis. These are useful in understanding biomass processes and performing mass balances on biomass conversion processes. Elemental chemical composition, moisture, ash and volatiles are essential for thermal/ thermochemical conversions of biomass. Additional information on the polymeric composition of biomass is important for conversions like torrefaction, pyrolysis and gasification. A typical ultimate analysis of the biomass in terms of its basic elements, moisture (M) and inorganic constituents (ASH) can be written as: $C + H + O + N + S + ASH + M = 100\%$. The hydrogen or oxygen content in the ultimate analysis includes only the hydrogen and oxygen present in the organic components of the biomass it does not include the hydrogen and oxygen in the moisture. The moisture in the biomass is expressed separately as M. The ultimate analysis is useful in calculating the quantity of combustion air required to sustain the combustion reactions. Generally, sulfur and nitrogen content of biomass is very low. As a result, biomass produces minimal SO_X and NO_X pollutants. Particulate emissions of unburned carbon in the flue gases can present pollution problems [49, 51].

Lignocellulosic Material	C	H	N	S	O
Hardwood stems	48.6	6.2	0.4	—	41.1
Softwood stems	52.1	6.1	0.2	—	41.0
Almond shells	54.7	7.5	0.3	0.3	37.4
Corncobs	43.6	5.8	0.7	1.3	48.6
Grasses	48.3	5.7	0.8	—	39.4
Wheat straw	53.9	7.0	3.0	—	36.1
Flax straw	43.10	6.20	0.68	0.09	49.90
Hazelnut seedcoat	52.90	5.6	1.4		42.7
Sugar cane bagasse	45.13	6.05	0.3	—	42.77
Switch grass	48.6	5.5	0.5		39.5

Table 3.
Ultimate analysis (wt%) of common LCB [35].

4. Biomass conversion

Biomass conversion is the key step to produce heat, biofuels or chemicals from LCB. Biomass can be converted to these products through chemical, biochemical and thermochemical conversion processes. Selection of the conversion process depends on number of factors, including the desired form of end products (heat, biofuels or chemicals), biomass feedstock (type, quantity and characteristics) available, environmental standards, policy, economic conditions and specific factors related to the project. In most situations, the selection is based on two main factors, i.e., desired form of end products and the available feedstocks [5, 21, 52]. The moisture content of biomass primarily determines the conversion process for the selected biomass. Thermochemical conversions like pyrolysis, gasification or combustion are ideal for dry biomasses such as wood or straw. These processes are often unsuitable for wet biomass because of low energy density due to high moisture content. The high moisture content significantly increases the transportation cost and energy cost for drying. Wet conversion processes such as hydrothermal processing and biochemical processing (fermentation and anaerobic digestion) have gained growing attention and are more suitable to process high moisture content biomass like aquatic biomasses, sewage sludge, food waste and manures. Despite consuming less energy, biochemical conversion processes require more time compared to thermochemical conversion processes. Subsequently, much attention has been placed on hydrothermal processing, which is more cost-effective than conventional thermal drying followed by thermochemical conversion. The additional parameters (like cost and feasibility of drying etc.) need to be considered when selecting the correct conversion process, primarily if the moisture content lies between wet and dry regions [6, 21, 53, 54]. Biomass has substantial quantities of free and bound water. Wet biomass requires drying before pyrolysis, gasification or combustion; hence, additional energy and facilities are needed. Drying biomass outside the pyrolysis reactor is beneficial (produce pyrolysis vapor with high calorific value and bio-oil with low water content) [23].

The important biochemical conversion processes include anaerobic/aerobic digestion, fermentation and enzymatic or acid hydrolysis. In biochemical conversion, biomass molecules are broken down into smaller molecules by bacteria or enzymes. This process is much slower than the thermochemical conversion process but does not require much external energy. In anaerobic digestion, bacteria take oxygen from the biomass itself instead of atmospheric oxygen. The products of anaerobic digestion are biogas (a mixture of methane, carbon dioxide) and solid digestate. Only 5–10% of the feed into the digester is digested by the anaerobic bacteria. The digestate consists of remaining indigestible material. Aerobic digestion, commonly known as composting, takes place in the presence of oxygen. It uses different types of microorganisms that access oxygen from the air, producing carbon dioxide, heat and solid residue (compost). In fermentation, starch is converted into sugars using acids or enzymes. Then sugar is converted into ethanol or other chemicals with the help of yeast. The fermentation of lignocellulosic feedstock requires additional pretreatment (hydrolysis) to breakdown the cellulose and hemicellulose into simple sugars. Hydrolysis can be achieved by the use of acids, enzymes or hydrothermally [5, 21, 52].

The lignin is not converted and is left for thermochemical conversion. Major thermochemical conversion processes include combustion, gasification, pyrolysis, torrefaction and hydrothermal processing. These are further discussed in the following sections. Thermochemical conversions generally offer many advantages over biochemical conversions, such as handling a wide variety of feedstocks, better conversion efficiency, high energy efficiency and shorter reaction times. As a result, in recent years, thermochemical conversions have received greater attention for biofuels production [3, 21].

5. Thermochemical conversion

There are many types of thermochemical conversion processes through which biomass is converted to solid, liquid and gaseous products. Thermochemical conversion processes use high temperatures to breakdown the bonds of organic matter. Thermochemical conversion routes can be classified according to the oxygen content used in the process, including combustion (complete oxidation), gasification (partial oxidation) and pyrolysis (thermal degradation in the absence of oxygen). Torrefaction is also performed in the absence of oxygen. Hydrothermal processing is an alternative route to process wet biomass using heat and pressure in the presence of water, which can also be considered a thermal degradation in the absence of oxygen. The typical products of thermochemical conversion of biomass are carbon-rich solid residue (biochar), condensable vapors (bio-oil or tar) and non-condensable gases. The distribution of products (biochar, bio-oil/tar and gases) depends primarily on the conversion process [21]. A brief description of biochar, bio-oil and gases are given below.

Biochar - Biochar is a porous carbonaceous material with a high degree of aromatization and strong antidecomposition ability. The physical, chemical and mechanical properties of biochar will depend on the feedstock material characteristics and pyrolysis conditions used for the production of biochar. It has a wide range of potential applications in various agronomic and industrial sectors. Biochar is used in agriculture to upgrade the soil quality, in waste treatment to remove organic contaminants, heavy metals and different types of dyes and pigments from textile industries and in power generation as a fuel. The most successful approach for high-yield biochar production is via slow pyrolysis [21].

Bio-oil - Bio-oil is a dark brown, free-flowing organic liquid mixture. It generally comprises of 15–35 wt% water (resulting from both the original moisture and as a pyrolysis product) and a mixture of organic compounds, such as acids, alcohols, ketones, aldehydes, phenols, ethers, esters, sugars, furans, alkenes, nitrogen compounds, miscellaneous oxygenates and solid particles. The final water content of bio-oils depends on the initial moisture content of biomass feedstock and water formation during pyrolysis. Water cannot be removed from bio-oil by conventional methods like distillation. Bio-oil has a low (15–20 MJ/kg) HHV in comparison to conventional petroleum fuel HHV of 42–45 MJ/kg due to the increased oxygen content (35–40 wt% on a dry basis). Bio-oil density is approximately 1200 kg/m^3; the viscosity ranges from 25 to 1000 cP (depending on the composition). It is acidic in nature (pH value of 2–4) due to the presence of organic acids such as formic and acetic acid and, hence, corrosive. Bio-oil has a large amount of oxygenated compounds and organic material; it is highly polar. As a result, bio-oil is hydrophilic. A distinct aqueous phase is only observed with bio-oil having water content in the range 30–45 wt% [23, 55]. Bio-oil will not mix with hydrocarbon liquids. Bio-oil has a complex mixture of oxygenated compounds that provide the potential and challenge for its utilization. It has a range of uses in energy applications, can be used in boilers for heat and power generation, cofired with natural gas/coal in power plants or blended with other fuels such as ethanol or gasoline. Bio-oil can be converted into fuels (ethanol and diesel) and chemicals through hydrocracking/hydroprocessing [56].

Non-condensable gases – Gases produced in biomass pyrolysis may consist of carbon dioxide (CO_2), carbon monoxide (CO), hydrogen (H_2), methane (CH_4), ethane (C_2H_6) and ethylene (C_2H_4), and small amounts of other gases, such as propane (C_3H_8), ammonia (NH_3), nitrogen oxides (NO_X), sulfur oxides (SO_X) and alcohols of low carbon numbers. The composition of the non-condensable gases will be determined by the pyrolysis temperature and the vapors condensing temperature. Lower pyrolysis temperatures (such as torrefaction) result in higher amounts of CO and CO_2, while higher pyrolysis temperatures result in increased content of CH_4 and H_2.

5.1 Combustion

Combustion is simply the burning of biomass in air. Chemically it is high-temperature exothermic oxidation of biomass in the presence of oxygen. Complete combustion of biomass involves the production of heat due to the oxidation of carbon and hydrogen of biomass to CO_2 and H_2O, respectively. The process consists of consecutive heterogeneous and homogeneous reactions. Biomass combustion basically depends on the properties of the feedstock and particle size, temperature and combustion atmosphere. Char (contains some organic carbon) and ash (typically includes inorganic oxides and carbonates) are the solid byproducts of combustion. Combustion temperatures are usually in the range of 700–1400°C [52, 57].

Energy stored in biomass can be converted into heat and power via combustion. The chemical composition and the combustion properties of biomass vary considerably depending on the biomass type. A wide range of biomass sources can be considered for combustion. Seasonal, regional variances and parts of the plant (bark, branches and leaves etc.) of the woody biomass (wood chips, wood pellets and waste woods etc.) can result in differences in the chemical composition of the feedstock. Straw is also considered as having potential as an alternative feedstock. Straw is essentially a waste product from agricultural crop production. This feedstock does not compete with agricultural products for the limited land resources. Besides wood and straw, a wide variety of waste products such as rice husks, wheat bran, peanut shells, coffee grounds, bagasse, etc., can be used as feedstock. These can be used as an inexpensive fuel to produce heat or electricity needed for industrial processes. The best quality fuels contain high amounts of carbon and hydrogen and low amounts of other elements (oxygen, nitrogen, sulfur and trace elements). Biomass usually contains higher levels of oxygen than fossil fuels. Impurities such as sulfur and nitrogen are associated with the emission of SO_X and NO_X. Trace elements such as potassium and sodium can cause fouling; chlorine leads to corrosion and silica causes excessive wear to milling equipment [52, 57].

Fresh woodchips can contain 50% moisture and leaves can be over 90% moisture. Most furnaces and boilers recommended biomass with less than 20% moisture. It is extremely difficult to maintain combustion with a moisture content of more than 55%. Higher levels of moisture affect combustion efficiency and increase the amount of smoke emitted. The water content also increases the combustion time of a biomass particle and, thus, extends the required residence time in a boiler/furnace. Torrefaction upgrades biomass by removing lighter volatiles and moisture. It improves the heating value of biomass, increases the hydrophobicity and stability and, thus, can be stored under the open sky [52, 57].

Combustion can be split into four stages: drying, pyrolysis (devolatilization), volatiles combustion and char combustion. When biomass particles enter a hot environment, moisture in the particles starts to evaporate. On further heating, volatile gases and tars are released from biomass particles, followed by the combustion of volatiles. The remaining char will essentially retain its original shape. In the char combustion stage, char reacts with oxygen to form mainly CO_2 (and CO due to incomplete combustion) and ash remains after combustion is completed. Detailed chemical reactions kinetics that takes place during biomass combustion are complex [52, 57, 58].

The initial combustion stage requires heat to evaporate moisture; hence, it is necessary to have biomass with minimal moisture content. Biomass has a significantly higher volatile matter content compared to coal and the fixed carbon to volatile matter (FC/VM) ratio is significantly low. Lower values of the FC/VM ratio leads to high ignition behavior. Biomass releases VM at a lower temperature and more rapidly than coal, thus, reducing ignition temperature compared to coal. Proper design of the air supply is important due to the faster release of VM in order not to delay combustion. Combustion of VM is fast compared to combustion of solid charcoal and a low ratio of FC/VM decreases the residence time in the boiler/furnace [52, 57, 58].

Incomplete combustion results in the formation of intermediates, including pollutants such as CO, CH_4 and particulate matter (PM). Ash handling, high emissions of NO_X, SO_X, CO_2 and particulate matter make combustion environmentally challenging. Biomass is more corrosive, tends to foul heating surfaces, ash from biomass tends to agglomerate. The boilers have to be redesigned to burn biomass properly. Depending on the condition and combustion properties of the biomass to be burned, different furnace designs and combustion parameters can be selected to ensure optimum efficiency. Direct combustion is currently the principal method of generating electricity around the world [52, 57, 58].

5.2 Gasification

Biomass gasification is a thermochemical process of converting solid biomass into a gaseous fuel known as synthesis gas or producer gas under a reduced oxygen atmosphere to avoid complete combustion [59, 60]. Gasification aims to maximize the conversion of biomass feed into useable gases. In the gasifier, feed is exposed to a high temperature atmosphere, which heats the biomass leading to thermal decomposition. In contrast to pyrolysis, the feed is brought into contact with a gasifying agent (air). At the gasifier temperature, reactions between oxygen and carbon take place. A mixture of many gases, primarily carbon monoxide and hydrogen, is released as the output product of the gasification process. The gas contains various percentages of CO, H_2, CH_4, CO_2, H_2O and N_2 depending on the quality of the biomass used and the way gasification is conducted. It also produces liquids (oils, tars, and other condensates) and solids (char, ash) from solid biomass feedstocks [5, 61, 62]. A simple way of representing the gasification reaction is shown below

$$Biomass + O_2(g) \rightarrow CO(g) + H_2(g) + CO_2(g) + CH_4(g) + Tar(l) + H_2O(l) + Char(s) + TraceSpecies \quad (1)$$

Gasification processes are designed to generate fuel or synthesis gases as the primary product that can be used in internal and external combustion engines as well as fuel cells, offering a viable solution to overcome energy demands. Currently, such syngas is used as fuel to generate heat and electricity or as a feedstock for many products in the petrochemical and refinery industries, like methanol, ammonia, synthetic gasoline, etc. The overall gasification process is endothermic, requiring energy input for the reactions to proceed, most of which operate between 600°C and 1500°C [59, 63]. The energy needed for this endothermic reaction is obtained by oxidation of part of the biomass through a direct heating (autothermal) or an indirect heating (allothermal) phase. The main operating parameters of gasification include type and design of gasifier, gasification temperature, flow rates of biomass and oxidizing agents (air or steam), type and amount of catalysts, and biomass type and properties [64]. In addition to the operating conditions of the gasifiers, the properties of biomass such as size, shape, density, chemical composition, energy content and moisture content affect biomass gasification. Gasifier reactors are simple in construction and their designs are generally categorized into the following types: downdraft, updraft, entrained flow, and fluidized bed.

If air is used as the gasifying agent, the producer gas is usually diluted by atmospheric nitrogen. As a result, producer gas has a relatively low calorific value of 4–6 MJ/m^3 (normal cubic meter) compared to the calorific value of natural gas of 39 MJ/m^3. Because of its low calorific value, larger volumes of producer gas are required to achieve a given energy output compared to natural gas. In some more applications, oxygen-enriched air, oxygen or even steam may be used as the gasifying agent, resulting in the production of syngas with higher calorific value in the range of 10–15 MJ/m^3 due to the absence of diluting nitrogen [16].

5.3 Pyrolysis

Pyrolysis is a thermal decomposition process that takes place in the absence of oxygen to convert biomass into three distinct product fractions: solid residue (biochar), condensable vapors resulting liquid product fraction (bio-oil) and non-condensable gaseous products. Once oxygen is removed, combustion cannot occur; instead, pyrolysis happens. Pyrolysis temperatures are usually between 300 and 700°C, depending on the pyrolysis process employed. Pyrolysis is the most promising technique to convert biomass into biochar and bio-oil. Lower pyrolysis temperatures and longer residence times tend to produce more biochar. High temperatures and longer residence times increase the production of gas. Moderate temperatures and short residence times tend to produce more liquids. Higher pyrolysis temperatures tend to produce a higher proportion of aromatic carbon [65, 66].

Depending on the operating conditions (heating rate, solid residence time and temperature), pyrolysis processes are classified as torrefaction, slow (conventional) pyrolysis, intermediate pyrolysis, fast pyrolysis, flash pyrolysis, and microwave pyrolysis. Various operating conditions are used in these processes; residence time can vary from less than 1 second to hours, heating rate can vary from less than 1°C/s to more than 1000°C/s and temperature ranges from 300 to 700°C or higher. As each type of pyrolysis produces different proportions of the three types of products (biochar, bio-oil and gas), careful selection of the pyrolysis process is essential to obtain the final desired product [21].

The primary conversion of biomass during the pyrolysis process can be described by three pathways; char formation, depolymerization and fragmentation. Char formation is generally favored by intra- and intermolecular rearrangement reactions resulting in higher thermal stability of the residue. This pathway is characterized by the formation of benzene rings and the combination of these rings into an aromatic polycyclic structure. These rearrangement reactions are generally accompanied by the release of water or non-condensable gas (devolatilization). Depolymerization is a dominant reaction during the initial stages of pyrolysis, characterized by the breaking of polymer bonds. This occurs when the temperature is sufficiently greater than the activation energies for the bond dissociation. Depolymerization increases the concentration of free radicals. It is followed by stabilization reactions to produce monomer, dimer and trimer units. These volatile molecules are condensable at ambient conditions and found in the liquid fraction. Fragmentation consists of breaking polymer bonds and even monomer bonds result in the formation of non-condensable gases and a range of organic vapors that are condensable at ambient conditions [55, 58, 67].

The decomposition of cellulose, hemicelluloses and lignin releases a mixture of condensable vapors and non-condensable gases. The condensable vapor contains (apart from water vapor) methanol, acetic acid, acetone (all three mainly from hemicellulose), hydroxyacetaldehyde, anhydrous monosaccharides (both mainly from cellulose), phenols and heavier tars (from lignin decomposition). The heavier, water-insoluble tars contain larger molecular fragments obtained after splitting the ether and C-C bonds in lignin. The resulting complex mixture, once condensed, is referred to as bio-oil. The main parameter that determines the degree of devolatilization of the biomass is the pyrolysis temperature. Yang et al. [68] observed great differences among the pyrolysis behaviors of the three components, cellulose, hemicellulose and lignin. These three forms of polymers are responsible for most of the physical and chemical property modification during the pyrolysis process. The mechanisms of pyrolysis of these polymers are chemically different from biomass species to species. Cellulose and hemicellulose decompose over a narrower temperature range, whereas lignin degrades over a wider temperature range than cellulose and hemicellulose. Pyrolysis of lignin is known to produce more biochar than pyrolysis of cellulose and hemicellulose. Biomass pyrolysis consists of three main stages: (a) initial evaporation of moisture, (b) primary decomposition and (c) secondary reactions (oil cracking and repolymerization). At the initial heating stage, when the biomass temperature is increased to about 100°C, the mass of biomass decreases due to the evaporation of free water. Bound water is then removed in heating the biomass to temperatures up to 160°C. At this stage, the heating value of pyrolysis gases is negligible. Thermal decomposition of biomass begins with devolatilization/ decomposition of extractives at temperatures <220°C. Hemicellulose is the least stable polymer and breaks down first at temperatures of 220 to 315°C with maximum mass loss at 268°C [23, 38, 55, 67, 68].

The reactions are endothermic between 180 and 270°C, sometimes becoming exothermic at temperatures above 280°C. The nature of pyrolytic decomposition reactions explains this phenomenon. Devolatilization and decomposition in pyrolysis is not a single step reaction and a difference can be made between primary and secondary reactions. The gas and vapor products of primary conversion are unstable under pyrolysis temperatures and, with sufficient residence time, can undergo secondary reactions such as cracking and/or repolymerization of primary volatile compounds. Cracking reactions consist of the breaking of volatile compounds into lower MW molecules. Repolymerization involves combining volatile compounds into higher MW molecules, which may not be volatile under pyrolysis temperatures. Repolymerization reactions become effective at later stages of pyrolysis, leading to the formation of char. It also results in the formation of secondary char. Primary char can act as a catalyst to the secondary reactions. Primary reactions are highly endothermic, while secondary reactions are exothermic and result in the production of secondary char and non-condensable gases at the expense of volatiles in the vapor phase. Decomposition of vapors to coke and secondary vapors has been suggested as the reason for exothermicity. The extent of secondary decomposition reactions determines the overall exothermicity of the pyrolysis reaction and the overall char (primary and secondary) yield. The occurrence of primary and secondary reactions in the thermal decomposition of biomass highlights the fundamental difference between fast pyrolysis and slow pyrolysis [38, 55, 67].

Cellulose has a high degree of polymerization and exhibits higher thermal stability. It decomposes in the temperature range 315 to 400°C. The secondary reactions continue to occur within the solid matrix with further increasing of the temperature. At temperatures above 400°C, the less volatile components are gradually driven off from solid char residue resulting in higher fixed carbon content and lower volatile carbon content of the solid char residue. As the temperature increases above 600°C, the condensable vapor components undergo cracking and polymerization reactions, resulting in a lower bio-oil yield. Lignin is the most difficult component to pyrolyse, which decompose in a wide temperature range from 160 to 900°C, the rate of lignin degradation reactions is slower than cellulose and hemicellulose [23, 38, 55, 68].

The combination of low heating rate and longer residence time allowed for repolymerization reactions to maximize biochar yield. A low temperature, high heating rate and short gas residence time would be required to maximize bio-oil yield. A high temperature, low heating rate and long gas residence time would be preferred to maximize the gas yield. As a result of high heating rates and short residence times, fast pyrolysis tends to yield higher proportions of bio-oils. In contrast, slow pyrolysis produces higher proportions of biochars because of slow heating rates and longer residence times. Pyrolysis requires relatively dry feedstock (usually moisture content <30 wt%, but moisture contents of ~10 wt% are preferred) and ground to different particle sizes based on the type of pyrolysis. Feedstock with high moisture content consumes more energy to account for increasing heat of vaporization during the heating of biomass towards the pyrolysis temperature. Additionally, the gases and vapors produced in pyrolysis using a high moisture feedstock are diluted with steam and have a lower calorific value [21, 61].

The molar H/C and O/C ratios of LCB are approximately 1.5 and 0.7, respectively. During pyrolysis, the biomass undergoes devolatilization and the solid portion gets enriched in carbon. The H and O are preferably removed over C and the H/C and O/C ratios tend to decrease as biomass undergo its transformation into biochar. The H/C and O/C ratios are used to assess the degree of aromaticity and maturation. Low-temperature chars have high H/C and O/C ratios, the values close to the original biomass. After pyrolysis, a significant decrease in the H/C and O/C atomic ratios is reported and it decreased straightly with increasing pyrolysis temperature. When the pyrolysis temperature is below 500°C, the reduction in H/C and O/C is mainly attributed to major decomposition reactions of biomass, including dehydration (water removal), decarboxylation (CO_2 removal) and decarbonylation (CO removal). Above 500°C, the H/C ratio decreases drastically compared to the O/C ratio, which indicated direct dehydrogenation and demethanation of the chars occurred [38, 69].

5.3.1 Torrefaction

Torrefaction is a pretreatment for upgrading biomass primarily for energy production. Torrefaction, a mild or incomplete form of pyrolysis, involves heating the feedstock to temperatures of 200 to 300°C at slow heating rates (less than 1°C/s) in the absence of air under atmospheric pressure conditions. The residence time depends on the particle size and ranges between several seconds and an hour. Torrefaction removes water as well as superfluous volatiles and partly decomposes the biopolymers (cellulose, hemicelluloses, and lignin) by giving off organic volatiles. It tends to yield higher proportions of solid (torrefied biomass) in addition to liquid and non-condensable gaseous products [21]. Decomposition can be further subdivided into (a) drying, (b) depolymerization and recondensation, (c) limited devolatilization and carbonization, (d) extensive devolatilization and carbonization. Hemicellulose is the least stable of three major polymers, soften between 150 and 200°C and break down via various dehydration, deacetylization and depolymerization reactions at processing temperature range 200–300°C. Predominant hardwood hemicellulose is xylan, while predominant softwood hemicellulose is glucomannan. Xylan tends to break down more quickly than glucomannan at lower temperatures. Therefore, hardwood has a higher mass loss (higher breakdown of hemicellulose) than softwood when treated at the same temperature. This indicates that different biomass species have different torrefaction kinetics. The cellulose and/or lignin degradation during torrefaction is small. A greater degradation has been reported at temperatures above 270°C. Different biomasses have different physical properties such as porosity, specific heat capacity, thermal conductivity, particle size distribution etc., which results in non-homogenous torrefied biomass. A narrower particle size distribution is required for efficient torrefaction and optimized product quality [23, 70].

As specified by the EBC [71], the molar oxygen to carbon (O/C) ratio of biochar should be less than 0.4. But torrefied biomass tends to have higher oxygen to carbon (O/C) ratio than the ECB specification of biochar. Therefore, torrefied biomass cannot be referred to as biochar. Torrefied biomass has physicochemical properties in between that of raw biomass and biochar. Torrefaction is a pretreatment method that is used primarily for moisture removal and densification of biomass, which will reduce the cost of transportation and increase the heating value of biomass. Torrefaction has higher conversion efficiencies compared to slow pyrolysis. Torrefied biomass yield has a maximum of 35 wt% (on a weight basis of dry, ash-free biomass feedstock). A typical mass yield from torrefaction is 70–80% and energy yield is 80–90%. The lower heating value (LHV) of torrefied biomass is about 20.4 MJ/kg compared to the LHV of charcoal between 28 and 33 MJ/kg. Torrefaction increases the hydrophobicity, stability, grindability and reduces biodegradability compared to the untreated biomass feedstock. Torrefied biomass can be stored long-term without degradation [21].

The heating value of torrefied biomass on a weight basis increases compared to its original biomass. Its heating value on a volume basis is not necessarily increased as torrefied biomass has relatively low bulk density. Torrefied biomass can be pelletized or briquetted to account for low bulk energy density. Consequently, bulk energy densities between 14.9 and 18.4 GJ/m^3 can be achieved.

5.3.2 Slow pyrolysis

Slow pyrolysis is characterized by moderate temperatures (300–550°C), slow heating rates (0.1–0.8°C/s) and longer residence time (5–30 min or even 25–35 h) [56]. Slow pyrolysis aims at maximizing the yield of biochar by promoting secondary reactions, which is achieved by longer vapor residence times. The biochar produced in slow pyrolysis consists of both primary and secondary char. The slow heating rate with moderate pyrolysis temperatures also promotes the production of biochar. Biochar yield and physicochemical properties depend on the feedstock properties and pyrolysis conditions such as processing temperature, heating rate and pyrolysis environment. In addition, moisture content and particle size also significantly affect biochar yield [28, 56, 72].

The biomass composition plays a significant role in the resulting biochar yield and the physicochemical nature of the biochar. Biomass cell wall constituents (cellulose, hemicellulose and lignin) behave differently in terms of decomposition and devolatilization in pyrolysis. As lignin decomposes at lower reaction rates and contains aromatics, it is known to contribute to high biochar yields in slow pyrolysis. Consequently, if high biochar yield is required, then lignin-rich biomass feedstocks are preferable. The extractives in biomass will evaporate and end up in the vapor phase or may be cracked, thereby contributing to char and non-condensable gas formation [38]. High ash contents in the biomass affect the slow pyrolysis process and the physicochemical properties of biochar in multiple ways. Alkaline and earth alkaline metals exhibit catalytic activity in pyrolysis. They catalyze secondary reactions of primary vapor components favoring higher yields of non-condensable gases and biochar or they may catalyze different primary decomposition reactions altogether. Biomass containing more minerals yields less biochar [67, 72–74]. Most of the constituents in ash (mainly the alkaline and heavy metals) are non-volatile within the range of temperatures typically employed in slow pyrolysis processes. Thus, ash remains in the solid biochar product, potentially affecting the use of biochar in downstream processes. If biochar (with high ash content) is used in combustion or gasification processes, slagging and equipment fouling occurs. Chlorine and sulfur lead to higher corrosion. Some downstream applications, such as soil amendment, can benefit from higher ash contents in biochar (nutrient recycling). High ash content biochar can be treated with leaching, which involves soaking of biochar in hot water or hot dilute acid. Washing with pure water at 80°C is sufficient to remove about 90% of the potassium found in biochar, produced at 550°C. The leaching process has been successfully applied to treat potassium and chlorine of biochar. Extra dewatering and drying processes are required after leaching in addition to handling and treatment of leachate, which may contain heavy metals extracted from biochar [38].

The process variables that affect the biochar yield and properties include pyrolysis temperature, heat transfer (to and in the biomass), biomass residence time and operating pressure. The primary biomass constituents (hemicelluloses, cellulose and lignin) undergo decomposition and devolatilization over different temperature ranges. Decomposition of hemicellulose occurs at temperatures below 300°C. The resulting gas and vapor products include non-condensable gases (such as CO, CO_2, H_2 and CH_4), water vapor and low MW oxygenated organic compounds (mainly acetic acid, formic acid, methanol, acetone and furfural). Extensive devolatilization of lignin and cellulose occurs at temperatures higher than 300°C (slow pyrolysis temperatures). Peak devolatilization occurs at temperatures around 500°C, yielding typical vapor products including levoglucosan, hydroxyacetaldehyde, acetic acid and hydroxymethyl furfural (HMF) all originate from cellulose [38, 74, 75].

Phenolic compounds (both in monomeric and oligomeric form) are typically liberated from lignin [76]. Further increase in temperatures causes secondary vapor phase cracking reactions to dominate, yielding additional non-condensable gases and secondary biochar. The higher the pyrolysis temperature, the lower the biochar yield. The additional secondary char formation occurring at higher temperatures is offset by further devolatilization of the primary biochar. The net effect is a decreasing biochar yield with increasing slow pyrolysis temperature. Slow pyrolysis uses low heating rates resulting in biomass conversion being rate limited by heat transfer. Additionally, the slow devolatilization resulting from a slow heating rate ensures maximum secondary char formation. The heat transfer is of less critical importance in slow pyrolysis (compared to fast pyrolysis) as very long biomass residence times are used [38]. Biochar yield increases with increasing residence time, decreasing pyrolysis temperature and lower heating rate. With increasing residence times, vapors are restrained and reacted with solid-phase extensively for more biochar yield [28, 56, 72].

Slow pyrolysis favors biochar production at the expense of bio-oil production. The biochar, bio-oil and gas yields are typically 35 wt%, 30 wt% and 35 wt%, respectively. The overall slow pyrolysis process can generally be exothermic due to the extensive occurrence of secondary reactions. Slow pyrolysis can accept a wide range of particle sizes (5–50 mm). Large biomass particles are frequently used when rapid heating rates are not required or when the desired product is biochar [21, 55].

5.3.3 Intermediate pyrolysis

In intermediate pyrolysis, the reaction is faster than slow pyrolysis but slower than fast pyrolysis. Intermediate pyrolysis differs from fast pyrolysis in terms of heat transfer to biomass feed. It occurs in the temperature range of 450–550°C and even lower temperatures are used (350°C). The heating rates are much lower than fast pyrolysis (100 to 500°C/min) with residence time ranging from 10 to 30 s and produce less biochar than slow pyrolysis [65, 77]. Intermediate pyrolysis occurs at controlled heating rates, thus, inhibiting the formation of high MW tars and yielding products (biochar, bio-oil and gases) with different product qualities. In intermediate pyrolysis, the biomass particles sizes and shapes are less critical than in fast pyrolysis. It can process a wider variety of biomass, larger particles up to pellets, chips and dust and also material with a water content of up to 40%. High cooling rates are needed for the vapors to reduce thermal post-decomposition reactions. A higher amount of bio-oil than slow pyrolysis can be produced through intermediate pyrolysis. More controlled chemical reactions occur and, thus, the reaction conditions offer a broad range of variation for process optimization. The typical product distribution of intermediate pyrolysis is 40–60% of bio-oil, 20–30% non-condensable gases and 15–25% biochar [77].

The biochars produced by intermediate pyrolysis have a high carbon and low volatile contents. The bio-oil produced by intermediate pyrolysis has a high calorific value; the oil fraction easily separates into organic and aqueous phases. Compared to bio-oil generated from fast pyrolysis, the liquid fraction from intermediate pyrolysis has some beneficial characteristics, including low tar yield, improved viscosity and heating value. These characteristics may result from a relatively long residence time and contact with biochar [78]. Intermediate pyrolysis at 400°C yielding 35% solid (biochar), 46% liquid and 19% gaseous products has been reported. The liquid fraction separates into an aqueous phase (38% with 50% water; HHV of 7 MJ/kg) and organic phase (8% pyrolytic lignin, phenols, etc., HHV of 24 MJ /kg) [79].

5.3.4 Fast pyrolysis

The fast pyrolysis typically involves high heating rates (10–1000°C/s), short residence times (0.5–2 s) [56] and temperatures between 450 and 550°C. It decomposes biomass quickly to generate liquid (bio-oil), solid (biochar) and gaseous products. The bio-oil, biochar and gas yields are typically 60–70 wt%, 12–15 wt% and 13–25 wt% on a dry basis, respectively [55].

Fast pyrolysis suppresses secondary reactions from taking place by having short vapor residence times (rapid removal and quenching of the condensable primary volatile vapors) and maintaining high biomass heating rates, thereby maximizing the yield of condensable vapors (bio-oil). In this way, secondary reactions of cracking and repolymerization are prevented. The intermediate products of flash degradation of hemicellulose, cellulose and lignin are rapidly quenched and condensed to bio-oil before further reactions break down higher MW components into gaseous products. This freezing of intermediates results in bio-oil containing many reactive species, contributing to its unusual characteristics. Condensable vapors are formed by rapidly and simultaneously depolymerizing and fragmenting cellulose, hemicellulose and lignin fractions with a rapid increase in temperature. Rapid quenching traps many of these products that would further react (depolymerize, decompose, degrade, crack or condense with other molecules) to form more non-condensable gases if the residence time at high temperature was extended [61, 80].

The distribution of products (bio-oil, biochar and gases) depends on the biomass composition and rate and duration of heating. If bio-oil is the product of interest, the optimum pyrolysis temperature range is 425–600°C, with the peak temperature below 650°C to strike a balance between thorough devolatilization and minimal secondary cracking of vapors. The optimum yield in vapor products translates to the quantity of bio-oil formed. However, the peak temperature can be up to 1000°C if gas production is of primary interest [5]. Woody biomass (poplar, sawdust, forest and wood residue) produces the highest bio-oil yield (around 75%). The second highest bio-oil yield is from energy crops (reed), followed by agricultural residues (wheat straw, flax straw etc.). This is due to the higher cellulose and hemicellulose in wood than energy crops and agricultural residues. Product (bio-oil, biochar and syngas) yields in fast pyrolysis are affected by the feed particle size. The heat transfer rate decreases with increasing particle size, thus, increasing biochar yield and decreasing bio-oil and syngas yield. Smaller particle size is better for internal heat transfer, which increases bio-oil yield. Both pyrolysis temperature and feedstock particle size need to be optimized for maximum bio-oil yield [56, 80].

A finely ground biomass feed (usually <1 mm) is required to achieve very high heat transfer rates and thereby very high heating rates, which reduce the mass and heat transfer limitations. The biochar yield in fast pyrolysis is generally low as only primary char is being produced (secondary reactions are suppressed) and high reaction rates also minimize biochar formation. The overall fast pyrolysis process is highly endothermic due to the absence of secondary reactions. Fast pyrolysis prefers biomass with low moisture content (< 10 wt%) in order to minimize the water in the product bio-oil. Low moisture content also facilitates grinding the feed to give sufficiently small particles to ensure rapid heating and fast pyrolysis [65].

5.3.5 Flash pyrolysis

Flash pyrolysis aims to maximize the liquid yield (bio-oil). It is characterized by high temperatures, higher heating rates (> 1000°C/s) and shorter residence times (< 0.5 s). Very fine particles of biomass feed (< 0.2 mm) are usually required. Flash pyrolysis is extremely fast, thus, leading to a reduced time for processing of the feedstock. It occurs in the temperature range of 800–1000°C [77]. The product containing condensable and non-condensable gas is cooled, thus increasing the liquid yield while reducing biochar production. The main product distributions of flash pyrolysis are similar to that of fast pyrolysis. The small particle sizes of biomass feed result in small particles of biochar. The liquid (bio-oil) yield is typically 75–80 wt% and biochar yield is 12–13 wt%. Biochar particles need to be removed because it can catalyze the polymerization of some of the products and increase bio-oil viscosity. Special reactors, such as appropriately designed fluidized bed or entrained flow reactors, are typically required [21].

5.3.6 Microwave-assisted pyrolysis

In conventional heating, heat is transferred to the material surface (by convection, conduction and radiation) and subsequently from the surface to the interior of the material by conduction as a result of temperature gradients. On the other hand, microwave energy is delivered directly into materials within an electromagnetic field. The electromagnetic field enters the material and generates thermal energy throughout the penetration depth by dielectric heating through interaction with polarizable dipoles present in the material and heat the material from inside. Microwave heating requires a material with a high dielectric constant. The dielectric constant is a measure of the ability of a material to absorb microwave energy. Biomass has a relatively low dielectric constant. As a result, microwave pyrolysis requires catalysts as well as microwave absorbers to improve the heating. The presence of water in biomass may increase the heating rate of microwave pyrolysis due to the high dielectric constant of water in comparison with biomass. Microwave-assisted pyrolysis usually operates in the temperature range of 400–800°C [72, 81].

Some of the advantages of microwave-assisted pyrolysis over conventional pyrolysis include uniform heating throughout, rapid heating rate, cleaner products due to no agitation, volumetric and selective heating. Microwave heating provides ease of operation by instant on/off control and improves the yield and quality of the products. It reduces the formation of hazardous products and minimizes pollutants emission, making the technique environmentally friendly [72, 81]. The other advantages include high heating efficiency as heating is in situ, the ability to handle wet biomass without drying and the ability to pyrolyse large biomass particles. The disadvantage of microwave-assisted pyrolysis is that it requires electricity which is expensive high-quality energy compared to the heat generated by the combustion of pyrolysis gases and vapors in conventional pyrolysis. Microwave pyrolysis generally needed pre-treatment and catalysts before heating [21].

5.4 Hydrothermal processing

Most biomass materials are wet and have moisture contents range up to 95 wt%. Biomass with more than 30 wt% moisture content requires energy costly drying operation prior to pyrolysis, which is one of the leading technical barriers in using wet biomass. For high moisture content biomass, the heat of moisture evaporation is greater than the heat available from the biomass, thus becoming a net energy consumption. Wet biomass, typically with 70 wt% or more water, can be converted using hydrothermal processing, which involves applying heat and pressure to convert biomass in the presence of water into carbonaceous biofuel. In hydrothermal processing, water plays an active role as a solvent and reactant. It uses subcritical or supercritical water to convert biomass into end products in the absence of atmospheric oxygen. Hydrothermal processing is a promising technique to convert wet biomass into carbonaceous solids at relatively high yields by omitting the energy-intensive drying before or during the process. Hydrothermal processing can be classified into three processes: hydrothermal carbonization (HTC), hydrothermal liquefaction (HTL) and hydrothermal gasification (HTG) based on reaction parameters such as temperature, pressure and residence time [21].

5.4.1 Hydrothermal carbonization

When biomass feedstock in water is heated at low temperatures (< 200°C) in a sealed vessel at autogenous pressure, mostly solids (called hydrochar) are formed in a process known as HTC. The decomposition temperature of hemicellulose is usually around 160°C in subcritical water, while cellulose and lignin decompose in the temperature range 180 to 200°C and above 220°C in subcritical water, respectively. The three products of HTC are hydrochar (solid fraction), aqueous solution (bio-oil mixed with water) and a small volume of gas (mainly CO_2). HTC aims to maximize the yield of hydrochar. The product distribution and characteristics will mostly depend on three factors: type of biomass, the pH and the maximum temperature used. The reaction time and the solids concentration in biomass water mixture also has a relatively smaller influence. A minimum HTC temperature of 160°C is needed for the hydrochar formation from glucose. The overall HTC reaction extent is negligible below these processing conditions. The maximum hydrochar yield is obtained at 200°C and then it decreases gradually. A decreasing trend of hydrochar formation with the increase in temperature is due to the higher temperatures favoring gasification reactions. Consequently, part of hydrochar is lost in the formation of volatile compounds. The process energy requirements and final product composition depend on the input biomass feed and the process conditions. The main product of HTC is hydrochar which can be easily separated from the liquid fraction due to its high hydrophobicity and homogeneous properties. The solid fraction can be used to produce dried hydrochar pellets for energy production and mono sugars can be recovered from the liquid fraction. HTC generates less harmful gases such as CO and CO_2 and produces hydrochar mass yields varying from 35% up to 80%, with hydrochar carbon content similar to lignite. Residence time varies from minutes up to several hours [21, 70, 82, 83].

5.4.2 Hydrothermal liquefaction (HTL)

At elevated temperatures (between 200 and 350°C) and pressures (5–20 MPa), HTL takes place and the biomass feedstock is mainly converted into a liquid product (aqueous soluble). Free radical and ionic reactions are considered to be the main reactions in HTL [21, 84]. The dielectric constant of water decreases rapidly with increasing temperature. At ambient conditions, the dielectric constant (a measure of hydrogen bonding) of water is about 80 F/m. It drops to about 27 F/m at 250°C and 5 MPa and to about 14 F/m at 350°C and 25 MPa. Water starts to display less polar behavior due to decreasing number of hydrogen bonds. The dissociation of water also increases with the increase of temperature. The ionic product of water (pK_w) decreases from 14 at 25°C to 11 at 250°C. The viscosity of water decreases with increasing temperature. Thus, mass transfer is enhanced and any mass-transfer-limited chemical reactions are accelerated [84, 85].

5.4.3 Hydrothermal gasification (HTG)

HTG operates near and/or above the critical point of water at temperatures of 400–600°C and pressures of 23–45 MPa. The biomass is mainly converted into a mixture of non-condensable gases (H_2, CO, CH_4 and CO_2). HTG is capable of producing syngas enriched with H_2. The conversion efficiency is highly improved when water reaches the critical point (374°C and 22.1 MPa) [21]. HTG involves the splitting of biomass polysaccharides with supercritical water (SCW). Due to higher reaction temperature, HTG progresses at a faster rate and complete decomposition of biomass is achieved. This is a distinctive feature of HTG in comparison with other hydrothermal treatments (HTC and HTL). The difficulty of byproducts treatment is one of the problems with HTC and HTL. Undesirable byproducts produced by HTC and HTL are occasionally dissolved in the water phase. HTG typically decomposes biomass into gas with a conversion higher than 0.8. The amount of organic compounds in the liquid fraction is low; hence, post-treatment is unnecessary or easily carried out. HTG usually requires wet biomass; other biomasses can also be used. Conventional thermal gasification technologies are available when biomass is not wet. Conventional thermal gasification cannot be effectively employed when the feedstock is wet or has a high moisture content (> 80%). Conventional thermal gasification is achieved by partial oxidation using air. Syngas is partially diluted with nitrogen in addition to tar production. Syngas from HTG does not contain tars; even if produced, they remain in the liquid phase and are not diluted with nitrogen [21, 86].

6. Conclusions

Biomass is a sustainable source of energy. Natural biomass is bulky and usually has a high moisture content which complicates the direct use of biomass as a fuel. Biomass is a complex mixture of polymers consisting of three key elements carbon, oxygen and hydrogen. In addition, there are macronutrients, micronutrients, trace elements and other heavy metals. The three major components of LCB are cellulose, hemicellulose and lignin. Other compounds present in LCB include inorganic compounds and organic extractives. The components of biomass behave differently during thermochemical conversion.

The proximate and ultimate analysis is vital for describing biomass and thus product distribution. Thermochemical conversion is an increasingly feasible route to use biomass sources to fulfill energy needs. Designing effective thermochemical conversion systems that result in proper combinations of biomass feedstocks and conversion technologies is vital. It involves the design of efficient supply chains and conversion technologies suitable to manage variations in biomass properties. It needs to be cost-effective while minimizing harmful emissions and maintaining product quality.

Different thermochemical conversions of biomass are employed to yield varying amounts of gaseous, liquid and solid products of interest. Torrefaction aims to maximize the yield of solid product, torrefied biomass. It is performed at longer residence time, slow heating rates and low temperatures (<300°C). The improved properties of torrefied biomass include increased energy density, hydrophobicity, grindability and reduced biodegradability compared to the initial biomass feed, thus, allowing cheaper transport, handling and processing of the torrefied biomass. Torrefied biomass has a higher O/C ratio than biochar. Biochar is the desired product in slow pyrolysis, and yield is maximized by conversion at longer residence time, slow heating rates, and moderate temperatures. Biochar applications include use as a soil amendment, as an adsorbent or as fuel in power generation. In fast pyrolysis, bio-oil is the desired product and yield is maximized by conversion at short vapor residence times, high heating rates and around 500°C. The gap between biomass feedstock and a usable product can be bridged by incorporating additional processes such as drying and grinding of biomass feed, palletization or briquetting of biomass feed and/or product and mineral leaching of products to treat high ash content. These additional processes have cost implications.

High moisture content is a major barrier in biomass handling and processing. Moisture content has a significant impact on the biomass conversion process. The main problem with the pyrolysis of high moisture biomass is that it requires a large amount of energy to evaporate moisture. Hydrothermal processing is a useful processing technique for biomass feedstocks with high moisture as it does not require the drying and/or dewatering processes, thereby reducing the operating costs.

Acknowledgements

The authors would like to thank the Office of Applied Research, Higher Colleges of Technology (HCT), UAE, for financial support.

Author details

Meegalla R. Chandraratne* and Asfaw G. Daful
Higher Colleges of Technology, Abu Dhabi, United Arab Emirates

*Address all correspondence to: mchandra@hct.ac.ae

IntechOpen

© 2021 The Author(s). Licensee IntechOpen. This chapter is distributed under the terms of the Creative Commons Attribution License (http://creativecommons.org/licenses/by/3.0), which permits unrestricted use, distribution, and reproduction in any medium, provided the original work is properly cited.

References

[1] Bhat AH, Dasan YK, Khan I. Extraction of Lignin from Biomass for Biodiesel Production. In: Hakeem KR, Jawaid M, Alothman OY (eds) *Agricultural Biomass Based Potential Materials*. New York: Springer, 2015, pp. 155-180.

[2] Kataki R, Chutia RS, Mishra M, et al. Feedstock Suitability for Thermochemical Processes. In: Pandey A, Bhaskar T, Stöcker M, et al. (eds) *Recent advances in thermochemical conversion of biomass*. Oxford: Elsevier, 2015, pp. 31-74.

[3] Okolie JA, Nanda S, Dalai AK, et al. A review on subcritical and supercritical water gasification of biogenic, polymeric and petroleum wastes to hydrogen-rich synthesis gas. *Renew Sustain Energy Rev* 2020; 119: 109546.

[4] Chandra R, Takeuchi H, Hasegawa T. Methane production from lignocellulosic agricultural crop wastes: A review in context to second generation of biofuel production. *Renew Sustain Energy Rev* 2012; 16: 1462-1476.

[5] Basu P. *Biomass Gasification, Pyrolysis and Torrefaction: Practical Design and Theory*. Second. Oxford: Elsevier, 2013.

[6] McKendry P. Energy production from biomass (part 1): Overview of biomass. *Bioresour Technol* 2002; 83: 37-46.

[7] Pilon G, Lavoie JM. Pyrolysis of switchgrass (*Panicum virgatum* L.) at low temperatures within N_2 and CO_2 environments: Product yield study. *ACS Sustain Chem Eng* 2013; 1: 198-204.

[8] Ruan R, Zhang Y, Chen P, et al. Biofuels: Introduction. In: Pandey A, Larroche C, Dussap C-G, et al. (eds) *Biofuels: Alternative Feedstocks and Conversion Processes for the Production of Liquid and Gaseous biofuels*. Oxford: Academic Press, 2019, pp. 3-43.

[9] Ernst M, Walter A, Schurr U. Biomass Production. In: Kaltschmitt M, Themelis NJ, Bronicki LY, et al. (eds) *Renewable Energy Systems*. 2013, pp. 510-521.

[10] Faaij APC. Biomass Resources, Worldwide. In: Kaltschmitt M (ed) *Energy from Organic Materials (Biomass)*. New York: Springer, 2019, pp. 299-350.

[11] Panwar NL. Biomass for Domestic and Agro-Industrial Applications. In: Botannini LF (ed) *Wood: Types, Properties and Uses*. New York: Nova Science Publishers, 2011, pp. 81-109.

[12] Faaij APC. Technical and Economic Potentials of Biomass until 2050: Regional Relevance for Energy Security. In: Brauch HG, Spring ÚO, Grin J, et al. (eds) *Facing Global Environmental Change: Environmental, Human, Energy, Food, Health and Water Security Concepts*. Berlin: Springer-Verlag, 2009, pp. 379-394.

[13] IEA. *World energy outlook 2020*. Paris, 2020.

[14] Timilsina G, Shah KU. Energy Technologies for Sustainable Development Goal 7. In: Adenle AA, Chertow MR, Moors EHM, et al. (eds) *Science, Technology and Innovation for Sustainable Development Goals: Insights from Agriculture, Health, Environment and Energy*. New York: Oxford University Press, 2020, pp. 36-63.

[15] Carpenter NE. *Chemistry of Sustainable Energy*. Boca Raton, FL: CRC Press, 2014. DOI: 10.1201/b16687.

[16] Champagne P. Biomass. In: Letcher T (ed) *Future Energy: Improved, Sustainable and Clean Options for our Planet*. Oxford: Elsevier, 2008, pp. 151-170.

[17] Corro G, Pal U, Bañuelos F, et al. Generation of biogas from coffee-pulp and cow-dung co-digestion: Infrared

studies of postcombustion emissions. *Energy Convers Manag* 2013; 74: 471-481.

[18] Wolf JP. Biomass combustion for power generation: an introduction. In: Rosendahl L (ed) *Biomass Combustion Science, Technology and Engineering*. Cambridge: Woodhead, 2013, pp. 3-8.

[19] Li R, Chen J. Planning the next-generation biofuel crops based on soil-water constraints. *Biomass and Bioenergy* 2018; 115: 19-26.

[20] Cowie A, Woolf D, Gaunt J, et al. iochar, carbon accounting and climate change. In: *Biochar for Environmental Management: Science, Technology and Implementation*. Oxon: Routledge, 2015, pp. 763-794.

[21] Daful AG, Chandraratne MR. Biochar Production From Biomass Waste-Derived Material. In: Hashmi S, Choudhury IA (eds) *Encyclopedia of Renewable and Sustainable Materials*. Oxford: Elsevier, 2020, pp. 370-378.

[22] Gundekari S, Mitra J, Varkolu M. Classification, characterization, and properties of edible and non-edible biomass feedstocks. In: Hussain CM, Sudarsanam P (eds) *Advanced Functional Solid Catalysts for Biomass Valorization*. Oxford: Elsevier Inc., pp. 89-120.

[23] Nachenius RW, Ronsse F, Venderbosch RH, et al. Biomass Pyrolysis. In: Marin GB, West DH, Li J, et al. (eds) *Advances in Chemical Engineering, Volume 42. Chemical Engineering for Renewables Conversion*. Oxford: Elsevier, 2013, pp. 75-139.

[24] Sanchez J, Curt MD, Robert N, et al. Biomass resources. In: Lago C, Caldés N, Lechón Y (eds) *The Role of Bioenergy in the Emerging Bioeconomy: Resources, Technologies, Sustainability and Policy*. Oxford: Elsevier Inc., pp. 25-111.

[25] Hodge BK. *Alternative Energy Systems and Applications*. Hoboken, NJ: John Wiley & Sons, 2017.

[26] Mansoori GA, Enayati N, Agyarko LB. *Energy (Sources, Utilization, Legislation, Sustainability, Illinois as Model State)*. Danvers, MA: World Scientific, 2016.

[27] Quader MA, Ahmed S. Bioenergy With Carbon Capture and Storage (BECCS): Future Prospects of Carbon-Negative Technologies. In: Rasul M, Azad A, Sharma S (eds) *Clean Energy for Sustainable Development Comparisons and Contrasts of New Approaches*. Oxford: Elsevier, 2017, pp. 467-481.

[28] Tripathi M, Sahu JN, Ganesan P. Effect of process parameters on production of biochar from biomass waste through pyrolysis: A review. *Renew Sustain Energy Rev* 2016; 55: 467-481.

[29] Wang S, Luo Z. *Pyrolysis of Biomass*. Berlin: Walter De Gruyter GmbH, 2017.

[30] Vassilev S V., Baxter D, Andersen LK, et al. An overview of the chemical composition of biomass. *Fuel* 2010; 89: 913-933.

[31] Vassilev S V., Baxter D, Andersen LK, et al. An overview of the organic and inorganic phase composition of biomass. *Fuel* 2012; 94: 1-33.

[32] Wang S, Dai G, Yang H, et al. Lignocellulosic biomass pyrolysis mechanism: A state-of-the-art review. *Prog Energy Combust Sci* 2017; 62: 33-86.

[33] Godin B, Agneessens R, Schmit T, et al. Chemical Composition and Biofuel Potentials of a Wide Diversity of Plant Biomasses. *Energy and Fuels* 2013; 27: 2588-2598.

[34] Akhtar J, Amin NAS. A review on process conditions for optimum bio-oil yield in hydrothermal liquefaction of biomass. *Renew Sustain Energy Rev* 2011; 15: 1615-1624.

[35] Dhyani V, Bhaskar T. A comprehensive review on the pyrolysis of lignocellulosic biomass. *Renew Energy* 2018; 129: 695-716.

[36] Potumarthi R, Baadhe R, Bhattacharya S. Fermentable Sugars from Lignocellulosic Biomass: Technical Challenges. In: Gupta V, Tuohy M (eds) *Biofuel Technologies: Recent Developments.* Berlin: Springer-Verlag, 2013, pp. 1-27.

[37] Dhyani V, Bhaskar T. Pyrolysis of Biomass. In: Pandey A, Larroche C, Dussap C-G, et al. (eds) *Biofuels: Alternative Feedstocks and Conversion Processes for the Production of Liquid and Gaseous Biofuels*. Oxford: Elsevier, 2019, pp. 217-244.

[38] Ronsse F, Nachenius RW, Prins W. Carbonization of Biomass. In: Pandey A, Bhaskar T, Stöcker M, et al. (eds) *Recent advances in thermochemical conversion of biomass*. Oxford: Elsevier, 2015, pp. 293-324.

[39] Pant KK, Mohanty P. Biomass, Conversion Routes and Products – An Overview. In: Hornung A (ed) *Transformation of Biomass Theory to Practice*. Chichester: John Wiley & Sons, 2014, pp. 1-30.

[40] Sannigrahi P, Ragauskas AJ, Tuskan GA. Poplar as a feedstock for biofuels: A review of compositional characteristics. *Biofuels, Bioprod Biorefining* 2010; 4: 209-226.

[41] Karimi K, Shafiei M, Kumar R. Progress in Physical and Chemical Pretreatment of Lignocellulosic Biomass. In: Gupta V, Tuohy M (eds) *Biofuel Technologies: Recent Developments.* Berlin: Springer-Verlag, 2013, pp. 53-96.

[42] Zheng Y, Shi J, Tu M, et al. Principles and Development of Lignocellulosic Biomass Pretreatment for Biofuels. In: Li Y, Ge X (eds) *Advances in Bioenergy -Vol 2*. Oxford: Elsevier Ltd, pp. 1-68.

[43] Nanda S, Mohammad J, Reddy SN, et al. Pathways of lignocellulosic biomass conversion to renewable fuels. *Biomass Convers Biorefinery* 2014; 4: 157-191.

[44] Tursi A. A review on biomass: Importance, chemistry, classification, and conversion. *Biofuel Res J* 2019; 6: 962-979.

[45] Agbor VB, Cicek N, Sparling R, et al. Biomass pretreatment: Fundamentals toward application. *Biotechnol Adv* 2011; 29: 675-685.

[46] Demirbas A. *Biorefineries: For Biomass Upgrading Facilities*. London: Springer, 2010. DOI: 10.1007/978-1-84882-721-9.

[47] Dheeran P, Reddy L. Biorefining of Lignocelluloses: An Opportunity for Sustainable Biofuel Production. In: Kumar S, Sani R (eds) *Biorefining of Biomass to Biofuels: Opportunities and Perception, Biofuel and Biorefinery Technologies 4*. Cham, Switzerland: Springer, 2018, pp. 1-23.

[48] Amidon TE, Wood CD, Shupe AM, et al. Biorefinery: Conversion of woody biomass to chemicals, energy and materials. *J Biobased Mater Bioenergy* 2008; 2: 100-120.

[49] Basu P. *Biomass Gasification, Pyrolysis and Torrefaction: Practical Design and Theory*. 2nd ed. London: Elsevier, 2018. DOI: 10.1016/B978-0-12-812992-0.00003-0.

[50] Nunes LJR, Matias JCDO, Catalão JPDS. *Torrefaction of Biomass for Energy Applications- From Fundamentals to Industrial Scale*. Oxford: Elsevier, 2018. DOI: 10.1016/b978-0-12-809462-4.00001-8.

[51] Jameel H, Keshwani DR. Thermochemical Conversion of Biomass to Power and Fuels. In: Cheng JJ (ed) *Biomass to Renewable Energy Processes*.

Boca Raton, FL: Taylor & Francis, 2018, pp. 375-421.

[52] Adams P, Bridgwater T, Lea-Langton A, et al. Biomass conversion technologies. In: Thornley P, Adams P (eds) *Greenhouse gas balance of bioenergy systems*. London: Elsevier, 2018, pp. 107-139.

[53] Jafri N, Wong WY, Doshi V, et al. A review on production and characterization of biochars for application in direct carbon fuel cells. *Process Saf Environ Prot* 2018; 118: 152-166.

[54] Maniscalco MP, Volpe M, Messineo A. Hydrothermal carbonization as a valuable tool for energy and environmental applications: A review. *Energies* 2020; 13: 04098. DOI: 10.3390/en13164098.

[55] Kan T, Strezov V, Evans TJ. Lignocellulosic biomass pyrolysis: A review of product properties and effects of pyrolysis parameters. *Renew Sustain Energy Rev* 2016; 57: 1126-1140.

[56] Roy P, Dias G. Prospects for pyrolysis technologies in the bioenergy sector: A review. *Renew Sustain Energy Rev* 2017; 77: 59-69.

[57] Mando M. Direct combustion of biomass. In: Rosendahl L (ed) *Biomass Combustion Science, Technology and Engineering*. Cambridge: Woodhead, 2013, pp. 61-83.

[58] Ranzi E, Faravelli T, Manenti F. Pyrolysis, Gasification, and Combustion of Solid Fuels. In: Marin G, West D, Li J, et al. (eds) *Advances in Chemical Engineering, Volume 49, Thermochemical Process Engineering*. Oxford: Elsevier, 2016, pp. 1-94.

[59] Cerinski D, Baleta J, Mikulčić H, et al. Dynamic modelling of the biomass gasification process in a fixed bed reactor by using the artificial neural network. *Clean Eng Technol* 2020; 1: 100029.

[60] Luo X, Wu T, Shi K, et al. Biomass Gasification: An Overview of Technological Barriers and Socio-Environmental Impact. In: Yun Y (ed) *Gasification for Low-grade Feedstock*. London: IntechOpen, 2018, pp. 3-18.

[61] Demirbas A. Biorefineries: Current activities and future developments. *Energy Convers Manag* 2009; 50: 2782-2801.

[62] Neubauer Y. Biomass gasification. In: Rosendahl L (ed) *Biomass Combustion Science, Technology and Engineering*. Cambridge: Woodhead, 2013, pp. 106-129.

[63] Safarian S, Saryazdi SME, Unnthorsson R, et al. Gasification of woody biomasses and forestry residues: Simulation, performance analysis, and environmental impact. *Fermentation* 2021; 7: 1-14.

[64] Kumar A, Jones DD, Hanna MA. Thermochemical biomass gasification: A review of the current status of the technology. *Energies* 2009; 2: 556-581.

[65] Bridgwater A V. Review of fast pyrolysis of biomass and product upgrading. *Biomass and Bioenergy* 2012; 38: 68-94.

[66] Harris Z, Milner S, Taylor G. Biogenic Carbon-Capture and Sequestration. In: Thornley P, Adams P (eds) *Greenhouse gas balance of bioenergy systems*. London: Elsevier, 2018, pp. 107-139.

[67] Collard FX, Blin J. A review on pyrolysis of biomass constituents: Mechanisms and composition of the products obtained from the conversion of cellulose, hemicelluloses and lignin. *Renew Sustain Energy Rev* 2014; 38: 594-608.

[68] Yang H, Yan R, Chen H, et al. Characteristics of hemicellulose, cellulose and lignin pyrolysis. *Fuel* 2007; 86: 1781-1788.

[69] Krull ES, Baldock JA, Skjemstad JO, et al. Characteristics of Biochar: Organo-chemical Properties. In: Lehmann J, Joseph S (eds) *Biochar for Environmental Management*. London: Earthscan, 2009, pp. 53-65.

[70] Lam PS, Tooyserkani Z, Naimi LJ, et al. Pretreatment and Pelletization of Woody Biomass. In: Fang Z (ed) *retreatment Techniques for Biofuels and Biorefineries*. Berlin: Springer-Verlag, 2013, pp. 93-116.

[71] EBC. *European Biochar Certificate - Guidelines for a Sustainable Production of Biochar. Version 6.3E*. European Biochar Foundation, Arbaz, Switzerland. 2012. DOI: 10.13140/RG.2.1.4658.7043.

[72] Yu KL, Lau BF, Show PL, et al. Recent developments on algal biochar production and characterization. *Bioresour Technol* 2017; 246: 2-11.

[73] Patwardhan PR, Satrio JA, Brown RC, et al. Influence of inorganic salts on the primary pyrolysis products of cellulose. *Bioresour Technol* 2010; 101: 4646-4655.

[74] Ronsse F, Bai X, Prins W, et al. Secondary reactions of levoglucosan and char in the fast pyrolysis of cellulose. *Environ Prog Sustain Energy* 2012; 31: 256-260.

[75] Ronsse F, Dalluge D, Prins W, et al. Optimization of platinum filament micropyrolyzer for studying primary decomposition in cellulose pyrolysis. *J Anal Appl Pyrolysis* 2012; 95: 247-256.

[76] Mohan D, Pittman CU, Steele PH. Pyrolysis of Wood/Biomass for Bio-oil: A Critical Review. *Energy & Fuels* 2006; 20: 848-889.

[77] Bhaskar T, Pandey A. Advances in Thermochemical Conversion of Biomass – Introduction. In: Pandey A, Bhaskar T, Stöcker M, et al. (eds) *Recent advances in thermochemical conversion of biomass*. Oxford: Elsevier, 2015, pp. 3-30.

[78] Corton J, Blanco-Sanchez PP, Khan Z, et al. The Impact of Biomass Feedstock Composition and Pre-treatments on Tar Formation during Biomass Gasification. In: Singh LK, Chaudhary G (eds) *Advances in Biofeedstocks and Biofuels*. Hoboken, NJ: John Wiley & Sons, 2016, pp. 33-54.

[79] Dahmen N, Henrich E, Kruse A, et al. Biomass Liquefaction and Gasification. In: Vertes A, Qureshi N, Blaschek H, et al. (eds) *Biomass to Biofuels: Strategies for Global Industries*. Cham, Switzerland: John Wiley & Sons, 2010, pp. 91-122.

[80] Bridgwater A. Fast pyrolysis of biomass for the production of liquids. In: Rosendahl L (ed) *Biomass Combustion Science, Technology and Engineering*. Cambridge: Woodhead, 2013, pp. 130-171.

[81] Li J, Dai J, Liu G, et al. Biochar from microwave pyrolysis of biomass: A review. *Biomass and Bioenergy* 2016; 94: 228-244.

[82] Heidari M, Dutta A, Acharya B, et al. A review of the current knowledge and challenges of hydrothermal carbonization for biomass conversion. *J Energy Inst* 2019; 92: 1779-1799.

[83] Titirici MM, Funke A, Kruse A. Hydrothermal Carbonization of Biomass. In: Pandey A, Bhaskar T, Stöcker M, et al. (eds) *Recent advances in thermochemical conversion of biomass*. Oxford: Elsevier, 2015, pp. 325-352.

[84] Singh R, Prakash A, Balagurumurthy B, et al. Hydrothermal Liquefaction of Biomass. In: Pandey A, Bhaskar T, Stöcker M, et al. (eds) *Recent*

advances in thermochemical conversion of biomass. Oxford: Elsevier, 2015, pp. 269-291.

[85] Toor SS, Rosendahl L, Rudolf A. Hydrothermal liquefaction of biomass: A review of subcritical water technologies. *Energy* 2011; 36: 2328-2342.

[86] Matsumura Y. Hydrothermal Gasification of Biomass. In: Pandey A, Bhaskar T, Stöcker M, et al. (eds) *Recent advances in thermochemical conversion of biomass*. Oxford: Elsevier, 2015, pp. 251-267.

Section 2

Pyrolysis for Material Science Applications

Chapter 9

Functional Materials Synthesis and Physical Properties

Tatsuya Senzaki, Michiaki Matsukawa, Takanori Yonai, Haruka Taniguchi, Akiyuki Matsushita, Takahiko Sasaki and Mokoto Hagiwara

Abstract

A citrate pyrolysis technique is a unique route to prepare reactive precursor mixtures through an ignition process of concentrated aqueous solution. This procedure enables to synthesize highly homogeneous and fine powders for functional materials. The double-chain based superconductor $Pr_2Ba_4Cu_7O_{15-\delta}$ and double perovskite photocatalytic semiconductor $Ba_2Tb(Bi,Sb)O_6$ were synthesized by using the citrate pyrolysis technique. For the present sample with a reduction treatment for 72 h, a sharp superconducting transition appeared at an onset temperature $T_{c,on}$=26 K accompanied by a zero-resistance state at $T_{c,zero}$=22 K. The superconducting volume fraction estimated from the magnetization measurement showed an excellent value of $\sim$ 58%. Both reduction treatment in a vacuum and subsequent quenching procedure are needed to realize higher superconductivity due to further oxygen defects. The polycrystalline samples for $Ba_2Tb(Bi_{1-x},Sb_x)O_6$ ($x = 0$ and 0.5) were formed in the monoclinic and cubic crystal structures. We conducted the gaseous 2-propanol (IPA) and methylene blue (MB) degradation experiments under a visible light irradiation, to evaluate photocatalytic activities of the powder samples. For the Sb50% substituted sample, the highest performance of MB degradation was observed. The effect of Sb-substitution on the photocatalytic degradation of MB is in direct contrast to that on the IPA decomposition under visible light irradiation. The enhanced photocatalytic properties in the citrate samples are attributed to their morphology, where fine particles are homogeneously distributed with a submicron order.

Keywords: citrate pyrolysis technique, double-chain based superconductor, pressure effect, double-perovskite oxide, photocatalytic semiconductor

1. Introduction

A citrate pyrolysis technique is similar to nitrate combustion synthesis methods [1] and a unique route to prepare reactive precursor mixtures through an ignition process of concentrated aqueous solution including stoichiometric amounts of metal ions. For 80 K-class high- T_c cuprate superconductors such as $YBa_2Cu_4O_8$, high-quality single-phase polycrystalline samples have been successfully prepared at ambient oxygen pressure [2, 3]. We believe that the present technique is a powerful tool to fabricate highly homogeneous and fine crystalline grains for functional materials, in comparison to conventional solid-state reaction methods.

IntechOpen

High- T_c copper-oxide superconductors discovered as far have close relationship with two-dimensional CuO_2 planes. In quasi one-dimensional (1D) Cu_2O_3 ladder system without CuO_2 planes, it has been reported in [4] that the application of external pressure under 3 GPa causes bulk superconductivity at T_c = 12 K. For Pr-based copper oxides including insulating CuO_2 planes, we have reported that $Pr_2Ba_4Cu_7O_{15-\delta}$ (Pr247) with metallic CuO double-chain structure achieves a superconducting-state with a higher T_c (15 K) after a reduction treatment [5].

In the subsequent microscopic research on Pr247 [6], nuclear quadrupole resonance observations have resolved that the newly discovered superconductivity occurs along the CuO double-chain block. The Pr-based cuprates, $PrBa_2Cu_3O_{7-\delta}$ (Pr123) and $PrBa_2Cu_4O_8$ (Pr124), have identical crystal structures as Y-based high-T_c superconductors, $YBa_2Cu_3O_{7-\delta}$ (Y123) and $YBa_2Cu_4O_8$ (Y124), respectively. Both the Pr123 and Pr124 compounds share insulating CuO_2 planes and exhibit no superconductivity [7, 8]. The CuO single chains in Pr123 follow semiconducting property but Pr124 has a metallic conduction along the CuO double chain block [9]. For Pr124, it is hard to control the carrier density of doped double chains, because it is thermally stable against high temperature heat treatment.

For Pr247 intermediate compound existing between Pr123 and Pr124 phases, there are alternate stacks of CuO single-chain and double-chain blocks along the *c*-axis such as {-S-D-S-D-} sequence [10, 11] (see **Figure 1**). Here, CuO single-chain and double-chain blocks along the *b*-axis are abbreviated as S and D, respectively. Under thermal control of the oxygen content along the semiconducting CuO single chains in Pr247, it is possible to investigate the physical characters of the metallic CuO double chains. In oxygen defect polycrystalline sample, we succeeded in the appearance of superconductivity at an onset temperature $T_{c,on}$ of $\sim$15 K [5].

The B-site substituted perovskite oxides $A_2B'B''O_6$ have been widely studied because of their attractive physical properties and potential applications [13] (see **Figure 2a**). Some of the double perovskite compounds such as A_2FeMoO_6 exhibiting negative tunneling magnetoresistance effect at room temperature are of great interest with a wide range of applications in magnetic devices [14]. Furthermore, multiferroic double perovskite oxides have an effective coupling between spontaneous ferroelectric polarization and ferromagnetic ordering, which is considered to be promising materials from view points of physics and its applications [15]. Recently, a series of double perovskite oxides $Ba_2Ln'Bi''O_6$ (Ln:lanthanides) has been examined on the view point of photocatalytic semiconductors for hydrogen generation by water splitting and are taken as alternative materials for TiO_2 oxide [16, 17]. In particular, Ba_2PrBiO_6 is found to possess highly photocatalytic performance, which is probably close to the valence mixing state [18, 19]. Furthermore, a previous study on the magnetic properties of the Ba_2PrBiO_6 compound revealed that the average valence of Pr ions is an intermediate state between trivalent and tetravalent [20].

As for the key factors to fabricate visible light driven photocatalysts, it is desirable to control the energy band gap between the valence and conduction bands of their semiconductors to utilize a wide range of visible light [21, 22]. Accordingly, we think that the Sb substitution at the B site of the parent material Ba_2PrBiO_6 is an effective approach to adjusting the band gaps. For high photocatalytic activity, we avoid charge recombination between electron and hole and try to promote the photogenerated charge separation in the photocatalystic materials [21, 23]. In our research [24], it has been demonstrated that the valence mixing states between Pr^{3+} and Pr^{4+} are closely related to the phenomena of charge separation.

For our further understanding of the enhanced effect of the mixed valence states at B-site ions of the double perovskite compound on the photocatalytic

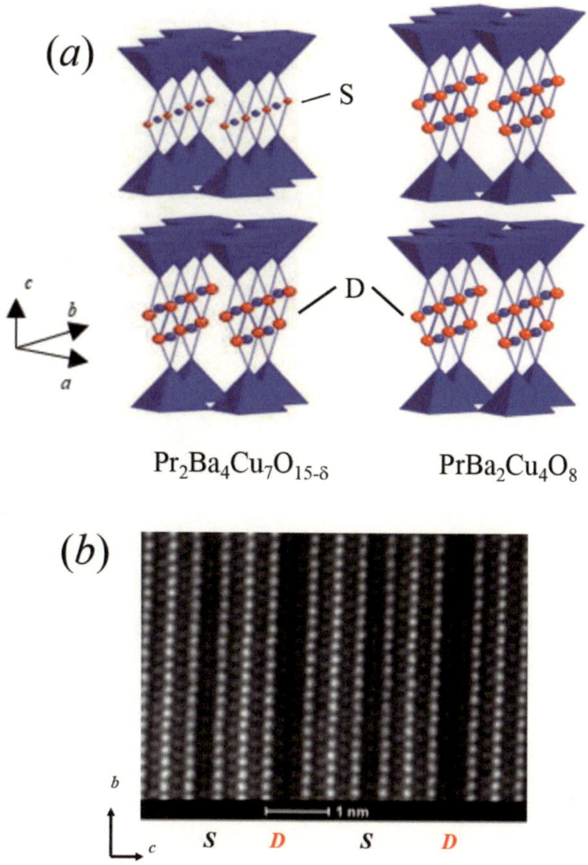

Figure 1.
(a) Typical crystal structure of $Pr_2Ba_4Cu_7O_{15-\delta}$ (Pr247) with CuO single-chain and double-chain blocks stacked along the c-axis. Here, S and D denote CuO single-chain and double-chain blocks along the b-axis. For comparison, the crystal structure of Pr124 is shown on the right hand side. (b) TEM image of superconducting Pr247. CuO single-chain and double-chain blocks are alternately stacked along the c-axis such as {-S-D-S-D-} sequence [12].

performance, we demonstrate the 2-propanol decomposition and methylene blue (MB) degradation under the irradiation of visible right for $Ba_2Tb(Bi,Sb)O_6$ samples prepared by the citrate pyrolysis technique. MB aqueous solution is adopted as the model pollutant [25].

2. Experiments

2.1 Double-chain based superconductor

2.1.1 Synthesis of $Pr_2Ba_4Cu_7O_{15-\delta}$ samples

First of all, flow chart of citrate pyrolysis technique for the synthesis of $Pr_2Ba_4Cu_7O_{15-\delta}$ (sample #1) is illustrated in **Figure 3b**. We synthesized high-quality polycrystalline samples of $Pr_2Ba_4Cu_7O_{15-\delta}$ by using a citrate pyrolysis technique [2, 3]. Stoichiometric quantities of high purity Pr_6O_{11}, $Ba(NO_3)_2$, and CuO were mixed and thoroughly ground. The mixture was dissolved in a nitric acid solution at 50–60°C. After adding citric acid to the resultant solution, we then neutralized it by adding aqueous

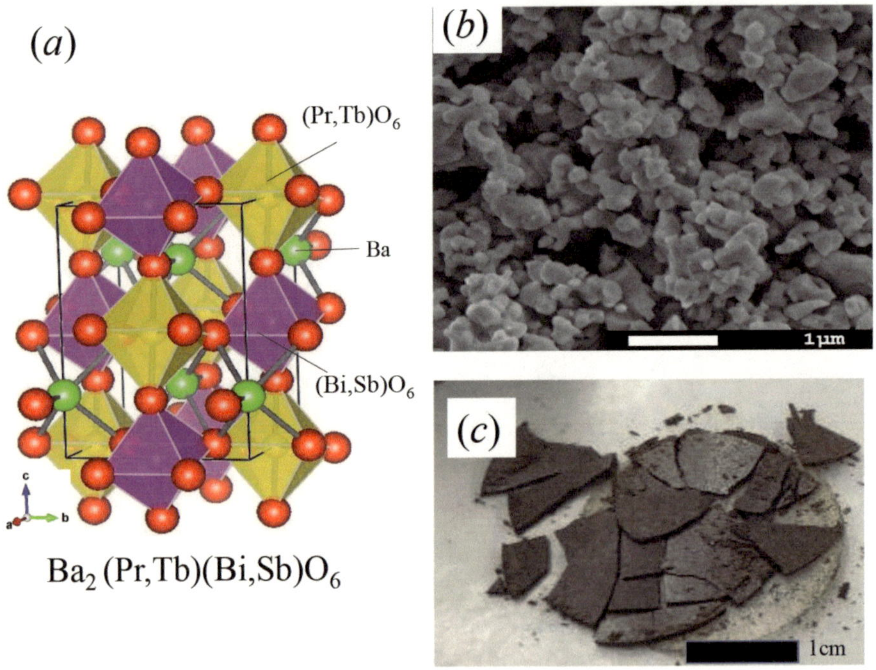

Figure 2.
(a) Typical crystal structure of double perovskite $Ba_2(Pr,Tb)(Bi,Sb)O_6$. (b) SEM image of Ba_2TbBiO_6 polycrystalline film fabricated from the single-phase powders by an electrophoretic deposition technique. (c) Photograph of the pelletized precursors of Ba_2TbBiO_6 after a citrate pyrolysis procedure.

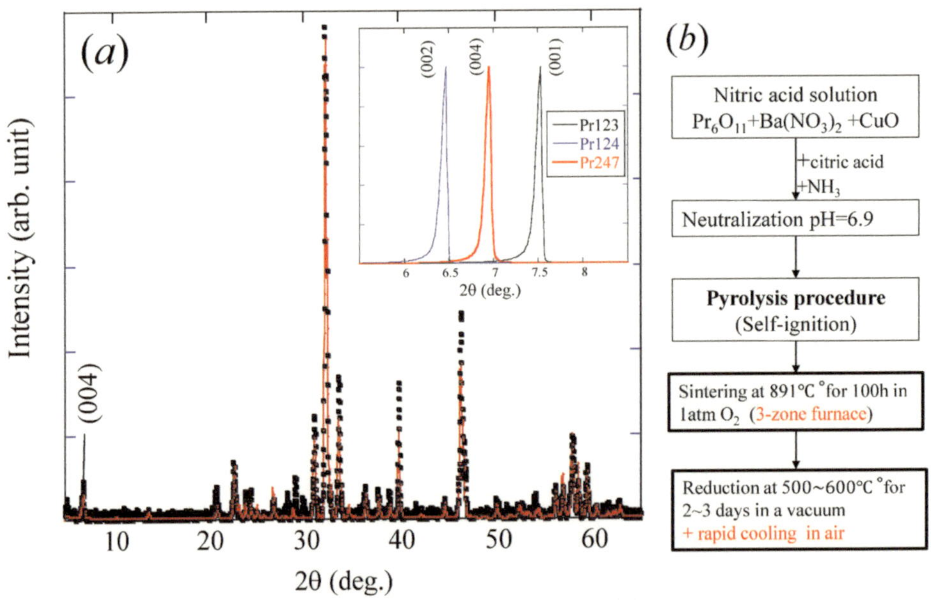

Figure 3.
(a)X-ray diffraction patterns of as-sintered polycrystalline $Pr_2Ba_4Cu_7O_{15-\delta}$. The (004) peak corresponds to one of typical Miller indexes of Pr247. The calculated curve is obtained using the lattice parameters in the text. Inset shows characteristic peaks at $2\theta \sim 7°$, indicating the formation of Pr123, Pr124, and Pr247 phases. (b) Flow chart of citrate pyrolysis technique for the synthesis of $Pr_2Ba_4Cu_7O_{15-\delta}$ (sample #1). After post-annealing of the as-sintered sample in a vacuum, quenching procedure is needed to obtain higher superconductivity due to further oxygen defects.

ammonia. After the solution was dried up under stirring on halogen lamp heater block, the porous black products were formed through the self-ignition process of it. Finally, the precursors were ground into fine powders and they were annealed under ambient pressure of flowing oxygen gas at 891°C ±0.5°C for an extended period over 100~120 h. For the present citrate pyrolysis synthesis procedure, we adopted the electric tube furnace with three zone temperature controlled system, to achieve the temperature uniformity within 1°C. In our previous study [26], we realized homogeneous distributions of the superconducting grains and improved weak links between their superconducting grains in the sintering procedure by using the 3 zone furnace. The oxygen in the as-sintered sample was removed by reduction treatment in a vacuum at 500~600°C, yielding a superconducting material. In particular, sample #1 was quenched in air from 300°C down to room temperature and samples #2–1 and #2–2 were slowly cooled in the electric furnace, as listed in **Table 1**. As mentioned below, this quenching procedure is needed to obtain higher superconductivity due to further oxygen defects. Typical dimensions of the pelletized rectangular sample were $4 \times 3 \times 1\ mm^3$.

2.1.2 Structural and physical properties of $Pr_2Ba_4Cu_7O_{15-\delta}$ samples

X-ray diffraction measurements on the produced samples were carried out at room temperature with an Ultima IV diffractometer (Rigaku) using Cu-Kα radiation. We evaluated the lattice parameters from the x-ray diffraction patterns using the least-squares fits.

The local crystal structure of the Pr247 sample was revealed by high-resolution transmission electron microscopy (TEM) using a JEOL3010 microscope operated at 300 kV at Tohoku University, to examine alternative stacking along the *c*-axis between CuO single-chain and double-chain blocks. The electric resistivity as a function of temperature was measured by the *dc* four-terminal method using a Gifford-McMahon cryocooler (Sumitomo heavy Industries).

We performed Hall coefficient measurements on the 48-h-reduced samples with the five-probe technique using a physical property measuring system (PPMS, Quantum Design), to check the sign of carriers in the present Pr247 superconductor. For determination of an onset temperature of superconductivity and superconducting volume fraction at low temperatures, the *dc* magnetization was performed under zero-field cooling (ZFC) in a commercial superconducting quantum interference device (SQUID) magnetometer (MPMS, Quantum Design).

2.1.3 Pressure effect on transport properties of $Pr_2Ba_4Cu_7O_{15-\delta}$ samples

It is well known that application of external pressure on oxide superconductors changes doped carrier densities, causing a positive or negative dependence of their

Sample no.	Reduced condition	δ	$T_{c,on}$ (K)	$T_{c,zero}$ (K)	Experimental data
#1	500°C 48 h, 600°C 24 h quenched	0.72	26	22	ρ(T), M(T)
#2-1[a]	500°C 48 h furnace cooled	0.56	26.5	15	TEM
#2-2[b]	500°C 48 h furnace cooled	0.56	26.5	15	R_H, pressure effect

In details, see the corresponding text and references. δ denotes oxygen deficiency.
[a]*Ref. [12].*
[b]*Ref. [27].*

Table 1.
Sample details of $Pr_2Ba_4Cu_7O_{15-\delta}$ used in the experiments.

superconducting transition temperature. We carried out the temperature dependence of the resistivity and the magneto-resistance (MR), under a maximum pressure of 2 GPa.

The MR effect of Pr247 sample was measured at low temperatures as a function of applied field up to 14 T using a 15 T-SM superconducting magnet at Institute for Materials Research, Tohoku University. The electric current *I* was excited in the longitudinal direction to the sample and the external magnetic field *H* was applied in the transverse direction to it ($H \perp I$). We measured the influence of applied pressure on the electric resistivity using a hybrid piston cylinder-type CuBe/NiCrAl cell under hydrostatic pressure up to 2.0 GPa, where inner and outer cylinders were made of NiCrAl and CuBe alloys, respectively. A mixture of Fluorinert FC-70 and FC77 (1:1) was used as a pressure transmitting medium in the experiments.

2.2 Double perovskite photocatalytic semiconductor

2.2.1 Synthesis of $Ba_2Tb(Bi,Sb)O_6$ samples

The citrate pyrolysis technique as mentioned above was applied to synthesis of $Ba_2Tb(Bi,Sb)O_6$ compounds. After mixing stoichiometric quantities of high purity Ba $(NO_3)_2$, Tb_4O_7, Bi_2O_3, and Sb, the resultant mixture was dissolved in a nitric acid solution at 70–80°C. Furthermore, adding citric acid to the solution, the neutralizing process was in progress by adding aqueous ammonia to it while the pH value of the solution reached $\sim$ 6.9. The transparent solution was dried under stirring on the halogen lamp hot plate, and the self-ignition process occurred in the 0.5 L beaker, resulting in the formation of the porous products. Finally, the precursors were ground into fine powders and they were annealed in air at 900–1000°C for 48-96 h, in order to synthesize the $Ba_2Tb(Bi,Sb)O_6$ double perovskite phase (samples #T1 and #T2).

For scanning electron microscope (SEM) measurements, Ba_2TbBiO_6 polycrystalline film on Ag substrate was fabricated from the single-phase powders by an electrophoretic deposition technique. The SEM image revealed the surface morphology and shape of the Ba_2TbBiO_6 powder sample.

2.2.2 Physical properties of $Ba_2Tb(Bi,Sb)O_6$ samples

We measured optical spectra by a diffuse reflectance method using a spectrophotometer (Hitachi U-3500) with the reference material of $BaSO_4$. The energy band gaps for the powder samples were estimated from the reflectance data on the basis of conventional Kubelka-Munk function [18, 28]. The *dc* magnetization measurement was performed over a wide range of temperatures under the magnetic field cooling process of 0.1 T using a SQUID magnetometer.

The effective magnetic moment (μ_{eff}) estimated from the the magnetic susceptibility data using the Curie–Weiss law gives rise to the average valence of the Tb ion at the B site, which is related to the ratio of the Tb^{3+} and Tb^{4+} ions.

2.2.3 Photocatalytic properties of $Ba_2Tb(Bi,Sb)O_6$ samples

We conducted the gaseous 2-propanol (IPA) and methylene blue (MB) degradation experiments, to evaluate photocatalytic activities of the powder samples (in detail, refer to [18, 29]). The visible light radiation experiment started after the IPA gas concentration reached constant under the dark condition. This conformation suggested that the IPA gas finished absorbing on the surface of powders. We used a 300 W Xe lamp equipped with UV and IR filtering functions (Cermax LX300F,

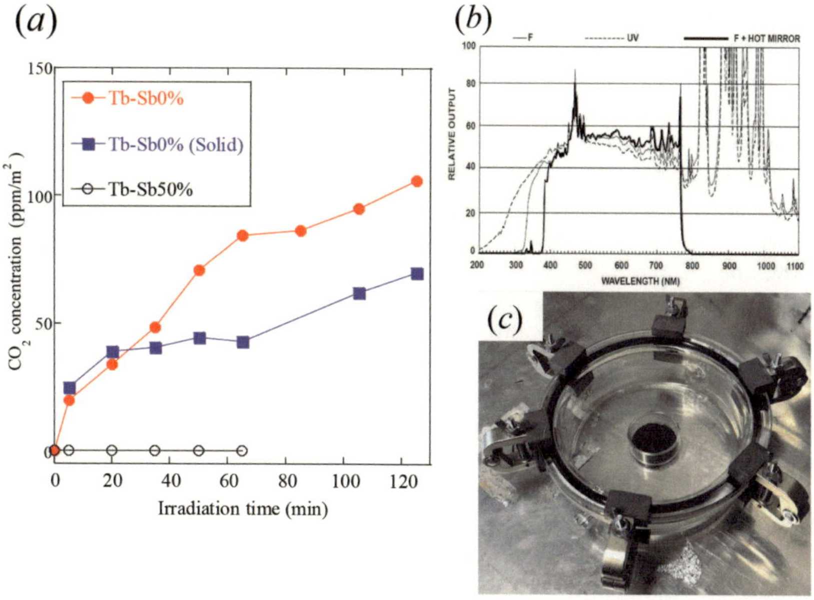

Figure 4.
(a) Photocatalytic activities of IPA decomposition vs visible light irradiation time for Ba_2TbBiO_6 citrate pyrolysis and solid-state samples. For comparison, the data for the Sb50% substituted sample are given. The gaseous concentration is normalized by the surface area of the powder samples. (b) Illuminating spectra of the Xe lamp limited in the visible wavelength range between 390 and 780 nm. (c) Photograph of 0.5-L glass reactor vessel for the IPA experiment. The powder sample (about 1 g) placed on the bottom of a small glass cell was set in its vessel.

Excelitas Technologies). The illuminating spectra of the Xe lamp covered the visible wavelength range from 390 to 780 nm (**Figure 4b**). We set the powder samples (about 1 g) placed on the bottom of a small glass cell in a 0.5-L glass reactor vessel (**Figure 4c**) and injected the dilute IPA gas (5 cc) into its vessel with a syringe.

It is well known that the IPA gas under photocatalytic reaction is finally decomposed into CO_2 [29]. Accordingly, the CO_2 concentration was measured as a function of irradiation time using a gas chromatography system (GC-2014, Shimazu Co.). The photocatalytic methylene blue degradation was carried out using 0.5 g of powders suspended in 50 mL of MB solution (10 ppm). The MB solution was stirred in dark for 30 min before starting visible light radiation. The bleaching of MB was measured using the UV–visible spectrometer (V550, JASCO Co.). The solution of about 3 mL was transferred from the 100 mL reactor vessel under light irradiation and it was then analyzed at the regular time interval, to determine the corresponding MB concentration.

3. Results and discussion

3.1 Double-chain based superconductor $Pr_2Ba_4Cu_7O_{15-\delta}$

3.1.1 Structural and superconducting properties of $Pr_2Ba_4Cu_7O_{15-\delta}$

X-ray diffraction data revealed that the as-sintered polycrystalline samples are an almost single phase with an orthorhombic structure (*Ammm*), as shown in **Figure 3a**. The lattice parameters of the as-sintered sample prepared using the three-zone controlled furnace are $a = 3.8919$ Å, $b = 3.9143$ Å, and $c = 50.7927$ Å.

These values fairly agree with the lattice parameters of Pr247 estimated by a previous study [30]. It was from gravimetric analysis estimated that the oxygen deficiencies in the 48 and 72 h reduced samples in a vacuum are $\delta =\sim 0.56$ and $\delta = \sim 0.72$, corresponding to samples #2–1, #2–2 and sample #1, respectively (**Table 1**). According to a variation in T_c as a function of the oxygen deficiency [31], the $T_{c,on}$ shows a rapid increase at $\delta \geq \sim 0.2$, then follows a stable increase with δ, and finally remains saturation around 26–27 K at $\delta \geq \sim 0.6$. Therefore, we conclude that the doped carrier concentrations in the present samples are located near the optimally doped region.

First of all, the temperature dependences of electric resistivities of the Pr123, Pr124, and as-sintered Pr247 compounds are shown in **Figure 5a**. The Pr123 and Pr124 samples exhibit semiconducting and metallic behaviors in ρ. The as-sintered Pr247 sample shows a weakly semiconducting property at high temperatures, it then reaches a maximum peak at intermediate tempertures around ~ 150 K, and finally it follows a metallic behavior with lowering T. It is well known that CuO single chains in Pr123 and CuO double chains in Pr124 show semiconducting and metallic behaviors, respectively. The reduction heat treatment on the as-sintered sample in a vacuum results in the appearance of superconducting state with T_c=22-27 K, accompanied by the strongly metallic properties over a wide range of temperature (**Figure 6a**). For the present Pr247 sample with a reduction treatment for 72 h, a sharp superconducting transition appears at an onset temperature $T_{c,on}$=26 K and then it achieves a zero resistance state at $T_{c,zero}$=22 K.

Furthermore, to check bulk superconductivity, we performed to measure low-temperature dependence of magnetic susceptibility χ of the 72-h-reduced superconducting sample measured at 5 mT under ZFC scan. **Figure 6b** exhibits diamagnetic signals below $T_{c,on}$ = 26 K for the 72-h-reduced sample. In addition, the superconducting volume fraction due to the shielding effect is estimated to be $\sim 58\%$ from the ZFC values at 5 K. In the inset of **Figure 6**, the magnified data are plotted to clarify the definition of $T_{c,on}$. We note that the quenching procedure in

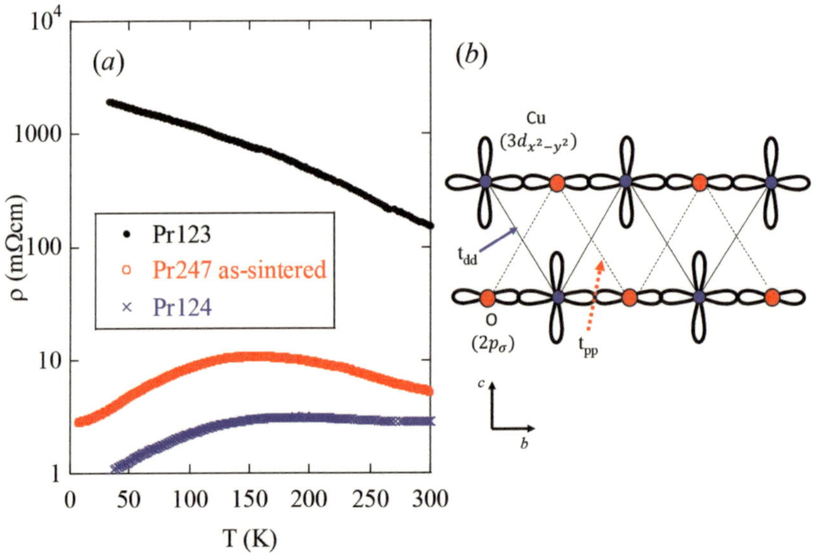

Figure 5.
(a) Temperature dependences of electric resistivities of the Pr123, Pr124, and as-sintered Pr247 compounds. (b) Schematic view of Cu($3d_{x^2-y^2}$) orbitals and O($2p_\sigma$) orbitals in a CuO double chain of $Pr_2Ba_4Cu_7O_{15-\delta}$. Here, t_{pp} and t_{dd} denote the hoping term between $2p_\sigma$ orbitals at the nearest neighbor oxygen sites and that between $3d_{x^2-y^2}$ orbitals at the nearest neighbor copper sites, respectively [32].

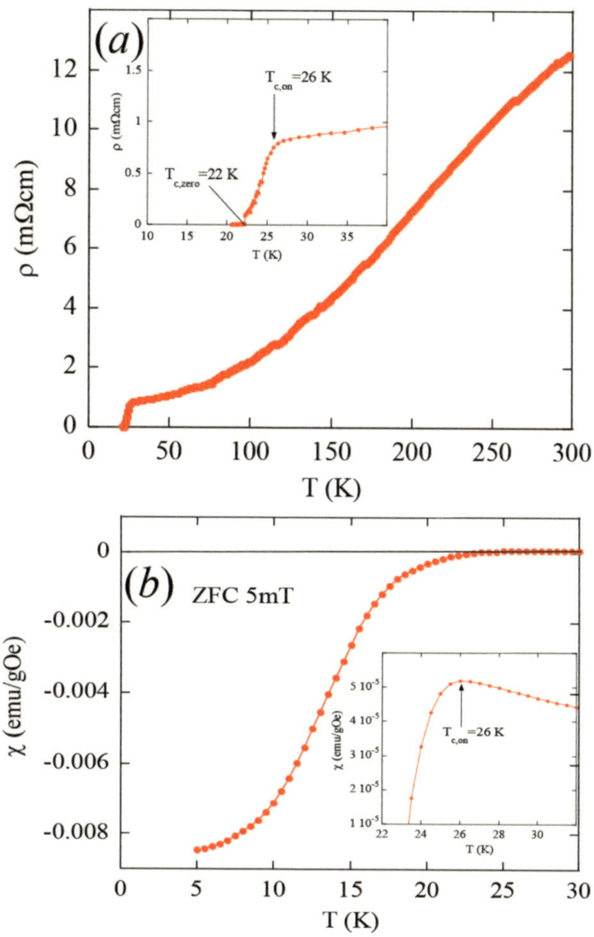

Figure 6.
(a) Temperature dependence of electric resistivity of the 72-h-reduced superconducting Pr247 compound. (b) Low-temperature dependence of magnetic susceptibility χ of the 72-h-reduced superconducting sample measured at 5 mT under ZFC scan. In the insets, the magnified data are plotted to clarify the definition of $T_{c,on}$.

air promotes a sharp superconducting transition appearing at temperatures between $T_{c,on}$=26 K and $T_{c,zero}$=22 K.

3.1.2 Hall and pressure effects of $Pr_2Ba_4Cu_7O_{15-\delta}$

Furthermore, in **Figure 7**, we show the temperature dependences of the Hall coefficients R_H for the as-sintered non-superconducting and 48-h-reduced superconducting samples of $Pr_2Ba_4Cu_7O_{15-\delta}$. For comparison, the R_H values of the as-sintered sample are taken from our previous work [33], which are similar, in magnitude and temperature dependence, to R_H of Pr124 with a metallic CuO double-chain block. (**Figure 1a**) For the 48-h-reduced sample, the R_H data exhibit negative values in the limited temperature range between 30 and 100 K, accompanied by electron doping due to the reduced heat treatment in a vacuum. Moreover, we estimate $R_H = -1.1 \times 10^{-3}$ cm^3/C at 30 K, which is in good agreement with the published data [33].

In **Figure 7b**, we show the temperature dependences of electrical resistivities of the 48-h-reduced superconducting sample under hydrostatic pressures up to 2.0 GPa. The application of external pressure on the 48-h-reduced sample suppressed the superconductivity with increasing the applied pressure. In the case

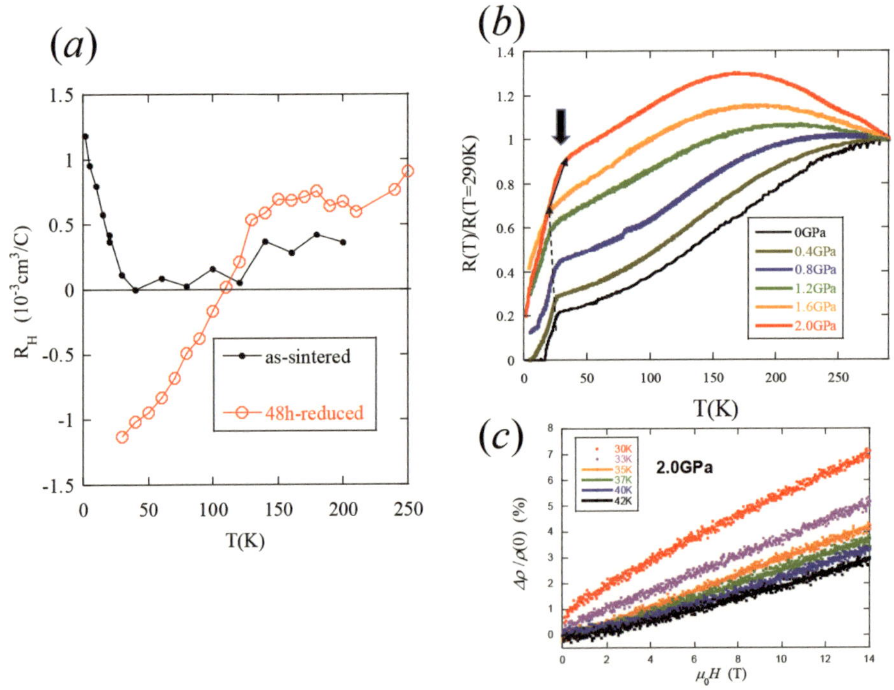

Figure 7.
(a) Temperature dependences of Hall coefficients for the as-sintered non-superconducting and 48-h-reduced superconducting Pr247 compounds. For comparison, the as-sintered data are cited from our previous paper [34]. (b) Temperature dependences of electrical resistivities of the 48-h-reduced superconducting Pr247 under various pressures up to 2.0 GPa. (c) Magneto-resistance effect (up to 14 T) of the 48-h-reduced superconducting Pr247 for temperatures close to 30 K under a maximum pressure of 2.0GPa.

of applied pressures above 0.8 GPa, the zero-resistance state vanished and the high-temperature metallic properties were transferred to the semi-conducting behaviors, accompanied by a rapid increase in ρ. The onset temperature of superconducting transition $T_{c,on}$ declined gradually from 26.5 K at ambient pressure through 24.1 K at 0.8 GPa down to 18.0 K at 1.6 GPa. However, the onset temperature was enhanced up to $\sim$ 30 K under a maximum pressure of 2.0 GPa.

The electronic phase diagram between normal and superconducting phases of a CuO double chain model has been clarified using the Tomonaga-Luttinger Liquid theory [32]. In the case of a shrinkage of the lattice spacing along *c*-axis between the two single chains of a CuO double-chain block, we expect the enhancement of both carrier hopping energies, t_{pp} and t_{dd}. Here, we define the hoping term between $2p_\sigma$ orbitals at the nearest neighbor oxygen sites and that between $3d_{x^2-y^2}$ orbitals at the nearest neighbor copper sites, as t_{pp} and t_{dd}, respectively (see **Figure 5b**). If we apply the external pressure on Pr247 including the CuO double chain block, it is theoretically predicted that the pressure induced enhancement of the hopping terms will result in a phase transition from the superconducting to normal phase. This theoretical prediction is qualitatively in agreement with the negative pressure effect on the superconducting phase observed in Pr247 [34].

We examined the magneto-resistance effect (up to 14 T) of the 48-h-reduced superconducting Pr247 for temperatures close to 30 K under a maximum pressure of 2.0 GPa, to establish a phase boundary between the superconducting and normal states. In **Figure 7c**, the MR data around 30 K tend to increase according to the upward covey behaviors at low fields. On the other hand, the MR around $\sim$40 K shows weak increases in the downward convey forms which is related to the model of slightly warped Fermi surfaces. (in details, see [27]) This finding indicates that the applied magnetic field destroyed the superconducting isolated regions and enlarged further the normal-state majority phase, resulting in the observed MR phenomena near 30 K. The re-entrant superconducting behavior observed at 2.0 GPa is an open question to be resolved in future through the structure analysis under applied pressures [35].

3.2 Double perovskite photocatalytic semiconductor

3.2.1 Structural and valence-state properties of $Ba_2Tb(Bi,Sb)O_6$ samples

X-ray diffraction patterns of Ba_2TbBiO_6 are shown in **Figure 8**. For the parent Ba_2TbBiO_6 with a monoclinic structure (the space group $I2/m$), the lattice parameters were estimated from the x-ray diffraction data to be $a = 6.1099$ Å, $b = 6.0822$ Å, $c = 8.5939$ Å and $\beta = 89.888°$, which fairly agree with previous data [20]. The peak intensity of (101) reflection (the inset of **Figure 8**) is responsible for the degree of B-site ordering in the double-perovskite crystal structures. When the B site ordering is assumed to be ~70 %, the tiny profiles around the (101) peak are well fitted by the least squared method using the RIETAN-FP program [37].

For the $Ba_2Pr(Bi,Sb)O_6$ system, the polycrystalline samples for $x < 0.5$ are formed in almost single phases of the monoclinic structure, while the $x \geq 0.5$ samples crystallize in a cubic structure with the space group $Fm\bar{3}m$ [24]. Substitution of the smaller Sb^{5+} (0.60 Å) ion at the Bi^{5+} (0.76 Å) site causes a monotonic decrease in the lattice parameters. For $Ba_2TbBi_{0.5}Sb_{0.5}O_6$, we obtain the lattice parameters $a = 8.4511$ Å and $\alpha = 90°$. **Figure 8b** shows tolerance factor vs. Sb content (x) for $Ba_2^{2+}Tb^{3+}Bi_{1-x}^{5+}Sb_x^{5+}O_6$ and $Ba_2^{2+}Tb^{4+}M1_{1-x}^{4+}M2_x^{4+}O_6$ with $M1^{4+} = Bi_{0.5}^{3+}Bi_{0.5}^{5+}$ and

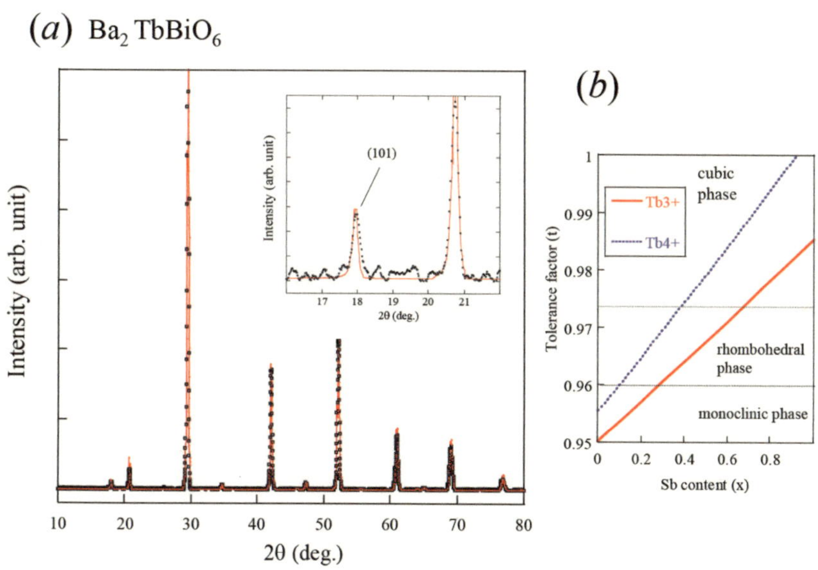

Figure 8.
(a)X-ray diffraction patterns of Ba_2TbBiO_6. Inset shows the enlarged diffraction data. The emergence of (101) reflection indicates B-cation ordering which is characteristic of the ordered double-perovskite structure. (b) Tolerance factor vs Sb content for $Ba_2TbBi_{1-x}Sb_xO_6$. The solid and dotted lines denote $Ba_2^{2+}Tb^{3+}Bi_{1-x}^{5+}Sb_x^{5+}O_6$ and $Ba_2^{2+}Tb^{4+}M1_{1-x}^{4+}M2_x^{4+}O_6$ with $M1^{4+} = Bi_{0.5}^{3+}Bi_{0.5}^{5+}$ and $M2^{4+} = Sb_{0.5}^{3+}Sb_{0.5}^{5+}$. The crystallographic phase diagram consisting of the monoclinic, rhombohedral and cubic phases is given as a function of tolerance factor in ref. [36].

$M2^{4+} = Sb^{3+}_{0.5}Sb^{5+}_{0.5}$. The tolerance factor of double perovskite compounds $Ba_2(Pr,Tb)(Bi,Sb)O_6$ is given by the following equation,

$$t = \frac{r_{Ba} + r_O}{\sqrt{2}\left(\frac{r_{Tb}+r_M}{2} + r_O\right)}, \tag{1}$$

where r_{Ba}, r_O, r_{Tb}, and $r_{M=(Bi,Sb)}$ are the ionic radii of the respective ions (in details, refer to [24]). The crystallographic phase diagram consisting of the monoclinic, rhombohedral and cubic phases is given as a function of tolerance factor in ref. [36]. The crystal structures obtained for the x = 0 and x = 0.5 samples are almost consistent with the phase diagram reported. Assuming the tetravalent state of Tb^{4+}, we obtain the value of t = 0.97914, indicating the stability of a cubic structure for the x = 0.5 sample. The microstructures and pelletized precursors for the Ba_2TbBiO_6 parent sample prepared by the citrate pyrolysis method are shown in **Figure 2b** and **c**. The crystalline grains of the citrate sample have an average size ranging from about 0.2 to 0.5 micron. On the other hand, the grain diameters of the solid-sate sample are distributed on a micron order scale and about one-order grater than those of the former. The citrate pyrolysis process fabricates uniformly dispersed grains with sub micron size compared with the solid-state preparation technique (see [24]).

The magnetic susceptibility data for the $Ba_2Tb(Bi_{1-x},Sb_x)O_6$ compounds (x=0 and 0.5) were measured as a function of temperature under a magnetic field of 0.1 T. (not shown here) The effective magnetic moments (μ_{eff}) are estimated from the magnetization data using the Curie–Weiss law. For the parent and x = 0.5 citrate samples, we obtain μ_{eff} = 8.91 μ_B and 8.86 μ_B, as listed in **Table 2**. Moreover, we try to evaluate the ratio of the Tb^{3+} and Tb^{4+} ions using the equation,

$$\mu^2_{eff} = y\mu^2_{eff}\left(Tb^{3+}\right) + (1-y)\mu^2_{eff}\left(Tb^{4+}\right) \tag{2}$$

where $\mu_{eff}(Tb^{3+})$ = 9.72 μ_B and $\mu_{eff}(Tb^{4+})$ = 7.94 μ_B. For the parent and x = 0.5 citrate samples, we obtain that the ratio of Tb^{3+} and Tb^{4+} ions is 0.52 : 0.48 and 0.49 : 0.51, respectively. The mixed valence state expected from the magnetic data qualitatively consists with the above discussion on the stability of cubic structure for the x = 0.5 sample. The magnetic data suggest that about half of Re ions (Re = Pr and Tb)

Sample no.	Composition	Synthetic method	Crystal symmetry	μ_{eff} (μ_B)	Experimental data
#T1	Ba_2TbBiO_6	citrate pyrolysis	monoclinic	8.91	IPA, MB, Opt.
#T2	Ba_2TbMO_6 M = $Bi_{0.5}Sb_{0.5}$	citrate pyrolysis	cubic	8.86	IPA, MB, Opt.
#T3	Ba_2TbBiO_6	solid state	monoclinic	9.07	IPA, MB, Opt.
#P1	Ba_2PrBiO_6	citrate pyrolysis	monoclinic[a]	3.08[a]	MB
#P2	Ba_2PrSbO_6	citrate pyrolysis	cubic[a]	3.0[a]	MB
#P3	Ba_2PrBiO_6	solid state	monoclinic[b]	3.15[b]	MB

The effective magnetic moments μ_{eff} were estimated from the magnetization data using the Curie–Weiss law. IPA, MB and Opt. denote gaseous 2-propanol decomposition, methylene blue degradation, and optical measurements, respectively.
[a]Ref. [24].
[b]Ref. [38].

Table 2.
Sample details of $Ba_2(Pr,Tb)(Bi,Sb)O_6$ used in the experiments.

are oxidized to the tetravalent state over the whole range of Sb substitution [24]. In a previous analysis of X-ray photoemission spectroscopy [39], it has been shown that a predominant peak of Pr^{3+} coexists with a smaller shoulder structure of Pr^{4+}, giving further evidence for the mixed valence state of the Pr ion.

3.2.2 Photocatalytic and optical properties of $Ba_2Tb(Bi,Sb)O_6$ samples

Next, the visible-light induced decomposition experiment using IPA gas was carried out, to examine photocatalytic activities of the powder samples.

In **Figure 4a**, the temporal variation of evolved CO_2 concentration after visible light irradiation are shown for the Ba_2TbBiO_6 citrate and solid state samples. For comparison, the data for the Sb50% substituted $Ba_2Tb(Bi_{0.5},Sb_{0.5})O_6$ sample are also given. The CO_2 concentrations for both the citrate and solid state x = 0 samples rapidly rise at the initial 20 min and then show a gradual increase at further illumination time. For the Sb50% substituted sample, no clear evolution of CO_2 was detected. It is expected that the heavy substitution of Sb ion at the B-site causes the band gap opening and considerably reduces the formation of electron–hole pairs, resulting in a strong suppression of the photocatalytic reaction processes. In our previous study [38], we investigated the influence of the band gap opening due to the Sb substitution on the basis of first-principles electric structure calculation. The Sb substitution at Bi site removes the Bi-orbitals and makes the corresponding band gap enlarged. The photocatalytic activity exhibits strong dependence of the Sb substitution, which is associated with the enhancement of the band gap energy. It is true that the photocatalytic behavior for the solid state sample is similar to that of the citrate sample. However, the evolved CO_2 quantities of the x=0 citrate sample are about twice as large as the data of the corresponding solid state sample, as shown in **Figure 4a**. The improved photocatalytic activity of the Ba_2TbBiO_6 citrate sample is attributed to its morphology, where fine polycrystalline grains with a sub micron order are homogeneously dispersed.

Furthermore, we demonstrated photocatalytic degradation of methylene blue (MB) vs. visible light irradiation time for the end-member samples with $Ba_2(Pr,Tb)BiO_6$ and Ba_2PrSbO_6 compositions. For comparison, the MB data of the Ba_2PrSbO_6 solid-state sample are given. Here, the MB degradation rate (%) is given by $(C(0) - C(t))/C(0) \times 100$, where the peak intensities located around λ = 665 nm at the initial and final concentrations at different time intervals are defined as $C(0)$ and C(t), respectively. **Figure 9a** shows the MB degradation rate after the visible light irradiation for Ba_2PrBiO_6 and Ba_2PrSbO_6 samples. In **Figure 9b** and **c**, the typical absorbance spectra and corresponding MB solutions for Ba_2PrSbO_6 at different irradiation time intervals are also displayed. The MB degradation rates under visible light irradiation show rapid increases due to Sb-substitution. For the Sb50% substituted $Ba_2Tb(Bi_{0.5},Sb_{0.5})O_6$ sample, the highest performance of MB degradation was observed. The citrate parent sample of Ba_2PrSbO_6 exhibits higher degradation in comparison with the data of the solid-state sample with the identical composition. For Ba_2PrBiO_6, the effect of Sb-substitution on the photocatalytic degradation of MB is in direct contrast to that on the IPA decomposition under visible light irradiation.

Finally, we carried out the optical measurements on the $Ba_2Tb(Bi_{1-x},Sb_x)O_6$ (x=0 and 0.5) powder samples by the diffuse reflectance method. In the first step, we transformed the observed reflectance data to the absorption coefficient α_{KM} by applying the conventional Kubelka-Munk function. In the next step, extrapolating the tangent line to the ε_p axis near the band edge, we evaluate the energy band gap from the intersection following the equation

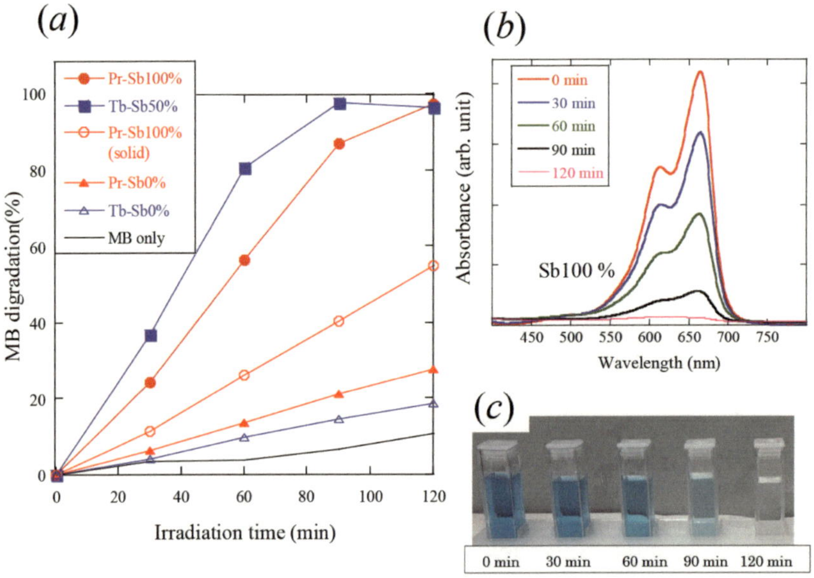

Figure 9.
(a) Photocatalytic degradation of methylene blue (MB) vs visible light irradiation time for $Ba_2(Pr,Tb)(Bi,Sb)O_6$ citrate pyrolysis samples. For comparison, the MB data of the Ba_2PrSbO_6 solid-state sample are given. (b) Absorbance spectra as a function of visible light irradiation time for the MB degradation in the case of Ba_2PrSbO_6 citrate pyrolysis sample. (c) Photocatalytic variations of the MB solutions at different irradiation time intervals.

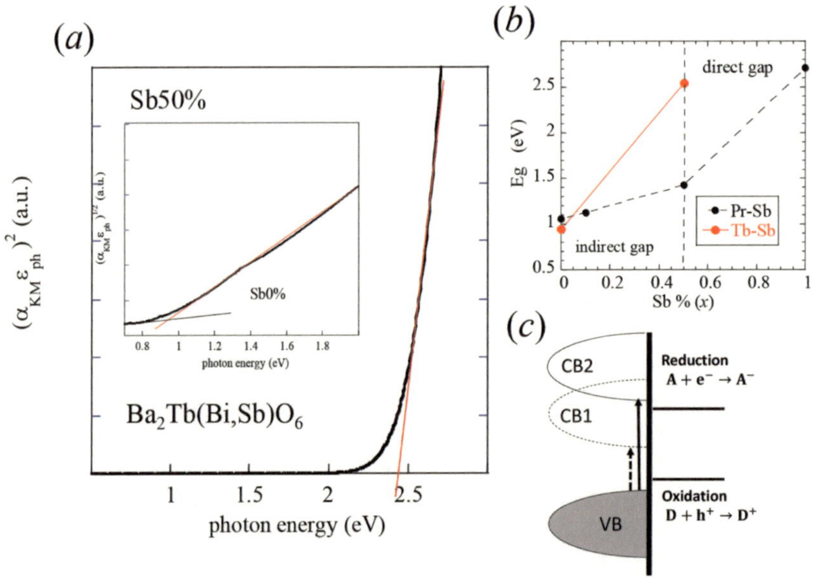

Figure 10.
(a) Optical properties of $Ba_2Tb(Bi_{1-x},Sb_x)O_6$ (x=0 and 0.5). $(\alpha_{KM}\varepsilon_p)^2$ vs ε_p for the $x = 0.5$ sample are plotted as a function of ε_p. The inset shows plots of $(\alpha_{KM}\varepsilon_p)^{1/2}$ vs ε_p. For the $x = 0$ sample, Eg is estimated from an intersection point of base line and straight line by extrapolation. (b) Band gap energies vs Sb content (x) for $Ba_2Tb(Bi_{1-x},Sb_x)O_6$ (x=0 and 0.5). The data of $Ba_2Pr(Bi,Sb)O_6$ are cited from our previous work [24]. (c) Schematic illustration between the semiconductor band positions and the the surface redox reactions in photocatalysis. The valence band (VB) is located to facilitate oxidation, but the lower conduction band (CB1) is not sufficiently positioned to facilitate reduction. The higher conduction band (CB2) is suitable for promoting effective reduction process.

$$(\alpha\varepsilon_p)^n \propto (\varepsilon_p - E_g), \tag{3}$$

where α, ε_p, and E_g are the absorption coefficient, the photon energy, and the band gap energy. The exponents, $n = 2$ and $n = 1/2$, are responsible for direct and indirect optical transitions, respectively [18, 19]. In **Figure 10a**, the absorption coefficient, $(\alpha_{KM}\varepsilon_p)^2$ is plotted as a function of ε_p for the x=0.5 sample. In the inset of **Figure 10a**, $(\alpha_{KM}\varepsilon_p)^{1/2}$ vs. ε_p are shown for the x=0 sample. We obtain that E_g = 0.92 eV at x=0 and 2.45 eV at x=0.5, assuming indirect and direct photon transitions, respectively. The magnitude of the energy band gap is enhanced due to the Sb substitution.

In **Figure 10c**, schematic view between the semiconductor band positions and the the surface redox reactions in photocatalysis is presented. The valence band (VB) is located to facilitate oxidation, but the lower conduction band (CB1) is not sufficiently positioned to facilitate reduction. The higher conduction band (CB2) is needed to promote reduction process. Applying the present band model to the photocatalytic activities observed for $Ba_2Tb(Bi,Sb)O_6$, the parent compounds with smaller energy band gap of nearly 1 eV are responsible for the lower energy level of conduction band (CB1) edge. On the other hand, from the effect of band gap opening due to Sb substitution, we believe that the CB2 band model is valid in the heavily Sb substituted samples. The former is closely related to facilitate oxidation reactions at the surface for the IPA decomposition process. In the latter compounds with the larger energy band gaps of $\sim$2.5 eV, the MB degradation is strongly promoted in contrast to that of the former compounds. These findings indicate that the conduction band edge of the heavily Sb substituted samples is optimized.

4. Conclusions

A citrate pyrolysis technique is a unique route to prepare reactive precursor mixtures through an ignition process of concentrated aqueous solution including metallic ions of stoichiometric composition. This procedure enables to synthesize highly homogeneous and fine powders for functional materials. The double-chain based superconductor $Pr_2Ba_4Cu_7O_{15-\delta}$ and double perovskite photocatalytic semiconductor $Ba_2Tb(Bi,Sb)O_6$ were synthesized by using the citrate pyrolysis technique.

The TEM image of the 48-h-reduced Pr247 revealed that CuO single-chain and double-chain blocks are alternately stacked along the c-axis such as {-D-S-D-S-} sequence. For the present Pr247 sample with a reduction treatment for 72 h, a sharp superconducting transition appeared at an onset temperature $T_{c,on}$=26 K accompanied by a zero-resistance state at $T_{c,zero}$=22 K. The superconducting volume fraction estimated from the magnetization measurement reached an excellent value of $\sim$ 58%. Both reduction treatment in a vacuum and subsequent quenching procedure are needed to realize higher superconductivity due to further oxygen defects. For the 48-h-reduced sample, the R_H data exhibited negative values in the limited temperature range between 30 and 100 K, accompanied by electron doping due to the reduced heat treatment in a vacuum. The re-entrant superconducting behavior observed at 2.0 GPa is an open question to be resolved in future through detailed structure analysis under applied pressures.

The polycrystalline samples for $Ba_2Tb(Bi_{1-x},Sb_x)O_6$ ($x = 0$ and 0.5) were formed in the monoclinic and cubic crystal structures. The magnetic data suggest that about half of Tb ions are oxidized to the tetravalent state over the wide range of Sb substitution, which is a common trend in the case of $Ba_2Pr(Bi,Sb)O_6$. For Ba_2Tb

$(Bi_{1-x},Sb_x)O_6$, we estimated E_g = 0.92 eV at x=0 and 2.45 eV at x=0.5, assuming indirect and direct photon transitions, respectively.

We conducted the gaseous 2-propanol (IPA) and methylene blue (MB) degradation experiments under a visible light irradiation, to evaluate photocatalytic activities of the powder samples. The band gap opening due to the heavy Sb substitution suppresses the formation of electron–hole pairs, causing a decrease of the photocatalytic reaction processes in the IPA decomposition. The MB degradation rates under visible light irradiation show rapid increases due to Sb-substitution. For the Sb50% substituted $Ba_2Tb(Bi_{0.5},Sb_{0.5})O_6$ sample, the highest performance of MB degradation was observed. In the $Ba_2(Pr,Tb)(Bi,Sb)O_6$ system, the effect of Sb-substitution on the photocatalytic degradation of MB is in direct contrast to that on the IPA decomposition under visible light irradiation. We conclude that the citrate pyrolysis samples of $Ba_2Tb(Bi,Sb)O_6$ exhibit excellent performances in comparison with the data of the solid-state samples with the identical composition. The enhanced photocatalytic properties in the citrate samples are attributed to their morphology, where fine particles are homogeneously distributed with a sub micron order.

Acknowledgements

This work was supported in part by MEXT Grands-in-Aid for Scientific Research (JPSJ KAKENHI Grants No. JP19K04995). We thank A. Sato, T. Teramura, and Y. Nakarokkaku for their supports in experiments.

Author details

Tatsuya Senzaki[1], Michiaki Matsukawa[1*], Takanori Yonai[1], Haruka Taniguchi[1], Akiyuki Matsushita[2], Takahiko Sasaki[3] and Mokoto Hagiwara[4]

1 Faculty of Science and Engineering, Iwate University, Morioka, Japan

2 National Institute for Materials Science, Tsukuba, Japan

3 Institute for Materials Research, Tohoku University, Sendai, Japan

4 Kyoto Institute of Technology, Kyoto, Japan

*Address all correspondence to: matsukawa@iwate-u.ac.jp

IntechOpen

© 2021 The Author(s). Licensee IntechOpen. This chapter is distributed under the terms of the Creative Commons Attribution License (http://creativecommons.org/licenses/by/3.0), which permits unrestricted use, distribution, and reproduction in any medium, provided the original work is properly cited.

References

[1] L. A. Chick, L. R. Pederson, G. D. Maupin, J. L. Bates, L/ E. Thomas, G. J. Exarhos, Materials Letters, vol. 10, 6-12, 1990.

[2] K. Koyama, A. Junod, T. Graf, G. Triscone, and J. Muller, Physica C, vol.185-189, 66-70, 1991.

[3] M. Hagiwara, T. Shima, T. Sugano, K. Koyama, and M. Matsuura, Physica C, vol. 445-448, 111-114, 2006.

[4] M. Uehara, T. Nagata, J. Akimitsu, H. Takahashi, N. Mori, and K. Kinoshita, J. Phys. Soc. Jpn. 65 (1996) 2764.

[5] M. Matsukawa, Y. Yamada, M. Chiba, H. Ogasawara, T. Shibata, A. Matsushita, and Y. Takano, Physica C 411 (2004) 101.

[6] S. Watanabe, Y. Yamada and S. Sasaki, Physica C 426-431 (2005) 473.

[7] L. Soderholm, K. Zhang, D. G. Hinks, M. A. Beno, J. D. Jorgensen,C. U. Segre, I. K. Schuller, Nature 328 (1987) 604.

[8] S. Horii, Y. Yamada, H. Ikuta, N. Yamada, Y. Kodama, S. Katano,Y. Funahashi, S. Morii, A. Matsushita, T. Matsumoto, I. Hirabayashi, and U. Mizutani, Physica C 302 (1998) 10.

[9] T. Mizokawa, C. Kim, Z. -X. Shen, A. Ino, T. Yoshida, A. Fujimori, M. Goto, H. Eisaki, S. Uchida, M. Tagami, K. Yoshida, A. I. Rykov, Y. Siohara, K. Tomimoto, S. Tajima, Yuh Yamada, S. Horii, N. Yamada, Yasuji Yamada, and I. Hirabayashi, Phys. Rev. Lett. 85 (2000) 4779.

[10] P. Bordet, C. Chaillout, J. Chenavas, J. L. Hodeau, M. Marezio,J. Karpinski, and E. Kaldis, Nature 334 (1988) 596.

[11] Y. Yamada, S. Horii, N. Yamada, Z. Guo, Y. Kodama, K. Kawamoto,U. Mizutani, and I. Hirabayashi, Physica C 231(1994)131.

[12] T. Konno, M. Matsukawa, H. Taniguchi, J.Echigoya, A. Matsushita, M. Hagiwara, T. Miyazaki, K. Sano, Y. Ono, Y. Yamada, T. Sasaki, and Y. Hayasaka, Physica C521-522 (2016) 13.

[13] S. Vasala, and M. Karppinen, *Prog.* Solid State Chem., vol. 43, pp. 1-31, 2015.

[14] K. -I. Kobayashi, T. Kimura, H. Sawada, K. Terakura, and Y. Tokura, *Nature*, vol. 395, pp. 677, 1998.

[15] R. Ramesh, and N. A. Spaldin, *Nature Materials*, vol. 2, pp. 21, 2007.

[16] A. Fujishima, and K. Honda, *Nature*, vol. 238, pp. 37, 1972.

[17] H. W. Eng, P. W. Barnes, B. M. Auerand, and P. M. Woodward, *J. Solid State Chem.*, vol.175, pp. 94, 2003.

[18] T. Hatakeyama, S. Takeda, F. Ishikawa, A. Ohmura, A. Nakayama, Y. Yamada, A. Matsushita, and J. Yea, *J. Cer. Soc.* Jpn., vol. 118, pp.91-95, 2010.

[19] A. Matsushita, T. Nakane, T. Naka, H. Isago, Y. Yamada, and Yuh Yamada, *Jpn. J. Appl. Phys.*, vol. 51, pp. 121802-1-5, 2012.

[20] William T. A. Harrison, K. P. Reis, A. J. Jacobson, L. F. Schneemeyer, and J. V. Waszczak, *Chem. Mater.*, vol. 7, pp. 2161-2167, 1995.

[21] X. Chen, S. Shen, L. Guo, and S. Mao, *Chem. Rev.* vol.110, 6503-6570, 2010.

[22] D. Sudha, and P. Sivakumar, Chemical Engineering and Processing, vol. 97, 112–133, 2015.

[23] B. Ohtani, *Catalysts*, vol.3, 942-953, 2013.

[24] A. Sato, M. Matsukawa, H. Taniguchi, S. Tsuji, K. Nishidate, S.

Aisawa, A. Matsushita, and K. Zhang, Solid State Science, 107, 10635, 2020.

[25] A. Housa, H. Lachheb, M. Ksibi, E. Elaloui, C. Guillard, J.M. Herrmann, Applied Catalysis B31, 145, 2001.

[26] K. Honnami, M. Matsukawa, T. Senzaki, H. Taniguchi, K. Takahashi, T. Sasaki, M. Hagiwara, Physica C585 (2021) 1353869.

[27] M. Kuwabara, M. Matsukawa, K. Sugawara,H. Taniguchi, A. Matsushita, M. Hagiwara, K. Sano, Y. Ono, and T. Sasaki, J. Phys. Soc. Jpn. 85 (2016) 124704.

[28] J. W. Tang, Z. G. Zou, and J. H. Ye, *J. Phys. Chem.* C, vol. 111, pp. 12779-12785, 2007.

[29] T. Murase, H. Irie, and K. Hashimoto, *J. Phys. Chem. B*, vol. 108, 15803-15807 (2004)

[30] Y. Yamada and A. Matsushita, Physica C 426-431 (2005) 213.

[31] M. Hagiwara, S. Tanaka, T. Shima, K. Gotoh, S. Kanda, T. Saito, and K. Koyama, Physica C 468 (2008) 1217.

[32] K. Sano, Y. Ono, and Y. Yamada, J. Phys. Soc. Jpn., 74 (2005) 2885.

[33] J. Tada, M. Matsukawa, T. Konno, S. Kobayashi, M. Hagiwara, T. Miyazaki, K. Sano, Y. Ono, and A. Matsushita, J. Phys. Soc. Jpn. 82 (2013) 105003.

[34] F. Ishikawa, K. Fukuda, S. Sekiya, A. Kaeriyama, Y. Yamada and A. Matsushita, J. Phys. Soc. Jpn., 76 (2007) Suppl. A 92.

[35] H. Taniguchi, A. Nakayama, M. Matsukawa, S. Nakano, M. Hagiwara and T. Sasaki, J. Phys. Soc. Jpn. 90 (2021) 015001.

[36] S. Otsuka, and Y. Hinatsu, J. Solid State Chem., vol. 227, pp. 132-141, 2015.

[37] F. Izumi and K. Momma, Solid State Phenom. 130 (2007) 15.

[38] H. Taniguchi, M. Matsukawa, K. Onodera, K. Nishidate, and A. Matsushita, *IEEE Transactions on Magnetics*, 55(2), (2019) 2400404.

[39] K. Onodera, T. Kogawa, M. Matsukawa, H. Taniguchi, K. Nishidate, A. Matsushita, and M. Shimoda, *J. Phys. Conf. Ser.*, vol. 969, pp. 012122-1-6, 2018.

Chapter 10

Aerosol Spray Pyrolysis Synthesis of Doped $LiNi_{0.5}Mn_{1.5}O_4$ Cathode Materials for Next Generation Lithium-Ion Batteries

George Ganas, George Karagiannakis, Aitor Eguia-Barrio, Miguel Bengoechea, Iratxe de Meatza and Georgia Kastrinaki

Abstract

The autonomy of next generation Electric Vehicles relies on the development of high energy density automotive batteries. $LiMn_{1.5}Ni_{0.5}O_4$ (spinel structure) is a promising active cathode material in terms of charge rate capability, theoretical capacity, cost and sustainability being a cobalt-free material. In the current study pristine and doped (Fe, Al, Mg) $LiMn_{1.5}Ni_{0.5}O_4$ particles were synthesized by an Aerosol Spray Pyrolysis pilot scale unit in a production rate of 100 gr h^{-1} and were evaluated for their electrochemical activity in Half Coin Cell form. The doped particles were characterized in terms of their surface area, particle size distribution, crystallite size, morphology and ion insertion of the doping element into the $LiNi_{0.5}Mn_{1.5}O_4$ lattice by Raman spectroscopy. The mixed oxide particles had homogeneous composition which is an inert characteristic of aerosol spray pyrolysis synthesis. The electrochemical activity of the material is attributed both to the nanoscale structure, by successful dopant ion insertion into the spinel lattice as well as to optimization of carbon and spinel particle interface contact in the microscale for increase of electrode conductivity.

Keywords: Li-ion batteries, aerosol spray pyrolysis, LNMO, spinel, cathode materials

1. Introduction

The unavoidable consumption of the global fossil fuel reserves in conjunction with the increasing environmental issues, have rendered the storage of electrical energy produced from renewable sources of greater importance than ever before [1]. Rechargeable Lithium-Ion batteries (LIB) have gained considerable attention among various energy storage technologies and are currently widely employed or considered to be deployed in electric devices, electric vehicles (EVs) and grid energy storage systems due to their relatively high energy density, high voltage, better cycle life and environmental friendliness [2–5]. Pyrolysis methods are extensively studied for the synthesis of anode and cathode materials for LIBs [6–11], either as mixed oxide particles [12, 13] or carbon-based structures [14], thereby providing a scalable and sustainable process for the production of electrode materials. The demand for high specific power, energy capacity [15, 16] and safety within the battery lifetime

IntechOpen

has led to the importance of "going nano" [17] in order to optimize electrode electrochemical activity and stability towards lithium ion intercalation (insertion within the material lattice) in the framework of the necessary cycling operation.

Aerosol spray pyrolysis synthesis is an established and scalable method for the synthesis of metal oxides [18], metal [19] and carbon-based [20] materials. The atomized droplets undergo controlled evaporation of solvents and subsequent precipitation of precursor materials by regulating the operating conditions of the process to allow tuning of size distribution, morphology, porosity and to achieve uniform multicomponent composition of the synthesized particles and structures. Furthermore, the residence time of the droplets in the heated reactors is in the time frame of seconds, while further (i.e. post-synthesis) calcination can be performed in powder form. The precursor solutions for aerosol spray pyrolysis are mostly aqueous based, leading only to gaseous byproducts that can be in-line processed or captured, thereby contributing further to an environmentally sustainable and cost-effective process.

To meet the demanding battery standards, a wide variety of cathode materials has been thoroughly studied, such as the layered $LiMO_2$ [21–24], the olivine $LiMPO_4$ [25–27] and the spinel structure of LiM_2O_4. Among the materials of the latter LiM_2O_4 family, $LiNi_{0.5}Mn_{1.5}O_4$ (LNMO) is considered a very promising candidate as a cathode material in both environmental [28, 29] and operational terms, since it exhibits an operating charge/discharge voltage of ~4.7 V vs. Li/Li^+, a theoretical capacity of ~147 $mAh{\cdot}g^{-1}$ [30] and it does not require the use of cobalt. In this work we investigate the doping of Mg, Al and Fe through partial substitution of Ni in the LNMO structure (Mg-LNMO, Al-LNMO and Fe-LNMO respectively) via the Aerosol Spray Pyrolysis (ASP) synthesis technique for a production capacity of approximately 100 $g{\cdot}h^{-1}$. Subsequently, the doping effect on the electrochemical performance of LNMO half cells is assessed experimentally.

2. $LiNi_{0.5}Mn_{1.5}O_4$ spinel structure

LNMO can be formed in two different space groups as it is also exhibited in **Figure 1(a)** the single cubic *P4₃32* phase (ordered), where in the absence of Mn^{3+} the Mn^{4+} and Ni^{2+} ions are located at 12d and 4a sites of the lattice respectively, Li ions occupy the 8c sites and O ions the 8c and 24e sites and b) the face centered *Fd-3 m* phase (disordered) where Mn^{4+} and Ni^{2+} are randomly distributed at 16c sites of the lattice, while Li and O ions are located at 8a and 32e sites respectively [31]. $P4_332$ demonstrates a stoichiometric amount of oxygen atoms, while *Fd-3 m* presents oxygen vacancies, represented by the non-stoichiometric parameter δ. Thus, a $LiNi_{0.5}Mn_{1.5}O_{4-\delta}$ structure appears with certain Mn ions being reduced to Mn^{3+} due to charge balancing [32].

Despite the fact that LNMO exhibits promising electrochemical features, it may demonstrate significant capacity reduction during cycling due to Mn diffusion in the electrolyte mass, caused by various lattice modifications in conjunction with the Jahn-Teller effect [33]. Since the electrochemical performance of LNMO can be affected by the synthesis route and thus the stoichiometry, morphology, crystal structure and impurities [34], many synthesis methods have been employed in order to improve LNMO performance: hydrothermal route [35, 36], co-precipitation [37–40], sol–gel [41, 42], induction thermal plasma [43], molten salt [44–46], liquid phase self-propagating high-temperature synthesis (LPSHS) and aerosol spray pyrolysis [47].

The electrochemical performance of LNMO can be improved via cation doping [48–50] by Mg, Al and Fe. This approach contributes to LNMO lattice stabilization, increase of specific capacity, reduction of the $Li_xNi_{1-x}O$ phase which is formed in

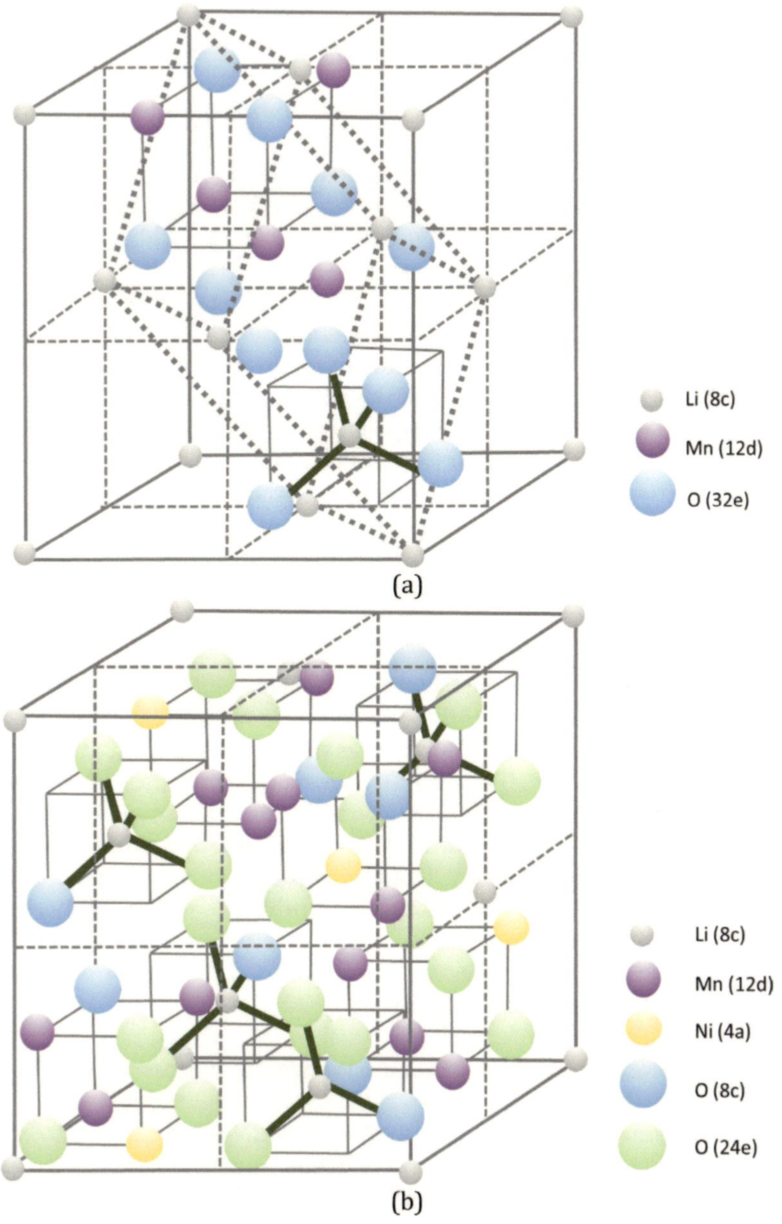

Figure 1.
Crystal phases of LNMO: (a) ordered P4332, (b) disordered Fd-3 m.

the case of the disordered *Fd-3 m* structure and increase of cycle life. Moreover, the doped compositions are less expensive, abundant and without scarcity or toxicity issues such as in the case of cobalt [51–58].

3. Experimental

3.1 Material synthesis

The LNMO material was synthesized by the Aerosol Spray Pyrolysis method [19], which utilizes the principles of bottom up synthesis in droplet micro-reactors.

A precursor solution is atomized into fine droplets which subsequently undergo evaporation of the solvent while transferred by a carrier gas, inside a heated tubular reactor. The advantages of the respective synthesis technique is the short production time of some seconds in the reactor zone and the scalability of the produced material quantities.

In the current synthesis campaigns, the precursor solution was a 1 M aqueous solution of $LiNO_3$ (Merck, ≥98%), $Ni(NO_3)_2{\cdot}6H_2O$ (Merck, ≥99%), $Mn(NO_3)_3{\cdot}4H_2O$ (Alfa Aesar, 98%), $Mg(NO_3)_2{\cdot}6H_2O$ (Sigma Aldrich, 98 + %, $Al(NO_3)_3{\cdot}9H_2O$ (Sigma Aldrich, ≥98%) and $Fe(NO_3)_3{\cdot}9H_2O$ (Merck, ≥99%), in molar ratios of Li:Ni:Mn:D: 1:0.5-x:1.5:x and the temperature of the heated reactor was 800°C. The collected powder was subsequently calcined at 850°C for 16 h (air atmosphere) in a muffle furnace in order to obtain the disordered phase of the LNMO material [11]. The temperature of the reactor controls the evaporation rate of the solvent and thus the chemical composition and morphology of the synthesized particles. The synthesis reactor is a pilot plant unit constructed and operated in CERTH's facilities. In this study, 4 different materials were produced; the pristine LNMO and 3 metal doped LNMO compositions using Mg, Al and Fe as additives, namely Mg-LNMO, Al-LNMO and Fe-LNMO, respectively.

3.2 Structural characterization

The identification of the phase structure was performed by XRD analysis using a Siemens D500/501 X-ray diffractometer with Cu Kα radiation between 5° and 80° at a scan rate of 0.040°/s. SEM/EDS analysis as well as the mapping of the materials were performed with a JEOL JSM-6300 microscope, while the TEM analysis was performed using a JEOL JEM 2010 high-resolution microscope. The ordered/disordered structure was identified with a Raman Renishaw microscope equipped with a 514 nm Argon laser of 50 mW. The specific surface area of the samples (BET method) was measured with the aid of a N_2 adsorption porosimeter (Micromeritics ASAP 2000, at 77 K, after degassing the samples at 250°C). Finally, the particle size distribution of the materials synthesized (powder form) was performed using a TSI PSD 3603 Particle Size Distribution Analyzer.

3.3 Electrode assembly and electrochemical characterization

The electrochemical performance of synthesized LNMO was evaluated at the laboratory scale by preparing electrodes and assembling half-coin cells (HCC). During the cathode formulation, an N-methyl-2-pyrrolidone (NMP) slurry was prepared, consisting in a first approach of 90 wt. % of active material with the rest 10 wt.% being carbon black (C-NERGY Super C65 from IMERYS Carbon & Graphite) and Poly-Vinylidene Fluoride (PVdF HSV900 from Arkema) as electronic conduction enhancer and binder, respectively, while different ratios followed (84/8/8, 80/10/10) when issues where observed on the slurry stability and electrical conductivity of the electrodes. After coating the slurry on an aluminum current collector foil (Showa Denko, 20 μm thickness) using the doctor blade technique to a targeted loading of 1.0–1.5 $mA{\cdot}cm^{-2}$, the resulted electrodes were dried at 120°C under vacuum overnight, and the coin cell assembly followed.

The electrochemical performance of the prepared electrodes has been analyzed by assembling and testing coin cells (CR2025, Hohsen), in HCC configuration using Li metal disk (50 μm thick, Albermale) as counter anode electrode. One layer of polyolefin Celgard 2325 separator was used in all coin cells with 50 μL of 1 M $LiPF_6$ EC/EMC electrolyte from Arkema. The cells were tested in a potential window of 3.5-5 V at room temperature in a working C-rate range between C/5 (5 h) and 5C (12 min). C-rate calculations were made based on a LNMO specific capacity of 135 $mAh{\cdot}g^{-1}$.

4. Physicochemical characterization results

Figure 2(a) depicts the XRD spectra of the synthesized LNMO materials. In all cases a desired well-crystallized $LiNi_{0.5}Mn_{1.5}O_4$ structure was obtained, with the characteristic peaks of the typical cubic structure at 19, 36, 44, 59, 64 2θ values; the higher resolution graph in **Figure 2(b)** referring to the pristine LNMO indicates two weaker peaks at ~37.5 and ~ 43.5 2θ values corresponding to the $Li_xNi_{1-x}O$ phase.

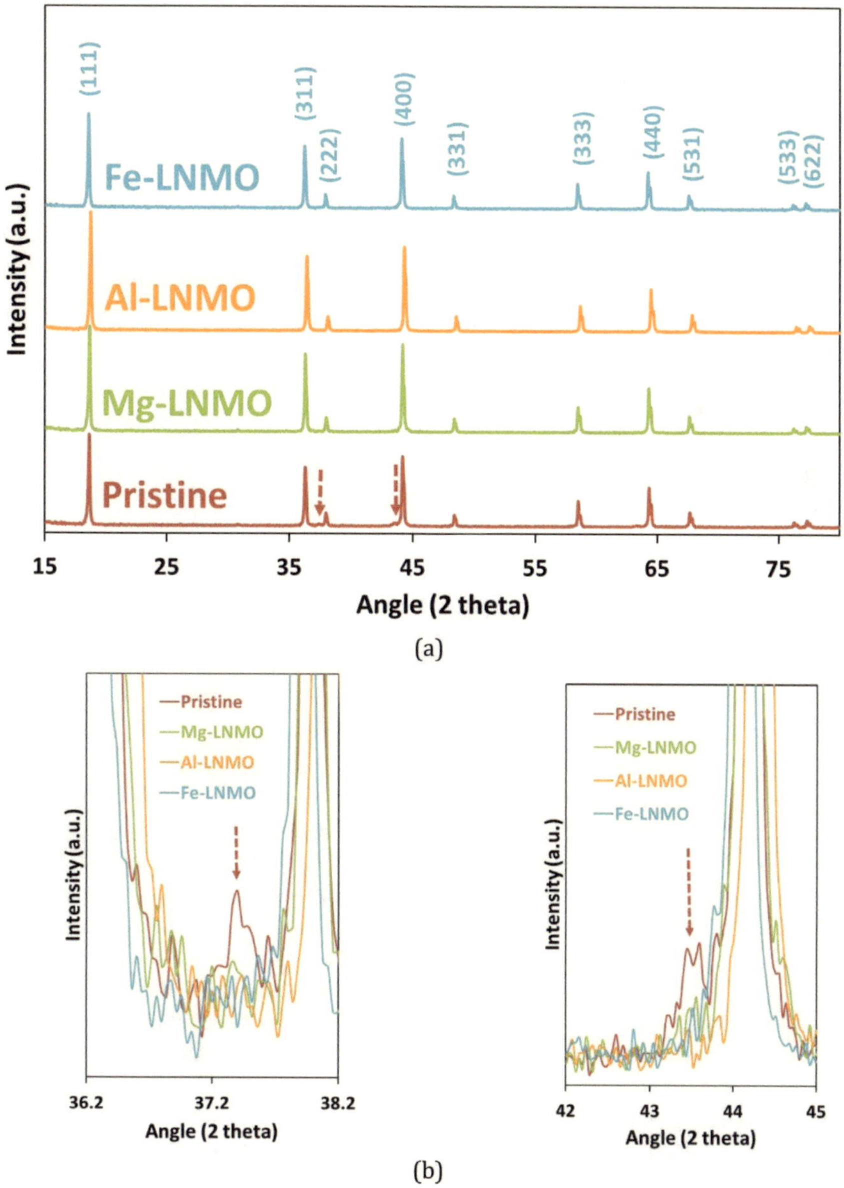

Figure 2.
(a) XRD of the pristine and doped $LiNi_{0.5}Mn_{1.5}O_4$ with Fe, Al and Mg materials (b) higher resolution XRD graph depicting the additional $Li_xNi_{1-x}O$ of the pristine sample.

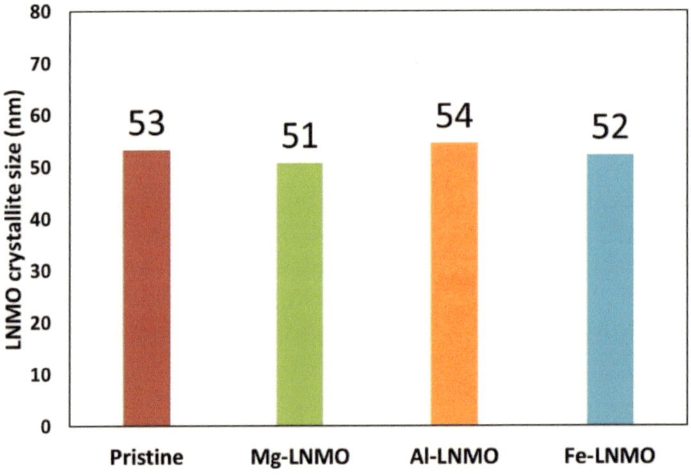

(a)

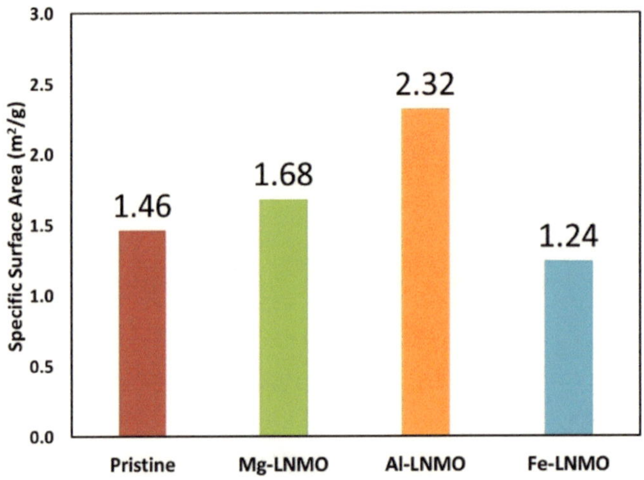

(b)

Figure 3.
Crystallite size (a) and specific surface area (b) of the doped LNMO materials.

On the contrary, in the case of doped LNMO compositions these peaks are not present, assuming, therefore, the elimination of the undesired $Li_xNi_{1-x}O$ phase.

Crystallite size was calculated by the Scherrer's formula from the FWHM of the *(111)* $LiNi_{0.5}Mn_{1.5}O_4$ peak; the calculated sizes in **Figure 3(a)** exhibit crystallites with an average size of ~53 nm for all the materials. Doping seems to affect the specific surface area in a different manner; in **Figure 3(b)** the pristine sample exhibits a 1.46 $m^2 \cdot gr^{-1}$ specific surface area which increases slightly to 1.68 $m^2 \cdot gr^{-1}$ for Mg and significantly to 2.32 $m^2 \cdot gr^{-1}$ for the Al sample, while in accordance to its crystallite size deviation the Fe sample exhibits lower surface area than the pristine at 1.24 $m^2 \cdot gr^{-1}$. The different surface areas can be related to the morphology exhibited by SEM in **Figure 4**, where spherical particles seem to exhibit a non-dense, macroporous structure. Particles synthesized by aerosol synthesis are expected to have a spherical morphology based on the aerosol process principle [19]. SEM images for the pristine and the doped particles depict multiple particulate aggregates forming the macroscopic spherical particles. The EDS mapping images show a homogeneous dispersion of the Ni, Mn and of the respective Mg, Al and Fe dopants per case,

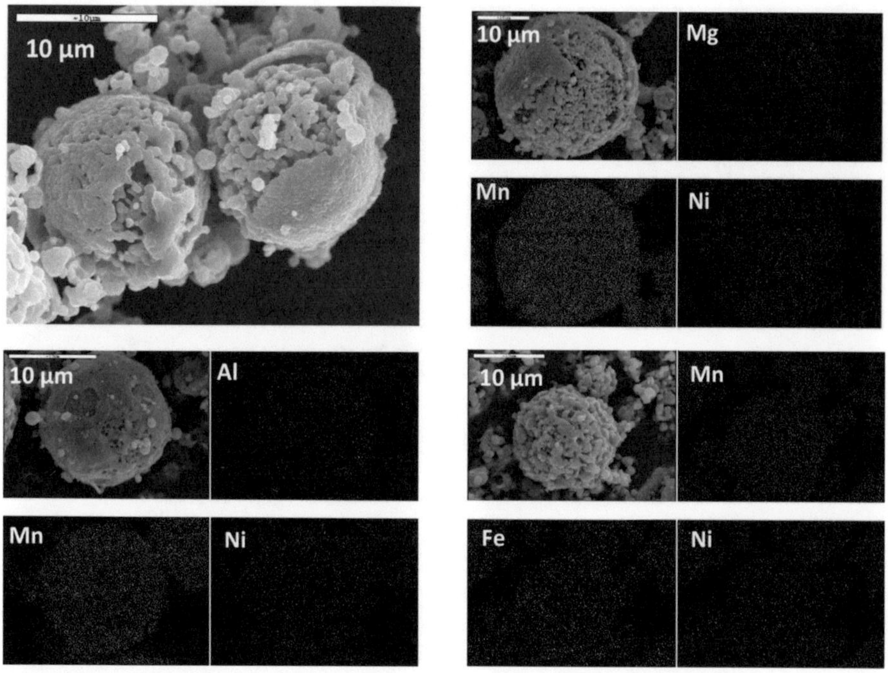

Figure 4.
SEM images and EDS mapping analysis for the pristine and doped LNMO particles.

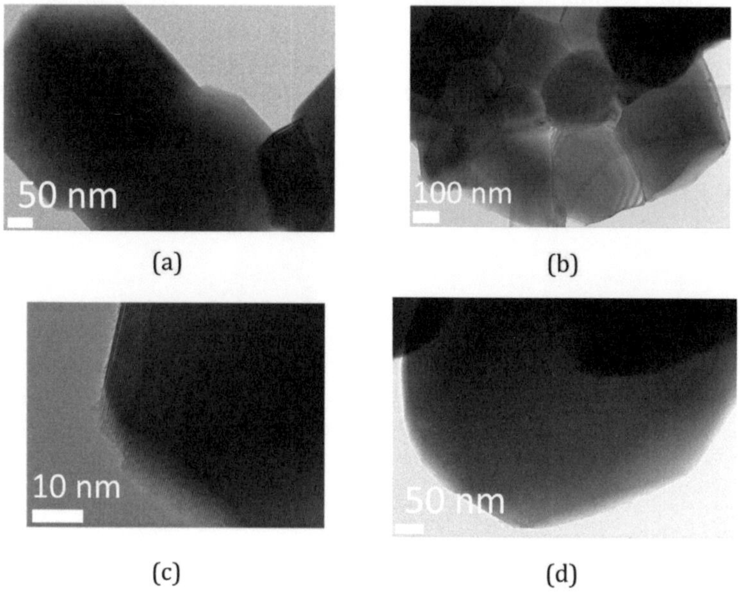

Figure 5.
TEM images depicting the crystallites of the LNMO of (a) pristine, (b) Mg-LNMO, (c) Al-LNMO and (d) Fe-LNMO.

in consistency with the XRD patterns which also confirm the homogeneity of the synthesized structures.

The morphology of the polycrystalline material obtained from the TEM images in **Figure 5** shows the aggregated morphology in the nanoscale, revealing similar morphological characteristics for all three doped compositions.

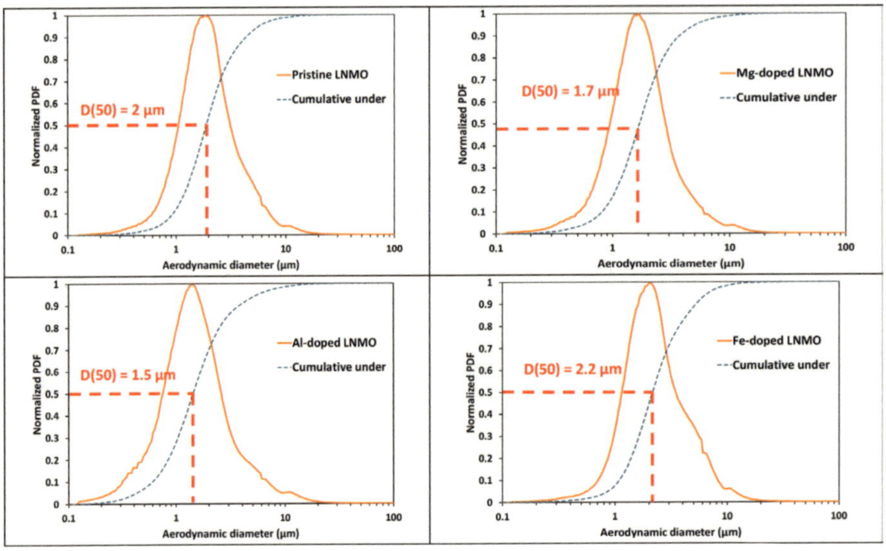

Figure 6.
Particle volume probability density function distribution of the pristine and doped LNMO particles measured by an aerodynamic diameter measuring technique.

Concerning the particle size distribution analysis, the mean particle diameter of pristine material was 2 μm; doping with Al and Mg led to smaller particle sizes while Fe doped had the higher particle size distribution at 2.2 μm, as observed in the normalized particle volume probability density function (PDF) plots of **Figure 6**.

Raman spectra was obtained in order to evaluate lattice differences of the pristine and doped LNMO materials. The spectra at **Figure 7** exhibits the 636 cm^{-1} peak, which is assigned to the Mn-O stretching vibration of the A_{1g} mode shifted to slightly higher wavenumbers due to the insertion of the Ni^{2+} ions in the spinel lattice [59]. The decrease of Ni amount (doped compositions), shifts the corresponding peak to the nominal location at 625 cm^{-1} attributed to the symmetric stretching

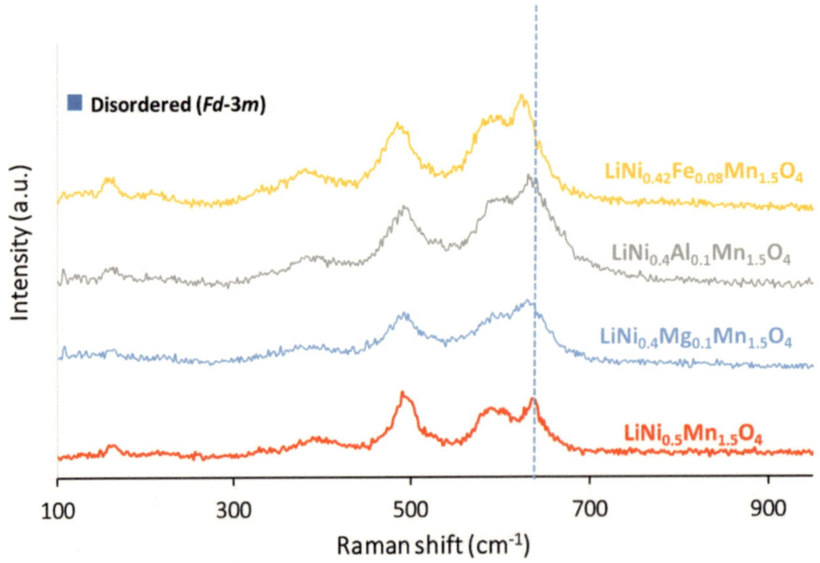

Figure 7.
Raman spectrum of the doped LNMO materials.

vibration of the Mn-O of the MnO_6 groups. The broadening of the two peaks at 582 and 605 cm^{-1}, which cannot be separated in the disordered structures, is attributed to the $F_{(3)2g}$ and the intensity of the shoulder located at 580 cm^{-1} is enhanced upon nickel substitution.

5. Electrochemical characterization results

The respective LNMO doped active materials were coated on aluminum foils using different formulations of active material, carbon and binder – as explained below – in order to study the material ratios and process parameters for slurry preparation and battery performance optimization. The coated material was also tested under different calendering densities, by varying the calendering pressure and final active layer thickness. The coated electrodes were analyzed for their electrochemical performance in the form of Half Coin Cells (HCC) with a metallic Lithium foil as counter electrode and evaluating their discharge specific capacity. Preliminary electrochemical tests were performed on all four materials using HCCs (3 cells per formulation/material). Discharge C-rate varied from C/5 to 10C on comparable electrodes prepared with a formulation AM/Carbon/PVdF ratio of 90/5/5, a real loading ~1.5 $mAh \cdot cm^{-2}$ and density of 2.5 $g \cdot cm^{-3}$.

As depicted in **Figure 8**, a profound specific discharge capacity decay was observed in both pristine and Mg-doped LNMO during cycles 1 to 5 (C/5 low rate and near-zero values were recorded until cycle 32 at high C-rates. Al-doped LNMO demonstrated high specific capacity values (~120 $mAh \cdot g^{-1}$) at low C-rates i.e. C/5, C/2, 1C while at higher discharge C-rates of 2C, 5C, 8C and 10C the specific capacity demonstrated values of ~110, ~90, ~50 and ~ 20 $mAh \cdot g^{-1}$ respectively. The discharge specific capacity almost regained its initial high values when C-rate was decreased at C/2 during cycle 24 and withstood high charge rates (~80 mAh/g at 3C charge, cycles 29–31). On the other hand, Fe-doped LNMO demonstrated similar behavior during cycles 1–8, however the performance was clearly poorer than the Al-doped LNMO composition at 2C and 5C (i.e. ~90 and ~ 18 $mAh \cdot g^{-1}$ respectively) while it dropped to zero during fast discharge at 8C and 10C. During the C-rate

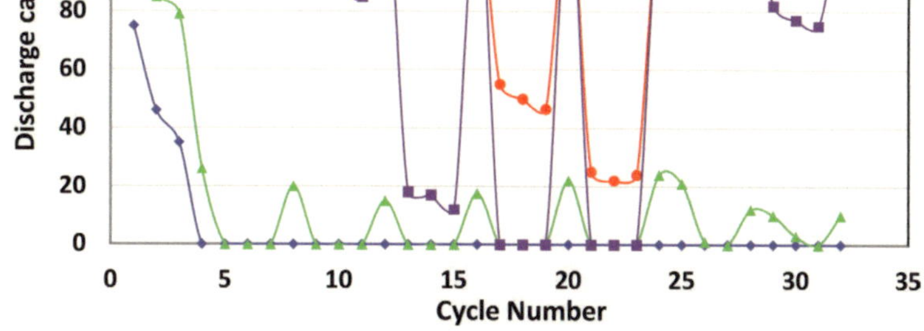

Figure 8.
Specific capacity of the different doped LNMO electrodes at different discharge rates (1C to 10C) and charge rates (1C, 3C).

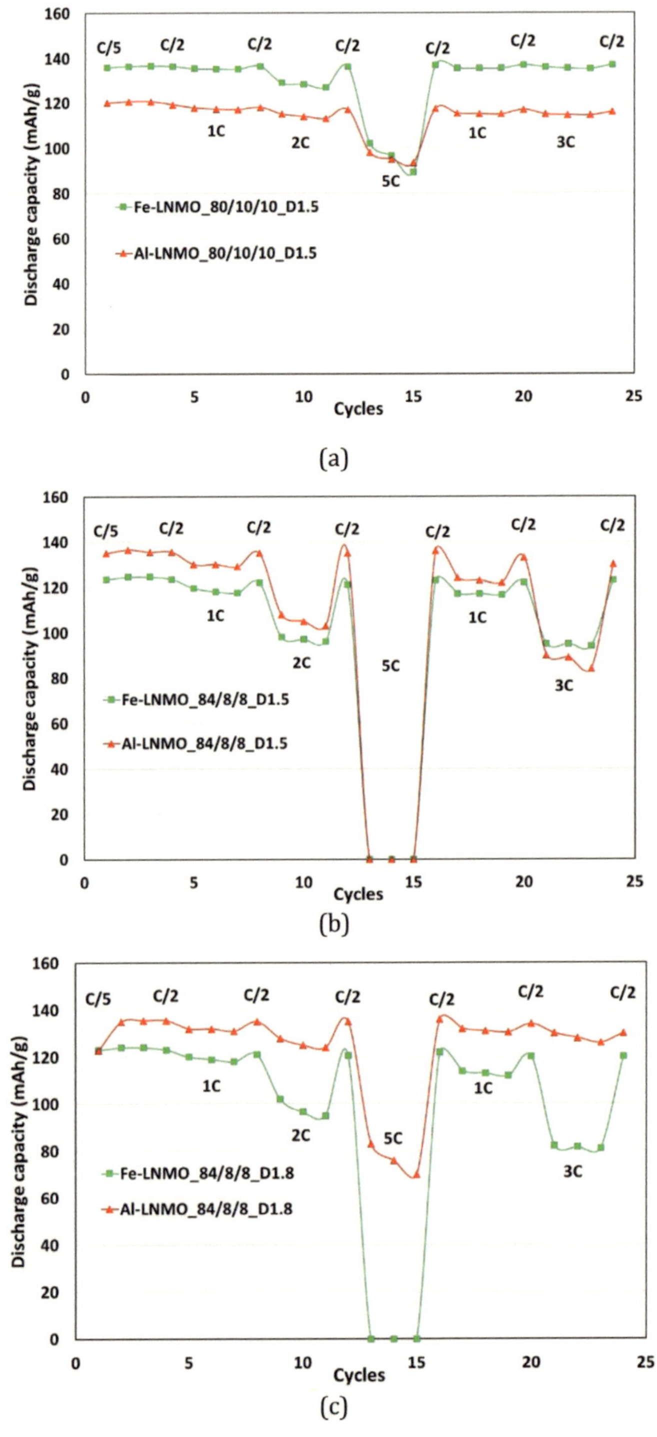

Figure 9.
Specific capacities of iron and aluminum doped LNMO for the same target loading of ~1.0 mAh·cm^{-2} under different electrode formulations and densities (a) 80/10/10_D1.5, (b) 84/8/8_D1.5 and (c) 84/8/8_D1.8 (D in gcm^{-3}).

reduction at cycle 24 (C/2), an increase in specific discharge capacity to the initial values was observed but again performance was inferior to the Al-doped LNMO case throughout cycles 25–32.

Based on the abovementioned electrochemical results, Fe-doped and Al-doped LNMO were chosen for further testing in HCCs in order to study the effect of the AM/Carbon/PVdF and electrode density on the electrochemical performance.

The effect of electrode formulations with higher conducting carbon content to improve electrical conductivity and the calendering density on electrochemical performance of Fe-doped and Al-doped LNMO is depicted in **Figure 9(a)–(c)**. The materials were tested with high C black content, an AM/carbon/PVdF ratio 80/10/10 ratio, that limited the achievable loading to 1 $mAh \cdot cm^{-2}$ and a density of 1.5 $g \cdot cm^{-3}$. **Figure 9(a)** depicts low C-rates of C/5 to 1C during cycles 1–8; Fe-doped LNMO demonstrated higher values of specific discharge capacity compared to Al-doped LNMO (~138 and 118 $mAh \cdot g^{-1}$ respectively), while similar performance in the range 80–100 $mAh \cdot g^{-1}$ was observed in the higher C-rate of 5C (cycles 13–15). Increasing the amount of active material to optimize the energy density of the active layer while maintaining slurry stability, achieved an AM/carbon/PVdF ratio of 84/8/8. As shown in **Figure 9(b)**, the Al-doped LNMO performance demonstrates a capacity improvement while a capacity reduction is observed in the case of Fe-doped LNMO. During cycles 1–4 (C/5 and C/2 rates) the specific capacity reaches values ~137 and 125 $mAh \cdot g^{-1}$ for Al-doped and Fe-doped LNMO respectively, with further decrease at 1C rate (130 and 120 $mAh \cdot g^{-1}$ respectively) and 2C rate (105 and 95 $mAh \cdot g^{-1}$ respectively). Both materials demonstrate zero capacity at 5C rate. On subsequent charge at 1C during cycles 17–19, a decrease in Al-doped LNMO capacity is observed compared to 1C discharge cycles 5–7 while insignificant change in Fe-doped LNMO was observed in the course of the same cycles. During 3C charge (cycles 21–23), Fe-doped LNMO performed better than Al-doped LNMO with a specific capacity of ~95 vs. 85 $mAh \cdot g^{-1}$ for the Al-doped LNMO composition.

Finally, by further increase of the density to 1.8 $g \cdot cm^{-3}$, a significant improvement of Al-doped LNMO performance was observed, especially at the higher C-rates of 2C, 3C and 5C in **Figure 9(c)**. On the contrary, no improvement was observed in the case of Fe-doped LNMO, as the capacity values remained almost the same as for 1.5 $g \cdot cm^{-3}$ densification. The enhanced electrochemical activity of the Al-doped LNMO at the 1.8 $g \cdot cm^{-3}$ density comparing to that of 1.5 $g \cdot cm^{-3}$ shows the importance of the physicochemical properties of the overall electrode, beginning from the nanoscale of Al ion doping in the LNMO lattice to the macroscale attributed to the density of the coated electrode; the conductivity (Li-ion mobility and electrical conductivity) of the cathode increases when carbon and LNMO particle interfaces are in better contact by calendaring.

6. Conclusions

Aerosol Spray pyrolysis is a sustainable and promising route for the synthesis of anode and cathode nanostructured materials, which is exploited by research laboratories and is employed for industrial production. In the current work, $LiNi_{0.5}Mn_{1.5}O_4$ spinel was synthesized at a production rate of 100 gr hrs^{-1} and was studied for three different doping compositions of Mg, Fe and Al; $LiNi_{0.4}Mg_{0.1}Mn_{1.5}O_4$, $LiNi_{0.42}Fe_{0.08}Mn_{1.5}O_4$ and $LiNi_{0.4}Al_{0.1}Mn_{1.5}O_4$ by substituting Ni molar ratio in the spinel lattice. The aqueous precursor solutions were sprayed at a heated tubular reactor at 800°C, while further calcination of the collected powder was carried out at 850°C for 16 hrs [11]. The Raman spectra and XRD diagrams depict successful synthesis of the mixed structure, while SEM showed spherical particles with partially hollow morphology. The electrochemical evaluation was performed for different electrode formulations and loading/densities in half coin cells at different C-rates. Fe- and Al-doped materials for the electrode formulation

AM/C/PVdF = 90/5/5 exhibited high specific capacity values (~120 $mAh \cdot g^{-1}$) at low C-rates, while at higher discharge C-rates of 2C, 5C, 8C and 10C the specific capacity demonstrated values of ~110, ~90, ~50 and ~ 20 $mAh \cdot g^{-1}$ respectively, showing interesting activity for further study, while pristine and Mg- doped material showed very low capacity at rates exceeding 1C. Fe- and Al- LNMO were further studied at lower active material ratios; namely 80/10/10 ratio using a density of 1.5 $g \cdot cm^{-3}$, where Fe-doped LNMO demonstrated higher values of specific discharge capacity compared to Al-doped LNMO at low C-rates of C/5 to 1C (~138 and 118 $mAh \cdot g^{-1}$ respectively), while similar performance in the range 80–100 $mAh \cdot g^{-1}$ was observed for the higher C-rate of 5C. Increasing the amount of active material to 84/8/8 to increase electrode energy density for practical battery application, improved the capacity for the Al-doped case, while a capacity reduction was observed for the Fe-doped LNMO. The current work shows the potential of Fe and Al doped LNMO synthesized by ASP as promising materials for cobalt free cathodes for the development of high voltage next generation Li-ion batteries.

Acknowledgements

We would like to thank the European Commission for funding this work through the HORIZON2020 eCAIMAN project (Grant No: GV-1-2014-653331). Dimitrios Zarvalis from CERTH is highly acknowledged for his contribution for the scaling up of the LNMO manufacturing process.

Conflict of interest

The authors declare no conflict of interest.

Abbreviations

LNMO	$LiNi_{0.5}Mn_{1.5}O_4$
XRD	X-Ray Diffraction
SEM	Scanning Electron Microscopy
EDS	Energy Dispersive Spectroscopy
TEM	Transmission Electron Microscopy
BET	Brunauer–Emmett–Teller
FWHM	Full Width Half Maximum
HCC	Half Coin Cell
AM	Active Material
PVdF	Polyvinylidene fluoride
LIB	Rechargeable Li-ion batteries
EV	Electric Vehicles
EMC	Ethyl Methyl Carbonate
NMP	N-methyl-2-pyrrolidone

Author details

George Ganas[1], George Karagiannakis[1], Aitor Eguia-Barrio[2], Miguel Bengoechea[2], Iratxe de Meatza[2,3] and Georgia Kastrinaki[1]*

1 Chemical Process and Energy Resources Institute, CERTH, Thessaloniki, Greece

2 CIDETEC, Basque Research and Technology Alliance (BRTA), Donostia-San Sebastián, Spain

3 Department of Organic and Inorganic Chemistry, Universidad del País Vasco (UPV/EHU), Bilbao, Spain

*Address all correspondence to: georgiak@certh.gr

IntechOpen

© 2022 The Author(s). Licensee IntechOpen. This chapter is distributed under the terms of the Creative Commons Attribution License (http://creativecommons.org/licenses/by/3.0), which permits unrestricted use, distribution, and reproduction in any medium, provided the original work is properly cited.

References

[1] Kuthanapillil M. Shaju and Peter G. Bruce, Nano-LiNi0.5Mn1.5O4 spinel: a high power electrode for Li-ion batteries, Dalton Trans., 2008, 5471-5475

[2] Chen G., Yang J., Tang J., Zhou X., RSC Adv. 5 (2015) 23067-23072.

[3] J.B. Goodenough, Y. Kim, Challenges for rechargeable Li batteries, Chem. Mater. 22 (2010) 587-603.

[4] V. Etacheri, R. Marom, R. Elazari, G. Salitra, D. Aurbach, Challenges in the development of advanced Li-ion batteries: a review, Energy Environ. Sci. 4 (2011) 3243-3262

[5] D. Andre, S.-J. Kim, P. Lamp, S.F. Lux, F. Maglia, O. Paschos, B. Stiaszny, Future generations of cathode materials: an automotive industry perspective, J. Mater. Chem. 3 (2015) 6709-6732

[6] Jung D. S., Ko Y. N., Kang Y.C., Park S.B., "Recent progress in electrode materials produced by spray pyrolysis for next-generation lithium ion batteries", Advanced Powder Technology 25 (2014) 18-31

[7] Zhu Y., Choi S. H., Fan X., Shin J., Ma Z., Zachariah M. R., Choi J. W., Wang C., "Recent Progress on Spray Pyrolysis for High Performance Electrode Materials in Lithium and Sodium Rechargeable Batteries", Adv. Energy Mater. 2017, 1601578

[8] Jang H., Seong C., Suh Y., Kim H., Lee C. Synthesis of Lithium-Cobalt Oxide Nanoparticles by Flame Spray Pyrolysis, Aerosol Science and Technology (2004), 38:10, 1027-1032, DOI: 10.1080/027868290524016

[9] Abram C, Shan J., Yang X., Yan C., Steingart D., Ju Y., "Flame Aerosol Synthesis and Electrochemical Characterization of Ni-Rich Layered Cathode Materials for Li-Ion Batteries", ACS Appl. Energy Mater. 2019, 2, 1319–1329

[10] Madero J. E., Li J., Shen K.-Y., Wojtak J., Axelbaum R.L., "An approach to low-temperature flame spray pyrolysis for the synthesis of temperature-sensitive materials: Application to $Li_{1.2}Mn_{0.54}Ni_{0.13}Co_{0.13}O_2$", Applications in Energy and Combustion Science 5 (2021) 100020

[11] Ganas, G., Kastrinaki, G., Zarvalis, D., Karagiannakis, G., Konstandopoulos, A.G., Versaci, D., Bodoardo, S., "Synthesis and Characterization of LNMO Cathode Materials for Lithium-ion Batteries", Materials Today: Proceedings 5 (2018) 27416-27424

[12] Shimoi N., "Synthesis of anode active material particles for lithium-ion batteries by surface modification via chemical vapor deposition and their electrochemical characteristics", Adv. Powder Technol. (2017)

[13] Ogihara T., Ogata N., Katayama K., Azuma Y., Mizutani N., Electrochemistry 68 (2000) 162-166

[14] Pateya T.J., Büchelb R., Nga S.H., Krumeich F., Pratsinis S.E., Nováka P., Flame co-synthesis of $LiMn_2O_4$ and carbon nanocomposites for high power batteries", Journal of Power Sources 189 (2009) 149-154

[15] Saft M., Chagnon G., Faugeras T., Sarre G., Morhet P., J. Power Sources 80 (1999), 180-189

[16] Y. Tanjo, T. Abe, H. Horie, T. Nakagawa, T. Miyamoto, K. Katayama, Soc. Automot. Eng. (Spec. Publ.) SP-1417 (1999) 51-55

[17] Bresser D., Paillard E., Copley M., Bishop P., Winter M., Passerini S., "The

importance of "going nano" for high power battery materials", Journal of Power Sources 219 (2012) 217-222

[18] Lorentzou S., Kastrinaki G., Pagkoura C., Konstandopoulos A.G. "Oxide Nanoparticles for Hydrogen Production from Water-splitting and Catalytic Oxidation of Diesel Exhaust Emissions", 2011, Nanoscience and Nanotechnology Letters, 3 (4)

[19] Kastrinaki G., Lorentzou S., Karagiannakis G., Rattenbury M., Woodhead J., Konstandopoulos A.G. (2018) "Parametric Synthesis Study of Iron Based Nanoparticles via Aerosol Spray Pyrolysis route", Journal of Aerosol Science, 115, pp. 96-107

[20] Nie P., Xu G., Jiang J., Dou H., Wu Y., Zhang Y., Wang J., Shi M., Fu R., Zhang X., "Aerosol-Spray Pyrolysis toward Preparation of Nanostructured Materials for Batteries and Supercapacitors", Small Methods 2018, 2, 1700272

[21] T.B. Reddy, D. Linden, Linden's Handbook of Batteries, fourth ed., McGraw-Hill, New York, 2011.

[22] M. Winter, J.O. Besenhard, M.E. Spahr, P. Novák, Adv. Mater. 10 (1998) 725-763.

[23] J.M. Tarascon, M. Armand, Nature 414 (2001) 359,

[24] J.B. Goodenough, Electrochemical cell with new fast ion conductors, US patent No. 4302518A, 1981, US.

[25] M. Wakihara, O. Yamamoto, Lithium ion batteries: fundamentals and performance, Wiley-VCH, New York, 1998.

[26] C. Miao, P.F. Bai, Q.Q. Jiang, X.Y. Wang, J. Power Sources 246 (2014) 232-238.

[27] D. Zhao, Y.L. Feng, Y.G. Wang, Y.Y. Xia, Electrochim. Acta 88 (2013) 632-638.

[28] D. Liu, W. Zhu, J. Trottier, F. Gagnon, C.M. Julien, J.B. Goodenough, K. Zaghib, RSC Adv. 4 (2014) 154-167

[29] R. Santhanam, B. Rambabu, J. Power Sources 195 (2010) 5442-5451

[30] Patoux, L. Daniel, C. Bourbon, C. Pagano, S. Jouanneau, S. Martinet, J. Power Sources 189 (2009) 344-352

[31] S.H. Oh, K.Y. Chung, S.H. Jeon, C.S. Kim, W.I. Cho, B.W. Cho, J. Alloys Comp. 469 (2009) 244-250

[32] M.G. Kim, J. Cho, Adv. Funct. Mater. 19 (2009) 1497-1514

[33] N.P.W. Pieczonka, Z. Liu, P. Lu, K.L. Olson, J. Moote, J.-H. Kim, J. Phys. Chem. C 117 (2013) 15947-15957.

[34] L. Wang, D. Chen, J. Wang, G. Liu, W. Wu, G. Liang, Powder Technology 292 (2016) 203-209

[35] Y. Xue, Z. Wang, F. Yu, Y. Zhang, G. Yin, J. Mater. Chem. A 2 (2014) 4185-4191.

[36] Qin, X., Gong, J., Guo, J., Zong, B., Zhou, M., Wang, L., Liang, G., Journal of Alloys and Compounds 786 (2019) 240-249

[37] Y.J. Gu, Y. Li, Y. Fu, Q.F. Zang, H.Q. Liu, J.X. Ding, Y.M. Wang, H.F. Wang, J. Ni, Electrochim. Acta 176 (2015) 1029-1035

[38] Shen, Y., Ju, X., Zhang, J., Xie, T., Zong, F., Xue, D., Lin, X., Zhang, J., Li, Q., Materials Chemistry and Physics 240 (2020) 122137

[39] Ren, X., Wang, Y., Xiao, Q., Lei, G., Li, Z., Material Letters 248 (2019) 97-100

[40] Li, Y., Wang, J., Yao, Q., Wang, Z., Zhou, H., Deng, J., Int. J. Electrochem. Sci., Vol. 14, 2019

[41] H.B. Lin, Y.M. Zhang, H.B. Rong, S.W. Mai, J.N. Hu, Y.H. Liao, L.D. Xing, M.Q. Xu, X.P. Li, W.S. Li, J. Mater. Chem A 2 (2014) 11987-11995

[42] Rahim, A.S, Aziz, N., Nor, A.M., Osman, Z., Molecular Crystals and Liquid Crystals, 695 (1) (2019) 10-18

[43] Sone, H., Yoshida, S., Tanaka, M., Watanabe, T., Journal of the Ceramic Society of Japan 128 (9) 635-640 2020

[44] G. Liu, X. Kong, H. Sun, B. Wang, Ceram. Int. 40 (2014) 14391-14395

[45] L. Wen, Q. Lu, G. Xu, Electrochim. Acta 51 (2006) 4388-4392.

[46] J.H. Kim, S.T. Myung, Y.K. Sun, Electrochim. Acta 49 (2004) 219-227

[47] Zarvalis D., Gkanas G., Kastrinaki G., Karagiannakis G., Konstandopoulos A.G., Eguia-Barrio A., Bengoechea M., de Meatza I., Ahniyaz A., Bourbon C., Fabre F., Peralta D, "LNMO cathode materials for high-voltage, next -generation automotive Li-ion cells", Proceedings of 7th Transport Research Arena (TRA 2018), Proceedings of 7th Transport Research Arena (TRA 2018).

[48] H. Zhang, Y. Xu, D. Liu, X. Zhang, C. Zhao. Electrochim. Acta 125 (2014) 225-231

[49] Q. Liu, S. Wang, H. Tan, Z. Yang, J. Zeng. Energies 6 (2013) 1718-1730.

[50] S. Bao, W. Zhou, Y. Liang, B. He, H. Li. Mater. Chem. Phys. 95 (2006) 188-192.

[51] C. Locati, U. Lafont, L. Simonin, F. Ooms, E. Kelder. J. Power Sources 174 (2007) 847-851.

[52] U. Lafont, C. Locati, W.J.H. Borghols, A. Lasinka, J. Dygas, A.V. Chadwick, E.M. Kelder. J. Power Sources 189 (2009) 179-184.

[53] M. Wagemaker, F.G.B. Ooms, E.M. Kelder, J. Schoonman, G.J. Kearley, F.M. Mulder. J. Am. Chem. Soc. 126 (2004) 13526-13533.

[54] S.T. Myung, S. Komaba, N. Kumagai. J. Electrochem. Soc. 148 (2001) A482-A489.

[55] Y.K. Sun, C.S. Yoon, C.K. Kim, S.G. Youn, Y.S. Lee, M. Yoshio, I.H. Oh. J. Mater. Chem. 11 (2001) 2519-2522.

[56] G.B. Zhong, Y.Y. Wang, Z.C. Zhang, C.H. Chen. Electrochim. Acta 56 (2011) 6554-6561.

[57] Asxaxs R. Alcantara, M. Jaraba, P. Lavela, J.M. Lloris, C.P. Vicente, J.L. Tiradoz. J. Electrochem. Soc. 152 (2005) A13-A18.

[58] J. Liu, A. Manthiram, J. Phys. Chem. C 113 (2009) 15073

[59] Zhang, X.; Cheng, F.; Zhang, K.; Liang, Y.; Yang, S.; Liang, J.; Chen, J. Facile polymer-assisted synthesis of $LiNi_{0.5}Mn_{1.5}O_4$ with a hierarchical micro–nano structure and high rate capability. RSC Adv. 2012, 2, 5669-5675.

Chapter 11

Accelerate the Aging of Polymer as Energy-Saving Method Prior to the Pyrolysis Process

Jin Hu

Abstract

Pyrolysis technology has appeared a long time ago, but it has not been widely recognized and used in the area of waste plastic/rubber management. The key reason is that the high energy consumption of pyrolysis is the most important problem that plagues the further development of pyrolysis. Prior to the pyrolysis of waste plastics/rubbers, in addition to dehydration and drying can save part of the energy consumption of pyrolysis, in this study, we have first reviewed and discussed high irradiance exposure to artificially accelerate the aging process of waste plastics/rubbers as pretreatment. The results from our preliminary experiments show that the pyrolysis process of the plastic that has undergone UV accelerated aging was speed up, accordingly to achieve saving energy in thermal cracking.

Keywords: pretreatment, natural aging, artificially accelerated aging, energy saving

1. Introduction

Pyrolysis is the thermochemical decomposition of organic matter into non-condensable gases, condensable liquids, and a solid residual coproduct, biochar or charcoal in an inert environment (i.e., in the absence of oxygen) [1].

In a singular pyrolysis reactor, oxygen must be excluded otherwise more of the gas, oils, and char will burn, thereby losing products and reducing efficiency. Pyrolysis has historically been implemented to produce useful substances such as methanol, acetone, acetic acid, and creosote from wood in predominantly batch process retorts prior to petrochemical production routes [2].

Nowadays, pyrolysis has great potential to convert waste such as plastic/rubber/biomass into valuable products, such as fuels, power, heat and other valuable chemicals and materials to achieve maximum economic and environmental benefits. For instance, liquid oil produced from different types of plastic waste had higher heating values (HHV) in the range of 41.7–44.2 MJ/kg similar to that of conventional diesel. Therefore, it has the potential to be used in various energy and transportation applications after further treatment and refining [3].

Municipal Solid Waste (MSW) is mainly derived from the disposal of general waste streams that include green waste, food waste, and miscellaneous products (ie, leather, textile, metal scraps), which can be separated as noncompostable materials [4]. Although most MSW ends up in unsightly landfill sites, a significant quantity has been used to produce different value-added products such as compost, feedstuffs, and biogas [5].

 IntechOpen

Most mixed MSW technologies attempt to treat large quantities of heterogeneous mixed waste streams. This can be appealing to governments which do not want to source separate waste and seek a single, technological solution. However, the approach of looking for a technology fix for mixed waste treatment presents unique challenges, and is not as successful as more comprehensive source separation strategies. Gasification, pyrolysis and plasma arc technologies are most applied for homogeneous material streams. The heterogeneous nature of MSW is not well suited to this type of technology [6].

Andrew and Jumoke's review [7] has shown that when appropriate system boundaries are applied, a MSW pyrolysis plant for self-sustaining energy from waste is thermodynamically unproven, practically implausible, and environmentally unsound. No practical examples of a self-sustaining MSW pyrolysis plant, using either gas, oil, or char were found.

For homogeneous material streams, an approach of energy saving is pretreatment of the waste, by removing wet organics and inert material while retaining the high-energy plastics in the waste stream. In the process of converting organic solid waste into energy, besides the major thermal cracking energy consumption, a variety of additional energy consumption would incur e.g. feedstock sorting, conditioning, drying, shredding, pyrolysis gas cooling/condensation, combustible gases cleaning...How to reduce the energy consumption in each link to achieve the ultimate goal of reducing total energy consumption and increasing energy efficiency in the entire process.

Moisture is present in all solid organic waste (even visibly dry material), existing at both surface and cellular level, therefore unless drying is set outside the system boundary it must be included in energy balances. Prior to the pyrolysis process occurs, removing this water from the solid organic waste is highly energy intensive due to the high latent and sensible enthalpy demands in both liquid and steam phases, and high enthalpy of vaporization [8]. These endothermic phenomena are known as parasitic enthalpy demand, and are well understood from extensive work with steam cycles. Therefore, dehydration and drying outside the pyrolysis system under controllable temperature and a good ventilation cycle are preference that has a much better effect than that in a relatively closed pyrolysis reactor and save more energy.

During the pyrolytic conversion, the process of transforming long-chain hydrocarbons into short-chain hydrocarbons needs an extraneous energy supply to drive the process so this is provided allothermally from electricity or by burning additional fuels. Generally speaking, catalytic pyrolysis is a major technique to obtain more oil or combustible gas at relatively low temperatures and low energy consumption. Apart from this, this study will first review the published experimental results of artificially speeding up the aging of raw materials for some common organic solid wastes under different conditions e.g. temperature, natural or artificial ultraviolet light. Second, to propose using this method in the pre-treatment of raw materials to reduce the energy consumption of thermal cracking from long-chain to short-chain, and finally shortening the thermal cracking time and reducing energy consumption.

2. Sources of solid waste

Polymer degradation can be caused by heat (thermal degradation), light (photodegradation), ionizing radiation (radio degradation), mechanical action, or by fungi, bacteria, yeasts, algae, and their enzymes (biodegradation). The deleterious effects of weathering on polymers generally has been ascribed to a complex set of processes in which the combined action of UV light and oxygen predominant.

The overall light-initiated process in the presence of oxygen generally is referred to as oxidative photodegradation or photooxidation. A pure thermal effect in possible because oxygen is always present and so the process is thermaloxidative degradation [9].

There are many different modes of polymer degradation. These are very similar since they all involve chemical reactions that result in bond scission.

2.1 Polypropylene

Polypropylene (PP), as a representative of modified plastics, is widely used in textiles, construction, and other industries, and can also be used to make fiber products, automotive plastic parts, woven bags, etc. Of these, PP is an attractive material for packaging due to its low cost, higher tensile strength, glossy and versatility [10].

However, since isotactic polypropylene is a spiral crystal and its unique molecular structure, its molecular chain is more susceptible to light, heat, oxidative degradation, and poor product cold resistance, resulting in its application range, especially as engineering materials and outdoor products are greatly restricted [11].

Gallo and co-workers have studied the natural weathering of both thin and thick polypropylene film samples [12–14]. The authors concluded that the degradation behavior of these PP films is different in its oxidation products and crystallinity. The structural rearrangement or chemical modification occurs mainly in a region nearer to the surface of the material during weathering. Hence, the thinner PP films were degraded easily than the thicker PP films or plates.

Rajakumar et al. [15] used mathematical models to predict attempted life time of PP. Their experimental and simulation results showed that the carbonyl growth is more affected by ultraviolet (UV) and cumulative total solar radiation for PP weathered during summer. The loss in tensile strength of PP weathered during summer is more dependent on the average temperature and the UV portion of the total solar radiation whereas, intensity of UV radiation has profound effect on the tensile strength of PP weathered during winter.

Meanwhile, the UV-induced degradation of PP has been investigated by many authors and reported in a number of reviews and research articles [16–18].

Ni et al. [19] have studied the effects of indoor temperature difference aging and ultraviolet light aging on the aging degree of PP samples by means of mechanical properties, capillary rheology and scanning electron microscopy.

Their experimental results show that the retention of elongation at break was 58.7% at 20°C and 6.7% at 100°C, respectively. And the greater the temperature difference, the faster the brittle fracture speed of PP and the greater the degree of fracture. The torque at a temperature difference of 100°C was reduced from 0.188 N·m to 0.099 N·m, the sample is degraded to a large extent and loses its value in use. The elongation at break under ultraviolet irradiation of 340 nm for 20 days was 15.3%, while under ultraviolet irradiation of 313 nm, which was 3.1%. The torque of PP samples aged under ultraviolet irradiation of 340 nm decreased from 0.188 N·m to 0.112 N·m for 20 days, while for samples under ultraviolet irradiation of 313 nm, the torque decreased to 0.084 N·m, respectively. At this time, the sample completely loses its use value. They concluded that different aging methods make different aging degrees for PP and the ultraviolet irradiation of 313 nm makes more serious aging for PP samples because of higher energy.

2.2 Polyethylene

According to the difference in production technology and physical and chemical properties, polyethylene molecules can be divided into low-density polyethylene (LDPE), high-density polyethylene (HDPE), medium-density polyethylene (MDPE) and linear low-density polyethylene (LLDPE) [20].

Among them, LDPE molecules are generated under high temperature and high-pressure conditions, and the reaction conditions are violent. The generated molecular chain has numerous branches, loose molecular arrangement, and relatively low crystallinity (that is, the percentage of crystalline regions in the polymer molecular chain), which is manifested as material density Small, poor strength, but with good flexibility, light transmission, and relatively easy to degrade; HDPE is produced by polymerization under lower temperature and pressure conditions through the action of a catalyst, with a few branches, dense molecular chain arrangement, and high molecular crystallinity. The material has high density, high mechanical strength, and the slowest degradation [21–23].

Polyethylene has a long molecular chain, large relative molecular mass, high crystallinity of the chain segment, strong hydrophobicity, and it is difficult to contact with biological or chemical substances or enter the microbial body to be catabolized; in addition, inside the polyethylene structural unit (–[CH2-CH2]n-), the physical and chemical properties of the CC and CH bonds of [CH2-CH2]n are stable, and higher energy or force is required to break the molecular bonds. The relative molecular weight is large, the molecular chain is long, the chain segment is high in crystallinity, and the hydrophobicity is strong [24–27]; these determine that the polyethylene molecule is difficult to degrade, so the degradation process under natural conditions is very slow [24].

Albertsson et al. [28] tracked and monitored the $14CO_2$ release of polyethylene film under soil landfill conditions through 14C labeling of polyethylene materials. The study found that the degradation rate of polyethylene film was only 0.2%~0.5% when buried in soil for 10 years. Ohtake et al. [29–31] further analyzed the surface molecular structure and molecular weight of polyethylene film and plastic bottles that have been buried in the soil for 32 years and calculated that the complete degradation of the polyethylene film with a thickness of 60 μm under the conditions of field soil landfilling is probably required 300 years.

The non-biological oxidative degradation of polyethylene refers to the oxidative cleavage of covalent bonds when the molecular chain of polyethylene is subjected to non-biological factors such as light, heat, and mechanical force higher than the energy of the covalent bond between the molecules. Under aerobic conditions, the process of rapidly reacting to generate unstable intermediate products such as peroxy or hydroperoxy, and further reacting to generate small molecules or low molecular weight substances such as aldehydes, ketones, acids, esters, carbon monoxide, etc. [32].

Chiellini et al. [33] analyzed the material molecular weight and surface properties and other indicators and confirmed that the decomposition rate of polyethylene film increased significantly with the increase of temperature (55, 70°C).

Briassoulis et al. [34] used artificial heating and ultraviolet radiation to speed up the aging of the residual polyethylene mulch film and backfilled it into the soil to observe the decomposition of the mulch film in its natural state. The results show that the polyethylene film without artificial accelerated aging treatment is not degraded significantly after being buried in the soil for 8.5 years, but after high temperature (50°C treatment for 800 h) and ultraviolet radiation (under 35~45 W·m-2 ultraviolet radiation, the lamp irradiated at a distance of 25 cm for 800 h). The polyethylene film was backfilled into the soil for 8.5 years and then completely decomposed into plastic micro-particles with a diameter of less than 1 mm, and the degradation process continued. The above results show that environmental factors such as light and heat can significantly promote the degradation reaction of polyethylene film.

Prakash Bhuyar et al. [35] selected plastic packaging sheet as LDPE sample (the low-density polyethylene) and garbage plastic sheet as HDPE sample (high-density polyethylene) for the degradation test under ultraviolet light. Such samples were placed in the Laminar Air Flow (LAF) under UV light of wavelength of 253.7 nm up to 30 days.

They found that the LDPE sheet after 30 days of UV-treated is more fragile and completely broken up into small parts compared to the LDPE sheet after 15 days of UV-treated. The longer the UV treatment applied to the LDPE sheet, the more crack and fragile the plastic sheet. The percentage weight loss for UV-treated LDPE was 87.5% for 30 days.

By comparison, they found HDPE sheet after 30 days of UV-treated is more crack and tear up into small parts compared to the HDPE sheet after 15 days of UV treated. The longer the UV treatment applied to the HDPE sheet, the more crack the HDPE sheet will be. The percentage weight loss for UV-treated HDPE was 21.6% for 30 days. This shows that the percentage weight loss for UV-treated LDPE is higher than the weight loss for UV-treated HDPE.

LU Lin et al. [36] have conducted the exposure test of medium density polyethylene (MDPE) for different time periods up to one year in Xisha (Paracel) Islands.

Their experimental results showed that during the aging period, the color abbreviation and hardness increased, while the gloss, the retention of tensile strength and that of elongation at break decreased. The aging evaluation showed a stepwise increasing tendency of aging rate with the extending of exposure time.

2.3 Polystyrene

Polystyrene (PS) is one of the most commonly used plastics at present. In recent decades, PS has been widely used through different ways of modification. Therefore, it is of important significance to study the aging behavior of PS. Polystyrene (PS) is a multipurpose polymer that is used in varied applications in rigid and foamed form. Polystyrene is manufactured by the addition polymerization of the styrene monomer unit. At room temperature, polystyrene is normally a solid thermoplastic, but can be melted at higher temperature for molding or extrusion, then resolidified. Styrene is an aromatic monomer, and polystyrene is an aromatic polymer [37].

PS is a widely used as thermoplastic. Its hardness, hydrophobic nature and chemical composition cause it to persist in nature without any decomposition for long period of time thus cause environmental pollution [38].

Polystyrene degrades very slowly in nature and the expanded polystyrene is not easily recyclable because of its lightweight and low scrap value. It is generally not accepted in curbside programs. Expanded polystyrene foam takes 900 years to decompose in the environment and has been documented to cause starvation in birds and other marine wildlife [39].

Shah et al. [39] have observed that the UV-irradiation having wavelength range of 365 nm has a profound effect on polystyrene sample. The reduction of the molecular weight of polystyrene shows the degradation of polystyrene macromolecules by irradiation. The increase in number of chain sessions per polymer with the increase irradiation time indicates in the increase in rate of degradation of polystyrene with irradiation time.

Zhang et al. [40] have conducted several UV accelerated weathering tests on PS samples in 4 different ways. They used UV 340 nm, irradiance 0.89 W/m^2. Among them, one way, the sample is unaged. The test method A, the sample was irradiated for 12 hrs at 60°C. The test method B, the sample was irradiated for 8 hrs at 60°C, then 4 hrs condensation at 50°C. The test method C, the sample was irradiated for 8 hrs at 60°C, then 3.75 hrs condensation at 50°C, 0.25 hrs water spray at ambient temperture (without irradiation) as the last step.

Zhang et al. [40] have investigated the thermogravimetric curve of PS samples which can reflect the aging degree of the sample because when the molecular weight decreases under the action of ultraviolet light, resulting in a corresponding decrease in the initial thermal decomposition temperature. The result is shown in the **Figure 1**. In the **Figure 1**, the PS sample aged by Method A has the lowest initial decomposition temperature, followed by B, and then C compared with unaged sample. It shows that the degree of influence of the three test conditions on the aging of the sample is method A>method B>method C. Their results show that UV radiation is the main factor for the chemical aging of PS, the temperature promotes the aging behavior, but humidity has little effect on the aging behavior.

2.4 Polystyrene foam

Polystyrene foam (EPS) is prepared by adding polystyrene (PS) resin as the main body and adding foaming agents and other additives. When EPS is exposed to the natural environment, under the influence of light, heat, oxygen, water, and other factors, the appearance will appear yellowing or even cracks, which will deteriorate its performance and shorten its service life...The aging behavior of EPS is the fundamental basis to determine service life and service environment.

Yao et al. [41] have found: with increasing time of UV accelerated aging, on the molecular chain of EPS some coloring groups that turned the color of the sample surface yellow were produced. With increasing aging time, cushioning efficiency decreased. When aging time was more than 100 h, there was a tendency to lose cushioning property. With increasing aging time, molecular weight of EPS

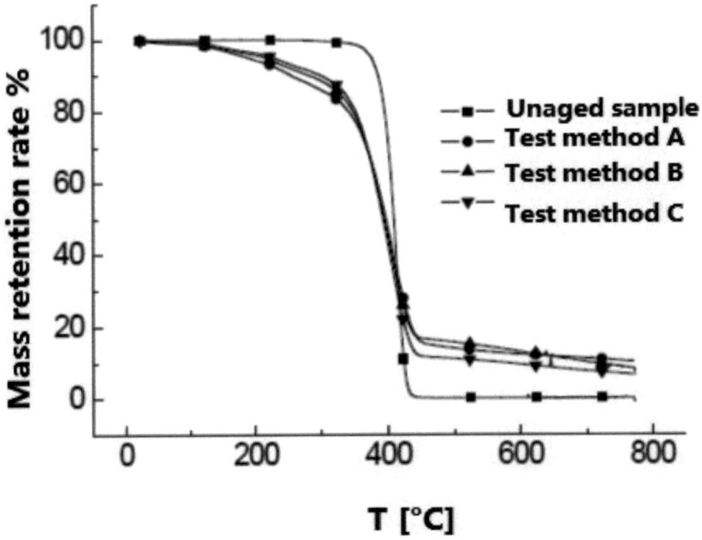

Figure 1.
TG curves of PS after UV aging test (regenerated figure from reference [40]).

decreased, the stability of the molecular structure of EPS was destroyed and photo-oxidative degradation occurs. The chemical degradation, resulting in molecular chain breaks and recombination phenomena, causing in the decline of mechanical properties.

2.5 Rubber

The rubber material is a polymer material with high elasticity and viscoelasticity. Among them, natural rubber (NR) is widely used in tires, conveyor belts, hoses, adhesive tapes, adhesives, sponges, and rubber products because of its excellent comprehensive properties. Production of various rubber products such as shoes, sports equipment, wires and cables, medical equipment, aircraft, and auto parts.

In the past 30 years, the average annual growth rate of global natural rubber consumption has exceeded 5%. The consumption has increased from 4.35 million tons in 1985 to 11.98 million tons in 2015 [42, 43].

In the actual storage and application process, rubber products are often in various environments and will be affected by oxygen, ozone, ultraviolet rays, and heat [44–47].

Weng et al. [47] have investigated Vulcanized natural rubber (NR) under quiescent thermal oxidation aging and high temperature fatigue loading with small strain amplitude. The vulcanized NR was simultaneously subject to high temperature (85°C) and cyclic loading. Under these conditions, their scanning electron microscope (SEM) images confirmed the appearance of nanoscale cracks and voids which are initiated by the combined impact of high temperature and cyclic loading. As the cracks evolve, nano-scale voids appear. They concluded that further evolution of the smaller voids leads to the larger spherical voids with their diameters varying between a few hundred nanometers and several microns.

In addition, seawater [48] will also affect rubber products. The above-mentioned various factors lead to the destruction of rubber composition and structure, gradually losing its original excellent performance, and even loss of use-value, the phenomenon of rubber aging [49]. "Aging" is a common problem of all polymer materials, especially natural rubber with unsaturated bonds on the major chain, which is more susceptible to aging by oxygen and free radicals. Once aging occurs, it will seriously affect its use value and application range, and even a major accident with dire consequences.

2.5.1 Aging mechanism of natural rubber

The aging process is: under the action of ultraviolet rays, the oxidation reaction of natural rubber starts from α-H, first is attacked by oxygen to generate hydroperoxide, and then further reacts to generate aldehydes and ketones, and at the same time the main chain is broken [50, 51]. The free radicals generated by chain scission may combine with double-free radicals, or they may be further attacked by oxygen to continue oxidation; in this process, double bonds addition reaction occurs, and the attack of free radicals on double bonds reduces the content of olefinic hydrogen. The early stage of the reaction is dominated by oxidative degradation, which is manifested as the surface of the aged sample is sticky, and cross-linking occurs in the later stage, the surface of the sample is dried and hardened, and cracks appear.

Ling Ding et al. [52]'s 1H-NMR and FT-IR absorption spectroscopy experimental results proved that the products obtained in the aging process of natural rubber in artificial ultraviolet light aging and natural aging are roughly the same. They found that natural rubber has poor resistance to ultraviolet light aging, and its molecular structure changes significantly when exposed to ultraviolet light for 2 h. Under natural aging conditions, the structure of natural rubber changes significantly after 30 days. Artificial UV aging for 7 days is equivalent to the effect of natural aging for 3 months.

3. Experimental setup

Because of the very slow degradation of polymer materials under natural conditions, therefore researchers from various countries have developed various artificial accelerated aging test methods to study the aging mechanism of polymer materials. The fluorescent ultraviolet lamp aging test has a high acceleration rate, and the aging resistance of the material can be obtained in a short time. It is one of the most commonly used artificially accelerated aging test methods for polymer materials.

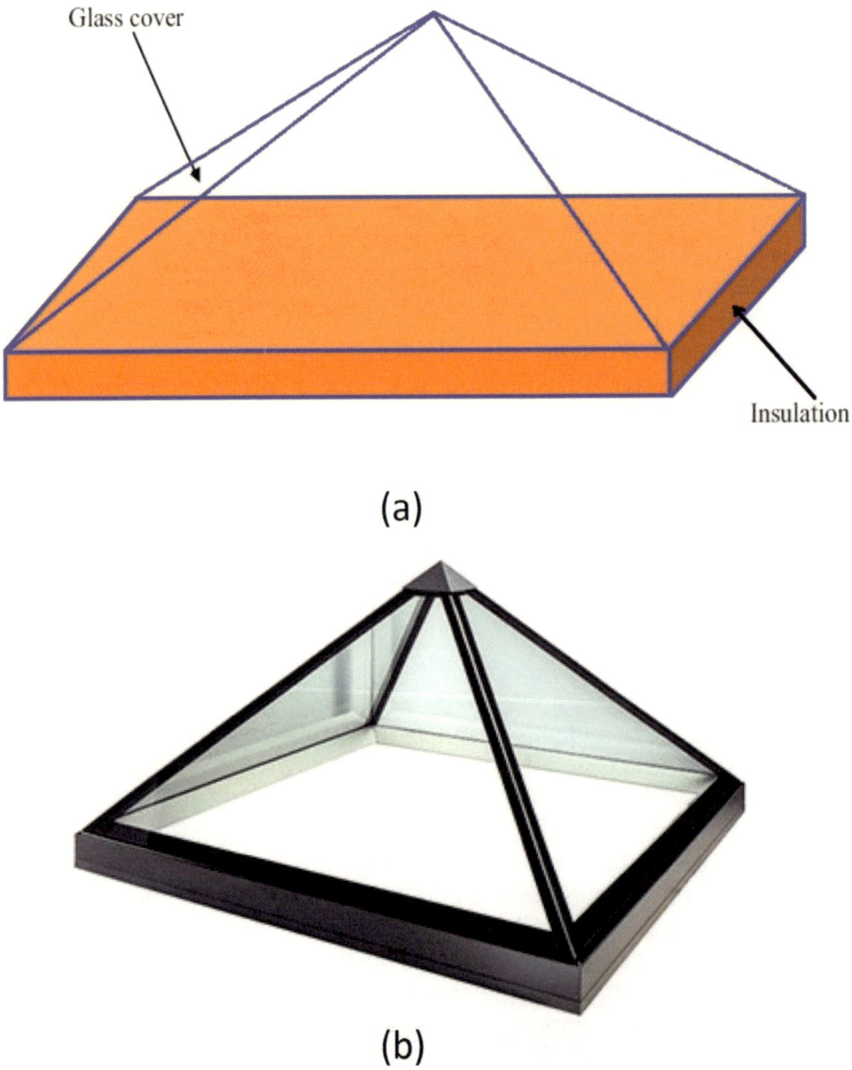

Figure 2.
a. a schematic diagram of the square pyramid test bench. b. the pyramid skylight structures.

In this study, we verify the effect of UV accelerated aging on pyrolysis through comparative experiments.

3.1 Raw materials

PP, PE, PS, EPS, Natural rubber sheets and PP granules for test are from Polyrocks Chemical Co., Ltd., Guangdong, China.

3.2 Experimental apparatus

Ultraviolet light has a wavelength of 10 to 400 nm. Among such ultraviolet rays, the part with a wavelength of 10 to 300 nm is absorbed by the atmosphere. The ultraviolet rays with a wavelength of 300 to 400 nm have a destructive effect, and their light energy is very large.

Various plastics are affected in different wavelength areas. The most influential wavelengths are 325 nm for polyester; 318 nm for polystyrene; 300 nm for polyethylene; 310 nm for polypropylene; 310 nm for polyvinyl chloride resin; and 322 nm to 364 nm for vinyl chloride-vinyl acetate copolymer [53].

UVA-340 can perfectly simulate the sunlight spectrum in the critical short-wavelength range, and the wavelength range is 315-400 nm. Its luminescence spectrum energy is mainly concentrated at the wavelength of 340 nm. UVA-340 fluorescent ultraviolet lamp can well simulate the short-wave ultraviolet (<365 nm) part of the spectrum. UVA-340 nm lamp was used in this study, with power 100w, lamp intensity: 30 mW/cm^2, made by Shenzhen Anhongda Optoelectronics Technology Co., Ltd.

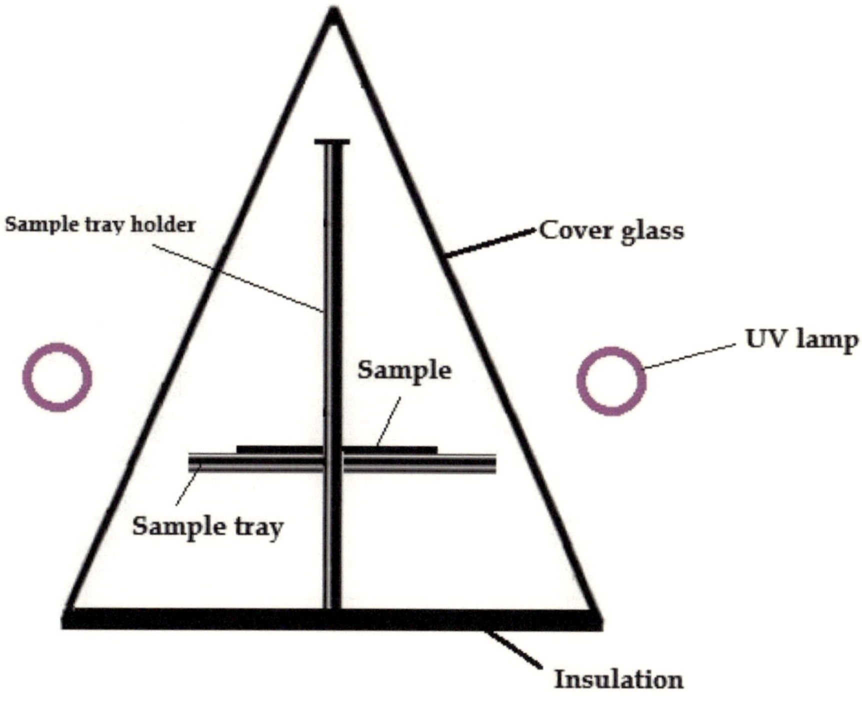

Figure 3.
Sectional view of the test bench.

Sample	Thickness (mm)	Photo
PP sheet	4	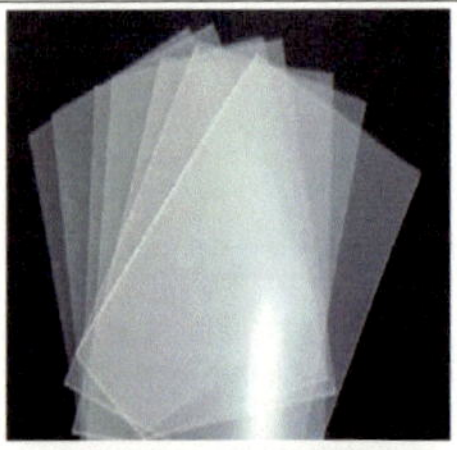
PE sheet	4	
PS sheet	4	
EPS foam sheet	4	
Natural rubber sheet	4	
PP Granules	N/A	

Table 1.
Samples for aging test.

As pyramid skylight has relatively higher Standard Visible Transmittance (Tvis) average than the most of other skylight [54].

Based on this, we choose two identical pyramid skylight structures (single-glazed with clear glass) as the experimental platform to receive the light source as shown in **Figure 2a** and **b**. Pyramid skylight was manufactured by Guangzhou Fineland Windows & Doors Co., Ltd.

A sample tray holder is placed in each pyramid skylight structure, and each sample tray is divided into multiple grids, which can separate different samples for simultaneous experiments, as shown in **Figure 3**.

3.3 Sample preparation

Cut PP, PE, PS, EPS, Natural rubber raw materials to prepare two sets of identical samples. Each sample size is 50 mm × 40 mm, the average thickness is 4 mm. PP granules 6 kg (**Table 1**).

4. Experimental results and discussion

For comparative testing, the samples will be performed simultaneously under two different conditions.

Natural aging experiment: A set of 5 sheet samples and 2 kg of PP granules were placed on the sample tray in one of the pyramid skylight structure. Under natural conditions, they were directly exposed to solar radiation. Because the pyramid skylight structure is closed, the impact of moisture on the material is not considered. The experiment time is 12 consecutive days.

Ultraviolet aging experiment: Place another set of 5 sheet samples and 2 kg of PP granules on a sample tray inside of another pyramidal skylight structure, and place them in a dark environment at a controlled constant temperature of 22°C. Four ultraviolet lamps were placed outside the four inclined triangle sides of the pyramid skylight structure. The position of the UV light source is higher than the sample tray and the distance between the sample and the ultraviolet lamp is 10 cm, as shown in the **Figure 3**. The sample was receiving 100% UV light for 12 consecutive days.

The rest 2 kg of PP granules are without any aging treatment.

4.1 Aging test results and discussion

The appearance changes of two groups of samples under different aging test conditions are shown in **Table 2**. In 12 days of natural aging, the color of PP sheet, EPS sheet, Natural rubber sheet and PP granules did not change. The PE sheet and PS sheet only turned yellow very slightly. It can be seen that the surface aging of the material is not obvious in the short-term irradiation of natural light. Normally, the aging test under natural conditions takes months up to years to see the obvious aging effect.

While under 12 consecutive days of UV light irradiation, PP granules very slightly turned yellow; PP sheet, EPS foam sheet turned light yellow; Natural rubber sheet turned yellow; PE sheet and PS sheet severely yellowed. Obviously, UV light has a significant influence on the aging of Natural rubber, PE and PS.

4.2 Pyrolysis test results and discussion

The pyrolysis experiments were conducted in a fully automatic 5 kW electromagnetic induction heating pyrolysis system, which was manufactured by Greenlina SA, Switzerland as shown in **Figure 4**. Under exactly the same experimental conditions without using any catalysts, three pyrolysis experiments were carried out on 2 kg PP granules without any aging treatment, 2 kg PP granules that had undergone natural aging and 2 kg PP granules that had been aging by

Polymer sample	Color before experiment	Color after natural aging	Color after UV aging
PP sheet	Milky	Milky	Light yellow
PE sheet	White	Very slight yellowing	Severe yellowing
PS sheet	Transparent	Very slight yellowing	Severe yellowing
EPS foam sheet	White	White	Light yellow
Natural rubber sheet	White	White	Yellow
PP granules	Milky	Milky	Very slight yellowing

Table 2.
Experimental results of two different aging methods.

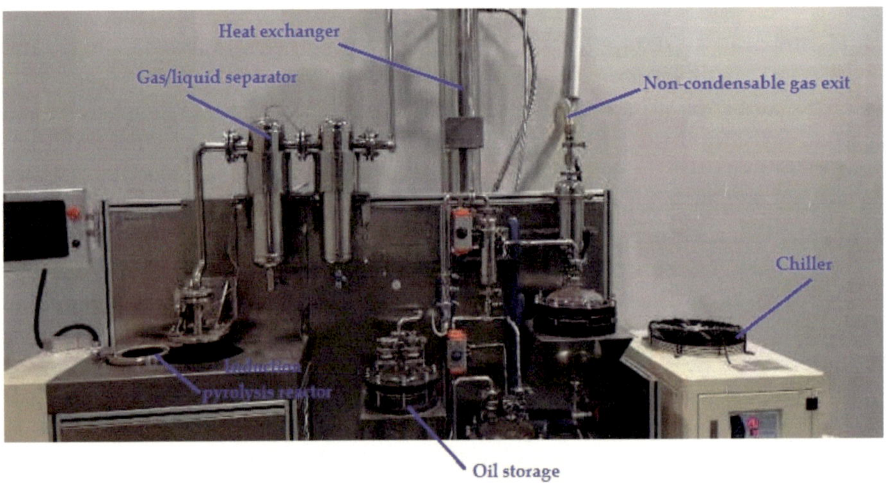

Figure 4.
Fully automatic electromagnetic heat induction pyrolysis system.

ultraviolet light respectively. The ambient temperature in the experiment was 27±0.5°C. The setting temperature of the chiller used for condensation was 3°C, and the start-up time of the chiller was completely the same in the two tests. The set temperature of the heating jacket on the separation tank was 120°C, and the heating jacket started to heat at the beginning of the experiment for both tests.

It is not difficult to see from the **Table 3** that without the aid of catalyst, under the same experimental conditions, there is not much difference in oil yield between the two different treatments of PP granules and the granules without any aging treatment. However, the PP granules that have undergone UV accelerated aging began to produce pyrolysis oil after about 17 minutes, and it took about 60 minutes to complete the entire pyrolysis process. The naturally aged PP granules started to produce pyrolysis oil after about 24 minutes, and it took about 77 minutes to complete the entire pyrolysis process. While the PP granules without aging treatment started to produce pyrolysis oil after about 25 minutes, and it took about 78 minutes to complete the entire pyrolysis process. Considering of reasonable deviation, short-term natural aging samples hardly improve the oil generation time and the time required to complete the entire pyrolysis process comparing with the sample without aging. Nevertheless, we have noticed that the pyrolysis (from long chains to short chains) of PP plastic granules after UV accelerated aging becomes much easier and faster. In other words, less energy was consumed to complete the entire pyrolysis process.

Sample	Mass (kg)	Set Power (kW)	Set Temperature (°C)	Time to start oil production (min)	Complete pyrolysis time (min)	Oil production (g)
PP granules (without aging)	2	4	350	~25	~78	1632
PP granules (Natural aging)	2	4	350	~24	~77	1640
PP granules (UVaging)	2	4	350	~17	~60	1657

Table 3.
The comparison test of pyrolysis.

Figure 5.
The pyrolysis oil of PP plastic granules.

Component	Aromatic %	Non-aromatic %
C_5-C_9	6.48	18.4
C_{10}-C_{13}	2.35	7.10
>C_{13}	13.18	52.49

Table 4.
Component in liquid fraction obtained from pyrolysis of PP granules using GC–MS analysis.

The pyrolysis-oil (**Figure 5**) from PP granules was analyzed by GC–MS (as shown in **Table 4**). The identified compounds were categorized according to the length of their carbon chain: C5–C9, C10–C13 and C>13. Those molecules with a chain length of C5–C9 represents to light hydrocarbons (gasoline fuel), which generally contain hydrocarbons between C5 and C9, while C>13 correspond to the heavy oils.

It was observed that pyrolysis of PP granules mainly yielded non-aromatic hydrocarbons with carbon number C>13 around 52.49% and aromatic hydrocarbons with carbon number C>13 around 13%. For PP granules, the resulting oils are mostly aliphatic hydrocarbons (alkanes and alkenes) with carbon number more than C13, which makes it suitable for use in diesel engines.

Compounds	Ratio (%)
C_9H_{18}	15.35
$C_{23}H_{46}$	15.18
$C_{18}H_{36}$	8.84
$C_{15}H_{30}$	8.68
$C_{12}H_{24}$	4.12

Table 5.
The most abundant compounds present in oils from the pyrolysis of PP granules as detected by GC–MS analysis.

The most abundant compounds present in oils from the pyrolysis of PP granules are present in **Table 5**. Pyrolysis of PP granules produced an oil containing C_9H_{18} (15.35%), $C_{23}H_{46}$ (15.18%), $C_{18}H_{36}$ (8.84%), $C_{15}H_{30}$ (8.68%) and $C_{12}H_{24}$ (4.12%).

5. Conclusions

Through review, we found that the heterogeneous nature of MSW is not well suited to gasfication/pyrolysis technology due to high energy consumption and low energy efficiency. Pyrolysis is suitable for homogeneous material streams (e.g., Waste plastics/rubber after sorting; numerous studies have shown that the short-term accelerated aging of polymers under a specific wavelength of artificial ultraviolet light is equivalent to the results of long-term natural aging of polymers.

In this study, 12 days of natural aging has almost no effect on the appearance of the experimental samples. While the artificial accelerated aging by the aid of 340 nm ultraviolet light in 12 days has a significant effect on the appearance of the experimental samples.

In the comparative pyrolysis experiments of PP granules under the same experimental conditions, the pyrolysis results confirmed that the PP granules undergone UV accelerated aging were faster than the short-term natural aging PP granules in the time of oil producing and completing the entire pyrolysis. The time was reduced about 1/4, thereby reducing total energy consumption. The performance of the PP granules undergone short-term natural aging is consistent with the performance of the PP particles that have not undergone any aging treatment, which confirms the short-term natural aging does not help the energy consumption of thermal cracking.

Pyrolysis is a high energy-consuming treatment method. In addition to the use of catalysts to greatly reduce the cracking temperature, it is recommended to use ultraviolet light to artificially accelerate the aging of the raw materials at the same time when the raw materials are dried/dehydrated in the pretreatment stage.

This additional pretreatment step can speed up the thermal cracking reaction, reduce energy consumption and increase thermal efficiency which has been confirmed by lab-scale experiments. While in some remote or power-deficient areas, it is not realistic to implement UV light to artificially accelerate aging during the pretreatment of raw materials. Where conditions are available, for large-scale pyrolysis production, there are currently many unknown factors such as: how much additional investment is needed as an additional step of pretreatment; the optimal irradiation frequency and intensity of artificial UV light for universal/mixed polymers; the shortest and most effective aging time required for artificial aging; the overall energy consumption of artificially accelerated aging by UV light; the overall energy saving during the pyrolysis after this pretreatment; whether it is cost-effective if the investment required for UV sped up aging compared with the benefits brought by energy-saving...all need to be further verified in the future study.

6. Outlook

SEM experiment of the sample after accelerated aging under ultraviolet light is need to carry out; at present, only the energy consumption comparison tests of PP plastic granules pyrolysis have been done due to time limitation. In the near future, more energy consumption comparison tests of different types of materials pyrolysis need to be carried out. Large scale experiments need to be carried out to estimate overall energy consumption and efficiency.

Acknowledgements

The author gratefully acknowledge the financial support provided by Greenlina SA Switzerland. Thanks to Polyrocks Chemical Co., Ltd., Guangdong for providing the experimental raw materials and cooperating with related experiments.

Declaration of competing interest

The author declares that he has no known competing financial interests or personal relationships that could have appeared to influence the work reported in this paper.

Author details

Jin Hu
Greenlina SA, Avenches, Switzerland

*Address all correspondence to: jin.hu@greenlina.com

IntechOpen

© 2021 The Author(s). Licensee IntechOpen. This chapter is distributed under the terms of the Creative Commons Attribution License (http://creativecommons.org/licenses/by/3.0), which permits unrestricted use, distribution, and reproduction in any medium, provided the original work is properly cited.

References

[1] Kaimin Shih, Environmental materials and waste: Resource recovery and pollution prevention, Apr 19, 2016, Academic Press, ISBN 9780128039069

[2] Garcia-Nunez, J.A., Palaez-Samaniego, M.R., Garcia-Perez, M.E., Fonts, I., Abrego, J., Westerhof, R.J.M., Garcia-Perez, M., 2017. Historical developments of pyrolysis reactors: A review. Energ. Fuel 31, 5751-5577

[3] Rashid Miandad, Mohammad Rehan, Mohammad A. Barakat, Asad S. Aburiazaiza, Hizbullah Khan, Iqbal M. I. Ismail, Jeya Dhavamani, Jabbar Gardy, Ali Hassanpour and Abdul-Sattar Nizami, Catalytic pyrolysis of plastic waste: Moving toward pyrolysis based biorefineries, Front. Energy Res., 19 March 2019 | DOI:10.3389/fenrg.2019.00027

[4] Hargreaves, J.C., Adl, M.S., Warman, P.R., 2008. A review of the use of composted municipal solid waste in agriculture. Agriculture, Ecosystems and Environment 123, 1e14.

[5] Vithanage, M., Wijesekara, S.S.R.M.D.H.R., Siriwardana, A.R., Mayakaduwa, S.S., Ok, Y.S., 2014. Management of municipal solid waste landfill leachate: A global environmental issue. In: Malik, A., Grohmann, E., Akhtar, R. (Eds.), Environmental Deterioration and Human Health. Springer, Netherlands.

[6] Neil Tangri and Monica Wilson, Waste Gasification & Pyrolysis: High Risk, Low Yield Processes for Waste Management, a Technology Risk Analysis GAIA March 2017.

[7] Andrew Neil Rollinson, Jumoke Mojisola Oladejo, 'Patented blunderings', efficiency awareness, and self-sustainability claims in the pyrolysis energy from waste sector, Resources, Conservation and Recycling 141 (2019) 233-242

[8] Twigg, M.V., 1996. Catalyst Handbook, 2nd ed. Manson, London, pp. 17-282.

[9] Feldman D (2002) Polymer weathering: Photo-oxidation. Journal of Polymer and the Environmental 10:163-173

[10] Roder H, Vogl O (1999) Prog Polym Sci 24:1205-1215

[11] Wang Wen, Xia Yu. Research and application of thermally conductive insulating materials. Journal of Insulation Materials, 2012, 45 (1): 19-24.

[12] Severini F, Gallo R, Ipsale S (1988) Polym Degrad Stab 22:185-194

[13] Gallo R, Severini F, Ipsale S, Fanti ND (1997) Polym Degrad Stab 55:199-207

[14] Gallo R, Brambilla L, Castiglionil C, Severini F (2006), J Macromol Sci Part A: Pure Appl Chem 43:535-554

[15] K. Rajakumar, V. Sarasvathy, A. Thamarai Chelvan, R. Chitra and C. T. Vijayakumar, Natural weathering studies of polypropylene, J Polym Environ (2009) 17:191-202. DOI:10.1007/s10924-009-0138-7

[16] Kotek J, Kelner I, Baldria J, Rabb M (2004) Eur Polym J 40:2731-2738

[17] Obadal M, Cermak R, Rabb M, Verneig V, Commereuc S, Fraisse F (2005) Polym Degrad Stab 88:532-539

[18] Shyichuk AV, Turton TJ, White JR, Syrotyhska ID (2004) Polym Degrad Stab 86:377-383

[19] NI Linggui, YIMIT Mamatjan, BEKEN Riza, JUMAHADER Asanhader,

CAO Liqin, Effect of Different Aging Methods on Aging Degree of Polypropylene, CHINA PLASTICS INDUSTRY, Vol 45, No. 1, Jan, 2017

[20] PAN Zu-ren. Polymer Chemistry [M]. Beijing: Chemical Industry Press, 2015

[21] Balasubramanian V, Natarajan K, Hemambika B, et al. High-density polyethylene(HDPE)-degradingpotential bacteria from marine ecosystem of gulf of Mannar, India [J]. Letter in Applied Microbiology, 2010, 51: 205-211.

[22] Fontanella S, Bonhomme S, Koutny M, et al. Comparison of the biodegradability of various polyethylenefilms containing pro-oxidant additives [J]. Polymer Degradation and Stability, 2010, 95(6): 1011-1021.DOI:10.1016/j.polymdegradstab. 2010.03.009

[23] YAN Hua, GUO Jun-jun, et al. Influence of density on degradation properties of polyethylene after photo-oxidation aging [J]. Polymer Bulletin, 2017(2): 46-55.

[24] Gopferich A. Mechanisms of polymer degradation and erosion [J]. Biomaterials, 1996, 17(2): 103-114

[25] Gewert B, Plassmann M M, Macleod M. Pathways for degradation of plastic polymers floating in themarine environment [J]. Environmental Science: Processes and Impacts, 2015, 17(9): 1513-1521.

[26] Krueger M C, Harms H, Schlosser D. Prospects for microbiological solutions to environmental pollutionwith plastics [J]. Applied Microbiology and Biotechnology, 2015, 99(21): 8857-8874. DOI:10.1007/s00253-015-6879-4

[27] Laycock B, Nikolić M, Colwell J M, et al. Lifetime prediction of biodegradable polymers [J]. Progress in Polymer Science, 2017, 71: 144-189. DOI:10.1016/j.progpolymsci.2017.02.004

[28] Albertsson A C, Karlsson S. The three stages in degradation of polymers-polyethylene as a modelsubstance, Journal of Applied Polymer Science, 1988, 35(5): 1289-1302.DOI:10.1002/app.1988.070350515

[29] Ohtake Y, Ashabe H, Murakami N, et al. Biodegradation of lowdensity polyethylene, polystyrene, polyvinyl-chloride, and urea-formaldehyde resin buried under soil for over 32 years [J]. Journal of Applied Polymer Science, 1995, 56: 1789-1796. DOI:10.1002/app.1995.070561309

[30] Ohtake Y, Kobayashi T, Asabe H, et al. Studies on biodegradation of LDPE: Observation of LDPE filmsscattered in agricultural fields or in garden soil [J]. Polymer Degradation and Stability, 1998, 60(1): 79-84. DOI:10.1016/S0141-3910(97)00032-3

[31] Ohtake Y, Kobayashi T, Asabe H, et al. Oxidative degradation and molecular weight change of LDPEburied under bioactive soil for 32~ 37 years [J]. Journal of Applied Polymer Science, 1996, 70: 1643-1659.

[32] Bonhomme S, Cuer A, Delort A, et al. Environmental biodegradation of polyethylene [J]. Polymer Degradation and Stability, 2003, 81: 441-452. DOI:10.1016/S0141-3910(03)00129-0

[33] Chiellini E, Corti A, D'Antone S, et al. Oxo-biodegradable carbon backbone polymers: Oxidativedegradation of polyethylene under accelerated test conditions [J]. Polymer Degradation and Stability, 2006, 91 (11): 2739-2747. DOI:10.1016/j.polymdegradstab. 2006.03.022

[34] Briassoulis D, Babou E, Hiskakis M, et al. Degradation in soil behavior of artificially aged polyethylenefilms with

pro-oxidants [J]. Journal of Applied Polymer Science, 2015, 132(30): 42289

[35] Prakash Bhuyar, Nurul Aqilah Binti Mohd Tamizi, Mohd Hasbi Ab. Rahim, Gaanty Pragas Maniam 1, Natanamurugaraj Govindan, Effect of ultraviolet light on the degradation of Low-Density and High-Density Polyethylene characterized by the weight loss and FTIR, Maejo International Journal of Energy and Environmental Communication, MIJEEC., 2019, 1-2 | 31

[36] LU Lin, SHI Yuye, GAO Jin, LI Xiaogang, Photo Degradat ion behavior of media dens ity polyethylene under Xisha (Paracel) islands environment, Journals of Materials Engineering, 2011(3): 45-54. doi:10.3969/j.issn.1001-4381.2011.03.011

[37] Ruoko (2012) Uv Light Induced Degradation of Polyethylene and Polystyrene–Spectroscopic and DSC Study. Tampere University of Technology, Msc thesis

[38] Singh B, Sharma N (2008) Mechanistic implications of plastic degradation. Polym Degrad Stab 93:561-584

[39] Syed Sikandar Shah, Imtiaz Ahmad, Muhammad Ishaq, Degradation study of used polystyrene with UV irradiation, Advanced Materials Science, 2017 doi: 10.15761/AMS.1000130

[40] Zhang Xiaodong, Jie Ganxin, Peng Jian, Feng Hao, Analysis on Influencing Factors of Aging of PS Using UV Accelerated Weathering Test, Plastics Science and Technology, Vol 38,No. 10, Oct 2010, 1005-3360(2010)10-0088-03

[41] YAO Peipei, LI Chen, XIAO Shengling Effect of ultraviolet aging on properties and structure of polystyrene, CIESC Journal, Vol 55, No. 11, Nov, 2014

[42] Qi D L, Wang X Q, Zhang Z Y, et al, Current situation of world natural rubber industry and its driving force by science and technology, Chinese Journal of Tropical Agriculture, 2013, 33(1): 61-66.

[43] Chen M W, The current situation and coping strategies of China's natural rubber industry Issues in Agricultural Economy, 2016(10): 91-94.

[44] Patel M, Skinner A R, Maxwell R S, Sensitivity of condensation cured polysiloxane rubbers to sealed and open-to air thermal ageing regimes, Polymer Testing, 2005, 24: 663-668.

[45] Zhang B L, Huang H H, Wang Y Z, et al, Aging characteristics of constant viscosity natural rubber, Polymer Materials Science and Engineering, 2012, 28(6): 24-27.

[46] Gu H S, Itoh Y, Aging behaviors of natural rubber in isolation bearings, Advanced Materials Research Vols. 163-167, 2010, 3343-3347.

[47] Gengsheng Weng, Guangsu Huang, Hangxin Lei, Liangliang Qu, Peng Zhang, Yijing Nie, Jingrong Wu, Crack initiation of natural rubber under high temperature fatigue loading, Journal of Applied Polymer Science, Vol. 124, 4274-4280 (2012).

[48] Gac P Y L, Arhant M, Davies P, et al., Fatigue behavior of natural rubber in marine environment: Comparison between air and sea water, Materials and Design, Volume 65, January 2015, Pages 462-467

[49] Weng G, Huang G, Lei H, et al., Crack initiation of natural rubber under high temperature fatigue loading, journal of applied polymer science, Journal of Applied Polymer Vol. 124, 4274-4280 (2012)

[50] N.S. Tomer, F. Delor-Jestin, R.P. Singh, J. Lacoste, Cross-linking

assessment after accelerated ageing of ethylene propylene diene monomer rubber, Polymer Degradation and Stability, Volume 92, Issue 3, March 2007, Pages 457-463.

[51] Xiang K W, Wang X A, Huang G S, et al., Thermal ageing behavior of styrene-butadiene random copolymer: a study on the ageing mechanism and relaxation properties, Polymer Degradation and Stability, Volume 97, Issue 9, September 2012, Pages 1704-1715

[52] Ling Ding, Zhihui Li, Hui Yang, Xiangpeng Wang, Lihua Liang, Jianyong Zhong, Yousi Zou, Aging Mechanism of Natural Rubber, POLYMER MATERIALS SCIENCE AND ENGINEERING, Vol. 34, No. 5, May 2018

[53] https://www.sohu.com/a/384028158_751728 (last access 07.08.2021)

[54] Jon McHugh, Rocelyn Dee and Mudit Saxena Heschong Mahone Group, Visible Light Transmittance of Skylights, project 5.3.4

Chapter 12

The Use of Biochar of High Growth Rate Plants to Agriculturally Remediate Heavy Metal Polluted Acidic Mine Wastes

Arturo Aguirre Gómez, Laura Virginia Nuñez Balderas, Claudia Hidalgo Moreno and Jorge Dionisio Etchevers Barra

Abstract

The chapter is meant to expose how a sound methodology can be instrumented to both, remediate acidic metal polluted mine wastes, taking advantage of the neutralizing power and high metal sorption affinity of biochar, and to utilize pyrolyzed material derived from high-rate growth plants (water hyacinth, *Eichhornia crassipes* Mart, and Eucalyptus, *Eucalyptus globulus* Labill), which have become of ecological relevance due to their unwanted proliferation over specific terrestrial, lacustrine or riverine environments. In addition, the proposal considers not only neutralizing the mine tailings and abating the toxic levels of specific heavy metals like Pb, Cd, Cu, Zn, etc., to fulfill the international and national standards and norms, but to conveniently combine biochar with widely used soil amendments to pass widely recognized biological tests of growth using heavy metal-sensitive plants. The approach addresses firstly: a) characterizing physiochemically mine tailings and biochar, in terms of their properties (metal speciation and contents, potential acidity and neutralization potential, chemical oxygen demand, heavy metal-biochar sorption-complexing affinities, among others), and secondly; b) creating a" fertile environment" by reconditioning, agriculturally, the heavy metal-polluted acidic mine waste to allow native vegetation, or other reforesting species, to regrow on the reclaimed site, based on the bioassay tests performances.

Keywords: biochar, mine tailings remediation, soil amendments, heavy metal phytotoxicity, heavy metal availability and speciation

1. Introduction

Multi-metallic mine wastes from mining activities have been produced in large quantities and disposed on soils for centuries all over the world, most of the times in a non-environmentally safe way. When exposed to environmental factors like rain and air mine tails undergo continuous chemical changes involving acid–base, redox, sorption-dissolution, and complexation equilibria increasing both, the (bio)-availability levels of specific metals to local biota, and the metal mobility, increasing the risk of extending the polluting effects to ground water and adjacent soils, affecting all forms of life. Biochar additions to soils and mine wastes have proved to be useful for mitigating the

IntechOpen

negative effects of metal phytotoxicity by augmenting the active surface and number of new sorption sites for metal retention and increasing pH of the once acidic metal-sulfidic mine wastes that would otherwise hamper the growth of vegetation. Among other beneficial effects of biochar additions, the increase of residence time of C in soils should be mentioned. The methodologic approach presented here shows a sound proposal of how a remediation strategy can be instrumented to agriculturally stabilize and reclaim acidic metal mine wastes and polluted soils. By taking advantage of many of the relevant biochar characteristics and properties that have proved to be instrumental when used, alone or in combination with fertilizers and other soil amendments (e.g., lime, gypsum, compost, etc.), to lower metal toxic levels, and modify the relevant physicochemical waste and soil characteristics and properties. Firstly, we focus on mitigating the phytotoxic effects of the bioavailable essential and toxic heavy metal (Cu, Zn, Pb, Cd, etc.) deposited on the acid mine tailing dams, and secondly, on creating a" fertile environment" by reconditioning agriculturally, the HM-polluted acidic mine waste, to allow metal sensitive higher plant species to grow, so that the more resistant native vegetation, or other reforesting species, can regrow on the reclaimed site, based on the successful bioassay tests performed. Pyrolyzed material derived from high-rate growth plants (water hyacinth, *E. crassipes* Mart); Eucalyptus, (*E. globulus* Labill), among others, have become of ecological relevance due to their unwanted proliferation in many terrestrial, lacustrine or riverine environments all over the planet, so that the proposal considers not only neutralizing the mine tailings and abating the toxic levels of specific heavy metals (HM) like Pb, Cd, Cu, Zn, etc., to fulfill the international and national standards and norms, but to conveniently combine biochar based on these type of plant materials, with widely used soil amendments to pass any of the widely recognized biological tests of growth using HM-sensitive plants, earthworms (*Eisenia fetida*/*Eisenia andrei* by dermal and alimentary uptake), acute toxicity test, etc., among others. The approach addresses firstly: a) characterizing physiochemically mine tailings, in terms of their relevant physical and chemical parameters and properties, i.e., metal speciation and contents, potential acidity, chemical oxygen demand, metal-biochar sorption-complexing affinities, neutralization kinetics, etc., to compare and establish definitive results of doses, etc., and secondly; b) establishing conditions for biochar production and pyrolysis so that optimal neutralization potential and surface for HM-sorption interactions can optimally be obtained.

2. Mine waste and metalliferous acid mine drainage

Mining extraction of metals from sulfidic materials produce considerable levels of potential acidity which eventually, if not prevented and neutralized, will generate the so-called metalliferous acid mine drainage resulting in the production of acidic water and mobilization of heavy metals [1, 2]. The quantity of acid-forming minerals found in many mines of central Mexico around the neo-volcanic mountainous axis [3, 4] and around the world, include dominantly pyrite (FeS_2), galena (PbS), sphalerite (ZnS), pyrrhotites ($Fe_{1-x}S$), chalcopyrite ($CuFeS_2$), arsenopyrite (AsFeS), bornite (Cu_5FeS_4), and many other metallic sulfosalts. After oxidation, these minerals generate the H^+-producing redox and hydrolytic processes of the components (e.g., S, Fe, Mn, Zn, Cu, Pb, etc.), left behind in the mine tailings. Mine wastes and tailings, polluted sites and pristine-soils (for reference and comparison) must then be chemically characterized in their metal speciation and contents, besides other relevant physicochemical properties and parameters, i.e., pH, Electrical Conductivity (EC), Cation Exchange Capacity (CEC), Acid Neutralizing Capacity (ANC), Chemical Oxygen Demand (COD)-value, etc. before any remediation technique is planned or executed. Metal speciation and fractionation must also be carefully considered so that the so-called bioavailable, labile-toxic metal fractions and pools (soluble, exchangeable, DTPA-, acid-extractable, total metal content, etc.) must correlate well with the requirements of clean up and agrostabilization of sites.

2.1 Chemical characterization of mine wastes for remediation

Routinely acid–base account (ABA) is determined to specify the total quantity of acid-forming sulfur compounds in a sample relative to the neutralizing potential [1, 2]. Acid–base account methods are applied routinely to mine waste analysis of materials to assess the potential of generating acid upon redox weathering processes. Acid–base account is evaluated by determining the neutralization potential (NP) and potential acidity (PA) of a sample: ABA = NP–PA [5]. However, when common sulfide and sulfate minerals are subjected to ABA extraction methods, the ABA method fails to accurately distinguish the acid-forming from nonacid-forming minerals, resulting in errors in the determination of potential acidity mainly because the lack of consideration of all possible acid-producing species that very slowly oxidize. These errors can be subsequently visualized not only in the discrepancies between NP- and COD-values measured in a sample, but in the inaccurate liming requirements predicted to be applied, resulting not only in insufficient neutralizing amounts required and an excessive cost to cover the slowly released and underestimated potential acidity, but in the potential reclamation failure when the ABA measurement is underestimated. In this respect, the methods regularly used to determine the acidity potential, involve the acid–base treatment of a mine waste and polluted and pristine soil samples, without considering, on one hand, the kinetics or slow neutralization reactions that with time might occur, and the slow oxidation of other components which contribute to waste acidification [6]. In our proposal we measure and determine the neutralization-kinetic curve of the acidic mine waste with lime and biochar, or other amendments applied, so that the time of reaction under laboratory conditions can be considered and extrapolated to green house and field experiments. **Figure 1** shows such a neutralization curve for an acidic mine waste from Zimapan, Hidalgo State, Mexico, which showed initially very low pH values.

These discrepancies have been recorded [5–7] between the acidity potential (AP) and the COD-values. In a freshly sample taken from the site at Zimapan mine in Mexico, whose initial pH, AP (ABA) and neutralization potential (NP) values were 6.9, 348 kg_{CaCO3} ton^{-1} and 201 kg_{CaCO3} ton^{-1}, respectively, calculation of the acid–base balance produced a ratio of NP/AP < 1.2 (=0.58), revealing a potential for acid mine drainage generation resulting from the eventual oxidation of sulfides to sulfates

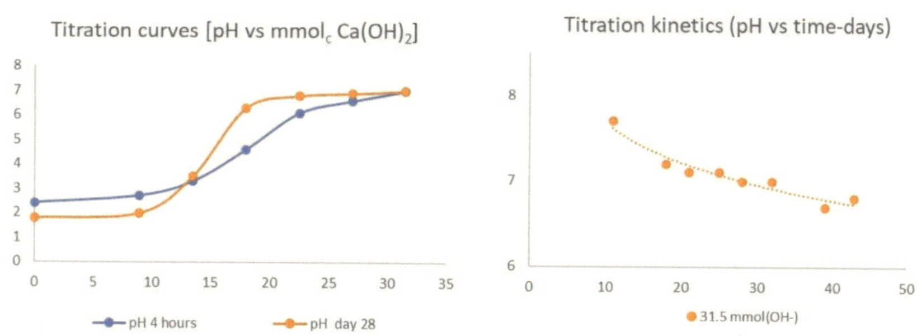

Figure 1.
Graphs show neutralization (pH-mmol$_c$ of OH^-) and kinetic (pH-time, in days) curves of an acid mine waste from Zimapan, Hidalgo State, Mexico with lime, $Ca(OH)_2$ whose initial pH value was 2.4. Solid:liquid mixture ratios were 1:3, in small polyethylene reactor flasks with cap. Ten flasks were prepared to which various dose of liming material was added at variable levels of 0–67.5 mmol$_c$ of $Ca(OH)_2$, which were kept incubated for the time period described. pH was monitored over time. As an example (see [6]) of curves, some flask-systems results were plotted [6].

according to reactions of the type e.g., S^{-II} + $^{3}/_{2}O_2$ + $H_2O \leftrightarrows SO_4^{2-}$ + $2H^+$, and the hydrolysis of the ferric ions, Fe^{3+}, ($Fe^{3+} + 3H_2O \leftrightarrows Fe(OH)_3 + 3H^+$). These values were found to be within the range of reported levels for mine tailings deposited in mining areas of central Mexico, whose acid–base balance usually lays between 0.1 and 0.9, and already present acid drainage production [8], as well as with values reported specifically for the mining area that are at 0.3 being considered as eventually acidic waste [9]. As a complement, the chemical oxygen demand, COD, corroborated an extra potential risk of acid drainage generation due to the slow oxidation processes of the mine residues. The COD-value of this sample was of 17 kg $K_2Cr_2O_7$ ton^{-1}, equivalent to 12% of S(-II) content (unpublished results), assuming that all the oxidable material was pyrite, and therefore to 376 kg_{CaCO3} ton^{-1} waste is calculated leading to an extra amount of liming material of 28 kg_{CaCO3} ton^{-1} (= COD-AP), expressed as potential acidity. Thus, despite being a recently neutralized the fresh waste, in the medium or long term, the deficit in the acid–base balance and the high COD value must be considered as indicators of the potential oxidation of metal sulfides. Besides, the kinetic factor would lead to a ΔpH/43 days of 0.9 units, which would correspond to an extra amount of an extra 4.5 $mmol_c$ (= 5.5 kg_{CaCO3} ton^{-1} waste) to compensate for a kinetic period of 60 days under laboratory conditions, resulting in total an extra amount of 22.8% more liming material. In close comparison with an acidified sample of the same site at Zimapan mine, calculations showed (see [6, 7]) original values of pH, AP, NP and COD-values, to be 2.2, 38.1kg_{CaCO3} ton^{-1}, 109 kg_{CaCO3} ton^{-1}, and 146 $kg_{K2Cr2O7}$ ton^{-1}, equivalent to still 1.2% S^{-II} content to be yet oxidable, if all materials were pyrite. Following similar calculations as above, the amount liming material required to neutralize this naturally oxidized mine waste would be estimated to be in the order of 37.5 $kg_{K2Cr2O7}$ ton^{-1} (plus the kinetic dose of 5.5 $kg_{K2Cr2O7}$ ton^{-1}), giving a total of 43 $kg_{K2Cr2O7}$ ton^{-1} [6, 7]. The eventual production of acid drainage conditions and the consequent availability and potential mobilization of the toxic metals contained in the waste material can only be prevented when the three aspects of analysis are considered: the ABA- and COD-value, and the kinetics of neutralization factor. From that it results clear that precise and full characterization of these determinations must be considered if a successful remediation of acidic mine tailings is planned. Besides the natural tendency of field researchers to apply lime on mine tailing dams, several studies have shown that some metal can be re-speciated causing some redissolution, as lime is a more soluble material than other precipitates, adsorbing or complexing amendments as phosphates, biosolids, compost, clays or biochar, alone or in combination, which has conducted many researchers to look for alternative ways of immobilizing metals in mine wastes [10–13]. We have carried out studies to prove this [7]. After applying lime, gypsum, phosphates and organic matter (compost) it was corroborated that lime and gypsum (results not shown) caused some redistribution and redissolution of metals. **Figure 2** shows for lime, when applied to local pristine soil-mine waste mixtures, ranging from 100:0% to 0:100% soil (s): mine waste (mw) mixtures. Adsorption experiments carried out on these soil-wastes emulate the various grades of pollution caused by the dispersion of the mine waste over clean unpolluted soil.

Graphs shows the effect that lime had over the desorption of the four studied metals. Curves show that as mixtures were limed at low, medium and high doses (where low, medium and high doses were chosen to suppress the summation of water-soluble, acid-extractable and DTPA-extractable heavy metals respectively), metals were, in all cases extracted in higher amounts as compared with the no-limed treatment, suggesting this amendment material serves well to neutralize the acidic mine waste, but fails to suppress the extractability of metals with respect to the non-limed fresh mine waste treatments. Results (not shown) for systems containing phosphates and organic matter (compost) effectively lowered the availability of these four metals, as other researchers have reported too [10–13]. Further experiments were planned

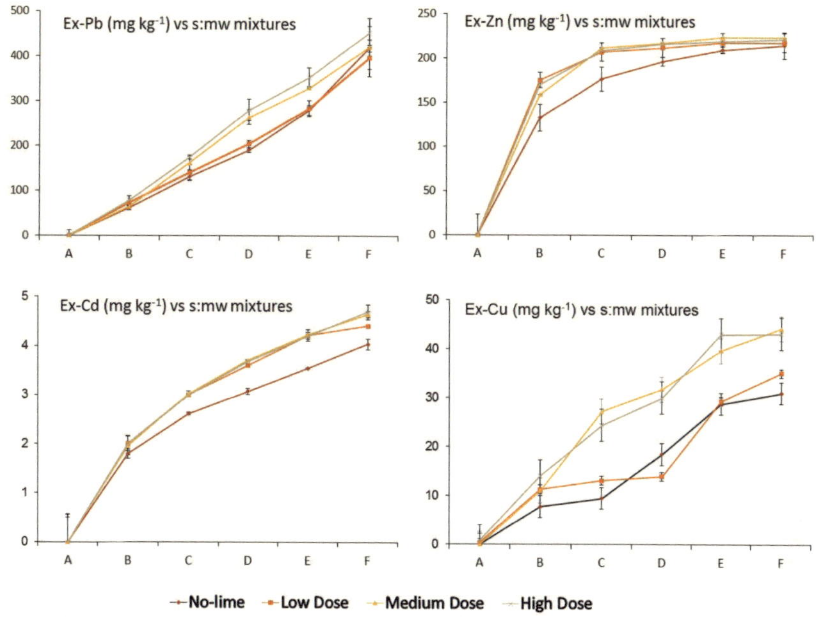

Figure 2.
Comparison of the relative effect of low, medium and high doses of lime versus the non-limed treatment, on the 0.2 N acetic acid (AcO)-extractable (Ex-) Pb, Zn, Cd and Cu levels for six experimental mixtures (mx) of soil (s)/mine waste (mw): A: 100% soil, B: 80:20 s:mw, C: 60:40 s:mw, D: 40:60 s:mw, E: 20:80 s:mw, F: 100% mw. Freshly deposited mine waste was sampled, showing a pH value of 7.37 (see [7]).

to continue with this research to test the effectiveness of biochar in combination with lime and phosphate [14, 15]. Results have been published elsewhere but conclusions are presented in our case of study included in Section 4 of this chapter.

2.2 Toxic heavy metals in mine tailings

To lessen HM toxicity hazards several studies involving physical temporary solutions [16, 17]; a wide variety of chemical treatments [18, 19]; and the not always feasible biological techniques [20, 21] have been conducted and published. Physical treatments include not only confining, solidification and vitrification procedures [16, 17] but S/S technologies which in many cases are considered not to be a permanent environmental solution as metals are not removed from contaminated media; always require future monitoring and; there is a questionable longevity of the effect of the ameliorating materials added plus requiring long-term management of resulting materials based on landfilling and soil caps to prevent erosion [16, 17]. On the other hand, chemical procedures used specially for polluted sites are generally based on precipitation [22], adsorption [23] or complexation processes [24]; this last to either mobilize and bioremove HM effectively. These physical and chemical procedures, if integrally and congruently used, should have an important impact on the abiotic and biotic factors, so that the toxicity of metals left after the treatments is reduced and an attractive growing media for phytoremediation can then be promoted and established. For soils most procedures used to diminish the mobility and leaching potential of HM have used chemical remediation techniques. Several studies report an effective immobilization of HM in polluted soils when applying lime, phosphate and/or organic matter amendments were applied, resulting in a reduced bioavailability of HM [25–28]. Addition of phosphate materials for instance was effective in immobilizing Pb [29]; whereas composts, biosolids, manure and peat materials [8–10, 30] were instrumental to reduce mobility of Cd, Pb and Zn, among other HM [31].

3. Agrostabilization of mine waste and polluted soils

Among the most promising remediation techniques to treat heavy metal polluted wastes and soil used are the application of lime [32], phosphates [29, 33], biosolids [34, 35], composts [36] and the application of amendments with biochar [12, 15, 37]. Although the application of biosolids and composts decreases the bioavailability of metals, their effect is variable depending on the metal, soil type, dose, type, and degree of humification of organic matter [38, 39]. Most of these techniques require constant applications to ensure their success. Other studies [40] have demonstrated that fresh composts increased the solubility of metals, except perhaps for Cu [7], so that pretreatments are required to stabilize organic manufacturing, which increase application costs [36]. These above studies considered the addition of ameliorating materials to reduce chemically the HM bioavailability, however if the addition of carefully selected agrochemical materials and doses is premeditated the remediation strategy should warrant in an initial step: 1) a stable and slow mobilization of bio-available HM; 2) the abatement and control of their toxic levels at least based on the threshold allowable levels dictated by the national and international standards and norms and; 3) a fertile media to sustain a consistent phytoremediation or phytomining activities, for instance through the use of HM sensitive indicator plant species according to standardized biological test methods [41, 42], or through the use of hyperaccumulator plants, respectively. Once this strategy works, and a well-nourished growing media is assured, a further important step should be funded in a studied introduction of perennial domesticated, native and/or undomesticated plant species whether HM hyperaccumulators or not. For acidic polluted mine wastes, studies should focus mainly on lowering the mobility of Cd, Cu, Pb, and Zn below official environmentally safe values to warrant a biologically clean and sustainable ecosystem. To reach this goal, at least two schemes must be visualized; First the addition of widely used agronomic materials consisting of lime ($Ca(OH)_2$); gypsum ($CaSO_4{\cdot}2H_2O$); P-fertilizer (KH_2PO_4) and/or compost, and more recently with much success biochar, to "treat" a gradient of soil-fresh mine tailings mixtures in order to assess the treatment effectiveness to lower both the most toxic metal species in solution, the free-metal ion chemical activity, (M^{2+}), of metals like Cd^{2+}, Cu^{2+}, Pb^{2+} and Zn^{2+} [43] and the so called bio-available (acid-extractable) HM fraction imposed by international standards and norms. Second, a bioassay should be carried out to find the conditions to pass sensitive indicator plant response tests in these "fertile" ameliorated media. These studies should prove to be useful in deriving soil-substrate quality criteria to establish specific cleanup strategies and/or verify the success of remediation processes. In order to evaluate metal toxicity/abatement effectively both, the bioavailable (acid-extractable) HM fraction and the chemical activity of the free metal ion, (M^{m+}) should be measured after incubation with the chemical-agronomic treatments mentioned above. Green house experiments and field trials are the next important steps to establish a well-founded remediation strategy.

3.1 Lime

The application of lime as ameliorating material alone or in combination with compost, initially increases pH values, lowers metal solubilities and buffers the changes of the acid–base equilibria. In addition, their low-cost and easily accessible and application aspects [10] favors at first instance their use, however, due to the transient nature of organic matter, its effectiveness might decrease afterwards [11]. Although excess liming may also lead to P-deficiency in plants, an excess of mineralization of labile carbon is also possible when lime is mixed with biosolids promoting the availability of P [12]. Several authors have tested this material in combination with-red mud (a by-product of aluminum manufacturing) [10], compost from water hyacinth [11], biochar and biosolids [12]. These studies coincide that the addition of lime is effective, but it is enhanced with mixed with other amendments. For example, Trippe, et al. (2021) (see [12]), showed that the above and belowground biomass of plants grown in mine tailings amended with lime, biochar and biosolids, was 3–4 times larger than the biomass of plants grown in tailings amended only with lime.

3.2 Phosphate fertilizers

The application of phosphates to metalliferous polluted wastes and soils is intended to form stable precipitates with metals, at the time of providing essential nutritional elements for the growth of the plant cover. For instance (see [44–46]). Compared to the control, they found that all P amendments significantly decreased relatively Pb phytoavailability up to 86%. The Pb immobilization was attributed to sorption and precipitates of Pb phosphate on calcite and phosphate phases and minerals. Water solubility of Cu and Zn was also lowered up to 80% and 69%, respectively. On the other hand, it has been put into discussion whether metals should be immobilized or not. Other studies have also addressed the remediation of heavy metals and metalloids of contaminated soils, to mobilize or to immobilize them by adding amendments or complexing substances, respectively. Of course, the final fate of metals becomes critical, whether they will be immobilized to mitigate the potential toxic effects on biota, or to increase their phytoavailability so that metals can be phytoextracted or leached them out from the system. In terms of adding phosphates to Pb-polluted environments, certainly it promotes Pb-immobilization, although the availability of other metals, still need to be addressed specifically, especially Cu.

3.3 Compost and biosolids

Some studies [38] have shown that the application of biosolids and compost decreases the bioavailability of some metals, but their effect is variable depending on the metal, soil type, dose, and degree of organic matter humification. However, some contrasting effects of manure and compost on soil pH, heavy metal availability and growth have also been recorded in soils contaminated by pyritic mine waste [39], so that in terms of specific metals, there are accounts signaling that some metals certainly are complexed and retain on organic material (e.g., Cu), but others not. The problem related to sewage sludge application arises when it contains high concentrations of potentially toxic heavy metals. The release of heavy metals associated with sewage sludge is strongly influenced by soil pH, cation exchange capacity, organic matter, and mobility and speciation of specific metals. Excessive application of sewage sludge to soil has been found to increase the bioavailability of heavy metals, but the low doses of sewage sludge did not cause a significant increase in heavy metal concentrations [47]. Although application of biosolids provides essential nutrients for plant growth, a careful assessment of the characteristics of sewage sludge, however, is required prior to land application. When sludge contains high concentrations of toxic elements, amendments may have a potential toxic impact. Co-application of organic matter, liming, phosphates, biochar, etc. may minimize the potential negative impact of land application of biosolids too. This last study [47] also showed that sewage sludge pyrolysis decreased the plant-availability, the mobile forms and the risk of leaching of Cu, Ni, Zn and Cd. Important to notice that the leaching of Cu, Ni and Zn were lower in the soil treated with biochar that in sewage sludge treatment. When mixed biosolids and biochar amendments increased soil respiration with respect to the control soil, being lower in the case of biochar than when biosolid was added alone. On the contrary, biochar amended samples reduced Ni, Zn, Cd and Pb plant availability, when compared to sewage sludge amended samples. It results clear then, that biosolids, at low dose have a good influence in soil than biochar alone, in that particular case, however, it also results clear that sewage sludge when mixed with biochar can have better results to lower metal availability for plants. Beneficial effects can be obtained when biochar is incorporated to the application of amendments.

3.4 Clays

Clays have been tested to remediate heavy metals contaminated soils too. Their use has been recommended because of the low cost, clay materials can easily be obtained, and result innocuous to the environment. Among these, bentonite, montmorillonite, attapulgite [48], palygorskite, sepiolite [49], illites and kaolinites [50] can also be mentioned. It has been observed that sepiolite can increase the pH of systems due to its alkaline nature, it is rich in oxides and carbonates of Ca, Mg, Al and Si, and it improves microbial activity in soils contaminated with mining residues [51]. Nevertheless, the main characteristic clays impose on treatments are related to those factors that influence the adsorption of heavy metals on clays. According to Otunola and Ololade, 2020 [50] these are: a) increase of cation exchange capacity (CEC) and the specific surface area of the clay, which are the determining properties of sorption; b) longer contact times between the clay and the contaminated site, both related to the type of metal and the pH value of the site; c) moderate dose of clay, being 4 to 8% to remove more than 70% of heavy metals [52]; d) increase of pH of the system or the site, as it influences the leaching of cationic metals. The best absorption they mentioned was between pH values of 6 to 8 [53]. The increase of pH generates negatively charged sorption sites on soil colloids and on the surface of organic matter which favors adsorption of the HM's [54]; and e) the temperature that influences the amount of adsorbed metal, being lower when temperature is greater than 36°C [49]. Nowadays, methods have been designed to modify the adsorption capacity, although it is not yet known which is the best modification method for each type of clay and the required dose [50]. These are the following modification methods: a) Forming organic complexes with clay, where the clay is mixed with compost or residual sludge, which can rise the sorption of metals up to six times more due to the upturn in CEC [55], this is an environmentally friendly method, however the presence of organic matter can affect the immobilization of metals [56]; b) Modification with nano-zero valent iron (nZVI), small particles are used as binders that increase the active adsorption sites on the clay surfaces by impregnation, followed by filtering, washing and drying [48]. This can also be done by chemical synthesis, with H_2 or Fe^{2+}, which are reduced at high temperatures [57]; c) Thermal modification that consists of heating the clay to temperatures of 200–1000°C for a time of 3–12 hours, the process dehydrates and dehydroxylates clays. The procedure reduces the mass, increases the porosity and the cation exchange surface but the disadvantage is that the time and temperature must be determined for each type of clay [58]; d) Modification with acids to remove impurities as Ca, Mg K and metal oxides present in the clay, and modify the adsorption surface [59]. Although the method increases the cost of the amendment, and it can release new agents into the environment [60]. However, the application of pristine or modified clays as remediation methods, require more field studies to determine the ecotoxicity and biocompatibility. Since the change in pH, redox potential, temperature and CEC of the soil strongly influences the mobility of metals. In addition, clays are not selective and have partial binding capacities [61].

3.5 Biochar

Biochar as a solid product of biomass pyrolysis has shown to present large specific surfaces for sorption [62] and remotion of heavy metals from mine tailing

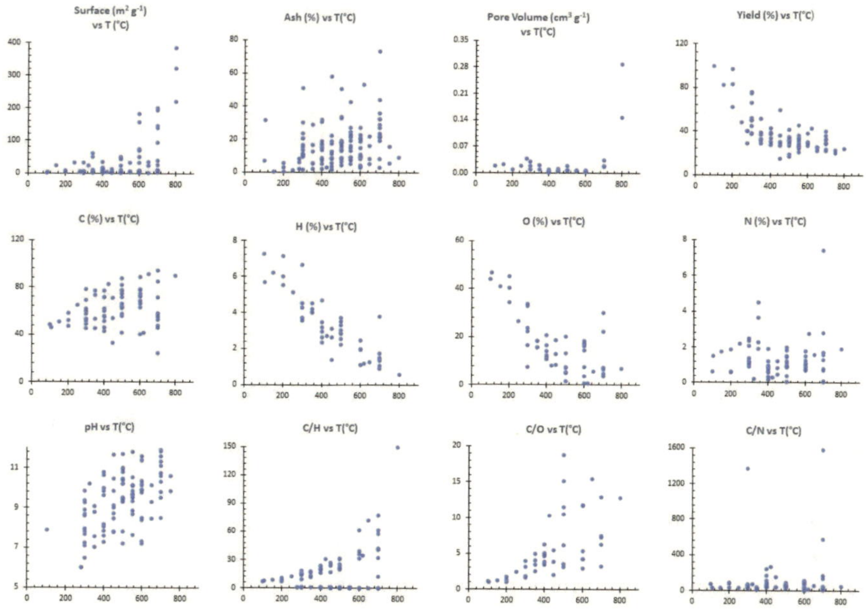

Figure 3.
Biochar characteristics and properties as affected by temperature of charring. Tendencies and general behavior of relevant parameters usually measured and reported are shown vs. temperature (°C). Data plotted show Specific surface (m^2 g^{-1}); Ash contents (%); Pore Volume (cm^3 g^{-1}); Yield (%); C-content (%); H-content (%); O-content (%); N-content (%); pH; and C/H; C/O; and C/N ratios. Data compiled from [75–103].

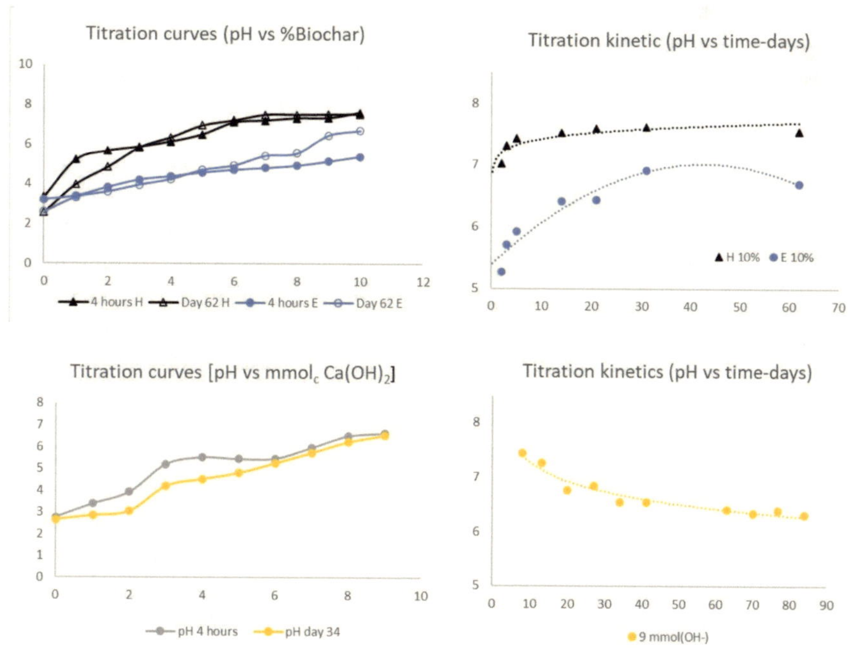

Figure 4.
Graphs show neutralization (pH-% biochar) and kinetic (pH-time in days) curves of an acid mine waste from Zimapan, Hidalgo State, Mexico with biochar of water hyacinth, Eichhornia crassipes Mart, (H) and Eucaliptus, Eucalyptus globulus Labill, (E) whose initial pH was ca. 2.9. Solid:liquid ratio was 1:3, processed in small polyethylene reactor flasks with cap. Ten flasks were prepared to which various dose of liming material was added at variable levels of 0–10% biochar, which were kept incubated for the time period described. pH-values were monitored over time. As an example of curves, some flask-systems results are plotted [14, 15].

sites and polluted soils [12].; being a porous material [63], it has been reported to improve soil physical properties [64]; to provide nutrients, allowing the development of microorganisms [62] too, and last but not least, increasing both, pH values which normally range between 5 to 12, depending on the material and the pyrolysis temperature (see **Figure 3**), and the alkaline neutralization potential (**Figure 4**).

3.5.1 Effects on the acidic mine waste remediation

Biochar can be used to modify soil structure, improve physicochemical properties and enhance nutrient intake by plants, modify habitat for microorganisms, adsorb pollutants and mitigate climate change by changing the availability of elements in soils, especially C [62]. Yuan et al., 2019 [62] also propose the use of biochar in alkaline-saline soils, in cultivation areas and industrialized regions. However, more long-term experiments are required to assess the environmental risk, in addition to considering more mechanisms to make progress in biochar production, when modifying biomass type, pyrolysis temperature, gas flow, retention times, all to increase the efficiency of using biochar as an amendment (see Section 4.1). The fissure intensity factor has also been evaluated [64, 65], with positive conclusions towards the reduction of cracks in this respect, because these affect permeability and water retention and balance in soils. Here water hyacinth (*Eichhornia crassipes*) and soil biochar were studied to know their effect on water content and soil suction in doses of 0 to 10% (w:w) for 3 months. More suction was observed when higher doses of biochar were added, when compared to unamended treatments, finding best doses close to 10% values for most soils, to increase permeability and water retention in soils. Zimmerman, 2010 [66] carried out laboratory experiments to model C-longevity in soils, for controlling pollution migration, increasing fertility and sequestering carbon. The author mixed pyrogenic or "black" carbon-lime-biosolids and biochar-lime-biosolids-inoculum in soils contaminated with mine tailings, the results were that even in the absence of inoculum, microorganisms made progress under the two systems studied. However, both systems generally decreased with increasing charring temperature. Recalcitrant organic carbon content of soil increased in the long run, showing a slow natural mineralization of up to 26% losses in 100 years, projecting half-lives of 10^2 to 10^7 years in the environment [66]. Unfortunately, there was found evidence to suggest that its stability was less than previously thought, being that biochar-carbon was considered to be extremely refractory. Biochar lability was found to be strongly controlled by the relative amount of a more aliphatic and volatile component, which later an inferred though came up to suggest that their losses could be utilized as an indicator of biochar C longevity. Final conclusions were made in terms of its practical value for considering biochar as a tool for soil remediation, amelioration, or atmospheric C sequestration. The stability of two biochar materials of rye grass and pine wood has also been evaluated at 350°C under oxic conditions. Markers of lipid extracts [67], and the composition of n-alkanes and n-fatty acids was used as indicators of stability to microorganisms. Pyrolysis breaks down these molecules and reduces them to shorter chains. Lignin seems to "survive" the process, it is also mentioned that the first day's microorganisms were more predisposed to degrade the biochar of grass than that of wood. Among some of the disadvantages of this amendment it can be found that: it can cause health problems when inhaling the dust [68]; requires high temperatures to be used as a remediation method for liming and adsorbing various metals [69]. The potential risk of using biochar and establishing vegetation creates these and some extra problems, since these may also acidify the soil due to the release of acids that again might solubilize the metals in the biochar, what has called for the attention to make a review of several cases in China where after a while some metals are released. Recent studies indicate that biochar applications could pose ecological and health risks by releasing toxic substances into soils or by inhalation of biochar dust during production or its application on the fields [67]. In China, biochar is commonly used for remediating agricultural soils. Processes for metal interactions with biochar-soil materials may involve adsorption-precipitation and/or complexation reactions, as affected by pH, CEC, organic matter contents, and other modified soil properties. Most field trials showed HM uptake by plants and concentrations of mobility-bioavailability HMs in soil that biochar addition reduced. However, there can also be adverse effects of applying biochar to contaminated soils. Toxic substances contained in biochar may be released to the environment, resulting in increased pollution, and impacting soil organisms and soil functioning, as there are no application standards of regulatory frameworks for biochar application, safety precautions, food safety and human health effects upon exposure, and long-term effects of biochar applications to HM contaminated soils on soil ecosystems. Biochar has been used mixed with mud at 300–900°C, with pyrolysis temperature affecting pH, specific surface, and functional groups, as well as the composition and content of the dissolved organic matter. The higher the temperature, the metal content decreased more (from 91.7% to 9.4% for Cu, 92–63% for Zn, 55.9% to 4.9% for Pb, and 78.2% to 12.5% for Cd, among other metal(loids)). The length of the shoots and roots, and the activity of microbes, in sandy and silty soils, were also increased. In general, biochar prepared at high temperatures was safe for their production or application. High doses of 3–9% (equivalent to more than 60 to 200 Mg ha^{-1}) were estimated to be required to remedy heavy metal contaminated soils [70]. Due to this it has been proposed to mix small amounts of biochar with lime or with compost to increase the pH of acid residues so that other properties in the soil can be improved [12, 71]. Miscanthus biochar (*Miscanthus giganteus*) was utilized in acid residues with low pH values ($pH < 3$), applying doses of 0 to 5% (w/w) with and without lime and fertilizers. To evaluate its effect, a bioassay carried out with Blue Wildrye (*Elymus glaucus* cv. 'Elkton'), showed that lime mixed with biochar greatly decreased the extractable metal levels. Higher doses of biochar also reduced the extractable concentration of metals and improved the microbial activity [71]. Currently, there are studies too, where biochar has also been modified by the addition of alkalis, oxidants (as O_3, H_2O_2, K_2MnO_4 and air), microwaves, CO_2 and steam, to improve its sorption capacity [72]. Biochar activation methods have been studied [73], utilizing water hyacinth to remove metals from polluted waters, which subsequently was pyrolyzed and applied to the soil with phosphates. According to authors [74] positive and negative effects on biochar activation were observed, depending on the method of activation, kind of bioassay and kind of soil too, however, the increase of biochar specific surface area caused a significant reduction of toxicity of water leachates due probably to the presence dissolved organic carbon in soils.

As noticed, biochar is widely recommended for remediation of heavy metals polluted environments due to its excellent immobilization effect, which may last for many years [67]. There is a general agreement, then, that the increase–decrease of biochar effectiveness with time, depends on the abiotic and biotic factors [56, 66–74].

3.5.2 Relevant characteristics of biochar as affected by temperature

According to all researchers and papers reviewed and mentioned above, many of the characteristics of biochar vary and depend mainly on temperature of charring, and type of initial material too, among other characteristics and conditions of biochar production. However, most of their important characteristics tend to be congruent with this. Nevertheless, some parameters and analyses reported in literature show good agreement and tendencies with temperature of charring, but some others not as can be seen from the curves plotted from a compiled data collected from references [75–103] (see **Figure 3**). Among the strategies that have been receiving greater attention, the use of biochar has been outstanding in this respect. The coal produced from the pyrolysis of biomass, compiled data with temperatures ranging from 250 to 750°C [104]. It is known that the material used for the preparation of the biochar, the pyrolysis temperature, the size of the waste to be pyrolyzed, the speed of pyrolysis, among others, influenced the physical, chemical and biological characteristics of the material obtained, such as ash content, specific surface, and pH-values, among others [104]. Biochar materials used in the retention of polar contaminants are obtained at high temperatures, because in this condition the specific surface area and sites for adsorption are increased [105]. Pyrolysis temperatures of 400 to 750°C caused the biochar produced to get values of pH > 7, higher values of ash contents and lower specific surfaces, than those generated at temperatures below 400°C [106]. However, biochar produced with straw, at high temperatures (>500°C), usually present high concentrations of the potentially toxic polycyclic aromatic hydrocarbons (PAHs) due to incomplete combustion. Therefore, it is necessary to carefully define the conditions of pyrolysis, to generate a non-toxic biochar suitable for the intended plants and soil microorganisms [67, 107]. There is information explaining the retention mechanisms that occur when biochar is used for the purpose of sequestrating organic and inorganic pollutants. Metals interact with biochar through electrostatic interactions, by cation exchange, adsorption by proton exchange and specific bindings [67, 108]. It is known that the adsorption of Pb by biochar is attributed to phosphate and carbonate in the aqueous phase of the pores of the material, whereas for Cu, pyrolysis temperature and increased pH, correlate well with metal adsorption. Uchimiya et al., 2012 [109] reported that biochar can adsorb up to 42,000 mg of Cu per kg^{-1} in aqueous solution at acidic pH values. If the pH value increases, the retention is due to the organic binders on the surface of the material and to the sorption-precipitation with carbonates and oxides which are also found in the pores.

Figure 3 shows that specific surface, pore volume, yield, C-, H-, O-content, pH and C/H, and perhaps C/O ratios show some tendencies as charring-burning temperature of materials increases. It must be mentioned that a variety of materials and burning conditions, besides temperature, vary, however, some very clear tendency behavior can be observed. Most interesting for the purposes of this chapter are the tendencies observed for pH, specific surface, yield and C-content and C/O ratios, as these variables will correspondingly stand for the acid–base account, sites for exchange capacity, and functional groups (represented by C and O-contents) which should favor remotion via sorption of HM present in the mine wastes. The production and application of biochar from plants with high growth rates and weeds, for instance, present areas of opportunity to remediate acid residues generated by mining activities, at the time of contributing to solve some of the unwanted proliferation of these plants on terrestrial and lacustrine, riverine, or irrigation and wastewater channels around urban areas all over the world. Water hyacinth plants present growth rates ranging from of 100 to 208 Mg ha^{-1} $year^{-1}$ [110, 111], which besides its high biomass yields, possesses strong adsorption capacities, due to its high cellulose content and carboxyl and hydroxyl functional groups [112]. These properties give to this material an extreme potential for use for remediating activities in metalliferous acidic mining wastes. For this reason, is that it has been used in wetlands, in solid dry form (as biochar), to remove toxic metals from aqueous solutions, wastewater and effluent treatments [113], and soil [114]. Eucalyptus trees on the other hand, also present high growth rates of the order of 25.5–40.2 Mg ha^{-1} $year^{-1}$ dry matter equivalent with cutting periods of 7 to 10 years [115]. Although not many studies can be found on this plant, it might easily be considered of special interest in this respect.

4. Biochar as a remediation amendment for acid mining wastes: A case of study

Zimapan mining district in Hidalgo State, Mexico generates mine wastes with high Cu, Pb and Zn contents. Water hyacinth (*Eichhornia crassipes*) and Eucalyptus (*Eucalyptus globulus* Labill) trees require attention as they have been proliferating excessively in lacustrine, riverine, irrigation- and wastewater-channels and in streets and parks of cities and urban areas of Mexico, and other parts of the world, creating enormous residues of plant material, branches and wood. Remediation treatments to solve both problems seem to converge into the single strategy presented here. The objectives of the research were: a) to evaluate biochar derived from water hyacinth, *E. crassipes* Mart (H) and Eucalyptus, *E. globulus* Labill (E) to treat acidic metalliferous mining wastes; and b) to compare biochar performances vs. lime (L), both alone and in combination with phosphate (F) fertilizers. Combined biochar and lime amendments with phosphates included were (FH), (EH) or with lime (FL), via running a root length growth bioassay using barley plants (*Hordeum vulgare* L), to correlate them with total, DTPA-extractable and water-soluble metal contents and pH as published in [14]. In this investigation four substrates were used to emulate pollution gradients, ranging from pristine soil to pure acidic mining waste. Treatments were: 100%-neutral pristine soil (M1) from the mine surroundings; 100%-acid mine residues (M4); and two mixtures of soil:residues (w/w) of 65:35 (M2) and 35:65 (M3). Substrates were treated with the amendment gradient doses (w/w) of up to 10% (H) or (E), lime, L (3.4%), and phosphate F (0.06%). Blanks (non-amended mixtures) were included. The study showed that water hyacinth (H) and eucalyptus (E) could be utilized as an acid mine residue treatment by converting them firstly into biochar. Biochar materials were produced from water hyacinth (H), *E. crassipes Mart*, and eucalyptus (E), *E. globulus* Labill (branches and leaves) collected from Lake Patzcuaro, Michoacan State, and from Texcoco, State of Mexico, respectively, both in Mexico. Briefly, plants were air dried and milled to pass a 3.5 mm mesh. Both plant materials were slowly pyrolyzed following conditions of temperature of 600°C, speed of 10°C min^{-1} and resilience time of 30 min, as described in Refs. [14, 15, 108]. The surface morphology of the both uncharred plant- and biochar-materials were examined using a scanning electron microscope and Jeol Scanning Electron Software for elemental analysis [14]. Images shown in **Figure 5** were taken with an

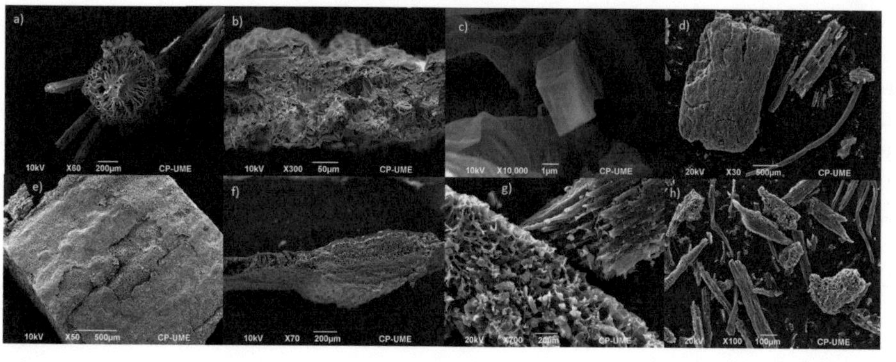

Figure 5.
Images of root (a) and leaf (b) cross-section of water hyacinth cut; $CaCO_3$ crystals in water hyacinth leaf (c); biochar particles of the bulb and root of the water hyacinth less than 2 mm (d); eucalyptus bark (e); cross-section of eucalyptus leaf (f); pyrolytic eucalyptus bark with crystals (g); and eucalyptus biochar particles greater than 2 mm (h). Scan images were taken with a JEOL Model JSM-6390 scanning electron microscope, at the electron microscopy unit (UME) of Postgraduate College, Montecillos-Campus (COLPOS), at Texcoco, Mexico State, Mexico [14].

EDX Oxford Instruments Model INCA-X-ACT probe. **Figure 5** includes the images of root- and leaf-cross-section of water hyacinth cut; $CaCO_3$ crystals in water hyacinth leaf; biochar particles of the bulb and root of the water hyacinth less than 2 mm; eucalyptus bark; cross-section of eucalyptus leaf; pyrolytic eucalyptus bark with crystals and eucalyptus biochar particles greater than 2 mm, respectively.

Alkalinity (ABA) determinations were carried out for biochar and lime treatments as reported by other works [1, 2, 5, 14, 15]. Results produced liming potentials for H and E equivalent to 16.4 g kg^{-1} (kg ton^{-1}) and 8.02 g kg^{-1} (kg ton^{-1}) of $CaCO_3$, respectively. The equivalent doses necessary to correct the acidity of the mine residues to bring them to pH 6.5 were 5% for H, and 9% for E as shown in the neutralization curves of biochar depicted in **Figure 4**. Lime dose to correct the acidity of M4 to reach pH of 6.5, was 9.2 $mmol_{(OH-)}$ for 10 g of M4, equivalent to 34 kg of $Ca(OH)_2$ ton^{-1} of the residues, amount that was found to be in the order of magnitude of other author's quantifications [5]. Worth to mention is that the pH stabilization time for lime was 34 days, a longer time in comparison to that required when biochar was used as a neutralizing agent. Extra details on the experiment and results can be found in Ref. [14, 15]. **Figure 4** shows the neutralization (titration) curves and kinetics of the acid mine waste for both studied biochar materials, as well as with lime. From curves it results clear that neutralization reactions under ideal (laboratory conditions) take several months to stabilize, which consequently will project to take even years for these materials to complete the neutralization reactions, depending of course on particle size (specific surface for reaction) and water content of substrate materials in the field, among others. We have discussed in the introduction section, some of the ABA method details to determine the balance between the acidity potential of the mine tailing material and the neutralization potential of bases naturally present in the soil. An extended discussion of the detail characterization of the acidity of mine materials, including the use of COD-determination and kinetics should be considered if a successful strategy to avoid the so-called acid mine drainage must be stopped. Calculations for neutralizing the mine waste with lime, were included in Section 2.1, and herein we present those of biochar as a liming material (see [14, 15]).

4.1 Biological tests

Biological methods to run biotoxicity assays have been widely used to test the ecological risk assessment of soils [41] and other ameliorated polluted substrate materials (wastes, biosolids, sludge, composts, etc.) [116–121]. These include measuring the inhibition of root growth [41, 42, 116, 118, 119–121]; the effects of chemicals on emergence and growth of higher plants [117, 119]; root length responses, germination viability; green house and field tests [122], using metal sensitive higher plants [41, 116–121] or mesofauna [123, 124] or on the emergence and growth of higher plants when evaluating the effects of pollutants on soil flora [117, 119]. However, it must be stated clearly, what the authors declare about the application and scope of these biological methods. As stated by the authors in many of the presented references, these bioassays are applicable to the comparison of soils of known and unknown quality and to the measurement of effects of materials (compost, sludge, waste) or chemicals deliberately added to the soil [41], and that as stated in references [116–120], that methods are not intended to be used as a measure of the ability of the soil [and other soil materials, wastes, biosolids, etc. to support sustained plant growth. Nevertheless, the importance and relevance of testing remediated environments should always rely on the fact that recovering native or reforesting vegetation, should be one, and not the unique target of the final goals of remediation strategies, to remediate sites and recover plant growth. Having said that, we present herein some of the more relevant results of the bioassays carried out with the neutralized and amended acidic mine tailings with water hyacinth (H) and Eucalyptus

(E) biochar materials as compared with lime (L), alone or mixed with mono-potassium phosphate (F). The biological test was applied according to the described methods in Ref.s [41, 42, 116–121]. The experiment and results are presented in Section 4.4.

4.2 Physicochemical characterization of soil and biochar materials

For the application of the bioassay, barley seeds (*Hordeum vulgare* L. var. Hope) were used and the applied procedure was carried out according to the methodology described by ISO-references [116, 117]. Physicochemical characterization values for pristine soil, mining waste and mixtures [M1 is 100% soil; M2 is 35:65% (M1:M4); M3 is 65:35% (M1:M4); and M4 is 100% acidic mining residue], and biochar materials (H) and (E) are shown in **Table 1**. Determined parameters included pH, Electrical Conductivity (EC), Saturation Percentage (SP), Cation Exchange Capacity (CEC), total-metal levels for Cu_T, Pb_T and Zn_T, and water soluble-Cu, Pb and Zn. Determinations were carried out according to procedures followed in reference [15].

Texture of the control soil (M1) was clayey, and its pH (=6.8) is considered neutral, while the M4 sieved at 2 mm presented an acidic pH of 3.3 (see **Table 1**). The mixtures of soil with the acidic mining residue (M4) raised the pH of both M2 and M3 substrates in greater proportion than in M4. The concentration of water soluble-Zn and -Pb in M2 and M3 samples were lower than the phytotoxic value reported by other authors [125, 126]. The pH value of each substrate decreased as the mine residue increased correspondingly. EC-values, the soluble Cu content, the extracted bases and the CEC increased from M2 to M4 (see **Table 1**), due to the dissolution of the carbonates present in the pristine soil [3] as promoted by the acidity of the mining residue. The pH values of the (H) and (E) biochar materials used in this study were, on an average, about 10 (**Table 1**), like those found by other authors [127–131] who reported that pH of biochar materials reached high alkaline values due to the pyrolysis temperatures used (> 500°C), contents greater than 2% of alkaline elements (Ca, Mg, K and Na), and losses of the functional groups of the source material (carboxyl, phenolics, lactone, etc.). Both the pH and neutralization potential values found in this study (equivalent to 16.4 and 8.02 g $CaCO_3$ kg^{-1} of biochar materials H and E, respectively), are directly related to the contents of exchangeable bases. These values are comparable too with those reported by other works [127]. The EC of the biochar were 0.3 and 0.6 dS m^{-1}, for E and H respectively. These results are similar to those obtained in wood and paper biochar produced at a temperature of 550 and 700°C [129], whose values ranged between 0.09 to 0.44 dS m^{-1} (ratio of 1:5 s:v). Other researchers [106, 132] have reported that some properties of biochar differ according to the pyrolysis method employed, and temperature, the source material, the amount applied and the climatic conditions, so that the variability in these conditions makes the results not entirely comparable and conclusive. The different and contrasting values of pH, neutralization potential, number of bases, EC, etc., might also vary depending on the analytical methods used, as well as the way in which these are processed, as demonstrated by Singh et al., 2010 [129]. **Table 1** also show the CEC-values measured by silver thiourea method (AgTU) [15], without removal of carbonates and soluble salts, which for E was 16.1 mmol(−) kg^{-1}, while for H it was 41.3 mmol(−) kg^{-1}. These authors [129] also recommend measuring CEC and base saturation, with the above method, but with prior removal of salts, since these can increase the pH of the extracting solution (pH > 8) promoting the precipitation of Ag with sulfides, which occurs at basic pH-values, leading to a possible overestimation of both parameters. The authors also determined CEC of an eucalyptus biochar, measured with AgTU without removal of carbonates and soluble salts and obtained values of 91 mmol(−) $kg^{-1,}$ however, when these carbonates and salts were removed, the CEC decreased by 60% (to 36.4 mmol(−) kg^{-1}). According to them, the actual CEC, if the effect

	pH	EC	SP	CEC	BS	Cu_T	Pb_T	Zn_T	Cu-soluble	Pb-soluble	Zn-soluble
						(mg kg^{-1})					
Mixture substrates											
M1	6.8 ± 0.2	334 ± 25	62 ± 0.1	27 ± 3	39 ± 1	76 ± 11	173 ± 3	379 ± 11	<DL	39 ± 5	<DL
M2	6.6 ± 0.2	415 ± 21	50 ± 0.4	46 ± 1	50 ± 1	210 ± 20	2384 ± 3	1090 ± 18	0.3 ± 0.04	64 ± 2	<DL
M3	4.7 ± 0.1	468 ± 70	38 ± 0.2	24 ± 6	21 ± 1	324 ± 29	4279 ± 3	1699 ± 24	1 ± 0.04	29 ± 2	5 ± 0.002
M4	3.3 ± 0.1	481 ± 43	35 ± 0.2	22 ± 5	18 ± 1	457 ± 38	6490 ± 3	2409 ± 31	11 ± 0.03	48 ± 6	29 ± 2
Biochar											
H	10.2 ± 0.1	604 ± 32	346 ± 7.0	42 ± 2	424 ± 3	199 ± 33	<DL	115 ± 20	1.2 ± 0.04	< DL	< DL
E	9.9 ± 0.1	277 ± 20	222 ± 0.1	16 ± 1	17 ± 1	213 ± 20	<DL	192 ± 9	1.5 ± 0.04	< DL	4 ± 2

Metal Detection Limit (DL): Cu_T = 10 mg kg^{-1}; Pb_T = 5 mg kg^{-1}; Zn_T = 8 mg kg^{-1}; and water-soluble Pb = 1 mg kg^{-1}; Zn = 0.1 mg kg^{-1} and Cu = 0.1 mg kg^{-1}. Where: $EC_{1:20}$ is the Electrical Conductivity (1:20 ratio, µS cm^{-1}); SP (%); Saturation Percentage; CEC is the Cation Exchange Capacity ($mmol_c$ kg^{-1}); BS is the Base Saturation $mmol_c$ kg^{-1}; Cu_T, Pb_T and Zn_T are the total metal levels.

Table 1.
Physicochemical characterization values and standard deviations of chemical properties determined on pristine soil (M1), mixture substrates (M2) and (M3), mine waste (M4) and biochar materials from water hyacinth (H) and eucalyptus (E). Parameters included are pH, Electrical Conductivity (EC) in dS m^{-1}, Saturation Percentage (%SP), Cation Exchange Capacity (CEC) $mmols_C$ kg^{-1}, Base saturation, BS [$mmols_C$ kg^{-1}], total-metal levels for Cu_T, Pb_T and Zn_T (mg kg^{-1}) and water soluble-Cu, Pb and Zn (mg kg^{-1}). Determinations were carried out according to [15].

of salts on the biochar of the study were eliminated, it would lead to CEC-values of 9.7 mmol(−) kg^{-1} for E and 24.8 mmol(−) kg^{-1} for H biochar.

4.3 Results of the root length bioassay

Figure 6 shows the results of the root length bioassay only for Cu, carried out with barley plants under the conditions mentioned at the start of the section. For more details of the experiments for other metals (like Pb and Zn) refer to [15]. Results show the effect of treatments on barley root length (*Hordeum vulgare* L. var. Hope), pH and soluble copper in the soil-acid mining residue gradient. The lowercase letters on each bar indicate statistical differences (Fisher's test, $P < 0.01$). Where the first letter and subscript indicate the mixture type [M1 is 100% soil; M2 is 35:65% (M1:M4); M3 is 65:35% (M1:M4); and M4 is 100% acidic mining residue] and the last letter(s) indicates the treatment. To be able to define a critical reference level ($P < 0.01$) of root growth length and using it as an indicator of phytotoxicity and efficacy of treatments, the average value of the root length of the pristine soil (M1) was used. This was the minimum mitigation value to be achieved. The M4 and M4F treatments had a growth below the critical level of the reference value (**Figure 6**), mainly due to the low pH of these substrates (3.2 and 3.7, respectively), that is, due to the phytotoxicity caused by the high chemical activities of the H^+ ions. The high activity of H^+ ions, caused by the hydrolytic oxidation of minerals such as pyrite (FeS_2), causes the release of free sulfuric acid [133]. In contrast, the response of the remaining treatments was equal to or greater than that

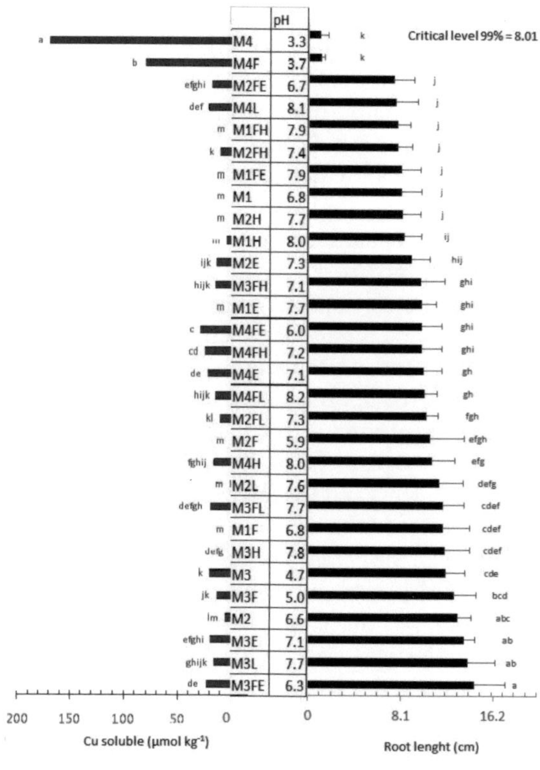

Figure 6.
Effect of treatments on barley root length (Hordeum vulgare L. var. Hope), pH and soluble copper in the soil gradient substrates-acid mining residue. The lowercase letters on each bar indicate statistical differences (Fisher's test, P < 0.01). Where the first letter and subscript indicate the mixture [M1 is 100% soil; M2 is 35:65% (M1:M4); M3 is 65:35% (M1:M4); and M4 is 100% acidic mining residue] and the last letter(s) indicates the treatment where F corresponds to phosphates, L to liming), H and E to biochar of water hyacinth and eucalyptus, respectively.

of M1, due, on the one hand, to the increase in pH, which favors the decrease of the chemical activity of H^+ ions and the phytotoxicity caused by the sorption of metals onto the solid phase [134]. In M1 the response was: M1F > M1E > M1H = M1 = M1FE = M1FH, where treatments with F and E were the ones with the greatest response. The rest of the treatments showed no significant differences. The effect of F treatment, in combination with biochar, was not significant with respect to M1. Karami et al. [135] showed that wood-derived biochar applied at doses of 20% (v:v), reduced phosphorus availability. In our case the bioassay showed this type of antagonism when phosphates were applied with H and E in M1. For the M2 substrate (pH = 6.6, 65% soil and 35% acid residue), the control treatment, i.e., the mixture without any additional treatment, promoted the highest root growth. The order of response was: M2 > M2L = M2F = M2FL = M2E = M 2H > M1 = M2FH = M2FE. It is evident that the presence of M1 soil in 2/3 parts of the substrate was sufficient to mitigate the phytotoxic effect of 1/3 of the acid residue. The addition of L, F, FL, E and H generated less root growth each time, respectively, with respect to the blank treatment. Treatment pH-values of those with neutralizing potential (L, H and E and mixtures of these with F) were greater than 7, while for phosphates pH was 5.9, having, as expected, no effect on pH. However, FH and FE treatments in this system decreased the response even more than the combination of lime-phosphate (FL) and that of lime (L). For the M3 (1/3 of M1 and 2/3 M4) the response order to the application of E and L, was: M3FE = M3L = M3E = M3F = M3 = M3H = M3FL > M3FH > M1. All these treatments showed better responses than the reference level shown by the pristine soil (M1). According to **Figure 6**, soluble-Cu was not detected in the pristine soil, but for the treatments of the same substrate (M3), this metal contributed evidently to the positive response. The pH values of the treatments ranged from 4.7 (M3) to 7.7 in M3L and M3FL. The correlation between the pH and the concentration of the soluble metal, and that between the soluble metal and the root length are related to the increased extractability of the soluble metal at low pH values (<4), while reducing root growth. The inferred relationship is indicative of the level of the phytoavailability of Cu, more than to the reduced phytotoxicity in the treatment. The increase in root length in the M2 and M3 substrates, compared to the reference materials (M1 and M4), was as expected, because soils naturally tend to reduce the mobility and mitigate the phytotoxicity of metals such as Pb, Zn and Cu, especially when metals are in high concentrations [134]. This seems to occur because there are sites where metals can be adsorbed and complexed with free organic and inorganic functional groups. However, the process can be accelerated if some amendments are used to agrostabilize the environment, such as phosphoric fertilizers, composts, biochar or biosolids, and/or liming materials, among others [136]. Also, the positive effect on root growth in M2 and M3 compared to M1 treatment is satisfactorily explained by the fertility contribution of phosphate (and potassium) from the F treatments and the micronutrients Cu and Zn coming from the mining residue. The addition to M4 to H, FL, E, FH and FE treatments enlarged the length of roots with respect to the blank. The order: M4H = M4FL = M4E = M4FH = M4 FE > M1 = M4L> > M4F = M4. Treatment response in this case showed equal or greater response in growth than M1, fact that can be explained by the pH increase caused by the basic materials added, and the increased phytoavailability of Cu, as compared with the other substrates.

5. Conclusions

The chapter presents in its introduction, a general view of the sound methodology that is proposed through the rest of the sections, to both, remediate acidic metal polluted mine wastes, taking advantage of the neutralizing alkalinity and high metal sorption affinity of pyrolyzed materials, and to utilize biochar derived from high-rate growth plants (water hyacinth, and eucalyptus, which have become of ecological relevance due to their unwanted proliferation over specific terrestrial, lacustrine or riverine environments. Relevant calculations are presented in the section two to analyze the discrepancies reported in literature among ABA methods, COD-redox measurements and kinetic factors, that may affect the determination of the right doses to neutralize with lime or pyrolyzed materials of alkaline reaction, the acidity of metalliferous mining wastes. In this section too, the status of the heavy metal problematic provoked by the sulfidic mine tailings deposits on dams is also considered, that in terms of dimensioning the relevant physicochemical characteristics to overcome when a remediation strategy is planned or put on action, when trying to mitigate and abate the toxic effects of HM on the biota. The section also addresses the importance of not only neutralizing the mine tailings but abating the toxic levels of heavy metals (HM) like Pb, Cd, Cu, and Zn, to fulfill the international and national standards and norms. Section three includes a review of the main amendment materials used, alone or in combination with biochar, to remediate heavy metal polluted sites. These references reviewed include lime, phosphates, compost, biosolids, clays and more precisely the role of biochar on these reclamation activities. A comprehensive compilation of characteristics of biochar as affected by temperature, is presented in **Figure 3**. Data plotted against charring temperature show behavior of specific surface; ash contents, pore volume, yield, C-, H-, O- and N-contents, pH; and C/H, C/O and C/N ratios. In Section 4 a case of study is presented to congruently exhibit how the strategy proposed was tested through following the methodology described herein. An experiment where biochar was combined with phosphate, and compared against the performance of lime, alone and combined with phosphate too, is exposed. Results of the root length bioassay only for Cu, carried out with barley plants under the conditions mentioned at the start of the section. For more details of the experiments for other metals (like Pb and Zn) refer to [15]. Results of the effect of soil-acid mining residue gradient were tested applying a bioassay on barley (*H. vulgare* L. var. Hope) plants and correlated with pH and soluble copper levels.

Author details

Arturo Aguirre Gómez[1*], Laura Virginia Nuñez Balderas[2], Claudia Hidalgo Moreno[3] and Jorge Dionisio Etchevers Barra[3]

1 Faculty of Superior Studies, Department of Chemistry, National Autonomous University of Mexico, Cuautitlan, Mexico

2 Faculty of Superior Studies, Department of Agricultural Sciences, National Autonomous University of Mexico, Cuautitlan, Mexico

3 Laboratory of Soil Fertility at Campus Montecillo, Ministry of Agriculture, Postgraduate College, Texcoco, Mexico

*Address all correspondence to: aag@unam.mx

IntechOpen

© 2021 The Author(s). Licensee IntechOpen. This chapter is distributed under the terms of the Creative Commons Attribution License (http://creativecommons.org/licenses/by/3.0), which permits unrestricted use, distribution, and reproduction in any medium, provided the original work is properly cited. CC BY

References

[1] Sobek, AA, Schuller, WA, Freeman, JR, and Smith RM. Field and laboratory methods applicable to overburdens and mine soils, Cincinnati, OH, US Environmental Protection Agency Environmental Protection Technology, EPA-600:1978;2-78-054.

[2] Schafer and Associates and Reclamation Research, Laboratory analytical protocol for the Streambank Tailings and Revegetation Study, Silver Bow Creek RI IFS, EPA Region VIll, Doc. No. SBC-STARS-LAP-F-Rl-1987;121-187.

[3] Moreno TR, Barbanson L, Careño AD. Neoformación mineralógica en residuos mineros (jales) del distrito minero Zimapán, Estado de Hidalgo, México. Minería y Geología. 2009;25(2):1-31.

[4] Armienta MA, Dótor A, Arcega-Cabrera FE, Talavera O, Aguayo A, Ceniceros N, Cruz O. Assessing water-sediment processes for metals in rivers polluted by mining to predict environmental impacts in developing countries. In: Webb B, De Boer D, Eds. Water Quality and Sediment Behavior of the Future: Predictions for the 21st century. Vol. 314. IAHS Pub; 2007;155-162.

[5] Jennings S.R. and Dollhopf D.J: Acid–base account effectiveness for determination of mine waste potential acidity. *Journal of Hazardous Materials* 1995;41:161-175.

[6] Cruz-Ruiz E. Effect of liming, manuring and fertilizing of acid mining wastes on the chemical availability and bioaccumulation of Cu, Cd, Pb and Zn in *Festuca arundinaceous*. M.Sc. [Thesis] on Earth Sciences Program at National Autonomous University of México (in Spanish) 2017.

[7] Guzmán RG. Strategy for the remediation of sites contaminated with Cu, Cd, Pb and Zn applying chemical-agronomic treatments to mining-metallurgical wastes and soils. M.Sc. [Thesis] of Environmental Engineering at National Autonomous University of Mexico, México 2012.

[8] Romero, FM, Gutiérrez, MR. Comparative study of the danger of mine wastes in two mining areas located in southern and central Mexico. Bulletin of the Mexican Geological Society: 2010;62,43-53.

[9] Reyes CVE, Veloz MA, Vargas M. Evaluation of alternatives for the reduction of the environmental impact due to historical mine wastes in the mining district of Zimapan. Autonomous University of the State of Hidalgo, Mexico. 2005; p 15.

[10] Gray CW, Dunham SJ, Dennis PG, Zhao FJ, McGrath SP. Field evaluation of in situ remediation of a heavy metal contaminated soil using lime and red-mud. Environ. Pollut. 2006;142(3):530-539.

[11] Singh J, Kalamdhad AS. Effects of lime on bioavailability and leachability of heavy metals during agitated pile composting of water hyacinth. Bioresource Technol. 2013;138:148-155.

[12] Trippe KM, Manning VA, Reardon CL, Klein AM, Weidman C, Ducey TF, Novak JM, Watts DW, Rushmiller H, Spokas KA, Ippolito JA, Johnson MG. Phytostabilization of acidic mine tailings with biochar, biosolids, lime and locally-sourced microbial inoculum: Do amendment mixtures influence plant growth, tailing chemistry and microbial composition? Appl. Soil Ecol. 2021:165:1-12.

[13] Cao X, Wahbi A, Ma L, Li B, Yang Y. Immobilization of Zn, Cu and Pb in contaminated soils using phosphate

rock and phosphoric acid. J. Hazard. Mater. 2009;164(2-3):555-564.

[14] Nuñez-Balderas LV. Agrostabilization of mining waste and contaminated soils with high heavy metal content. M.Sc. [Thesis]of Edaphology Program, Postgraduate College. Ministry of Agriculture, Mexico 2017.

[15] Núñez-Balderas LV, Aguirre-Gómez A, Hidalgo-Moreno C, Carrillo-Ávila N, Etchevers-Barra JD. Acid residues remediation from mines using biochar, monopotassium phosphate and lime. Rev. Int. Contam. Ambie. 2020;36(3)593-605.

[16] Dermont G., Bergeron M, Mercier G, Richer-Laflèche M. Soil washing for metal removal: A review of physical/chemical technologies and field applications. J. Hazard. Matter. 2008;152:1-31.

[17] Mulligan CN, Yong RN, Gibbs, BF. An evaluation of technologies for the heavy metal remediation of dredged sediments. J. Hazard. Mater. 2001;85:145-163.

[18] Leštan D, Luo CL, Li XD. The use of chelating agents in the remediation of metal-contaminated soils: A review. Environ. Pollut. 2008;153:3-13.

[19] Kumpiene J, Lagerkvist A, Maurice C. Stabilization of As, Cr, Cu, Pb and Zn in soil using amendments: A review. Waste Management. 2008;28:215-225.

[20] Haimi J. Decomposer animals and bioremediation of soils. Environ. Pollution. 2000;107:233-238.

[21] Nessner VK, Esposito E. Biotechnological strategies applied to the decontamination of soils polluted with heavy metals. E. Biotechnology Advances. 2010;28:61-69.

[22] Alpaslan B, Yukselen MA. "Remediation of Lead Contaminated Soils by Stabilization/Solidification," Water, Air and Soil Pollution, 2002;133:253-263.

[23] García SA, Alastuey A. Querol X. Heavy metal adsorption by different minerals: application to the remediation of polluted soils. Sci. Total Environ. 1999;242: 179-188.

[24] Gibert O, de Pablo J, Cortina JL, Ayora C. Municipal compost-based mixture for acid mine drainage bioremediation: metal retention mechanisms for permeable reactive barriers, Appl. Geochem. 2005;20: 1648-1657.

[25] Berti W, Cunningham S. In-place inactivation of Pb in Pb-contaminated soils Environmental Science and Technology. 1997:31:1359-1364.

[26] Castaldi P, Santona L, Melis P. Heavy metal immobilization by chemical amendments in a polluted soil and influence on white lupin growth. Chemosphere. 2005;60:365-371.

[27] Farrell M. Jones DL. Use of Composts in the Remediation of Heavy Metal Contaminated Soil. Journal of Hazardous Material, 2010;175:575-582.

[28] Chen ZS, Lee GJ, Liu JC. The effects of chemical remediation treatments on the extractability and speciation of cadmium and lead in contaminated soils. Chemosphere. 2000;41:235-242.

[29] McGowen SL, Basta NT, Brown GO. Use of diammonium phosphate to reduce heavy metal solubility and transport in smelter-contaminated soil. J. Environ. Qual. 2001;30:493-500.

[30] Basta NT, McGowen SL. Evaluation of chemical immobilization treatments for reducing heavy metal transport in a smelter contaminated soil. Environ. Pollut. 2004;127:73-82.

[31] Wang Z, Liu G, Zheng H, Li F, Ngo HH, Guo WS, Liu C, Chen L,

Xing B. Investigating the mechanisms of biochar's removal of lead from solution. Bioresource Technology, 2015;177:308-317.

[32] Bolan, N.S., D.C. Adriano, A.P. Mani, and A. Duraisamy.. Immobilization and phytoavailability of cadmium in variable charge soils. II. Effect of lime compost. Plant Soil. 2003;251:187-198.

[33] Bolan NS, Adriano DC, Mani AP, Duraisamy A, Arulmozhiselvan A. Immobilization and phyto-availability of cadmium in variable charge soils: I. Effect of phosphate addition. Plant Soil. 2003;250:83-94.

[34] Wang D, Felice ML, Scow KM. Impacts and interactions of biochar and biosolids on agricultural soil microbial communities during dry and wet-dry cycles. Applied Soil Ecology: 2020;152:103570.

[35] Smith, SR. A critical review of the bioavailability and impacts of heavy metals in municipal solid waste composts compared to sewage sludge. Environ. Int. 2009;35:142-156.

[36] Tandy S, Healey JR, Nason MA, Williamson JC, Jones DL. Remediation of metal polluted mine soil with compost: Co-composting versus incorporation. Environ. Pollut. 2009;157: 690-697.

[37] Beesley L, Moreno-Jimenez E, Fellet G, Melo L, Sizmur T. Biochar and heavy metals, 2° ed. In: J. Lehmann and S. Joseph. (Eds.), Biochar for environmental management. Science, technology and implementation, Routledge, New York, 2015; pp. 563-559.

[38] Singh, RP, Agrawal M. Potential benefits and risks of land application of sewage sludge. Waste. Manag. 2008;28:347-358.

[39] Walker DJ, Clemente R, Bernal ML. Contrasting effects of manure and compost on soil pH, heavy metal availability and growth of *Chenopodium album* L. in a soil contaminated by pyritic mine waste. Chemosphere 2004;57:215-224.

[40] Almas A, Singh BR, Salbu B. Mobility of cadmium-109 and zinc-65 in soil influenced by equilibration time, temperature, and organic matter. J. Environ. Qual. 1999;28:1742-1749.

[41] Hooper M, Anderson M. Soil toxicity and bioassessment test methods for ecological risk assessment. Toxicity Test Methods for Soil Microorganisms, Terrestrial Plants, Terrestrial Invertebrates, and Terrestrial Vertebrates." Prepared for Ecotoxicology Program Integrated Risk Assessment. Branch Office of Environmental Health Hazard Assessment. California Environmental Protection Agency, USA. 2009.

[42] EPS. Biological test method: Test for measuring emergence and growth of terrestrial plants exposed to contaminants in soil EPS1/RM/45. Environmental Protection Series. Environmental Protection Service. Canada. 2007.

[43] Aguirre-Gomez A, McBride MB, Norvell WA. A voltametric method for determining free metal activities in aqueous solutions: Part 2. Anodic Stripping voltammetry of Cd, Cu, Pb and Zn in synthetic and soils solutions. Inter. J. Environment Poll (IJEP). 2006;26:68-89.

[44] Cao X, Wahbi A, Ma L, Li B, Yang Y. studied successfully the immobilization of Zn, Cu and Pb in contaminated soils using phosphate rock and phosphoric acid. J. Hazard. Mater. 2009;164(2-3): 555-564.

[45] Bolan N, Kunhikrishnan A, Thangarajan R, Kumpiene J, Park J, Makino T, Kirkham MB, Scheckel K. studied the remediation of heavy

metal(loid)s contaminated soils. To mobilize or to immobilize? J. Hazard. Mater. 2014;266:141-166.

[46] Osborne lR, Baker LL, Strawn DG.. Lead immobilization and phosphorus availability in phosphate-amended, mine-contaminated soils. J. Environ. Qual. 2015;44 (1):183-190.

[47] Méndez A.; Gómez J, Paz-Ferreiro, Gascó G. Effects of sewage sludge biochar on plant metal availability after application to a Mediterranean soil. Chemosphere 2012;(89)11:1354-1359.

[48] Xu C, Qi J, Yang W, Chen Y, Yang C, He Y, Wang J, Lin A. Immobilization of heavy metals in vegetable-growing soils using nano zero-valent iron modified attapulgite clay. Sci. Total Environ. 2019;686:476-448.

[49] Hamid Y, Tang L, Hussain B, Usman M, Liu L, Ulhassan Z, He Z,Yang X. Sepiolite clay: A review of its applications to immobilize toxic metals in contaminated soils and its implications in soil–plant system. Environ. Technol. In no. 2021;23:15.

[50] Otunola B, Ololade O. A review on the application of clay minerals as heavy metal adsorbents for remediation proposes. Environ. Technol. In no. 2020;18:1-14.

[51] Abad-Valle P, Alvarez-Ayuso E, Murciego A, Pellitero E. Assessment of the use of Sepiolite to restore heavy metal polluted mine soil. Geoderma. 2016;280:57-66.

[52] Usman ARA, Kuzyakov Y, Stahr K. Effect of clay minerals on extractability of heavy metals and sewage sludge mineralization in soil. Chem. Ecol. 2004;20(2):123-135.

[53] Zhang G, Lin Y, Wang M. Remediation of copper polluted red soils with clay materials. J. Environ. Sci. 2011;23(3):461-467.

[54] Es-sahbany H, Berradi M, Nkhili S, Hsissou R, Allaoui M, Loutfi M, Bassir D, Belfaquir M, El Youbi MS. Removal of heavy metals (nickel) contained in wastewater-models by adsorption technique on natural clay. Mater. Today: Proc. 2019;13:866-875.

[55] He H, Ma L, Zhu J, Frost RL, Theng BK, Bergaya F. Synthesis of organoclays: A critical review and some unresolved issues. Appl. Clay Sci. 2014;100:22-28.

[56] Wang J, Shi L, Zhai L, Zhang H, Wang S, Zou J, Shen Z, Lian C. Chen Y. Analysis of the long-term effectiveness of biochar immobilization remediation on heavy metals contaminated soil and the potential environmental factors weakening the remediation effect: A review. Ecotoxicol. Environ. Saf. 2021;207:1-13.

[57] Zhang W, Qian L, Ouyang D, Chen Y, Han L, Chen M. Effective removal of Cr (VI) by attapulgite-supported nanoscale zero-valent iron from aqueous solution: Enhanced adsorption and crystallization. Chemosphere. 2019;221:683-692.

[58] Bojemueller E, Nennemann A, Lagaly G. Enhanced pesticide adsorption by thermally modified bentonites. Appl. Clay Sci. 2001;18(5-6):277-284.

[59] España VAA, Sarkar B, Biswas B, Rusmin R, Naidu R. Environmental applications of thermally modified and acid activated clay minerals: Current status of the art. Environ. Technol. Innov. 2019;13:383-397.

[60] Li J, Wang X, Zhao G, Chen C, Chan Z, Alsaedi A, Hayat T, Wang. Metal–organic framework-based materials: superior adsorbents for the capture of toxic and radioactive metal ions. Chem. Soc. Rev. 2018; 47(7): 2322-2356.

[61] Sarkar B, Rusmin R, Ugochukwu UC, Mukhopadhyay R,

Manjaiah KM. Modified clay minerals for environmental applications. In: Mercurio M, Sarkar B, Langella A, editors. Modified clay and zeolite nanocomposite materials. Amsterdam: Elsevier; 2019. p.113-127.

[62] Yuan P, Wang J, Pan Y, Shen B, Wu C. Review of biochar for the management of contaminated soil: preparation, application and prospect. Sci. Total Environ. 2019;659:473-490.

[63] Batista EMCC, Shultz J, Matos TTS, Fornari MR, Ferreira TM, Szpoganicz B, Freitas RA, Mangrich AS. Effect of surface and porosity of biochar on water holding capacity aiming indirectly at preservation of the Amazon biome. Sci. Rep. 2018;8:1-9.

[64] Bordoloi S, Gopal P, Boddu R, Wang Q, Cheng YF, Garg A, Sreedeep S. Soil-biochar-water interactions: Role of biochar from *E. crassipes* in influencing crack propagation and suction in unsaturated soils. J. Clean Prod. 2019;210:847-859.

[65] Ippolito JA, Cui L, Novak JM, Johnson MG. Biochar for Mine-land Reclamation. In: Ok YS, Tsang DWC, Bolan N, Novak JM editors. Biochar from Biomass and Waste: Fundamentals and Applications. Amsterdam: Elsevier; 2019. p. 75-90.

[66] Zimmerman AR. Abiotic and microbial oxidation of laboratory-produced black carbon (biochar). Environ. Sci. Technol. 2010;44:1295-1301.

[67] Knicker H, Hilscher A, La Rosa JM, Gonzalez-Perez JA, Gonzalez-Vila FJ. Modification of biomarkers in pyrogenic organic matter during the initial phase of charcoal biodegradation in soils. Geoderma. 2013;197:43-50.

[68] He L, Zhong H, Liu G, Dai Z, Brookes PC, Xu J. Remediation of heavy metal contaminated soils by biochar: Mechanisms, potential risks and applications in China. Environ. Pollut. 2019;252: 846-855.

[69] Xing J, Li L, Li G, Xu G. Feasibility of sludge-based biochar for soil remediation: Characteristics and safety performance of heavy metals influenced by pyrolysis temperature. Ecotoxicol. Environ. Saf. 2019;180:457-465.

[70] Phillips CL, Trippe KM, Whittaker G, Griffith SM, Johnson MG, Banowetz GM. Gasified grass and wood biochars facilitate plant establishment in acid mine soils. J. Environ. Qual. 2016;45(3):1013-1020.

[71] Novak JM, Ippolito JA, Ducey TF, Watts DW, Spokas KA, Trippe KM, Sigua GC, Johnson MG. 2018. Remediation of an acidic mine spoil: Miscanthus biochar and lime amendment affects metal availability, plant growth, and soil enzyme activity. Chemosphere. 2018;205:709-718.

[72] Zhang F, Wang X, Xionghui J, Ma L. Efficient arsenate removal by magnetite-modified water hyacinth biochar. Environ. Pollut. 2016;216:575-583.

[73] Mosa A, El-Ghamry A, Tolba M. Funtionalized biochar derived from heavy metal rich feedstock:phosphate recovery and reusing the exhausted biochar as an enriched soil amendment. Chemosphere. 2018;198:351-363.

[74] Koltowski M, Charmas B, Skubiszewska-Zieba J, Oleszczuk P. Effect of biochar activation by different methods on toxicity of soil contaminated by industrial activity. Ecotox. Environ. Saf. 2017;136:119-125.

[75] Yue, Y., Lin Q, Irfan M, Chen Q, Zhao X, Li G. Slow pyrolysis as a promising approach for producing biochar from sunflower straw. Bioresources. 2018;13:7455-7469.

[76] Silva MP, Lobos MLN, Piloni RV, Dusso D, Quijón MEG, Scopel AL,

Moyano EL. Pyrolytic biochars from sunflower seed shells, peanut shells and Spirulina algae: their potential as soil amendment and natural growth regulators. SN Appl. Sci. 2020;2:1-15.

[77] Deng S, Hu B, Chen T, Wang B, Huang J, Wang Y, Yu G. Activated carbons prepared from peanut shell and sunflower seed shell for high CO_2 adsorption. Adsorption. 2015;21:125-133.

[78] Zhang Z, Feng X, Yue X-X, An FQ, Zhou WX, Gao JF, Hu TP, Wei CC. 2015. Effective adsorption of phenols using nitrogen-containing porous activated carbon prepared from sunflower plates. Korean J. Chem. Eng. 2015;32(8): 1564-1569.

[79] Saleh ME, El-Refaey AA, Mahmoud AH. Effectiveness of sunflower seed husk biochar for removing copper ions from wastewater: a comparative study. Soil water Res. 2016;11:53-63.

[80] Silva MP, Moyano EL, Scopel AL. Potential applications of biochar and terpene enriched bio-oil produced from a semi-arid native asteraceae. J. Anal. Appl. Pyrolysis. 2017;126:39-49.

[81] Wang G, Dai Y, Yang H, Xiong Q, Wang K, Zhou J, Li Y, Wang S. A review of recent advances in biomass pyrolysis. Energy Fuel. 2020;34:15557-15578.

[82] Wang J, Zhao M, Zhang J, Zhao B, Lu X, Wei H. Characterization and utilization of biochars derived from five invasive plant species *Bidens pilosa* L., Praxelis clematidea, *Ipomoea cairica*, *Mikania micrantha* and *Lantana camara* L. for Cd^{+2} and Cu^{+2} removal. J. Environ. Manag. 2020;111746.

[83] Ahmad M, Moon DH, Vithanage M, Koutsospyros A, Lee SS, Yang JE, Lee SE, Jeon C, Ok YS. Production and use of biochar from buffalo-weed (*Ambrosia trifida* L.) for trichloroethylene removal from water. J. Chem. Technol. Biotechnol. 2014;89: 150-157.

[84] Tang J, Zhang S, Zhang X, Chen J, He X, Zhang Q. Effects of pyrolysis temperature on soil–plant-microbe responses to *Solidago canadensis* L.-derived biochar in coastal saline-alkali soil. Sci. Total Environ. 2020;38938.

[85] Mahmoud ME, El-Ghanam AM, Saad SR, Mohamed RHA. Promoted removal of metformin hydrochloride anti-diabetic drug from water by fabricated and modified nanobiochar from artichoke leaves. Sustain. Chem. Pharm. 2020;18:100336.

[86] Vilvanathan S, Shanthakumar S. Ni(II) adsorption onto *Chrysanthemum indicum*: influencing factors, isotherms, kinetics, and thermodynamics. Int. J. Phytoremediation. 2016;18:1046-1059.

[87] Yousaf B, Liu G, Abbas Q, Ali MU, Wang R, Ahmed R, Wang C, Al-Wabel MI, Usman ARA. Operational control on environmental safety of potentially toxic elements during thermal conversion of metal-accumulator invasive ragweed to biochar. J. Clean. Prod. 2018;195: 458-469.

[88] Fan L, Zhou X, Liu Q, Wan Y, Cai J, Chen W, Chen F, Ji L, Cheng L, Luo H. Properties of Eupatorium adenophora Spreng (Crofton weed) biochar produced at different pyrolysis temperatures. Environ. Eng. Sci. 2019;36:937-946.

[89] Yuan J, Xu R. The amelioration effects of low temperature biochar generated from nine crop residues on an acidic ultisol. Soil Use Manag. 2011;27(1):110-115.

[90] Angin D. Effect of pyrolysis temperature and heating rate on biochar obtained from pyrolysis of safflower

seed press cake. Bioresour. Technol. 2013;128:593-597.

[91] Samsuri AW, Sadegh-Zadeh F, Seh-Bardan BJ. Adsorption of As(III) and As(V) by Fe coated biochars and biochars produced from empty fruit bunch and rice husk. J. Environ. Chem. Eng. 2013;1:981-988.

[92] Ahmad M, Rajapaksha AU, Lim JE, Zhang M, Bolan N, Mohan D, Vithanage M, Lee SS, Ok YS. Biochar as a sorbent for contaminant management in soil and water: A review. Chemosphere. 2014;99:19-33.

[93] Sari NA, Ishak CF, Bakar R.A. Characterization of oil palm empty fruit bunch and rice husk biochars and their potential to adsorb arsenic and cadmium. Am. J. Agric. Biol. Sci. 2014;9:450-456.

[94] Sun J, Lian F, Liu Z, Zhu L, Song Z. Biochars derived from various crop straws: characterization and Cd(II) removal potential. Ecotoxicol. Environ. Saf. 2014;106:226-231.

[95] Naeem MA, Khalid M, Arshad M, Ahmad R. Yield and nutrient composition of biochar produced from different feedstocks at varying pyrolytic temperatures. Pak. J. Agric. Sci. 2014;51:75-82.

[96] Meng A, Zhang Y, Zhuo J, Li Q, Qin L. Investigation on pyrolysis and carbonization of *Eupatorium adenophorum* Spreng and tobacco stem. J. Energy Inst. 2015;88:480-489.

[97] Wang X, Zhou W, Liang G, Song D, Zhang X. Characteristics of maize biochar with different pyrolysis temperatures and its effects on organic carbon, nitrogen and enzymatic activities after addition to fluvo-aquic soil. Sci. Total Environ. 2015;538:137-144.

[98] Liu S, Meng J, Jiang L, Yang X, Lan Y, Cheng X, Chen W. Rice husk biochar impacts soil phosphorous availability, phosphatase activities and bacterial community characteristics in three different soil types. Appl. Soil Ecol. 2017;116:12-22.

[99] Bian R, Li L, Bao D, Zheng J, Zhang X, Zheng J, Liu X, Cheng K, Pan G. Cd immobilization in a contaminated rice paddy by inorganic stabilizers of calcium hydroxide and silicon slag and by organic stabilizer of biochar. Environ. Sci. Pollut. Res. 2016;23:10028-10036.

[100] Narzari R, Bordoloi N, Sarma B, Gogoi L, Gogoi N, Borkotoki B, Kataki R. Fabrication of biochars obtained fromvalorization of biowaste and evaluation of its physicochemical properties. Bioresour. Technol. 2017;242:324-328.

[101] Cao Y, Xiao W, Shen G, Ji G, Zhang Y, Gao C, Han L. Carbonization and ball milling on the enhancement of Pb(II) adsorption by wheat straw: competitive effects of ion exchange and precipitation. Bioresour. Technol. 2019;273:70-76.

[102] Shi J, Fan X, Tsang DCW, Wang F, Shen Z, Hou D, Alessi DS. Removal of lead by rice husk biochars produced at different temperatures and implications for their environmental utilizations. Chemosphere. 2019;235:825-831.

[103] Rizwan M, Lin Q, Chen X, Li Y, Li G, Zhao X, Tian Y. Synthesis, characterization and application of magnetic and acid modified biochars following alkaline pretreatment of rice and cotton straws. Sci. Total Environ. 2020;714:136532:

[104] Ippolito JA, Strawn DG, Scheckel KG, Novak JM, Ahmedna M, Niandou MAS. Macroscopic and molecular investigations of copper sorption by a steam-activated biochar. J. Environ. Qual. 2012;41:1150-1156.

[105] Kloss S, Zehetner F, Dellantonio A, Hamid R, Ottner F, Liedtke V,

Schwanninger M, Gerzabek MH, Soja G. Characterization of slow pyrolysis biochars: Effects of feedstocks and pyrolysis temperatures on biochar properties. J. Environ. Qual. 2012;41:990-1000.

[106] Novak JM, LimaI, King B, Gaskin JK, Steiner C, Das KC, Ahmedna M, Rehrah D, Watts DW, Busshner WJ, Schumber H. Characterization of design biochar produced at different temperatures and their effects on a loamy sand. Ann. Env. Sci. 2009;3:195-206.

[107] IBI. Standardized Product Definition and Product Testing Guidelines for Biochar That Is Used in Soil. International Biochar Initiative, United States; 2014.

[108] Tang, J., W. Zhu, R. Kookana, and A. Katayama. Characteristics of biochar and its application in remediation of contaminated soil. J. Biosci. Bioeng. 2013;6:653-655.

[109] Uchimiya M, Cantrell KB, Hunt PG, Novak JM, Chang S. Retention of heavy metals in a Typic Kandiudult amended with different manure-based biochars. J. Environ. Qual. 2012;41:1138-1149.

[110] Masto RE, Kumar S, Rout TK, Sarkar P, George J, Ram LC. Biochar from water hyacinth (Eichornia crassipes) and its impact on soil biological activity. Catena. 2013:111:64-71.

[111] Reddy, K.R. and J.C. Tucker. Productivity and nutrient uptake of Water Hyacinth, *E. crassipes* I. Effect of nitrogen source. Economic. Bot. 1983;37:237-247.

[112] Sindhu R, Binod P, Pandey A, Madhavan A, Alphonsa JA, Vivek N, Gnansounou E, Castro E, Faraco V. Water hyacinth a potential source for value addition: An overview. Bioresource Tecnol. 2017;230:152-162.

[113] Wang H, Xia W, Lu P. Study on adsorption characteristics of biochar on heavy metals in soil. Korean J. Chem. Eng. 2017;34(6):1867-1873.

[114] Sarkar M, Rahman AKML, Bhoumik NC. Remediation of chromium and copper on water hyacinth (*E. crassipes*) shoot powder. Water Resour. Ind. 2017;17:1-6.

[115] Lindenmayer, D.B., R.B. Cunningham, M.L. Pope, P. Gibbons and C.F. Donenelly.. Cavity size and types in Australian eucalypts from wet and dry forest types: A simple of rule of thumb for estimating size and number of cavities. For. Ecol. Manage. 2000;137:139-150.

[116] ISO 11269-1. Determination of the Effects of Pollutants on Soil Flora - Method for the Measurement Growth. International Organization for Standardization. Geneve, Switzerland, 1993.

[117] ISO 11269-2. Soil quality - Determination of the effects of pollutants on soil flora: Part 2: Effects of chemicals on the emergence and growth of higher plants. International Organization for Standardization. Geneve, Switzerland 2005.

[118] ISO 11269-1: Soil quality: Determination of the effects of pollutants on soil flora - Part 1: Method for the measurement of inhibition of root growth. International Organization for Standardization. Geneve, Switzerland, 2012.

[119] ISO 11269-2: Soil quality: Determination of the effects of pollutants on soil flora - Part 2: Effects of contaminated soil on the emergence and early growth of higher plants. International Organization for Standardization. Geneve, Switzerland, 2012.

[120] OECD Guidelines for the testing of Chemicals 208: Terrestrial Plant Test:

Seedling Emergence and Seedling Growth Test. 2006.

[121] Ginocchio R. Rodríguez P. Validation of the terrestrial biotic ligand model: Prediction of copper toxicity to barley in soils of north-central Chile. Centro de Investigación Minera y Metalúrgica. Chile. 2007

[122] Nasona P, Johnson RH, Neuschützc C, Alakangasa L, Öhlander B. Alternative waste residue materials for passive *in situ* prevention of sulfide-mine tailings oxidation: A field evaluation. J. Hazar. Mat. 2014;267:245-254.

[123] ISO 11268-1 Soil quality–Effects of pollutants on earthworms - Part 1: Determination of acute toxicity to *E. fetida/*. International Organization for Standardization. Geneve, Switzerland, 2012.

[124] ISO 11268-2 Soil quality - Effects of pollutants on earthworms - Part 2: Determination of effects on reproduction to *E. fetida/Eisenia Andrei*. International Organization for Standardization. Geneve, Switzerland, 2012.

[125] Macnicol RD, Beckett PHT. Critical tissue concentrations of potentially toxic elements. Plant Soil. 1985:85:107-129.

[126] Kabata-Pendias, A. and H. Pendias.. Trace elements in soils and plants, third ed. CRC Press, Boca Raton, USA, 2001; 403 pp.

[127] Berek AK, Hue NV. Characterization of biochars and their use as an amendment to acid soils. Soil Sci. 2016;181:412-426.

[128] Tan C, Yaxin Z, Hongtao W, Wenjing L, Zeyu Z, Yuancheng Z, Lulu r. Influence of pyrolysis temperature on characteristics and heavy metal adsorptive performance of biochar derived from municipal sewage sludge. Bioresour. Technol. 2014:164:47-54.

[129] Singh, B, Singh BP, Cowie AL. Characterization and evaluation of biochar for their application as a soil amendment. Aust. J. Soil Res. 2010:48:516-525.

[130] Gaskin JW, Steiner C, Harris K, Das KC, Bidens B. Effect of low temperature pyrolysis conditions on biochar for agriculture use. Trans. ASAE. 2008;51:2061-2069.

[131] Chun Y, Sheng G, Chiou CT, Xing B. Compositions and sorptive properties of crop residue-derived chars. Environ. Sci. Tech. 2004;38: 4649-4655.

[132] Scott HL, Ponsonby D, Atkinson CJ. Biochar: an improve of nutrient and soil water availability-what is the evidence? CAB-Reviews. 2014;9:1-19.

[133] Tordoff GM, Baker AJM, Willis AJ. Current aproaches to the revegetation and reclamation of metalliferous mine wastes. Chemosphere 2000;41:219-228.

[134] Jiang J, Xu R, Jiang T, Li Z. Immobilization of Cu (II), Pb (II) and Cd(II) by the addition of rice straw derived biochar to a simulated pollution Ultisol.. J. Hazard. Mater. 2012;229-230: 145-150.

[135] Karami N, Clemente R, Moreno-Jiménez E, Lepp NW, Beesley L. Efficiency of green waste compost and biochar soil amendments for reducing lead and copper mobility and uptake to ryegrass. J. Hazard. Mater. 2011;191:41-48.

[136] Adriano DC, Wenzel WW, van Gronsveld J, Bolan N. Role of assisted natural remediation in environmental cleanup. Geoderma 2004;122:121-142.

Chapter 13

Rejuvenator Obtained by Pyrolysis of Waste Tires for Use in Asphalt Mixtures with Added Reclaimed Asphalt

Lidija Ržek, Mojca Ravnikar Turk and Marjan Tušar

Abstract

Although in recent years, big progress has been made in the field of recovering waste tires, they still represent an unwanted waste and their production is constantly increasing. We can use waste tires as a raw material for a new product. In our study, multiple liquid products were produced by pyrolysis of waste tires. After extensive testing of their properties, we selected the most suitable pyrolytic product for the purpose of rejuvenation. Rejuvenators are designed to soften the old, brittle and stiff aged bitumen in reclaimed asphalt. Bitumen with its viscoelastic characteristics is the most important component of asphalt and dictates its behaviour. Commonly bitumen, after adding rejuvenator, becomes less viscous, more ductile and its coating properties are restored. By using a pyrolytic rejuvenator, the proportion of reclaimed asphalt added to the asphalt mixture was increased. The reuse of reclaimed asphalt and waste tires means a reduction in waste material and is therefore important for the preservation of the environment and sustainable development.

Keywords: pyrolyse, rejuvenator, bitumen, reclaimed asphalt, waste tires

1. Introduction

Bitumen combined with stone aggregate creates an asphalt mixture, which is the most common surface layer of the roads. Bitumen in asphalt mixtures ages already during the production, transport and installation of asphalt mixture and the process of ageing continues during the use of asphalt. Ageing processes are influenced by several factors, the most important of them are: temperature, UV radiation and oxygen exposure. Subsequently, bitumen becomes harder and more brittle, its viscosity increases, adhesion and cohesion deteriorate, which leads to the ravelling and the formation of cracks in asphalt mixture [1–3]. With the use of special additives, the so-called rejuvenator, bitumen restores its basic properties [4, 5]. Furthermore, rejuvenators allow mixing and installation at lower temperatures, which reduces production costs and energy consumption, therefore the production of asphalt is more environmentally friendly. The rejuvenator enables to re-install the aged bitumen from reclaimed asphalt (RA) into the fresh asphalt [6]. The first rejuvenator was used as early as 1960 [7]. Over the years, several different products have been used as rejuvenators: emulsions, oil components with a high content of maltene, soft bitumen with high penetration, tertiary amines, waste vegetable oils, waste motor oils and various combinations of these materials [8, 9]. A rejuvenator made from waste tires by a pyrolysis process, a pyrolytic rejuvenator, was developed in our study [10].

Pyrolysis in technical terms is a process in which the chemical decomposition of a substance takes place at elevated temperatures and without the presence of oxygen. Because of the increasing number of vehicles, demand for car tires is also increasing, and as a consequence their production is increasing (324 million tires were sold in 2019 in Europe [11]), resulting in an increasing number of waste tires at the end of their service life. European Directive on the landfill of waste 1999/31/EC [12] has prohibited the disposal of waste tires at landfill sites since 2006, as their improper storage may endanger human health and pose a threat to the environment, e.g. fire, the spread of rodents and insects and dangerous emissions that can be released into the air or the earth. In addition to this directive, waste management is also significantly influenced by the waste directive 2008/98/ES [13] which sets out a 5-step waste hierarchy, namely: prevention, preparing for re-use, recycling, other recovery (e.g. energy recovery) and disposal. Pyrolysis is a suitable process for the treatment of waste tires, as it has a small impact on the environment, and at the same time obtained products that can be used further. In the past, pyrolytic products from waste tires have been mainly used as additives to reduce the temperature sensitivity of bitumen [14–17]. The pyrolytic product from waste tires pyrolysis may contain high concentrations of polycyclic aromatic hydrocarbons (PAH). Therefore, the slow pyrolysis process of waste tires was performed at a relatively low temperature where the highest treatment temperature (HTT) did not exceed 500°C. Consequently, a product with low PAH content, i.e. lower than those reported in existing studies [18, 19] was obtained.

In the first phase of the presented study, several different pyrolytic products were developed [20]. Among them, the most appropriate for the purpose of rejuvenator was selected. In the second part, the influence of the pyrolytic rejuvenator on the properties of the non-aged and aged bitumen was examined. Results of the tests proved that the pyrolytic rejuvenator revived aged bitumen [21]. In the last phase, the pyrolytic rejuvenator was used in asphalt mixtures, to which the percentage of reclaimed asphalt was gradually increased [22].

2. Pyrolytic product for the purpose of rejuvenation

Pyrolytic products in the research were obtained by the process of slow pyrolysis. In the first phase, the conditions of the pyrolysis process e.g. the duration (from 10 min to 150min) and the pyrolysis temperature (from 280–500°C) were changed in order to obtain different pyrolytic products. In the second phase, the pyrolytic products were modified with various oils and crushed rubber. All composed pyrolytic products were similar to bitumen. Their properties and properties of their blends with reference bitumen were evaluated by standard European mechanical tests, which are usually used to determine the properties of bitumen. As reference bitumen and as a matrix of the blends a 50/70 penetration grade bitumen (B50/70) was used. All the blends were laboratory produced by adding a controlled quantity of the pyrolytic product to the bitumen. The blends of bitumen and pyrolytic product were commonly produced by mixing two of the components in ratios of 1:1 (labelling them B+No. of pyrolytic product). Only the blends of pyrolytic product No. 11 were also prepared in smaller concentrations (11_x%, where x represents a share of the pyrolytic product).

The softening properties were determined by using the Ring and Ball method (RB) according to the EN 1427 [23]. Fraass breaking point test according to the EN 12593 [24] was used to determine the brittleness of the products at low

	RB	Fraass breaking point	Force ductility				Mixing temperature (@η = 0.17 Pas)	Compaction temperature (@η = 0.26 Pas)
			T	Elongation	Force (max)	Energy $E'_{0,4}$-$E'_{0,2}$		
	[°C]	[°C]	[°C]	[mm]	[N]	[J/cm²]	[°C]	[°C]
B 50/70	50.5	−11	25	1500*	0.97	0.07	141.2	121.9
1	44.6	−14	25	236	0.1	0.00	153.6	127.1
2	59.1	2	25	187	1.01	0.06	192.9	159.9
3	54.6	−18	25	137	0.34	0.02	169.2	141.2
4	61.3	−12	25	257	1.41	0.10	185.6	156.8
5	60.6	−8	25	354	1.41	0.12	187.8	158.6
6	47.3	−20	25	232	0.27	0.02	169.3	138.8
7	45.9	−14	25	827	0.41	0.02	139.4	116.4
8	65.6	+1	25	1309	12.59	0.91	177.6	153.1
9	42.0	−20	15	145	0.14	0.01	190.6	150.3
10	30.4	−23	15	202	0.04	0.00	155.7	121.5
11	—	−1	15	—	—	—	202.3	154.4
12	43.9	−21	15	325	0.07	0.01	212.4	173.6
13	54.9	−21	15	322	0.29	0.05	255.5	207.5
14	37.5	−22	15	230	0.0	0.00	150.3	122.4
B + 1	43.3	−13	25	902	0.31	0.02	152.1	125.8
B + 2	46.4	−8	25	534	0.52	0.03	146.0	123.3
B + 3	47.1	−18	25	335	0.58	0.04	143.6	121.9
B + 4	46.9	−11	25	1500	0.96	0.06	146.6	123.3
B + 5	46.6	−10	25	1492	0.82	0.05	152.1	128.0
B + 6	44.8	−18	25	760	0.43	0.03	142.2	119.2
B + 7	45.7	−14	25	1500*	0.72	0.04	140.1	118.2
B + 8	52.3	−6	25	1500*	3.15	0.22	150.7	129.1
B + 9	38.1	−24	15	1500*	0.67	0.10	158.6	129.3
B + 10	37.2	−22	15	1500*	0.61	0.06	135.3	113.8
B + 11	—	−24	—	—	—	—	140.3	105.2
11_5%	42.8	−17	15	1343	2.6	0.23	183.3	117.2
11_10%	36.6	−23	15	746	0.45	0.03	134.8	111.8
11_20%	22.4	−23	—	—	—	—	129.6	102.5
B+12	40.2	−21	15	1326	0.10		176.8	143.5
B+13	45.4	−14	15	1500*	0.28	0.03	169.9	140.8
B+14	42.6	−19	15	1122	0.20	0.00	139.6	117.2

** Maximum elongation.*

Table 1.
Results of standard mechanical tests of pyrolytic products and their blends with reference bitumen [20].

temperatures. The tensile properties of the bitumen and pyrolytic products were determined by the force ductility method in accordance with EN 13589 [25].

For the production of asphalt, it is important to determine the optimum mixing temperature for specific bitumen, which is the temperature to be maintained at the asphalt plant during mixing. For laying of asphalt the compaction temperature, at which the mixture has to be compacted on sites has to be determined. The mixing and compaction temperatures for pure pyrolytic rejuvenators and bitumen blends were determined with the rotational viscometer Haake RS50. Both temperatures determine recommended viscosity of bitumen: 0.170 ± 0.02 Pas form mixing and 0.260 ± 0.03 Pas for compaction, respectively [26].

Results in **Table 1** show most pyrolytic products lowered the softening point of the blends in comparison with reference bitumen. That means the pyrolytic product could also lower the high softening point of the extracted bitumen in reclaimed asphalt. For our product, this is a good feature, as we want to use it as a rejuvenator. In addition, almost all pyrolytic products lowered the Fraass breaking point, meaning the temperature range is extended.

Results of final elongation at force ductility tests show that all pyrolytic products broke before they reached the maximum possible length (1500 mm, designated 1500* in **Table 1**) to the contrary of reference bitumen. The pyrolytic products 4, 5, 7, 8, 9, 10, and 11 (in adequate concentration) retained the elongation ability of the reference bitumen in the blends. All other pyrolytic products shortened the elongation of the blends. Except for the pyrolytic product no. 8, all other values of

	Viscosity		
	@T = 60°C	@T = 100°C	@T = 150°C
Label	**[Pas]**	**[Pas]**	**[Pas]**
B 50/70	209.37	2.03	0.12
B + 1	71.50	1.56	0.23
B + 2	104.19	1.67	0.17
B + 3	110.84	1.67	0.15
B + 4	92.59	1.59	0.17
B + 5	99.02	3.36	0.18
B + 6	61.04	1.31	0.14
B + 7	71.39	1.21	0.13
B + 8	234.85	2.60	0.21
B + 9	56.31	1.74	0.29
B + 10	30.91	0.90	0.12
B + 11	6.93	0.43	0.15
11_5%	73.03	1.35	0.12
11_10%	31.14	0.90	0.09
11_20%	9.01	0.53	0.09
B + 12	94.83	2.78	0.58
B + 13	144.06	3.33	0.45
B + 14	59.34	1.25	0.12

Table 2.
Results of viscosity measurements of the blends (pyrolytic products with reference bitumen) [21].

maximum force measurements were in the range around 1 N or even smaller. The ductility test was performed at 25°C for the pyrolytic products (and their blends) from 1 to 8. For others, the test temperature was lowered, to 15°C. Pyrolytic product 11 and its blends, B+11 and 11_20%, could not be tested even at the temperature of 15°C, so we did no performed test for those three samples.

The results of the mixing and the compaction temperatures of the pyrolytic products are higher than the reference bitumen's. Although this indicates that the pyrolytic products have a higher viscosity, also the homogeneity of the sample influenced the viscosity. Inhomogeneous samples (2, 9, 10, and 11) have higher viscosity. Blends of pyrolytic products 7, 10, 11 and 14 have lower mixing and compaction temperatures.

The viscosity of the blends was determined at three different temperatures and was measured at the constant shear rate. Results (**Table 2**) show that the viscosity of the bitumen and the blends is decreasing with increasing temperature. In general, the viscosity of all blends is lower than the viscosity of the references bitumen at 60 °C and 100°C. At 150°C viscosity of almost all blends was at least as high as the reference's bitumen; nevertheless, the absolute values of the viscosity were very small.

Based on the results of these tests we decided that out of the fourteen manufactured and modified pyrolytic products, the most suitable pyrolytic product for the role of a pyrolytic rejuvenator was number 14.

3. Effect of pyrolytic rejuvenator on non-aged and laboratory aged bitumen

After selecting the appropriate pyrolytic rejuvenator, the focus of the study was to evaluate its effect on paving grade bitumen. When reclaimed asphalt (with aged bitumen) is added to the new asphalt mixture, stone aggregate and a calculated amount of fresh (non-aged) bitumen are also added at the same time. Consequently, researches on non-aged bitumen and on laboratory aged bitumen were conducted. The short term ageing, to which bitumen is subjected during mixing, transport and installation of asphalt, is in laboratory simulated by rolling thin film oven test (RTFOT) method according to EN 12607 [27]. According to this method, bitumen is aged under the influence of high temperatures and constant air flow. The ageing conditions are not exactly the same as in asphalt production, but the ageing results are comparable [28]. Ongoing ageing of bitumen during road use was simulated by the pressure ageing vessel (PAV) method according to EN 14769 [29]. PAV ageing at elevated pressure and temperature was performed on RTFOT aged bitumen. PAV simulates ageing according to climatic conditions (temperature, UV, etc.), but cannot take into account variables in the asphalt, such as the proportion of air voids, the type of aggregate and the absorbency of the aggregate.

Blends of non-aged bitumen and pyrolytic rejuvenator were laboratory prepared in different concentrations: 3%, 5%, 10% and 20% rejuvenator based on the mass of the reference bitumen (**Table 3**). All prepared samples and concentrations of the blends are presented in **Table 3**.

In addition to before mentioned standard test on samples, determination of the sample's consistency by needle penetration test according to the EN 1426 [30] at 25°C was also performed. The elastic recovery of the samples was determined according to the EN 13398 [31]. According to this standard, a specimen was first elongated to 20 cm and then cut in the middle to obtain two halves of the thread. After the predetermined time (30 minutes) for recovery has elapsed, the shortening of the half threads was measured and expressed as the percentage of the elongation length.

To check the possible phase separation in the blend of bitumen and pyrolytic products, a storage stability test according to EN 13399 [32] was performed only for the blend of pyrolytic product No. 14 and reference bitumen. In the test, the sample of the blend is maintained in the vertical vessel at 180°C for three days. After the sample is cooled down, it is cut into three equal parts. The two ends (top and bottom) are further analysed to evaluate possible differences in characteristics. The affinity between pyrolytic rejuvenator and stone aggregate (limestone) was checked by standard rolling bottle method according to EN 12697–11 [33] and compared with the affinity of reference bitumen (non-aged) and the blend.

The properties of samples in the low-temperature range were characterized with the bending beam rheometer (BBR) according to EN 14771 [34]. Bending tests are suitable for testing brittle materials when measurements at tensile load do not provide insight into the properties of the material or are not feasible. The stress relaxation in bitumen is significantly slower at low temperatures, which can lead to the formation of cracks in the asphalt and loss of binder functionality. During the BBR test a bitumen beam is bent under a constant load and deformation of bitumen is measured. The flexural creep stiffness $S_m(t)$ is calculated at time t = 60 s. The characteristic parameter is also value m_{60}, the slope of the curve S (t) at t = 60 s, which indicates the relaxation capacity of the bitumen stress. We presumed that the adequate quality of bitumen at low temperatures is ensured by the maximum value of S_{60} = 300 MPa and the minimum value of m_{60} = 0.300.

The results of softening point, Fraass breaking point and penetration of reference bitumen (non-aged, 'B 50/70' and laboratory aged, 'B_PAV'), pyrolytic rejuvenator 'PR' and their blends are shown in **Table 3**. Comparison of penetration and Fraass breaking point between 'B 50/70' and pyrolytic rejuvenator 'PR' indicates on their different chemical composition. Pyrolytic rejuvenator shows a significantly lower value of softening point and Fraass breaking point and at the same time a much higher value of penetration.

	Label	Proportion of the reference bitumen	Proportion of the pyrolytic rejuvenator	RB	Penetration	Fraass breaking point
		[%](m/m)	[%](m/m)	[°C]	[1/10 mm]	[°C]
Non- aged bitumen	B 50/70	100	0	50.5	53	−11
	B50/70_3%	97	3	50.2	55	−11
	B50/70_5%	95	5	50.3	59	−20
	B50/70_10%	90	10	49.0	60	−16
	B50/70_20%	80	20	47.3	75	−16
	B50/70_50%	50	50	42.6	115	−19
Pyro-lytic rejuve-nator	PR	0	100	37.5	233	−22
Aged bitumen	B_PAV	100	0	70.4	21	−10
	B_PAV_3%	97	3	69.0	22	−6
	B_PAV_5%	95	5	67.7	22	−6
	B_PAV_10%	90	10	65.6	26	−7
	B_PAV_20%	80	20	61.6	34	−7

Table 3.
Tested blends of pyrolytic rejuvenator (labels, RB and penetration) [21].

Results on non-aged samples show the effect of the rejuvenator on the standard mechanical properties (penetration, softening point, Fraass breaking point) of bitumen. Penetration values increased, at the same time softening values decreased with increasing rejuvenator proportion. All added amounts of rejuvenator to non-aged bitumen decreased the values of the Fraass breaking point.

Original B 50/070 was RTFOT+PAV aged, simulating naturally aged bitumen in reclaimed asphalt. Then the same proportions of pyrolytic rejuvenator No. 14 (as in the case of non-aged bitumen) were added and the blends were tested.

A comparison of non-aged and aged bitumen blends shows the impact of laboratory ageing. Aged B 50/70 (B_PAV) and all bitumen blends became stiffer since the penetration of aged bitumen decreased and the softening point increased. Unexpectedly, all proportions of rejuvenator which were added to RTFOT+PAV aged B 50/70 decreased the values of the Fraass breaking point. It should be noticed that the repeatability of the Fraass breaking point test is 3°C.

Table 4 presents results of tensile properties of bitumen, pyrolytic rejuvenator and their blends. Non-aged bitumen and its blends with rejuvenator elongated to the maximum length (1500 mm). The elongation of aged bitumen was prolonged with the addition of a rejuvenator. When mixing materials such as bitumen, two consequences can be observed: the mixing effect (mostly linear change) and the structural-interaction effect (mostly nonlinear change) [21]. Maximum force decreased proportionally with the added rejuvenator indicating a linear change occurred. The effect of nonlinearity is not observed, as the elongation at maximum force, Fmax, is the same for all samples with non-aged bitumen, as well as for all samples with aged bitumen, regardless of the amount of rejuvenator added. The elongation of the non-aged bitumen was about 1.5 times greater than the elongation of the aged bitumen with rejuvenator, indicating that aged bitumen was not completely restored. The results show that due to the added rejuvenator, the mechanical properties of aged bitumen approached the values of non-aged bitumen, but a complete restoration was not achieved.

	Force Ductility at 25°C				Elastic recovery
	Elongation	Force Fmax	Elongation at Fmax	Energy	
Label	[mm]	[N]	[mm]	[J/cm^2]	[%]
B 50/70	1500*	0.97	17.18	0.07	/
B50/70_3%	1500*	1.58	7.22	0.11	13
B50/70_5%	1500*	1.27	7.21	0.10	13
B50/70_10%	1500*	0.99	7.21	0.07	12
B50/70_20%	1500*	0.71	7.11	0.05	12
PR	230	0.04	17.12	0.00	−38
B_PAV	184	21.66	12.01	1.34	30
B_PAV_3%	304	20.03	12.10	1.37	27
B_PAV_5%	469	17.08	12.40	1.28	21
B_PAV_10%	307	13.38	12.71	0.95	21
B_PAV_20%	370	8.27	12.89	0.62	22

** Maximum elongation.*

Table 4.
Tested blends of pyrolytic rejuvenator (ductility, elastic recovery) [21].

Temperature	Mixing temperature (η = 0.17 Pas)	Compaction temperature (η = 0.26 Pas)	BBR T at S_{60} = 300 MPa	BBR T at m_{60} = 0.300	Higher T (BBR)
Label	[°C]	[°C]	[°C]	[°C]	[°C]
B 50/70	141	122	−19.3	−20.8	−19.3
B50/70_3%	140	120	−19.6	−21.5	−19.6
B50/70_5%	140	120	−19.8	−22.0	−19.8
B50/70_10%	138	118	−20.7	−22.7	−20.7
B50/70_20%	139	118	−22.6	−24.5	−22.6
PR	150	122	−31.3	/	/
B_PAV	155	139	−16.1	−11.7	−11.7
B_PAV_3%	154	137	−16.9	−12.9	−12.9
B_PAV_5%	153	137	−17.5	−12.8	−12.8
B_PAV_10%	151	134	−18.3	−13.3	−13.3
B_PAV_20%	150	132	−19.6	−18.8	−18.8

Table 5.
Results of rheological tests [21].

The elastic recovery did not change significantly for non-aged bitumen regardless of the amount of rejuvenator. In the case of aged bitumen, the elastic recovery decreases with the amount of rejuvenator and thus approached the value of the reference aged bitumen. The elastic recovery of the rejuvenator had a negative value, so the sample was stretching after the test, meaning that the rejuvenator had no elastic properties and all energy was lost. We expected that due to rubber content in car tires some elasticity will remain in our product, but it is evident that all rubber from tires decomposed during the pyrolytic process.

Mixing and compaction temperatures (**Table 5**) of aged bitumen were higher than temperatures of non-aged bitumen, and in both cases, the temperatures decreased with the increasing amount of added rejuvenator.

S_{60} and m_{60} are criteria that determine the lower limit of the bitumen application temperature. When the conditions: $S_{60} \leq 300$ MPa and $m_{60} \geq 0.300$ are met at the same time, the bitumen shows sufficient low temperature resistance. **Table 5** shows the results, i.e. the minimum temperatures for all samples. For non-aged bitumen, the critical temperature was determined by S_{60}, while for aged bitumen the temperature at parameter m_{60} was decisive. With the addition of the rejuvenator, the critical temperature was lowered. Aged bitumen with the highest amount of rejuvenator additive achieved similarly low temperatures as the reference non-aged bitumen.

Storage stability was checked only on a blend with the highest amount of pyrolytic rejuvenator, B50/70_50%. The test results showed that although an immense quantity of the rejuvenator was added to the reference bitumen, the blend remained homogeneous. This was evident from the very small changes in the penetration and softening values of the binder in the upper and lower parts of the tube (**Table 6**). Results also show that due to storage at high temperature the characteristics (pen, RB) did not change much.

The result of the affinity test (**Table 7**) shows that after the first 6 hours there was no difference between the tested samples. The binder detached only slightly from the aggregate (limestone) in all samples. After one day of testing in the rolling

	Conditions	RB	Penetration
Label		[°C]	[1/10 mm]
B50/70_50%	Ambient	42.2	115
B50/70_50% upper part	72 h at 180°C	43.0	118
B50/70_50% lower part	72 h at 180°C	42.8	117

Table 6.
Results of storage stability test [21].

Affinity [%]	Time of test [h]				
Sample	0	6	24	48	72
B50/70	100	95	80	55	40
B50/70_50%	100	95	80	50	30
PR	100	95	90	55	45

Table 7.
Results of affinity test.

bottle, differences between the samples appeared more obvious. The pyrolytic rejuvenator had the best affinity with aggregate. After two days, the reference bitumen and rejuvenator covered the aggregate equally well, and after three days the aggregate was best covered with pyrolytic rejuvenator and worst with a mixture of reference bitumen and pyrolytic rejuvenator.

To verify the relationship between rheological and mechanical measured properties, presented in **Tables 3–5**, linear relationships between individual properties were examined. Linearity between properties was evaluated with statistical parameter R^2 (**Table 8**). The results show that there is no linear relationship between BBR measurements and empirical mechanical tests. Correlation between the parameters S_{60} and m_{60} and the Fraass breaking point was expected since all measurements were performed in the low temperature range. Similarly, we expected a relationship between the properties measured in the medium and high temperature range. The softening point temperature was compared with the mixing and compaction temperatures. It turned out that there is no linear correlation for the samples of non-aged bitumen, and there is a good linear dependence for the samples of aged bitumen. There is also a good relationship between the results of the two most basic mechanical tests, penetration and softening point, for both non-aged and aged paving grade bitumen samples.

	RB – penetration (Table 3)	RB – mixing temperature (Tables 3 and 5)	RB – compaction temperature (Tables 3 and 5)	BBR S60 - Fraass breaking point (Tables 3 and 5)	BBR m60 - Fraass breaking point (Tables 3 and 5)
Non-aged bitumen	0.91	0.00	0.02	0.10	0.20
Aged bitumen	0.94	0.91	0.96	0.06	0.00

Table 8.
The R^2 values for various properties of non-aged and aged bitumen with rejuvenator.

4. Asphalt mixtures with reclaimed asphalt and pyrolytic rejuvenator

Asphalt mixtures have to withstand dynamic loads as well as high and low temperatures without cracking or rutting. Several standardized laboratory tests enable the evaluation of these asphalt characteristics. Before implementation, characteristics of reclaimed asphalt (RA) had to be established. Several samples of reclaimed asphalt from the same stockpile were sieved into sub fractions and extracted bitumen was investigated. The results showed that the bitumen content in RA was 4.5%. Based on this, the required amount of fresh bitumen and rejuvenator for each asphalt mixture type AC 8 surf were calculated.

The bitumen content of all prepared asphalt mixtures was determined at 5% of the mass regarding the total asphalt mixture mass. Regarding to the established effects of pyrolytic rejuvenator on (non-aged and laboratory aged) bitumen, the addition of 20% of pyrolytic rejuvenator to the bitumen from reclaimed asphalt was established. Control asphalt mixture of fresh materials (0% RA) and asphalt mixtures with 20%, 40% and 60% of RA according to the weight of the stone aggregate were prepared in laboratory (set of asphalt mixture samples is described in **Table 9**). Each asphalt mixture with RA was produced without and with a pyrolytic rejuvenator. All together seven asphalt mixtures were prepared with the same B50/70 bitumen (from the same producer) as used in previous research.

All asphalt mixtures samples were mixed in a laboratory and compacted in accordance with EN 12697–30 [35]. Basic information about asphalt mixtures was gained by determining bulk density, EN 12697–6 [36], maximal density, EN 12697–5 [37], void content, EN 12697–8 [38], and indirect tensile strength (ITS), EN 12697–23 [39]. ITS is maximum tensile stress applied to a cylindrical specimen loaded diametrically until the break. Cylindrical specimens (nominal diameter 100 mm) were for ITS compacted by 50 impacts on each side at temperature 150°C. Water sensitivity tests were completed according to EN 12697–12 [40], in order to evaluate the effect of moister. Water sensitivity is expressed by ITSR – indirect tensile strength ratio. Two sets of four cylindrical specimens were prepared, compacted by 35 impacts on each side at temperature 150°C. One set of specimens was conditioned in water for three days; the other set was kept dry. The ratio between their ITS values expresses water sensitivity.

For determination of characteristics at low temperature thermal stress restrained specimen test (TSRST, EN 12697–46 [41]) was carried out. Asphalt

Sample	Proportion of the RA	Addition of the pyrolytic rejuvenator	Measured bitumen content	The bulk density	The maximal density	Void content
	[%] (m/m)	/	[%] (m/m)	[kg/m^3]	[kg/m^3]	[%] (V/V)
0% RA	0	No	5.0	2398	2478	3.2
20% RA	20	No	5.0	2403	2490	3.5
20% RA + rej.	20	Yes	5.0	2407	2492	3.4
40% RA	40	No	5.0	2407	2494	3.5
40% RA + rej.	40	Yes	5.0	2394	2492	3.9
60% RA	60	No	5.0	2425	2502	3.1
60% RA + rej.	60	Yes	5.0	2417	2501	3.4

Table 9.
Standard properties of asphalt mixtures [22].

mixtures were compacted in form of slabs, EN 12697–33 [42], from which prismatic specimens were cut. In TSRST, the specimen, whose length is held constant during the test, is subjected to a temperature decrease with a constant temperature rate. Due to the confined thermal shrinkage, cryogenic stress builds up in the specimen. The results of the tests are the progression of the cryogenic stress over the temperature range until break, σcry(T), and the failure stress, σcry,failure, at the failure temperature, Tfailure.

To check the behaviour of asphalt mixtures at elevated temperatures the formation of wheel tracking was checked according to EN 12697–22 [43].

Extracted bitumen from all asphalt mixtures was investigated according to before mentioned standard testing methods.

The siewing curve had to be adjusted for each asphalt mixture (**Figure 1**) because different shares of RA were added. In all figures addition of rejuvenator is designated with RA+rej. The curves overlap with each other well and bitumen from RA does not affect the siewing curve. The inhomogeneity of RA was successfully solved by presiewing RA into sub fractions (0/2 mm, 2/4 mm and 4/8 mm).

The densities were determined according to standard procedures and the results are given in **Table 9**. The difference between the densities and void content of individual asphalt mixtures was very small. Results confirm that mixtures were comparable.

The presence of water in the asphalt is expected. Moisture is one of the most important factors influencing the durability of asphalt. The combination of excess moisture and traffic load shortens the life of the asphalt. Moisture in asphalt causes two main destructive mechanisms: deterioration of adhesion and cohesion. Strong deterioration of adhesion can be observed as peeling of bitumen from the stone aggregate, and deterioration of cohesion is observed as softening of the binder, which leads to lower strength of asphalt. The water sensitivity of asphalt mixtures is expressed by the ITSR quotient, which represents the ratio of indirect tensile strength of wet and dry specimens, expressed as a percentage. **Figure 2** shows the results of the water sensitivity test. For all tested asphalt mixtures indirect tensile strength increased with the addition of RA and decreased only slightly when rejuvenator was added. Asphalt mixtures were less sensitive to water after the addition of RA, as the ITSR ratio increased. No significant effect on ITSR was observed with the addition of a rejuvenator.

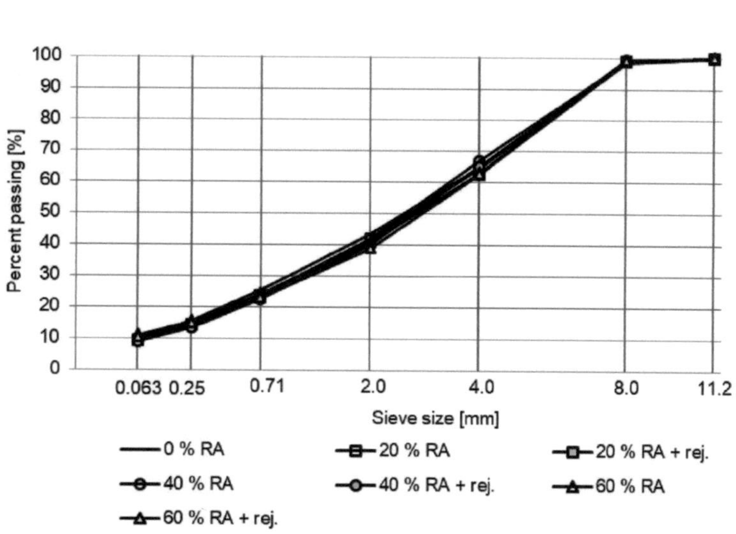

Figure 1.
The sieving curves of laboratory produced asphalt mixtures AC 8 surf [22].

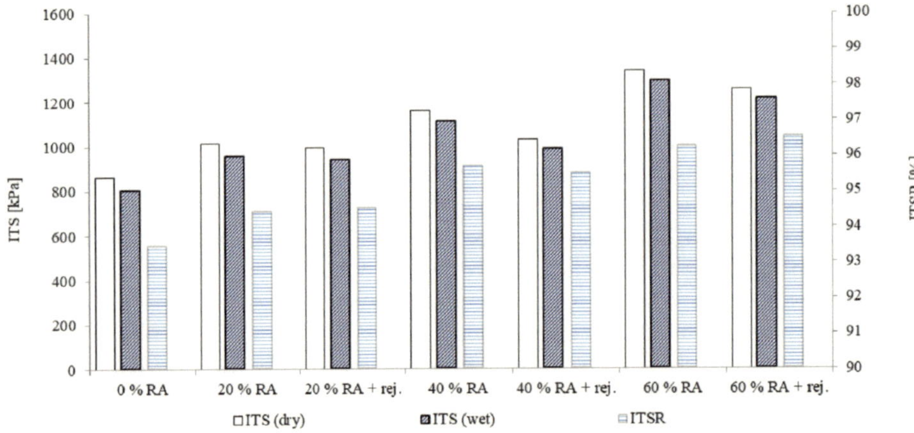

Figure 2.
The water sensitivity of the asphalt mixtures [1].

Testing of wheel track formation was performed at an elevated temperature of T = 50°C. The differences between the sample results (**Figure 3**) were not significant, indicating that the selected test temperature was too low.

Because the aged bitumen from RA increases the stiffness of the binder in the asphalt mixture, resistance to fatigue cracking is weakened in asphalt mixtures with a high amount of RA. In our study, the resistance to low temperature cracking was checked only on mixtures with the highest proportion of RA, as 60% of RA represents the worst conditions for asphalt resistance at low temperatures. The average results (three specimens were tested for each mixture) for three mixes are shown in **Figure 4**. Compared to the mixture with 0%RA, the failure temperature increased due to the added RA, which means deterioration of the mechanical properties of the asphalt mixture. The results for 60% RA mixture with the addition of pyrolytic rejuvenator show that the failure temperature decreased, and the resistance of the asphalt mixture with RA and the rejuvenator was therefore slightly better than the resistance of the basic mixture. The breaking stresses were similar for the three

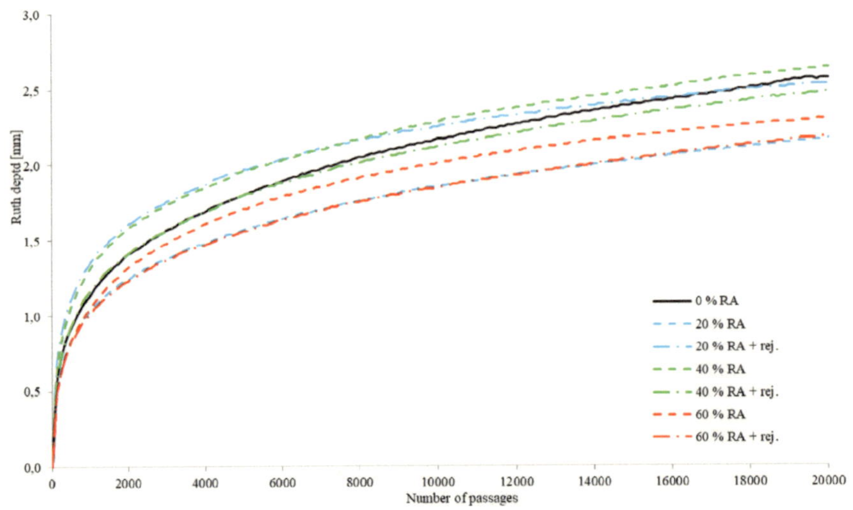

Figure 3.
Results of the wheel tracking test at T = 50°C.

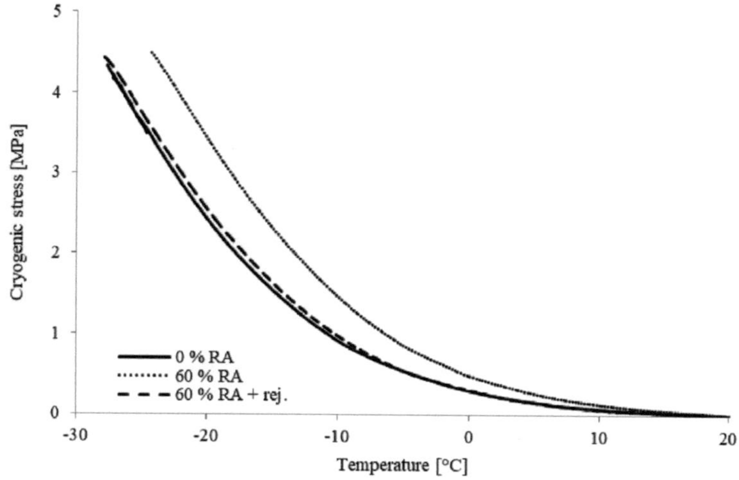

Figure 4.
Temperature dependency of cryogenic stress [1].

Label	RB [°C]	Fraass breaking point [°C]	Penetration [1/10 mm]
0% RA	58.6	−11	45
20% RA	59.4	−10	37
20% RA + rej.	58.2	−11	39
40% RA	60.8	−7	32
40% RA + rej.	58.3	−8	36
60% RA	63.3	−6	26
60% RA + rej.	60.7	−8	32
100% RA	65.3	−2	21

Table 10.
Results of standard mechanical tests of extracted bitumen [22].

tested mixtures, but the course of cryogenic stresses shows that the stresses were higher in the mixture with 60% RA than in the other two mixtures.

The extracted bitumen of asphalt mixtures was also investigated; the results are presented in **Table 10**. The highest value of the softening and Fraass breaking point was determined in bitumen from reclaimed asphalt (100% RA), meaning the bitumen was the most brittle and hard, as expected. The results showed that the addition of RA to asphalt mixtures without pyrolytic rejuvenator increased the softening point and Fraass breaking point and decreased penetration. However, the rejuvenator had a beneficial effect, as mixtures with the rejuvenator exhibited a lower softening point, Fraass breaking point and higher penetration than mixtures without pyrolytic rejuvenator. As expected, the bitumen became harder as the RA content increased, but the bitumen softened with the addition of a rejuvenator.

5. Conclusions

In the presented research, several pyrolytic products were tested. Based on the results, the pyrolytic product 14 was selected as the most suitable for the purpose of rejuvenating. It had a homogeneous structure, was solid at room temperature and it flowed at elevated temperatures. The indicator of suitability for use as a rejuvenator was the pyrolytic product's low softening and Fraass breaking point and high penetration value.

In mixtures with reference bitumen, the low Fraass breakpoint value was not maintained. The penetration of bitumen was increased by adding a pyrolytic rejuvenator and the softening point was decreased. The results of standard mechanical tests confirmed that the pyrolytic product softened the bitumen, which is the basic purpose of the rejuvenator.

The next part of the research was dedicated to the determination of the optimal proportion of rejuvenator that would regain the original properties of bitumen from reclaimed asphalt. This part of the research was performed on non-aged and laboratory aged bitumen. The addition of rejuvenator was limited at the upper limit, as the prepared pyrolytic product contained polycyclic aromatic hydrocarbons (PAHs). The results of decreased softening point and increased penetration show that the bitumen softened due to the rejuvenator. The Fraass breaking point increased in aged bitumen after the addition of a rejuvenator, indicating deterioration in the properties of the bitumen.

Despite the added rejuvenator the elongation in the ductility test was maintained for non-aged bitumen and increased for aged bitumen. However, in proportion to the increase in the rejuvenator, the maximum force in the sample decreased. The elastic recovery of non-aged bitumen did not change due to the rejuvenator, while in aged bitumen it decreased after the addition of the rejuvenator. The results of BBR confirmed our expectations, as resistance at low temperatures was increased by the addition of a rejuvenator. The addition of 20% of the pyrolytic product changed most properties of the aged bitumen in order to approach the characteristics of the non-aged bitumen. Although a complete recovery of the aged bitumen was not reached, the pyrolytic product can be successfully used as a rejuvenator. After the addition of pyrolytic product, the properties of aged bitumen were restored in the direction of the properties of the non-aged bitumen.

The asphalt mixtures were less water sensitive when RA and rejuvenator were added. Results proved RA improved the adhesion and cohesion of bitumen. Rejuvenator did not deteriorate the cohesion significantly, even though it made bitumen softer.

The results of the TSRST test revealed that RA deteriorates the properties of asphalt mixtures. However, low temperature cracking resistance improved with the addition of a rejuvenator. The improvement was made to such an extent that the asphalt mixture with the highest proportion of RA and rejuvenator had slightly better resistance than the control asphalt mixture.

On the basis of all the results, it can be concluded that pyrolytic rejuvenator enables the increase of RA share in asphalt mixture.

Author details

Lidija Ržek[1*], Mojca Ravnikar Turk[1] and Marjan Tušar[1,2]

1 Slovenian National Building and Civil Engineering Institute, Ljubljana, Slovenia

2 National Institute of Chemistry, Ljubljana, Slovenia

*Address all correspondence to: lidija.rzek@zag.si

IntechOpen

© 2021 The Author(s). Licensee IntechOpen. This chapter is distributed under the terms of the Creative Commons Attribution License (http://creativecommons.org/licenses/by/3.0), which permits unrestricted use, distribution, and reproduction in any medium, provided the original work is properly cited.

References

[1] Sanders P. Silvia Project Report: Review of Recycling and Rejuvenation Procedures. Sustainable Road Surfaces for Traffic Noise Control, European Commission, 2005.

[2] Lu X., Isacsson U. Construction and Building Materials: Effect of ageing on bitumen chemistry and rheology. 2001; 16;1: 15-22. DOI: 0.1016/S0950-0618 (01)00033-2

[3] Wu J. The influence of mineral aggregates and binder volumetrics on bitumen ageing [thesis]. Nottingham: University of Nottingham; 2009.

[4] Shen J., Amirkhanian S., Miller J. A. Journal of Materials in Civil Engineering: Effects of Rejuvenating Agents on Superpave Mixtures Containing Reclaimed Asphalt Pavement. 2007; 19, 5: 376-384. DOI: 10.1061/ (ASCE)0899-1561(2007)19:5(376)

[5] Ržek L., Tušar M., Slemenik Perše L. International Journal of Pavement Engineering: Modelling rheological characteristics of rejuvenated aged bitumen. 2020; DOI: 10.1080/10298436. 2020.1799205

[6] Tušar M., Avsenik L. Case Study: Increasing the Percentage of Recycled Asphalt. In: Torrenti J.-M., La Torre F., editors. Materials and infrastructures 2, London, ISTE; Hoboken: J. Wiley & Sons. 2016.p. 33-44. DOI: 10.1002/ 9781119318613.ch3

[7] Brownridge J. The Role of an Asphalt Rejuvenator in Pavement Preservation: Use and Need for Asphalt Rejuvenatio. In: Compendium of Papers from the First International Conference on Pavement Preservation, 13-15 April 2010; Newport Beach, California; 2010. p. 351-364

[8] Zaumanis M., Mallick R. B., Poulikakos L., Frank R. Construction and Building Materials: Influence of six rejuvenators on the performance properties of Reclaimed Asphalt Pavement (RAP) binder and 100% recycled asphalt mixtures. 2014; 71:. 538-550. DOI: 10.1016/j. conbuildmat.2014.08.073

[9] Karlsson R., Isacsson U. Journal of Materials in Civil Engineering: Material-related aspects of asphalt recycling – State-of-the-art. 2006; 18; 1: 81-92. DOI: 10.1061/(ASCE) 0899-1561(2006)18:1(81)

[10] Avsenik L., Tušar M. Građevinar: Analysis of possible use of pyrolytic products as binders in asphalt mixes. 2016; 3: 191-198. DOI: 10.14256/ JCE.1305.2015

[11] ETRMA The European Tire Industry: Facts and Figures 2020 Edition. Available: https://www.etrma. org/wp-content/uploads/2019/12/ Figures-leaflet-updated-front-2019- larger-NEW-LABEL.pdf. [Accessed: 2020-01-14].

[12] Council Directive 1999/31/EC of 26 April 1999 on the landfill of waste, 1999.

[13] "Directive 2008/98/Ec Of The European Parliament And Of The Council of 19 November 2008," 2008. [Online]. Available: https://eur-lex. europa.eu/legal-content/EN/TXT/HTM L/?uri=CELEX:32008L0098&from=SL. [Accessed: 2021-06-17].

[14] Chaala A., Ciochina O. G., Roy C. Resources, Conservation and Recycling: Vacuum pyrolysis of automobile shredder residues: use of the pyrolytic oil as a modifier for road bitumen. 1999; 26; 155-172. DOI: 10.1016/S0921- 3449(99)00003-8Get rights and content

[15] Yousefi A. A., Ait-Kadi A., Roy C. Fuel: Effect of used-tire-derived pyrolytic oil residue on the properties of

modifed asphalts. 2000; 75: 975-986. DOI: 10.1016/S0016-2361(99)00216-1

[16] Chaala A., Roy C., Ait-Kadi A. Fuel: Rheological properties of bitumen modified with pyrolytic carbon black. 1996; 75; 13: 1575-1583. DOI: 10.1016/0016-2361(96)00143-3

[17] Roy C., Chaala A., Darmstadt H. Journal of Analytical and Applied Pyrolysis: The vacuum pyrolysis of used tires: End-uses for oil and carbon black products. 1999; 51; 1-2: 201-221. DOI: 10.1016/S0165-2370(99)00017-0.

[18] Martínez J. D., Puy N., Murillo RGarcia. T., Navarro M. V., Mastral A. M. Renewable and Sustainable Energy Reviews: Waste tyre pyrolysis – A review. 2013; 23: 179-213. DOI: 10.1016/j.rser.2013.02.038

[19] Cunliffe A. M., Williams P. T. Journal of Analytical and Applied Pyrolysis: Composition of oils derived from the batch pyrolysis of tyres. 1998; 44; 2: 131-152. DOI: 10.1016/S0165-2370(97)00085-5

[20] Avsenik L., Tušar M. Pyrolytic product used as an alternative binder rejuvenator. In 6th Eurasphalt & Eurobitume Congress,1-3 June 2016; Prague, 2016.

[21] Avsenik L, Klinar D., Tušar M., Slemenik Perše L. Construction and Building Materials: Use of modified slow tire pyrolysis product as a rejuvenator for aged bitumen. 2016; 120: 605-616. DOI: 10.1016/j.conbuildmat.2016.05.140

[22] Ržek L., Ravnikar Turk M., Tušar M. Construction and Building Materials: Increasing the rate of reclaimed asphalt in asphalt mixture by using alternative rejuvenator produced by tire pyrolysis. 2020; 117177. DOI. 10.1016/j.conbuildmat.2019.117177

[23] EN 1427 :2007. Bitumen and bituminous binders - Determination of the softening point - Ring and Ball method.

[24] EN 12593:2007. Bitumen and bituminous binders - Determination of Fraass breaking point.

[25] EN 13589:2008. Bitumen and bituminous binders - Determination of the tensile properties of modified bitumen by the force ductility method.

[26] Mwanza A. D., Hao P., Xongyonyar X. International Journal of Civil Engineering: Determination of mixing and compaction temperatures shift for asphalt mastic at different type and content of mineral fillers. 2012; 2; 4: 159-166.

[27] EN 12607-1:2007. Bitumen and bituminous binders - Determination of the resistance to hardening under influence of heat and air - Part 1: RTFOT method.

[28] Airey G. D. Rheological Characteristics of Polymer Modified and Aged Bitumens [thesis]. Nottingham: University of Nottingham; 1997.

[29] EN 14769:2012. Bitumen and bituminous binders - Accelerated-long term ageing conditioning by a Pressure Ageing Vessel (PAV).

[30] EN 1426:2007. Bitumen and bituminous binders - Determination of needle penetration.

[31] EN 13398:2010. Bitumen and bituminous binders - Determination of the elastic recovery of modified bitumen.

[32] EN 13399:2010. Bitumen and bituminous binders - Determination of storage stability of modified bitumen.

[33] EN 12697-11:2012. Bituminous mixtures - Test methods - Part 11: Determination of the affinity between aggregate and bitumen.

[34] EN 14771:2012. Bitumen and bituminous binders - Determination of the flexural creep stiffness - Bending Beam Rheometer (BBR).

[35] EN 12697-30:2012. Bituminous mixtures - Test methods - Part 30: Specimen preparation by impact compactor.

[36] EN 12697-6:2012. Bituminous mixtures - Test methods - Part 6: Determination of bulk density of bituminous specimens.

[37] EN 12697-5:2010. Bituminous mixtures - Test methods - Part 5: Determination of the maximum density.

[38] EN 12697-8:2004. Bituminous mixtures - Test methods - Part 8: Determination of void characteristics of bituminous specimens.

[39] EN 12697-23:2004. Bituminous mixtures - Test methods - Part 23: Determination of the indirect tensile strength of bituminous specimens.

[40] EN 12697-12:2009. Bituminous mixtures - Test methods - Part 12: Determination of the water sensitivity of bituminous specimens.

[41] EN 12697-46:2012. Bituminous mixtures - Test methods for hot mix asphalt - Part 46: Low temperature cracking and properties by uniaxal tension tests.

[42] EN 12697-33:2007. Bituminous mixtures - Test methods - Part 33: Specimen prepared by roller compactor.

[43] EN 12697-22:2007. Bituminous mixtures - Test methods for hot mix asphalt - Part 22: Wheel tracking.

Chapter 14

The Biomass Waste Pyrolysis for Biopesticide Application

Mashuni and M. Jahiding

Abstract

The pyrolysis method has been used in various fields and has attracted the attention of many researchers so that this method can be applied to treat biomass waste. Pyrolysis of biomass occurs through heating a substance with limited oxygen so that the decomposition of complex compounds such as lignocellulose into simpler compounds occurs. The heat energy of the pyrolysis process encourages the oxidation of biomass so that complex carbon molecules break down into carbon and bio-oil. Pyrolysis of biomass for coconut shells, cashew nut shells, and cocoa pod husk was carried out at a temperature of 400–600°C with a flow rate of 6–7°C/min. The content of bio-oil compounds from its biomass based on the analysis of gas chromatography–mass spectroscopy obtained phenolic acid, pyrimidine derivatives, amines, carbamate acids, furans, esters derivatives, pyridine, ketones, furans, and aldehydes that can be used as active compounds for biopesticides.

Keywords: biomass, biopesticide, lignocellulose, pyrolysis

1. Introduction

Global output of biomass production from agricultural and forestry residues is estimated at 146 billion MT per year [1], most of which are disposed of in landfills or burned to ashes. The burning of these biomass residues can cause soil degradation by affecting soil biota. In addition, large amounts of particulates, volatile organic carbon and semi-volatile organic carbon compounds, ash, sulfate aerosols, and trace gases are also released into the atmosphere [2]. These pollutants contribute to greenhouse gas emissions, which can contribute to many serious environmental problems on a global scale, such as increasing global climate change, extinction of biodiversity, and socioeconomic severe and health problems. Therefore, it is essential to minimize the burning or wasting of plant biomass and instead develop low-cost pollution reduction and sustainable technologies to convert it into valuable bioproducts [3].

The chemical composition of biomass, both lignocellulosic and herbaceous, can be characterized by five main components: cellulose, hemicellulose, lignin, extractive/volatile, and ash. Cellulose and hemicellulose, combined with the third major component of biomass, lignin, make up more than 90% of lignocellulosic biomass and 80% of herbaceous biomass. Lignin is a complex array of phenolic compounds interwoven with the cellulose and hemicellulose fractions of the biomass structure. This interwoven property of lignin helps impart rigidity to lignocellulosic materials, such as trees [4]. Biomass can be converted into energy through thermal, biological, and physical conversion processes, such as direct combustion, pyrolysis, and gasification [5].

IntechOpen

Pyrolysis of biomass is the decomposition of chemical components of lignocellulosic by heating or incomplete combustion to be broken down into compounds with shorter chains [6]. Pyrolysis is a decomposition process or decomposition of compounds in raw materials in the presence of heat of combustion and limited oxygen so that gas, liquid, and charcoal are obtained, the amount of which is influenced by the type of material, method, and conditions of pyrolysis. Incomplete combustion of pyrolysis causes complex carbon compounds not to be oxidized to carbon dioxide in raw materials containing cellulose, hemicellulose, and lignin [7]. Conversion of agricultural residue biomass by the pyrolysis method into bio-oil is a potentially attractive technology to remove and process waste from agriculture and greenhouses into alternative sources of green energy and value-added chemicals [8]. Researchers at the Institute for Chemicals and Fuels from Alternative Resources (ICFAR) at the University of Western Ontario designed a highly automated rapid pyrolysis to convert biomass to bio-oil, gas, and biochar at the temperature of 250-800°C under nearly atmospheric pressure and in the absence of oxygen [9]. Bio-oil from biomass pyrolysis produces a complex mixture of chemicals including acids, ketones, furans, phenols, hydrosugars, and other oxygenates, which have antibacterial and antifungal properties against several pathogenic and carcinogenic bacteria as well as biopesticides [10–12].

2. Compositional of lignocellulosic biomass

In particular, lignocellulosic biomass, mainly composed of lignin, cellulose, and hemicellulose, has become an essential topic because lignocellulosic biomass does not compete with food and feed. There are several technologies for the conversion of lignocellulosic biomass into energy and chemicals [13]. As illustrated in **Figure 1**, the three main components are unevenly distributed in the cell wall as the skeleton, the connecting material, and the hard solid, respectively. Cellulose macromolecules regularly assemble to form tough microfibers that serve as the skeletal material of the cell wall, and the inner space is packed with hemicellulose linking materials and amorphous lignin [15]. Cellulose is linked to hemicellulose and lignin molecules

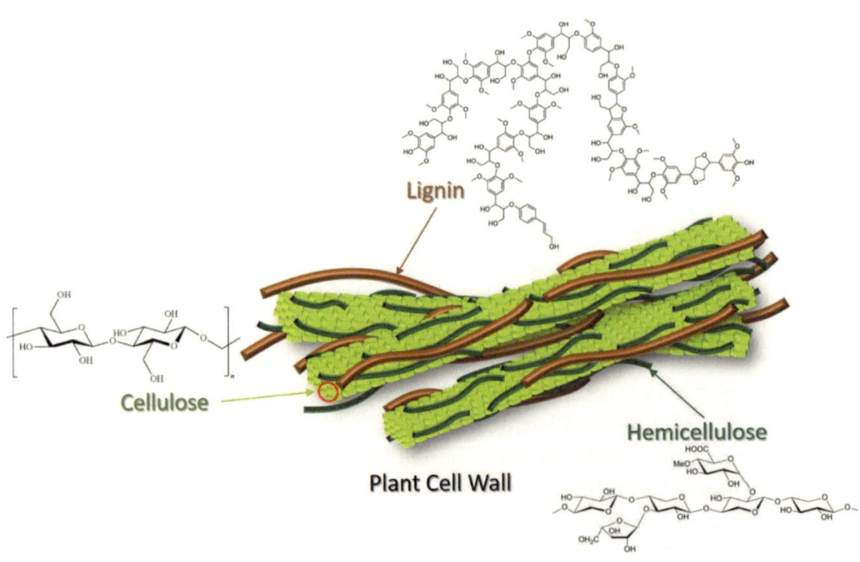

Figure 1.
The structure of lignocellulosic biomass consists of lignin, cellulose, and hemicellulose [14].

Biomass	Cellulose (%)	Hemicellulose (%)	Lignin (%)	References
Coconut shells (CS)	25.0	22.3	28.6	[17]
Cashew nut shells (CNS)	9.6	28.3	28.8	[18]
Cocoa pod husk (CPH)	26.1	12.8	28.0	[19]

Table 1.
The compositional variation lignocellulosic biomass from CS, CNS, and CPH.

mainly through hydrogen bonds, whereas the relationship between hemicellulose and lignin includes hydrogen and covalent bonds [16]. Carbohydrates and lignin are tightly bound in lignin-carbohydrate complexes, resulting in residual carbohydrates or lignin fragments in the extracted lignin or hemicellulose samples.

Lignocellulosic materials are mainly composed of cellulose (35–50%), hemicellulose (15–35%), and lignin (10–35%) [4]. The concentrations of the components mentioned vary with different plant species (as shown in **Table 1**). In addition to the three main components, a small fraction of extractives and inorganic ash are also present in the biomass as non-structural components, not cell walls or cell layers. Wood biomass contains significantly higher amounts of the three main components (»90%), while agricultural and herbaceous biomass contains more extractives and ash.

3. Optimization of parameter pyrolysis biomass

Thermal decomposition of organic matter in the absence of oxygen has been widely developed as a promising platform for producing fuels, preservatives, pesticides, and chemicals from various types of biomass. Pyrolysis produces charcoal, liquid, and gas products, which is highly dependent on the reaction conditions. Fast pyrolysis of biomass at a rapid heating rate and a short residence time of hot steam (<1 s) produces bio-oil with a yield of up to 75% of weight [12, 13].

Pyrolysis is a technology that converts lignocellulosic biomass into gaseous, liquid, and solid products by using heat under an inert atmosphere. Depending on the heating rate and residence time of the pyrolysis stream in the reactor, pyrolysis can be broadly classified into slow and fast pyrolysis. Slow pyrolysis involves thermal cracking of lignocellulosic biomass at low heating rates to produce a high-yield solid product known as biochar (or charcoal). In fast pyrolysis, high yields of liquid (bio-oil) products are obtained because the short residence time of the pyrolysis vapor in the reactor suppresses the secondary reaction, promoting the formation of gas and biochar [8, 20]. The reaction temperature for the pyrolysis of lignocellulosic biomass usually ranges from 500 to 800°C [21]. The physical properties characterization of the pyrolysis results of CS, CNS, and CPH biomass at temperatures of 400–600°C can be seen in **Table 2**.

During pyrolysis, biomass undergoes primary and secondary reactions involving heat and mass transfer mechanisms. The immediate response consists of decomposing lignocellulosic biomass, which leads to the formation of introductory and intermediate products. This intermediate species undergoes secondary cracking. The pathways for the first category include dehydration and charring reactions, while the second is decomposition and evaporation of intermediates. The pyrolysis products obtained in these competitive reactions are susceptible to operational variations and types of biomass [21]. Parameters play a significant role to determine the composition and properties of the pyrolysis products. Since biomass consists of cellulose, hemicellulose, and lignin, the degree of thermal fragmentation of

Biomass	Temperature (°C)	Density (g/cm^3)	pH	Flow rate (°C/min)	Yield (%)		
					Bio-oil	Char	Gas
CS	400	1.088	5	10 ± 1	39	28	33
	500	1.085			48	34	18
	600	1.083			38	25	37
CNS	400	1.078	3		38	24	38
	500	1.070			40	23	37
	600	1.087			43	22	35
CPH	400	1.083	5		36	32	32
	500	1.088			39	34	27
	600	1.084			37	36	27

Source: research by Mashuni et al. [22].

Table 2.
The operational parameters, physical characteristics, and yield (%) of the pyrolysis results of CS, CNS, and CPH biomass.

these components depends on the operating parameters [23]. **Table 2** shows the distribution of products obtained from various biomass and pyrolysis temperatures, indicating considerable flexibility that can change process conditions. During pyrolysis, many factors affect product properties such as type of biomass, residence time, age percentage of moisture in feed biomass, temperature, pressure conditions (atmosphere, vacuum), particle size, and heating rate of biomass so that pyrolysis efficiency also affects product composition. Optimization of reaction conditions can increase the yield of pyrolysis products to any of the three pyrolysis fuels such as pyrolysis oil, gas, or solid charcoal [20]. This parameter has a significant influence on the composition of the pyrolysis product.

4. Chemical compound composition of biomass pyrolysis bio-oil

The chemical composition of bio-oil derived from different biomass feedstocks is different. The difference can also be caused by other reaction conditions for the

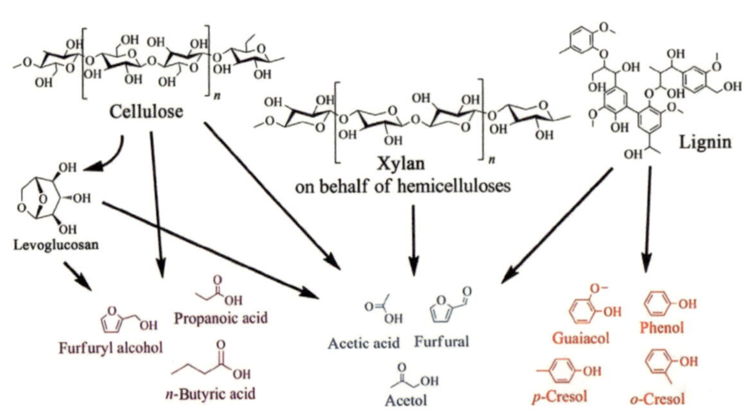

Figure 2.
The chemical formation pathways during pyrolysis of lignocellulosic biomass [25].

pyrolysis operation. This regard provides details of experimental results obtained with varying feedstocks of biomass and reaction conditions in pursuit of chemical production through pyrolysis of lignocellulosic biomass. During pyrolysis, the main components of lignocellulosic biomass will be thermally decomposed to produce valuable chemicals. Cellulose is a high-molecular-weight linear polymer consisting of D-glucose monomer units. The basic units are linked to each other by β-1,4-glycosidic bonds. Cellulose provides mechanical strength to plant cells and is generally degraded in the temperature range of 315–400°C [24]. Cellulose is classified as a homopolysaccharide, whereas hemicellulose is a heteropolymer referred to as a heteropolysaccharide. Hemicellulose consists of pentosan or hexosan, forming polymers such as xylan, glucan, xyloglucan, and glucomannan. Hemicellulose is usually degraded at 220–315°C. Lignin has a complex three-dimensional structure. The basic lignin units are three monolignol precursors, namely p-coumaryl alcohol, coniferyl alcohol, and sinapyl alcohol. Lignin degradation occurs over a wide temperature range covering 150–900°C [13]. **Figure 2** shows a simple pathway for the formation of chemicals produced during the pyrolysis of biomass.

Identified compound	Chemical formula	Peak area %		
		400°C	500°C	600°C
Phenol and its derivatives				
2-Methoxy-phenol	$C_7H_8O_2$	9.45	7.30	9.07
Phenol	C_6H_6O	20.81	23.88	21.92
2,6-Dimethoxy-phenol	$C_8H_{10}O_3$	11.82	9.50	11.54
3,4-Dimethyl-phenol	$C_8H_{10}O$		2.01	1.15
2-(Methylthio)-phenol	C_7H_8OS	3.08		
4-Butoxy-phenol	$C_{10}H_{14}O_2$	1.28	0.45	
Methylparaben	$C_8H_8O_3$		1.08	
1-(2,4-dihydroxyphenyl)-ethanone	C_8H_8O	10.25	0.93	
Amine and amides				
1-Methyl-2,5-pyrrolidinedione	C_5H_7NO	1.85	2.69	
Ketones				
3-Methyl-1,2-cyclopentanedion	$C_6H_8O_2$	8.83	8.96	7.66
Benzene, and its derivatives				
1,2,3-trimethoxybenzene	$C_9H_{12}O_3$	3.11	3.38	
1,4-Dimethoxy-benzene	$C_8H_{10}O_2$	5.10	2.93	5.08
(4-Methoxyphenyl)-hydrazine	$C_7H_{10}N_2O$	2.35	2.03	2.77
(4-Methoxyphenoxy)trimethyl-silane	$C_{11}H_{18}O_2Si$	3.89	6.09	1.13
Others				
3-Methyl-2-(2-oxopropyl)furan	$C_8H_{10}O_2$	4.45	1.53	4.38
(Z)-4-Methyl-5-(2-oxopropylidene)-5H-furan-2-one	$C_8H_8O_3$		4.38	13.06
Maltol	$C_6H_6O_3$	2.22	1.47	1.78

Source: research by Mashuni et al. [22].

Table 3.
The chemical compound composition of CS bio-oil.

Identified compound	Chemical formula	Peak area %		
		400°C	500°C	600°C
Phenol and its derivatives				
2-Methoxy-phenol	$C_7H_8O_2$	8.31		
Phenol	C_6H_6O	18.69	2.71	3.17
3-Methyl-phenol	C_7H_8O		2.31	4.55
2,6-Dimethoxy-phenol	$C_8H_{10}O_3$	10.52		
2-(2-Hydroxyethoxy)phenol	$C_8H_{10}O3$		4.92	
2-Ethoxy-4-methylphenol	$C_9H_{12}O_2$		5.22	
p-Cresol	C_7H_8O		5.76	
Catechol	$C_6H_6O_2$		29.29	
4-Ethyl-phenol	$C_8H_{10}O$		2.28	
1,2,3-Benzenetriol	$C_6H_6O_3$		6.12	
Pyridines and pyrimidines				
3-Methyl-pyridine	C_6H_7N		2.79	2.06
4-Hydroxy-6-methylpyrimidine	$C_5H_6N_3O$			1.20
Amine and amides				
N-Butylidene-N-oxide-methanamine	$C_5H_{11}NO$			10.42
5-Methyl-1H-1,2,4-Triazol-3-amine	C_3H_6N			1.54
N-Benzyl-2-phenylethanamine	$C_5H_{17}N$			16.54
Ketones				
3-Methyl-1,2-cyclopentanedion	$C_6H_8O_2$		8.85	13.78
Benzene, and its derivatives				
1,4-Dimethoxy-benzene	$C_8H_{10}O_2$	3.55	2.28	
4-Methyl-1,2-benzenediol	$C_{11}H_{12}O_4$		3.19	2.32
(4-Methoxyphenyl)-hydrazine	$C_7H_{10}N_2O$	2.33		
(4-Methoxyphenoxy) trimethyl-silane	$C_{11}H_{18}O_2Si$	5.32		
Picein	$C_{11}H_{18}O_7$	2.36		
Acids				
Dehydroacetic acid	$C_8H_8O_4$	3.34		
Hexanoic acid	$C_6H_{12}O_2$			2.95
Pentanoic acid	$C_5H_{10}O_2$		6.09	2.97
Others				
3-Methyl-2-(2-oxopropyl)furan	$C_8H_{10}O_2$		2.95	
(Z)-4-Methyl-5-(2-oxopropylidene)-5H-furan-2-one	$C_8H_8O_3$	11.30		
2(3H)-Furanone	$C_4H_4O_2$		10.16	
2,5-Furandicarboxaldehyde	$C_6H_4O_3$		2.95	
2,5-Diethoxytetrahydrofuran	$C_6H_{12}O$		7.80	
3,6-dimethyl-1,2,4,5-tetrazine	$C_4H_6N_4$		29.29	
Maltol	$C_6H_6O_3$	1.94	3.99	2.09
Source: research by Mashuni et al. [22].				

Table 4.
The chemical compound composition of CNS bio-oil.

Based on the results of the GC-MS analysis in **Table 3**, CS bio-oil contains phenol compounds and their derivatives, amines, amides, ketones, benzene, and furans. In **Table 4**, CNS bio-oil contains phenolic compounds and its derivatives, pyrimidines, pyridines, amines, amides, ketones, benzene, acids, and furans. In **Table 5**, CPH bio-oil contains compounds including phenols and its derivatives, amines, amides, ketones, benzene, acids, and furans. The thermal decomposition of cellulose produces carbonyl and furan compounds. Hemicellulose decomposition is similar to cellulose decomposition but produces acids. Meanwhile, the decomposition of lignin produces various types of phenolic compounds. Phenolic compounds contribute to the taste and color of bio-oil and have antibacterial properties. The

Identified compound	Chemical formula	Peak area %		
		400°C	500°C	600°C
Alkanes and alkenes				
2-Methoxy-butane	$C_5H_{12}O$		6.88	
5-(1-Methylpropyl)-Nonane	$C_{13}H_{28}$	7.96		1.95
Phenol and its derivatives				
2-Methoxy-phenol	$C_7H_8O_2$	2.14	24.19	16.82
Phenol	C_6H_6O	2.83		
3-Methyl-phenol	C_7H_8O			3.50
2,6-Dimethoxy-phenol	$C_8H_{10}O_3$	7.19	7.18	7.81
2-Methoxy-5-methylphenol	$C_8H_{10}O_2$		3.20	3.13
p-Cresol	C_7H_8O		3.86	2.06
Methyl-(2-hydroxy-3-ethoxy-benzyl)ether	$C_{10}H_{14}O_3$	1.28	1.20	2.37
Pyridines and pyrimidines				
3-Methyl-pyridine	C_6H_7N	4.06	2.08	1.36
2,5-Dimethyl-pyridine	C_7H_9N	16.38	9.29	
5-Methyl-pyrimidine	$C_5H_6N_2$		2.63	2.68
Amine and amides				
Guanidine	CH_5N_3	2.83		0.70
6-Aminonicotinamide	$C_6H_7N_3O$			6.25
Ketones				
1-(1H-pyrrol-2-yl)-ethanone	C_6H_7NO	1.17		
3-Methyl-1,2-cyclopentanedion	$C_6H_8O_2$	19.26	8.04	10.35
Benzene, and its derivatives				
1,2,3-Trimethoxybenzene	$C_9H_{12}O_3$	2.09		2.37
Others				
3-Methyl-2-(2-oxopropyl)furan	$C_8H_{10}O_2$	2.71	0.99	3.73
(Z)-4-Methyl-5-(2-oxopropylidene)-5H-furan-2-one	$C_8H_8O_3$	1.79	5.21	9.18
Maltol	$C_6H_6O_3$		6.15	5.82

Source: research by Mashuni et al. [22].

Table 5.
The chemical compound composition of CPH bio-oil.

dominant compounds in CS, SNS, and CPH bio-oils are phenolics and their derivatives. High phenolic content is generally applied in food systems as a browning agent or preservative, while it is applied as a pesticide in the agricultural industry. The use of pyrolysis liquid as a pesticide can be applied to all types of commercial biopesticides [26].

5. Application and effectiveness of biopesticide

Biopesticides are natural pesticides that can be obtained either from a microorganism or certain plant extracts including biomass waste. Utilization of biopesticides in plantation and agricultural management to inhibit or kill pests that cause disease in plants by using active components is found in microbes or certain plant extracts. The main advantages of biopesticides are specificity to target insects, effectiveness at low doses, biodegradability, and low toxicity compared to conventional chemical pesticides. Furthermore, synthetic pesticides are toxic, can kill various kinds of pests and beneficial insects, and cause environmental pollution.

Lignocellulosic biomass decomposition by pyrolysis method produces bio-oil, which can be applied as a biopesticide. The chemical components of bio-oil resulting from lignocellulosic pyrolysis are ammonia, hexane, alcohol, ketone, carbonyl, acetic, and phenolic acids [26–28]. One of the groups of compounds that have the most role in inhibiting microbial growth is phenolic compounds. Biomass contains lignin, a source of phenolic compounds either through the pyrolysis method, which functions as an antioxidant and can be used as a functional food ingredient [29]. Epidemiological studies show that phenolic compounds have potential effects to prevent chronic disease and have anticarcinogenic, anti-inflammatory, antimicrobial, antifungal [26, 30], and biopesticides [31, 32]. Bio-oil from the pyrolysis results can be used as a biofungicide because it contains active compounds that are effective as inhibitors for the growth of fungi and bacteria (as shown in **Tables 3–5**).

Phenol compounds are compounds that contain a hydroxyl group (–OH), which is directly attached to an aromatic hydrocarbon ring group. The activity of phenol compounds comes from the number of hydroxyl groups on the benzene ring. Classification of phenolic compounds contained in plants are simple phenols, benzoquinones, phenolic acids, acetophenone, naphthoquinones, xanthones, coumarin bioflavonoids, stilbenes, tyrosine derivatives, hydroxycinnamic acid, flavonoids, lignans, and tannins. Flavonoids and polyphenols can inhibit the growth of pathogenic bacteria and fungi [33].

Several researchers have tested the effectiveness of bio-oil from organic waste as a natural pesticide [34, 35]. The bio-oil of the CN exhibited antifungal activity against *Phytophthora plamivora* to 60% at 0.100% v/v concentration, whereas CN bio-oil at concentration 0.125% v/v showed inhibition of *Phytophthora palmivora* growth to 100%. In other words, 0.125% was the fungicidal concentration of the bio-oil. The CN bio-oil can inhibit *Phytophthora palmivora* growth since it contains a great amount of phenolic compounds and it is well known that phenolic compounds are one of the main inhibitors of fungal growth. This was supported by the fact that when the cell wall of fungi treated with phenol, it caused the shrinking, indicating the disruption of cell wall structure [11]. In addition, bio-oil also inhibits the growth of the fungus, *Promecotheca palmivora* is a pathogenic fungus that causes cacao pod rot disease [11]. In Mashuni et al.'s research, CNS bio-oil was used as a natural ingredient to improve the quality of cocoa seed supply during storage to replace synthetic pesticides. The bio-oil of CNS was used as a biofungicide to protect seeds from fungal attack by inhibiting the growth of its mitotoxin, where the results showed that bio-oil was better at inhibiting fungal growth on cocoa beans than synthetic fungicides [36].

The content of phenolic compounds contained in bio-oil can inhibit the growth of the fungus *Fusarium oxysporum*. The fungus *F. oxysporum* can cause damage to agricultural management, which is very detrimental to farmers. The fungus *F. oxysporum* damages red chilies, tomatoes, long beans, potatoes, cabbage, and several other crops that will cause the plants to wither [37]. Phenolic compounds that are high enough in bio-oil resulting from biomass pyrolysis can be applied to the inhibition of the growth of the fungus *F. oxysporum* already in the category of murder. The antifungal activity test against *F. oxysporum* was carried out by the dilution method. The CPH bio-oil shows antifungal activity with a minimum inhibitory concentration (MIC) of 1% (v/v) and a minimal killing concentration (MKC) of 9% (v/v) [32].

Biofungicide is a type of natural pesticides that is used to inhibit or kill fungi that cause disease in plants by using active components found in microbes or plants. The use of biofungicides is cheaper and environmental friendly than synthetic fungicides. Bio-oil from the pyrolysis of CN, CNS, and CPH is an alternative that can be used as a raw material for making biopesticides because it contains several active compounds that have the property of inhibiting and killing pathogenic microbes.

6. Conclusions

The abundance of the variability of biomass resources available today is a significant opportunity for the utilization of biomass. Biomass containing lignocellulose consisting of cellulose, hemicellulose, and lignin with good processing can be converted into useful active compounds. Biomass processing using the pyrolysis method has a positive impact because it produces bio-oil that can be used as a biopesticide, which does not involve chemical solvents in the process. The chemical compound content of bio-oil from CP, CNS, and CPH biomass pyrolysis based on the analysis of gas chromatography-mass spectroscopy (GC-MS) obtained phenolic acid, pyrimidine derivatives, amines, carbamate acids, furans, esters derivatives, pyridine, ketones, furans, and aldehydes that can be used as active compounds for biopesticides. The use of bio-oil as a biopesticide can reduce biomass waste and overcome the negative impacts of using synthetic pesticides so that it can have a positive impact on health and the environment.

Acknowledgements

This research was supported by the Ministry of Education, Culture, Research, and Technology of the Republic of Indonesia.

Author details

Mashuni[1*] and M. Jahiding[2]

1 Department of Chemistry, Halu Oleo University, Kendari, Southeast Sulawesi, Indonesia

2 Department of Physics, Halu Oleo University, Kendari, Southeast Sulawesi, Indonesia

*Address all correspondence to: mashuni2696@gmail.com; mashuni@uho.ac.id

IntechOpen

© 2021 The Author(s). Licensee IntechOpen. This chapter is distributed under the terms of the Creative Commons Attribution License (http://creativecommons.org/licenses/by/3.0), which permits unrestricted use, distribution, and reproduction in any medium, provided the original work is properly cited.

References

[1] Patel A, Agrawal B, Rawal BR. Pyrolysis of biomass for efficient extraction of biofuel. Energy Sources, Part A: Recovery, Utilization, and Environmental Effects. 2019;**42**(13): 1649-1661. DOI: 10.1080/15567036.2019.1604875

[2] Grewal A, Lord A, Gunupuru LR. Production, prospects and potential application of pyroligneous acid in agriculture. Journal of Analytical and Applied Pyrolysis. 2018;**135**:152-159. DOI: 10.1016/j.jaap.2018.09.008

[3] Muntean M, Guizzardi D, Schaaf E, et al. Fossil CO_2 Emissions of All World Countries—2018 Report. Luxembourg Publications; Office of the European Union; 2018. DOI: 10.2760/30158

[4] Williams CL, Emerson RM, Tumuluru JS. Biomass compositional analysis for conversion to renewable fuels and chemicals. In: Tumuluru JS, editor. Biomass Volume Estimation and Valorization for Energy. Rijeka: IntechOpen; 2017. pp. 251-270. DOI: 10.5772/65777

[5] Oyebanji JA, Okekunle PO, Lasode OA, Oyedepo SO. Chemical composition of bio-oils produced by fast pyrolysis of two energy biomass. Biofuels. 2018;**9**(4):479-487. DOI: 10.1080/17597269.2017.1284473

[6] Tsamba AJ, Yang W, Blasiak W. Pyrolysis characteristics and global kinetics of coconut and cashew nut shells. Fuel Processing Technology. 2006;**87**:523-530. DOI: 10.1016/j.fuproc.2005.12.002

[7] Wei X, Ma X, Peng X, Yao Z, Yang F, Dai M. Comparative investigation between co-pyrolysis characteristics of protein and carbohydrate by TG-FTIR and Py-GC/MS. Journal of Analytical and Applied Pyrolysis. 2018;**135**:209-218. DOI: 10.1016/j.jaap.2018.08.031

[8] Bridgwater AV. Review of fast pyrolysis of biomass and product upgrading. Biomass and Bioenergy. 2012;**38**:68-94. DOI: 10.1016/j.biombioe.2011.01.048

[9] Xu R, Ferrante L, Briens C, Berruti F. Flash pyrolysis of grape residues into biofuel in a bubbling fluid bed. Journal of Analytical and Applied Pyrolysis. 2009;**86**(1):58-65. DOI: 10.1016/j.jaap.2009.04.005

[10] Sandström L, Johansson A, Wiinikka H, Öhrman OGW, Marklund M. Pyrolysis of nordic biomass types in a cyclone pilot plant: Mass balances and yields. Fuel Processing Technology. 2016;**152**:274-284. DOI: 10.1016/j.fuproc.2016.06.015

[11] Mashuni YNA, Jahiding M, Kadidae LO, Djaila R, Hamid FH. Analysis of liquid volatile matters from coconut shell pyrolysis by GC-MS and its potential as antifungal agent. Asian Journal of Chemistry. 2020;**32**(7): 1728-1732

[12] Johansson A, Wiinikka H, Sandström L, Marklund M, Öhrman OGW, Narvesjö J. Characterization of pyrolysis products produced from different Nordic biomass types in a cyclone pilot plant. Fuel Processing Technology. 2016;**146**:9-19. DOI: 10.1016/j.fuproc.2016.02.006

[13] Azadi P, Inderwildi OR, Farnood R, King DA. Liquid fuels, hydrogen and chemicals from lignin: A critical review. Renewable and Sustainable Energy Reviews. 2013;**21**:506-523. DOI: 10.1016/j.rser.2012.12.022

[14] Jensen CU, Katerine Julie Guerrero R, Karatzos S, Olofsson G, Iversen SB. Fundamentals of hydrofaction™: Renewable crude oil from woody biomass. Biomass Convers

Biorefinery. 2017;7(4):459-509. DOI: 10.1007/s13399-017-0248-8

[15] Wang S, Luo Z. Pyrolysis of Biomass. China. De Gruyter; 2016. DOI: 10.1515/9783110369632

[16] Vorwerk S, Somerville S, Somerville C. The role of plant cell wall polysaccharide composition in disease resistance. Trends in Plant Science. 2004;**9**(4):203-209. DOI: 10.1016/j.tplants.2004.02.005

[17] Gonçalves FA, Ruiz HA, dos Santos ES, Teixeira JA, de Macedo GR. Valorization, comparison and characterization of coconuts waste and cactus in a biorefinery context using $NaClO_2$–$C_2H_4O_2$ and sequential $NaClO_2$–$C_2H_4O_2$/autohydrolysis pretreatment. Waste and Biomass Valorization. 2019;**10**(8):2249-2262. DOI: 10.1007/s12649-018-0229-6

[18] de Souza NB. Determination of Composition of Cashew (*Anacardium occidentale*) Nut Shell by Thermogravimetry. Vol. 2019. In: University of Applied Science. Jerman. 2019:2019.

[19] Lu F, Rodriguez-Garcia J, Van Damme I, et al. Valorisation strategies for cocoa pod husk and its fractions. Current Research in Green and Sustainable Chemistry. 2018;**14**:80-88. DOI: 10.1016/j.cogsc.2018.07.007

[20] Jahirul MI, Rasul MG, Chowdhury AA, Ashwath N. Biofuels production through biomass pyrolysis—A technological review. Energies. 2012;**5**:4952-5001. DOI: 10.3390/en5124952

[21] Akhtar J, Saidina N, Wood P. A review on operating parameters for optimum liquid oil yield in biomass pyrolysis. Renewable and Sustainable Energy Reviews. 2012;**16**(7):5101-5109. DOI: 10.1016/j.rser.2012.05.033

[22] Mashuni NAY, Jahiding M. Pengembangan Pestisida Alami Berbasis Liquid Volatile Matter dari Limbah Organik dalam Mengendalikan Hama dan Penyakit untuk Meningkatkan Produktivitas Petani Kakao. In: PENPRINAS MP3EI 2011-2025. Universitas Halu Oleo. p. 2017:2011-2025

[23] Wang S, Daia G, Yang H, Luo Z. Lignocellulosic biomass pyrolysis mechanism: A state-of-the-art review. Progress in Energy and Combustion Science. 2017;**62**:33-86. DOI: 10.1016/j.pecs.2017.05.004

[24] Yang H, Yan R, Chen H, HoLee D, Zheng C. Characteristics of hemicellulose, cellulose and lignin pyrolysis. Fuel. 2007;**86**:1781-1788. DOI: 10.1016/j.fuel.2006.12.013

[25] Zhang X, Yang G, Jiang H, Liu W, Ding H. Mass production of chemicals from biomass-derived oil by directly atmospheric distillation coupled with co-pyrolysis. Scientific Reports. 2013;**3**(1):1-7. DOI: 10.1038/srep01120

[26] Mashuni, Jahiding M, Kurniasih I, Zulkaidah. Characterization of preservative and pesticide as potential of bio oil compound from pyrolisis of cocoa shell using gas chromatography. In: International Conference on Chemistry, Chemical Process and Engineering (IC3PE). Yogyakarta. 2017;**1823**:1-8. DOI: 10.1063/1.4978081

[27] Ifa L, Yani S, Mandasini, Sabara Z, Sirajudin N, Rusnaenah A. Production of phenol from liquid smoke resulted by the pyrolysis of cashew nut shells. IOP Conference Series: Earth and Environmental Science. 2018;**175**(1):1-5. DOI: 10.1088/1755-1315/175/1/012033

[28] Jahiding M, Mashuni, Ilmawati WOS, et al. Characterization of coconut shell liquid volatile matter (CS-LVM) by using gas chomatroghaphy. Journal of Physics

Conference Series. 2017;**846**(1):1-7. DOI: 10.1088/1742-6596/846/1/012025

[29] Hii CL, Law CL, Suzannah S, Misnawi CM. Polyphenols in cacao (*Theobroma cacao* L.). Asian Journal of Food and Agro-Industry. 2010;**2**(04): 702-722

[30] Martínez R, Torres P, Meneses MA, Figueroa JG, Pérez-álvarez JA, Viuda-martos M. Chemical, technological and in vitro antioxidant properties of cocoa (*Theobroma cacao* L.) co-products. Food Research International. 2012;**49**:39-45. DOI: 10.1016/j.foodres.2012.08.005

[31] Mashuni, Jahiding M, Ilmawati WS, et al. Characterization of liquid volatile matter (LVM) of cashew nut shell using pyrolysis and gas chomatroghaphy. Journal of Physics: Conference Series. 2017;**846**:1-6. DOI: 10.1088/1742-6596/846/1/012026

[32] Mashuni P, Yanti NA, Jahiding M, Kadidae LO, Asis WA, Hamid FH. Penentuan Kandungan Fenolik Total Liquid Volatile Matter dari Pirolisis Kulit Buah Kakao dan Uji Aktivitas Antifungi terhadap Fusarium oxysporum. ALCHEMY Jurnal Penelitian Kimia. 2019;**15**(1):165-176. DOI: 10.20961/alchemy.15.1.24678.165-176

[33] Guerrero L, Castillo J, Quiñones M, et al. Inhibition of angiotensin-converting enzyme activity by flavonoids: Structure-activity relationship studies. PLoS One. 2012;7(11):1-11. DOI: 10.1371/journal.pone.0049493

[34] Mourant D, Yang DQ, Lu X, Roy C. Anti-fungal properties of the pyroligneous liquors from the pyrolysis of softwood bark. Wood and Fiber Science. 2007;**37**(3):542-548

[35] Mourant D, Riedl B, Rodrigue D, Yang DQ, Roy C. Phenol-formaldehyde-pyrolytic oil resins for wood preservation: A rheological study. Journal of Applied Polymer Science. 2007;**106**(2):1087-1094. DOI: 10.1002/app.26817

[36] Mashuni, Kadidae LO, Yanti NA, Jahiding M, Mappease, WON S. Evaluation of liquid volatile matter antifungal activities from pyrolysis of cashew nut shell on cocoa seeds. Global Journal of Pharmacology. 2019;**13**(1): 1-9. DOI: 10.5829/idosi.gjp.2019.01.09

[37] Jumjunidang, Edison, Riska, Hermanto C, Layu P. Fusarium Pada Tanaman Pisang di Provinsi NAD : Sebaran dan Identifikasi Isolat Berdasarkan Analisis Vegetative Compatibility Group. Jurnal Hortikultura. 2012;**22**(2):164-171

Chapter 15

Pyrolysis-GC/MS, A Powerful Analytical Tool for Additives and Polymers Characterization

Sonia Huertas de la Torre

Abstract

Analytical Pyrolysis technique hyphenated to GC/MS is used to obtain structural information of macromolecules. The composition and relative abundance of the pyrolysis products are characteristic for a given polymer and their determination allows the identification of materials that cannot be determined otherwise. In the present work different examples are presented in which different pyrolysis techniques are used, such as pulsed, fractional and sequential pyrolysis, working with polymer solutions, cold injection in split or narrow band mode or thermochemolysis. Specially useful is serial pyrolysis which is performed in two or more steps. Initially, volatile compounds are thermally extracted from the sample at a relatively low temperature (thermal desorption), and then the same sample is pyrolyzed at a higher temperature.

Keywords: pyrolysis-GC/MS, thermal desorption-GC/MS, polymer, additive

1. Introduction

Analytical Pyrolysis technique hyphenated to GC/MS is used to obtain structural information of macromolecules by GC/MS analysis of their thermal degradation products in the absence of oxygen [1–4]. By Py-GC/MS the polymers are converted into lower molecular weight products by the action of heat. The relative proportions of the produced products depend on the composition of the samples, the temperature and the time it is applied. The composition and relative abundance of the pyrolysis products are characteristic for a given polymer and their determination allows the identification of materials that cannot be determined otherwise. This technique can also provide quantitative analysis of polymer structure, including monomer composition, stereochemistry, tacticity, and molecular arrangements in homo and copolymers [5].

Fourier transform infrared (FT-IR) spectroscopy, nuclear magnetic resonance (RMN) and thermal analysis techniques such as thermogravimetric analysis and differential scanning calorimetry (DSC) are relevant techniques for the analysis of polymers. Py-GC/MS is a complementary technique which provides useful information that allows a deeper characterization and identification of these polymers.

Real cases often deal with the characterization of unknown samples. In these cases, it can be challenging to select appropriate instrument set-up and method

IntechOpen

parameter starting points. However, based on the experience, there are some general considerations on how approach method development in the most efficient way.

Py-GC/MS of polymeric materials is a very powerful technique as it allows the identification of polymer components from a complex pyrolyzate mixture using a library search approach [6–21]. Besides, there are commercial pyrolysis libraries that can be a really useful help.

In more complex cases simple mass spectral identification via library (e.g. NIST, Wiley, etc.) comparison will be of limited use. A database approach is very helpful for automating the extraction of meaningful information simplifying the data evaluation task [22]. This approach consists basically of three steps: measuring reference substances to generate a database, measuring unknown or difficult samples and using the database for the identification of these samples.

Using chemometric methods can help extract useful information in very complex samples [23, 24].

Py-GC/MS has a lot of benefits [25, 26] that makes this technique a really powerful one: Pyrolysis of polymeric materials as a sample introduction technique for gas chromatography allows studying materials and compounds that are not suitable for traditional GC/MS analysis; it allows studying polymeric structures from pure systems to multi-block polymers; it requires minimal sample preparation and solvent is not required for most applications, meaning that low concentration monomers, residual solvents, additives and crosslinking agents can be identified without adding additional contaminants.

Most of the instruments allow different pyrolysis techniques [27, 28], such as stepwise pyrolysis (the sample temperature is raised stepwise, and the pyrolysis products are recorded between each step), fractional pyrolysis (A pyrolysis where the sample is pyrolyzed under different conditions in order to investigate different sample fractions), sequential pyrolysis (the same initial sample is repeatedly pyrolyzed under identical conditions: final pyrolysis temperature, temperature-rise time and total heating conditions), cold injection in split or narrow band mode (the pyrolysis products can be transferred to the GC/MS with a cold injection system) or thermochemolysis (thermally assisted hydrolysis and methylation (THM)) to improve the chromatographic analysis of polar functional groups, such as carboxylic acids.

Besides it is possible to work with polymer solutions, subsequently venting the solvent, which enables the highly precise introduction of small amounts of polymer in solution.

Another advantage of this technique is that it requires small amount of sample [25], in the range of micrograms to milligrams, typically with 300 μg of polymer a complete characterization is possible.

The technique has also some limitations [28]. Some of them are: to obtain reliable comparisons between different laboratories reproducible experimental conditions are required; inhomogeneous samples can have variable results; it does not detect most inorganic components; it is a destructive technique.

Quality control in production, product development and forensic science are some of the key areas where py-GC/MS is applicable.

In this chapter some case studies are shown as real examples from the industry to highlight the technique's potential.

2. Practical examples

Some real examples will be presented to show the potential of this technique. The work presented in this section was carried with a pyrolyzer from Gerstel (see

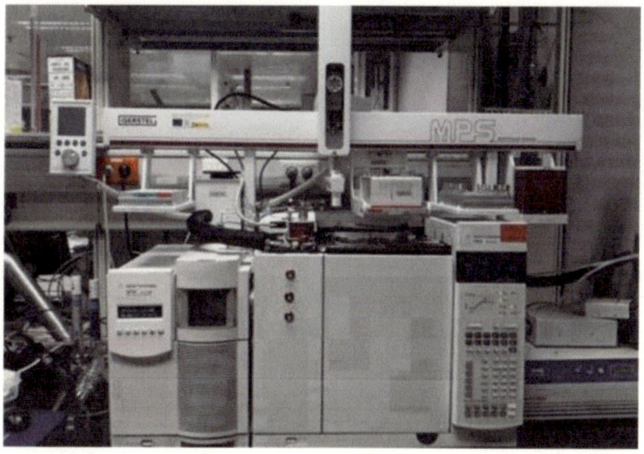

Figure 1.
GC/MS with a MultiPurpose sampler (MPS), thermal desorption unit (TDU) and pyrolysis module (PYRO).

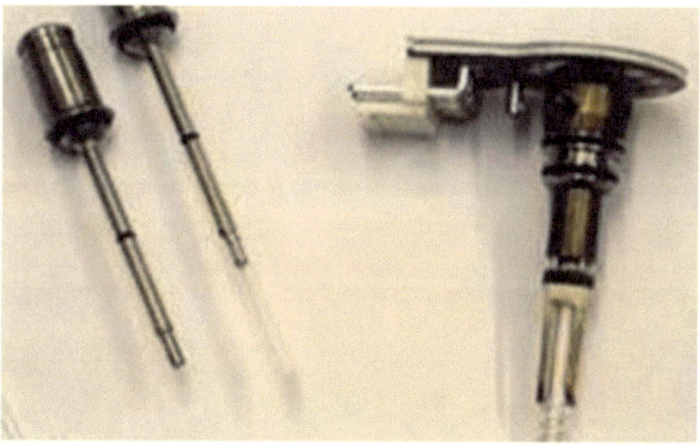

Figure 2.
Pyrolysis module with platinum filament, transport adapters and sample holders (Gerstel).

Figures 1 and 2), that can pyrolyze solids and liquids in a temperature range from 350°C up to 1000°C in a very flexible and automatic way [29], and determine thermal decomposition products in the GC/MS. All spectra were obtained under electron impact (EI) conditions. Ionization voltage: +70 eV. The MS was used in the full scan mode.

2.1 Chemical characterization of a synthetic foam by Py-GC/MS

The application of this technique to the analysis of polyurethanes offers information about the isocyanate and the polyester or polyol used in their formulation. Three reaction mechanisms have been proposed to account for the products arising from the thermal or pyrolytic degradation of the urethane portion of a poly(urethane) [30, 31]. The first one is the dissociation of the urethane linkage to form an isocyanate and an alcohol, the second is the formation of a primary amine, an olefin and carbon dioxide (CO_2) via a six-membered transition state and the third is the loss of CO_2 to produce a secondary amine. The three mechanisms are shown in **Figure 3**.

Methylene diphenyl diisocyanate (MDI) is a diisocyanate that exists in a pure or a polymeric form. The pure one is made up of three positional isomers, shown below (**Figure 4**):

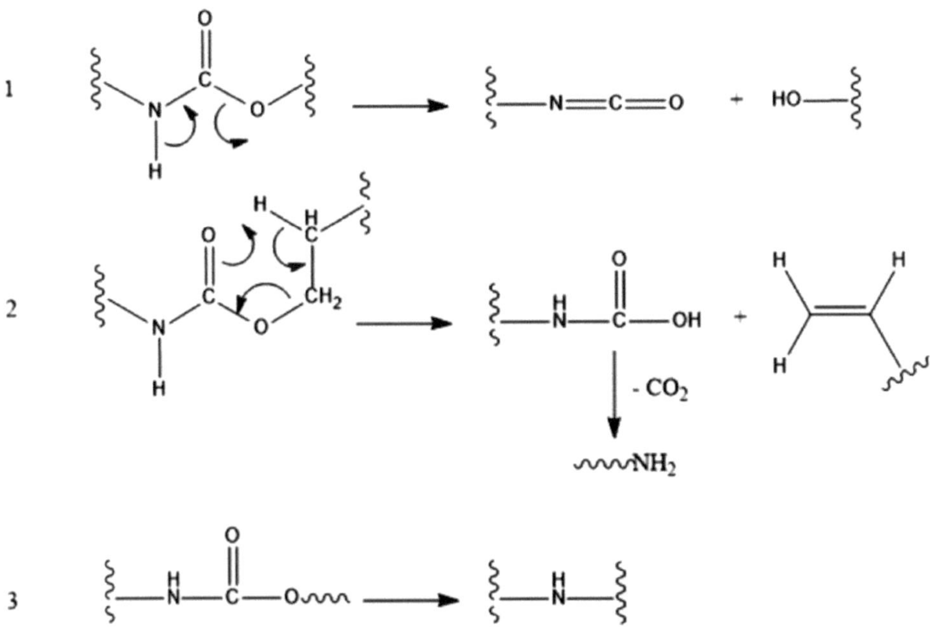

Figure 3.
Three reaction mechanisms proposed for the thermal or pyrolytic degradation of the urethane portion of a poly (urethane).

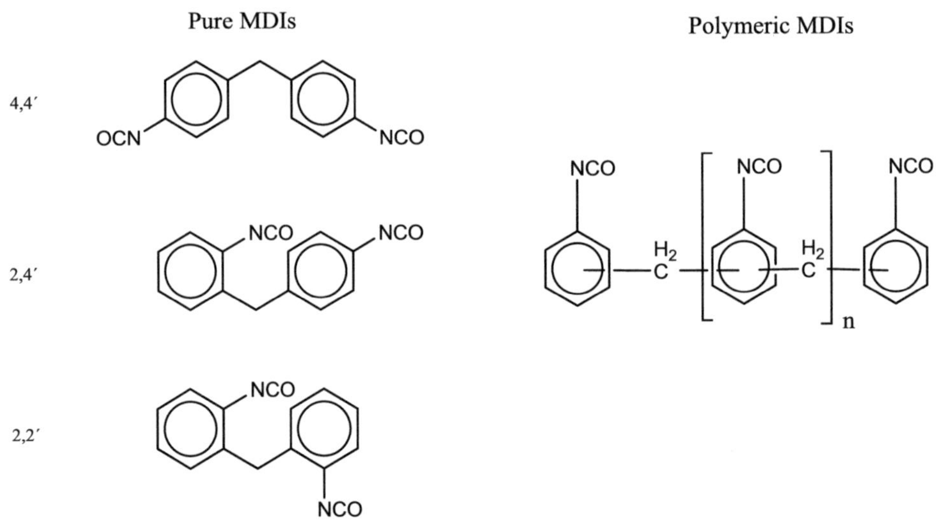

Figure 4.
Chemical structures of monomeric and polymeric MDI.

Figure 5 shows some possible products produced in the pyrolysis of a polyurethane made with MDI.

The foam was analyzed by py-GC/MS at a pyrolysis temperature of 600°C.

By studying the pyrogram (**Figure 6**) the following conclusions can be obtained: the sample is a polyurethane formulated from a polyether with styreneacrylonitrile (SAN) and MDI, and although none of the three isomers of MDI monomer was detected, its diamine, 4,4′-diaminodiphenylmethane (MDA), is present and also some other degradation products, such as p-tolyl-isocianate, and a phenyldiurea. The main identified products are shown in **Table 1**.

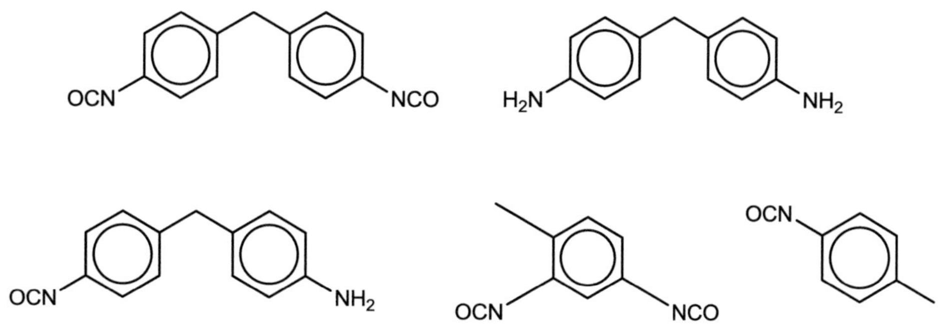

Figure 5.
Possible products structures.

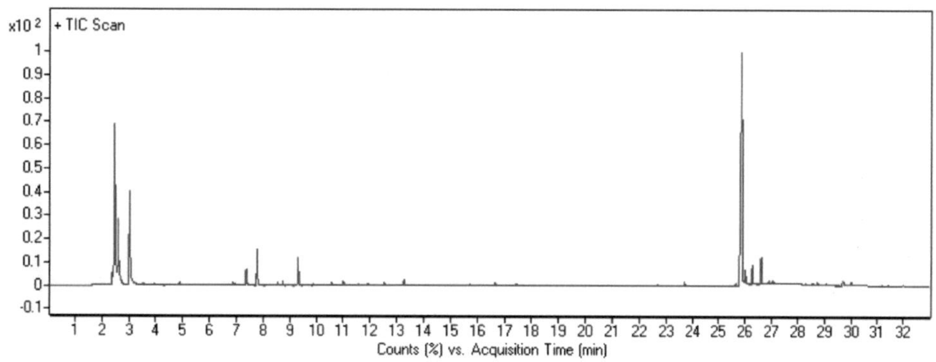

Figure 6.
Py-GC/MS of the PU foam.

RT (min)	Compound	Match
2,44	propylene	906
2,59	acetaldehyde	803
2,63	2-butene	683
3.00	acetone	870
3,35	acrylonitrile	877
7,34	diisopropyl ether	850
7,72	1-(1-methyletoxi)-2-propanone	928
9,27	styrene	969
12,50	p-tolyl-isocianate	847
23,68	palmitic acid	886
25,62	stearic acid	831
25,77	4,4′-diaminodiphenylmethane	963
25,97	1,1′-(2-methyl-1,4-phenylen)diurea (proposed compound)	

Table 1.
Identified products.

Some of the products detected were produced during pyrolysis mediated between the generated radicals, including monomers in the polymer matrix, as well as acrylonitrile and styrene.

Py-GC/MS provides very relevant information on the monomers used in the polyurethane formulation.

2.2 Chemical characterization of a paraffin inhibitor additive

A paraffin inhibitor additive was analyzed by py-GC/MS at a pyrolysis temperature of 600°C. The pyrogram is shown in **Figure 7**.

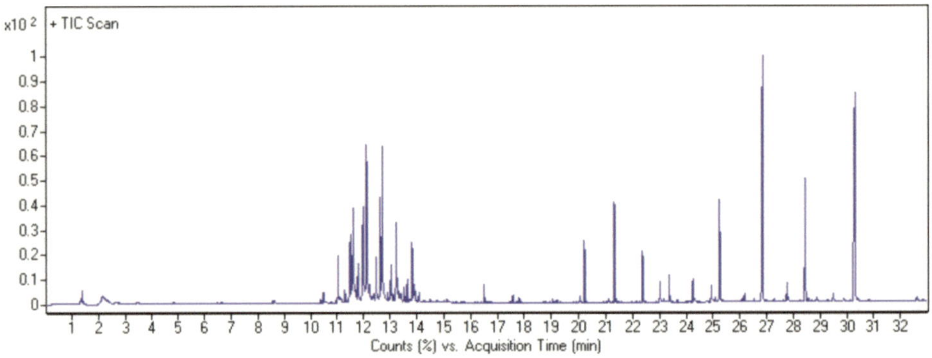

Figure 7.
Py-GC/MS of the paraffin inhibitor additive.

The following products were detected after the pyrolysis of the sample (See **Tables 2–4**):

Compounds	Area%
Aromatic fraction	57,4
Methacrylates	15,3
Acrylates	3,2
C3 and C4 olefin fraction	20,7
Methacrylic acid	0,6
Olefins	2,8

Table 2.
Relative composition expressed as area percentage.

Methacrylates fraction	RT (min)	Area%	Match
Octadecyl Methacrylate	27,47	47	933
Eicosanyl Methacrylate	28,42	11	861
Docosanyl Methacrylate	30,24	43	847

Table 3.
Relative composition of the methacrylate fraction expressed as area percentage.

Acrylates fraction	RT (min)	Area%	Match
Dodecyl Acrylate	20,20	61	931
Tetradecyl Acrylate	22,36	39	822

Table 4.
Relative composition of the acrylate fraction expressed as area percentage.

- Aromatic cut, in the range between the boiling points of benzene and methyl naphthalenes.
- Methacrylates of C16, C18, C20 and C22 alcohols (even with their olefins, product of their thermal degradation)
- Acrylates of C12 and C14 alcohols (even with their olefins, product of their thermal degradation)
- A light C3 and C4 hydrocarbons cut (olefins).
- Traces of methacrylic acid.

2.3 Chemical characterization of lubricity improvers additives

Samples were analyzed by py-GC/MS at a pyrolysis temperature of 600°C. The pyrograms are shown in **Figure 8**.

The pyrolysis of the second lubricity improver additive generates a complex mixture of products mainly with saturated and monounsaturated C16 and C18 hydrocarbon chains, they are generally quite branched, and they are presented in the following families of products, always in the form of a mixture of isomers:

- Olefins.
- Alkenyl-succinic anhydrides.

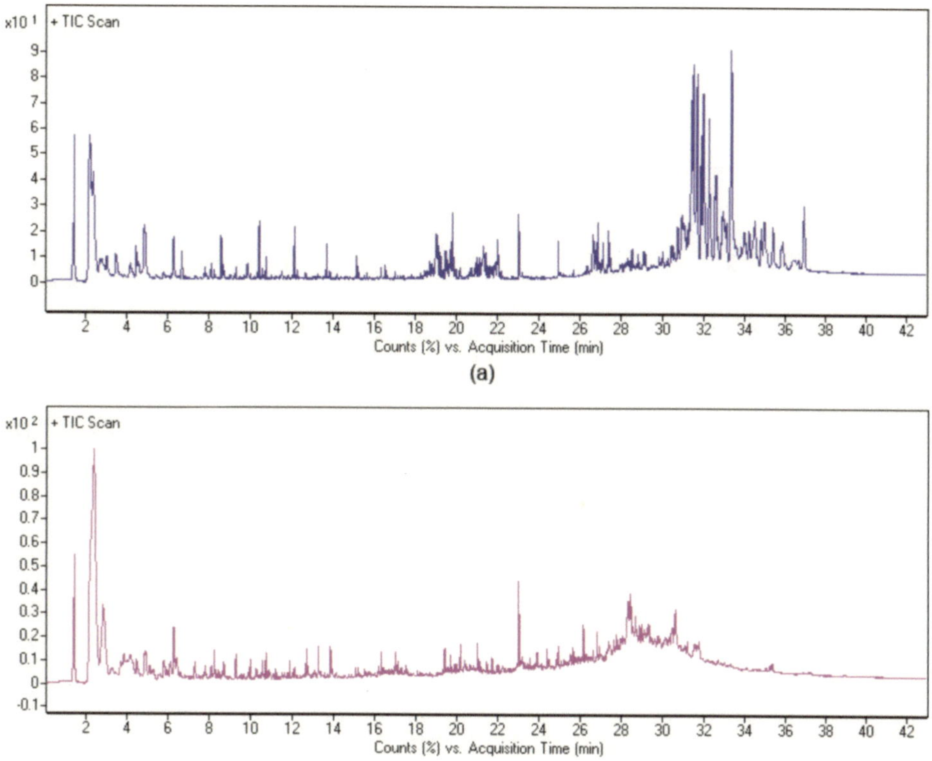

Figure 8.
a) Py-GC/MS of lubricity improver additive 1 and b) 2.

- Esters of palmitic and stearic acids.
- The rest of the detected products are:
    - o Mix of linear mono-olefins C6 to C13.
    - o Palmitic and stearic acids.
    - o Succinic anhydride.
    - o 2-hydroxyethanol acrylate.
    - o C4 fraction (butenes)
    - o Ethylene glycol.
    - o Traces of aromatic fraction (benzene-naphthalene)

The pyrolysis of the second lubricity improver additive produces species similar to that of the previous one, with the only difference in the length of the hydrocarbon chains, which in this case are mostly C14 and C16.

2.4 Chemical characterization of viscosity index improvers additives

Four viscosity index improvers additives were analyzed by Py-GC/MS at a pyrolysis temperature of 600°C. The pyrograms are shown in **Figures 9** and **10**.

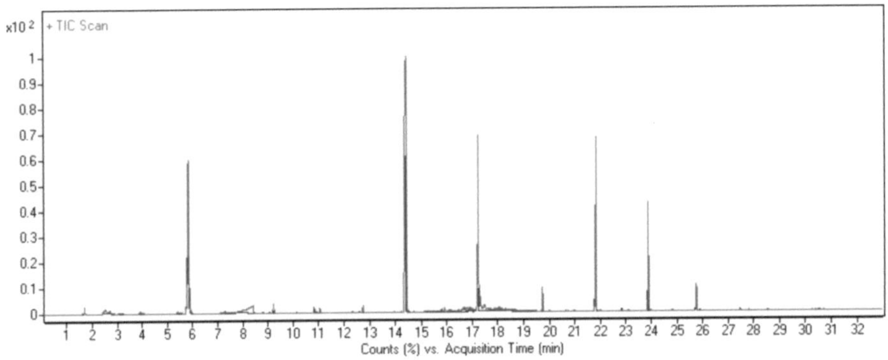

Figure 9.
Py-GC/MS of two of the viscosity index improvers additives.

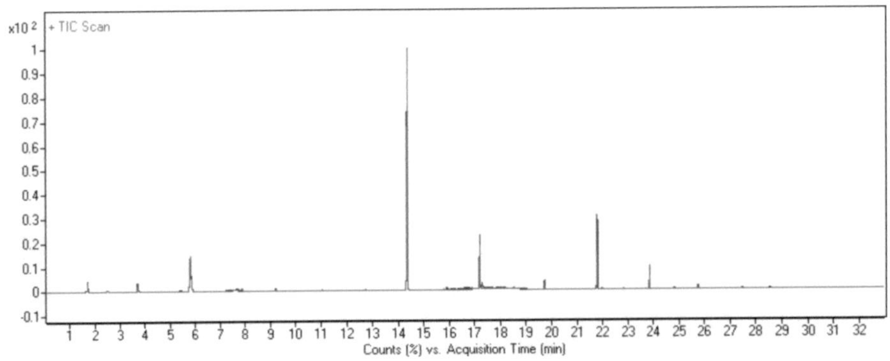

Figure 10.
Py-GC/MS of two of the viscosity index improvers additives.

The obtained chromatograms make it possible to identify the compounds from the pyrolysis of the polymeric fraction:

- Mono-olefins
- Methacrylates
- A mixture of hydrocarbons of mineral origin.

By normalizing the response of the detected alkyl-methacrylates (**Figure 11**), the comparative results that are shown in **Tables 5** and **6** are obtained.

Traces of methacrylic acid are also detected in all four samples.

Poly(methylmethacrylate) is formed by a polymerization reaction of methyl methacrylate.

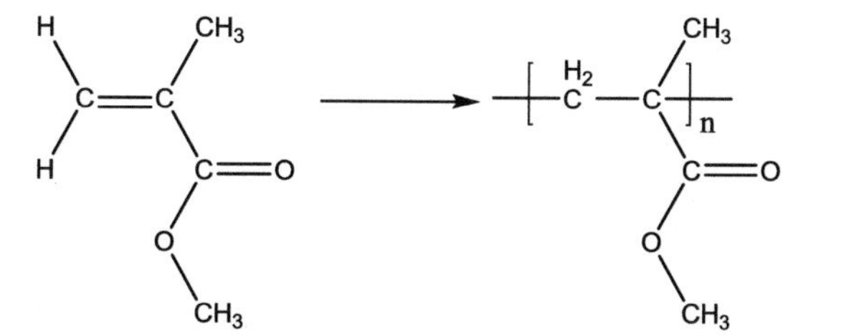

Figure 11.
Poly(methyl methacrylate) formation.

RT (min)	Compound	Additive 1	Match	Additive 2	Match
5,82	Methyl Methacrylate	43,40	907	51,72	890
10,83	Buthyl Methacrylate	1,51	925	0,31	702
20,65	Decyl Methacrylate	0,30	821	0,00	
21,81	Dodecyl Methacrylate	35,67	951	29,00	946
22,83	Tridecyl Methacrylate	0,52	851	0,00	838
23,86	Tetradecyl Methacrylate	14,67	894	15,39	900
24,80	Pentadecyl Methacrylate	0,40	869	0,09	806
25,73	Hexadecyl Methacrylate	2,97	905	3,32	881
27,47	Octadecyl Methacrylate	0,56	840	0,17	781
	Mineral Base (on the total)	6,90		3,37	

Table 5.
Relative composition expressed as area percentage.

RT (min)	Compound	Additive 3	Match	Additive 4	Match
5,82	Methyl Methacrylate	40,86	905	64,66	904
15,50	2-Ethylhexyl Methacrylate	1,34	787	0,00	
21,82	Dodecyl Methacrylate	42,31	947	25,16	943
23,86	Tetradecyl Methacrylate	13,08	894	8,54	887
25,73	Hexadecyl Methacrylate	2,41	899	1,64	854
	Mineral base (on the total)	16,77		17,88	

Table 6.
Relative composition expressed as area percentage.

The polymer fraction of the four samples can be described as mixtures of poly-methyl-methacrylate (PMMA) with poly-alkyl-methacrylates (PAMAs) in an approximate ratio 1:1, the latter mainly with C12, C14 and C16 radicals.

All samples show signals corresponding to hydrocarbon bases of mineral origin in the boiling point range between 1-dodecene and 1-octadecene.

Obtained data are based on:

Normalization of areas of the methacrylate fraction, for the distribution of the alkyl groups.

Signal of the mineral base by contribution of areas to the total ionic current diagram.

These data are not quantitative, so they can only be interpreted in comparative terms.

2.5 Determination of maleic anhydride (MAH) in polypropylenes and ethylene-butyl acrylate (EBA) copolymers to establish the relationship between free and grafted MAH in the polymeric matrix

Two samples of homopolymer polypropylene and three EBA samples were analyzed by TD-GC/MS and Py-GC/MS. These are two base resins that have been subjected to grafting processes with MAH at theoretical contents of around 1%. The samples have been analyzed with two different methodologies:

- Thermal desorption in two stages: 100°C and 320°C for each sample. Determination of the volatiles detected in the analysis at 100°C is carried out and the residual sample is subsequently subjected to a second thermal desorption at 320°C.

- Pyrolysis: samples are subjected to 340°C for a period of time, and then, the temperature of the pyrolyzer is increased to 800°C. The products are transferred directly to the chromatographic column without being cryogenically focused, thus avoiding that various fractions of the same compound emitted at different times during the heating of the pyrolyzer coexist in the liner.

With the first methodology the chromatograms that were obtained showed peaks corresponding to the presence of MAH. However, the sum of the two obtained signals is adjusted to MAH contents two orders of magnitude lower than the MAH used in the preparation of the samples, consequently, the MAH detected in the two cases must be assigned to the free MAH. So, the MAH that was detected in the analysis at 320°C cannot be consider as grafted MAH.

The tests carried out by pyrolysis (see **Figure 12**) show a single MAH signal per sample, this leads to think that the grafted MAH has not been detected as such. As it is part of the polymer chain, the MAH is incorporated by the double bond of its molecule, that is, as succinic anhydride units (**Figure 13**), pyrolysis does not make the grafting on

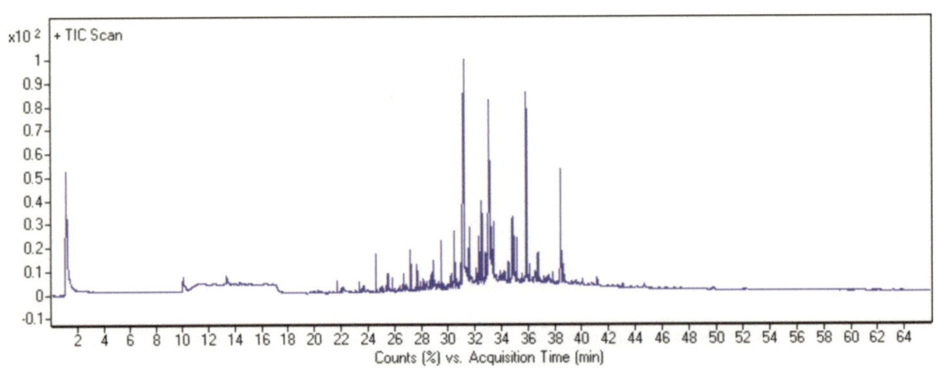

Figure 12.
Py-GC/MS (pyrolysis temperature of 850°C) of one of the grafted EBA samples.

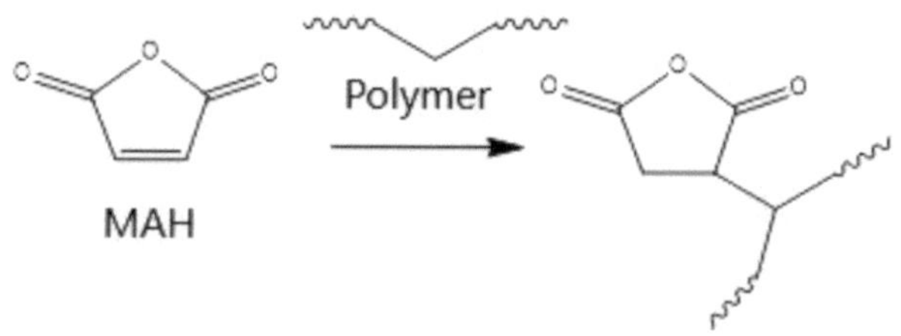

Figure 13.
Incorporation of MAH in the polymer.

the polymer reversible, derivatives of polymeric matrix and succinic anhydride are generated, which makes it impossible to directly estimate the fraction of MAH grafted.

Non-destructive spectroscopic (IR, NMR) techniques are recommended for this purpose.

2.6 Identification of deposits in filters of an Electrodeionization (EDI) unit for water purification in an industrial complex

In the filters of an EDI unit some deposits have been formed causing unit clogging.

The cause of the obstruction could be a phenolic epoxy paint that had been used recently. It was thought that over time some components of said paint dissolve in water and were the deposit formation cause.

A detailed analysis of the paint and the deposit was done in order to identify if the deposits formed correspond to any of its components and which component could be the cause.

The samples were analyzed by TD-GC/MS and Py-GC/MS.

Figures 14 and **15** show the thermal desorption (at a temperature of 350°C) and the subsequent pyrolysis (at a temperature of 650°C), respectively, of the paint sample.

With the analytical conditions that were tested, benzyl alcohol, phenol and phenolic derivatives were mainly detected. These compounds are compatible with a phenolic epoxy paint. See **Tables 7** and **8**.

2.7 Polyethylene and polypropylene artificial weathering

A method of analysis of Polyethylene, Polyethylene waxes and Polypropylene by thermal desorption and pyrolysis in series coupled to GC/MS was developed.

Optimal analysis conditions were sought for each matrix, testing different temperatures of thermal desorption and pyrolysis.

The objective is twofold. On the one hand, it is about obtaining information on the products that could be generated during a future biological and/or thermal

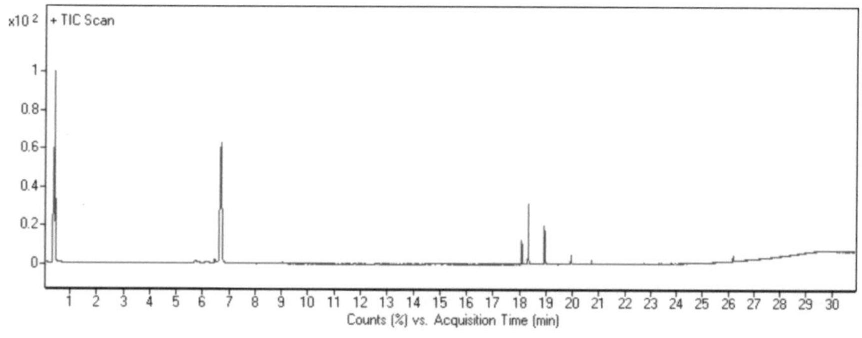

Figure 14.
DT-GC/MS of the paint sample (350°C).

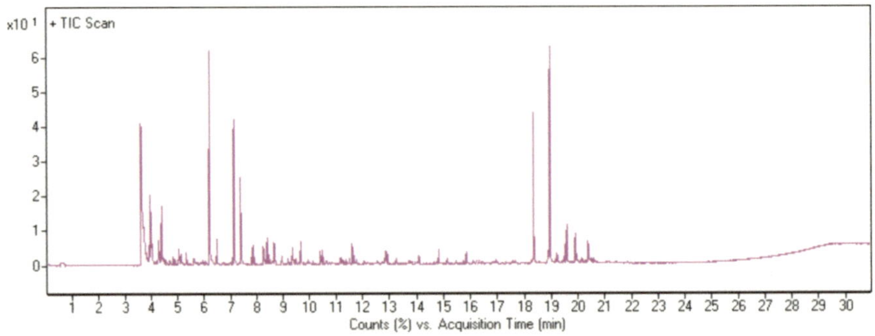

Figure 15.
Py-GC/MS of the paint sample (650°C).

RT(min)	Compound	Match	Area	Area%
5,74	Alkylbenzene C_9H_{12}	754	3,482,469	1,9
6,15	Alkylbenzene C_9H_{13}	803	2,707,299	1,5
6,69	Benzylic alcohol	943	133,976,378	73,2
18,09	Phenol, 2,2′-methylenebis-	910	6,308,961	3,4
18,33	Phenol, 2-[(4-hydroxyphenyl)methyl]-	929	15,308,848	8,4
18,94	Phenol, 4,4′-methylenebis-	892	11,162,787	6,1
19,95	Phenol, 4,4′-(1-methylethylidene)bis-	926	3,338,930	1,8

Table 7.
Identified compounds (DT-GC/MS) with area % > 1.

RT (min)	Compound	Match	Area	Area%
18,95	Phenol, 4,4′-methylenebis-	897	46,657,339	10,6
6,19	Phenol	929	38,255,006	8,7
3,64	C4 Olefin	744	29,689,766	6,7
18,34	Phenol, 2-[(4-hydroxyphenyl)methyl]-	922	28,302,868	6,4
3,62	Propane	677	26,265,730	6,0
7,13	Methylphenol	920	24,482,943	5,6
7,39	Methylphenol	917	18,073,779	4,1
3,97	Benzene	867	17,241,399	3,9
4,39	Toluene	876	13,477,989	3,1
3,74	Cyclopentadiene	710	11,580,548	2,6
19,58	4,4′-Ethylidenediphenol	731	7,498,180	1,7
4,27	Dimethylciclopentene	801	6,978,780	1,6
19,89	9,10-Anthracenedione, 1,4,4a,9a-tetrahydro-	704	6,006,814	1,4

Table 8.
Identified compounds (Py-GC/MS) with area % > 1.

degradation of polymers, and on the other hand, having information on the type of polymer and additives that they might contain.

Sample 1, sample 3 and sample 5 were a virgin polyethylene, a virgin polyethylenic wax and a virgin polypropylene sample, respectively. They were oxidized in an oven at 96°C for 28, 21 and 7 days, respectively. The corresponding oxidized samples were Samples 2, 4 and 6, respectively.

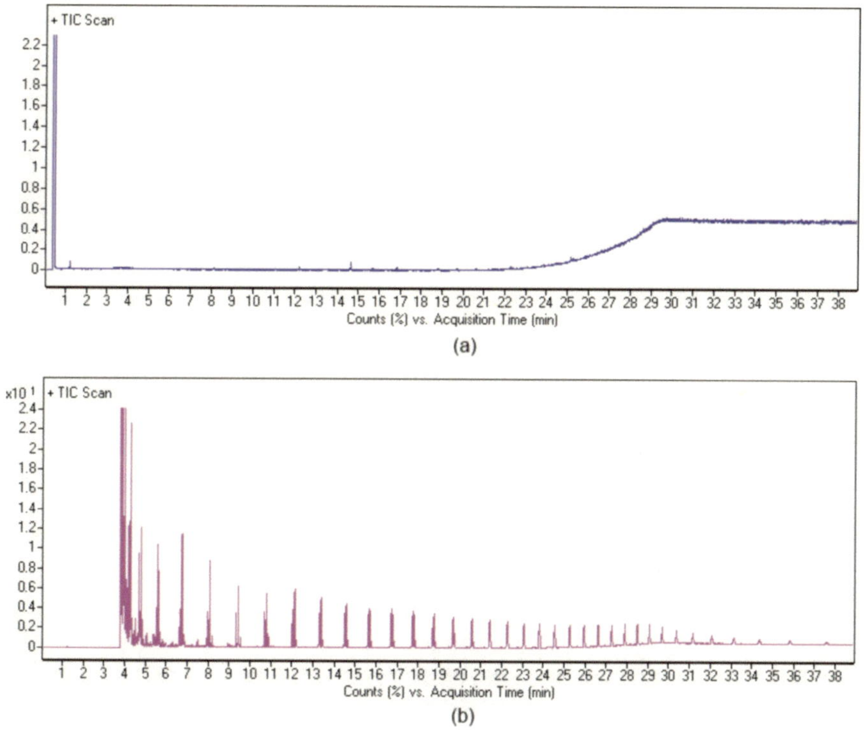

Figure 16.
a) DT- and b) Py-GC/MS of sample 1.

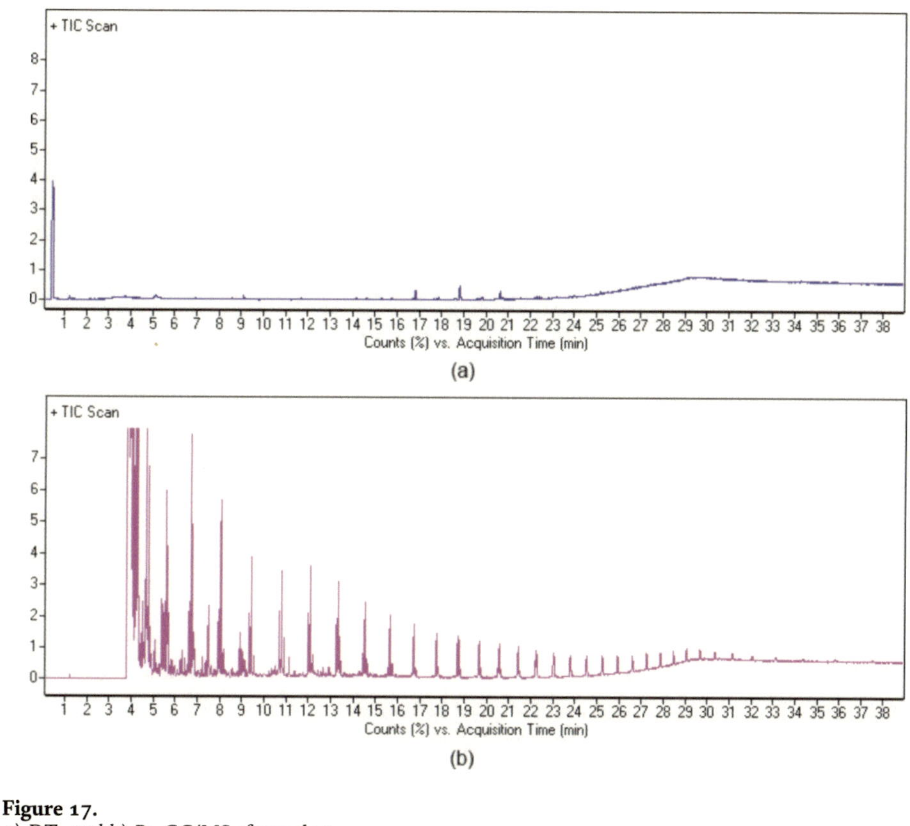

Figure 17.
a) DT- and b) Py-GC/MS of sample 2.

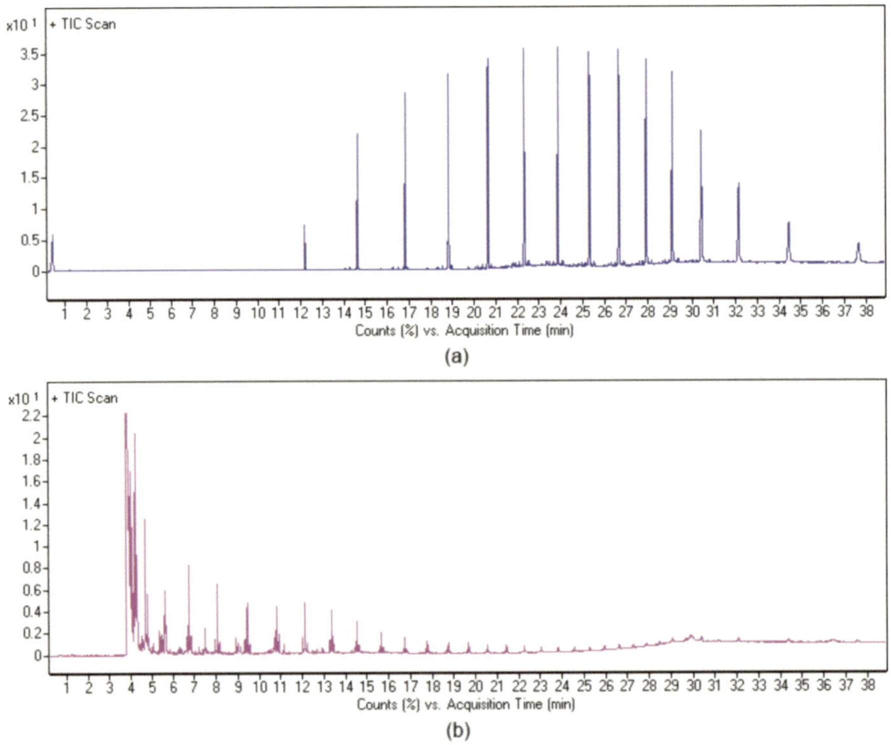

Figure 18.
a) DT- and b) Py-GC/MS of sample 3.

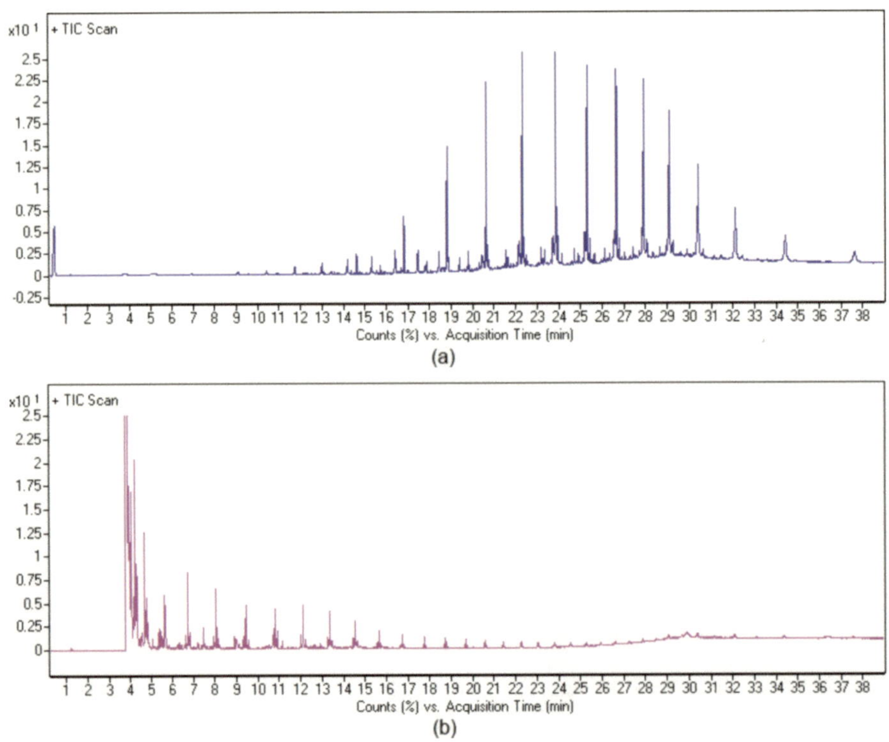

Figure 19.
a) DT- and b) Py-GC/MS of sample 4.

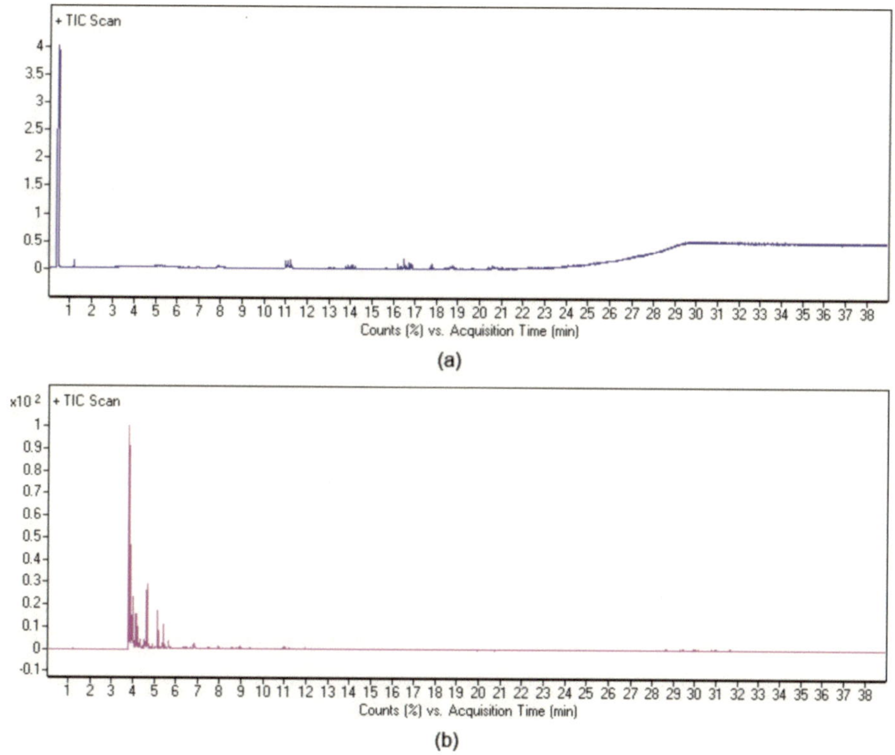

Figure 20.
a) DT- and b) Py-GC/MS of sample 5.

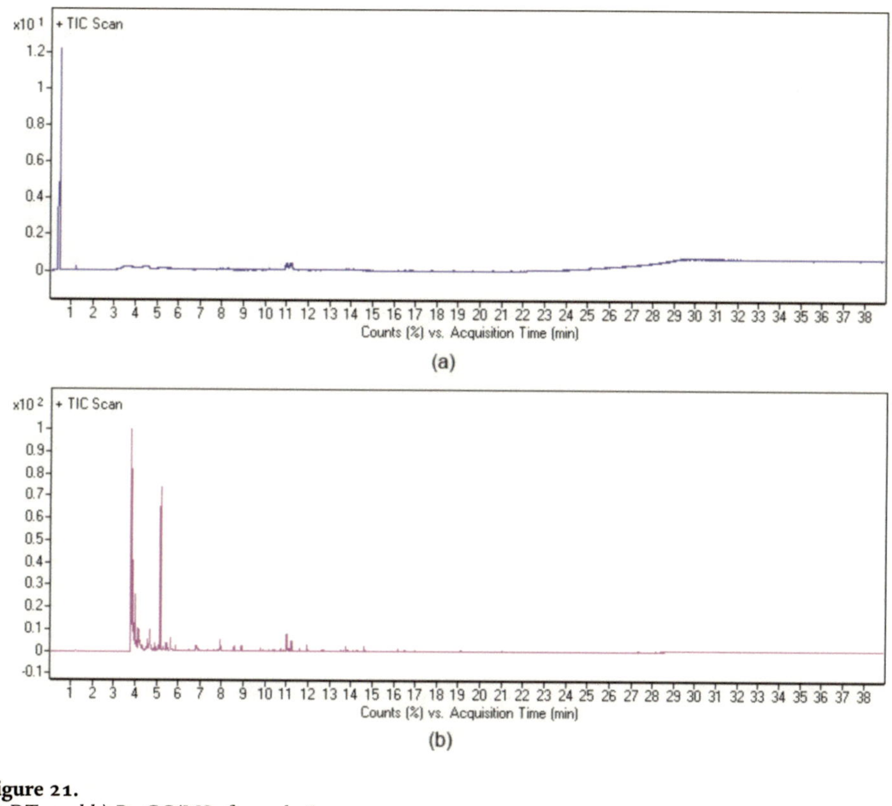

Figure 21.
a) DT- and b) Py-GC/MS of sample 6.

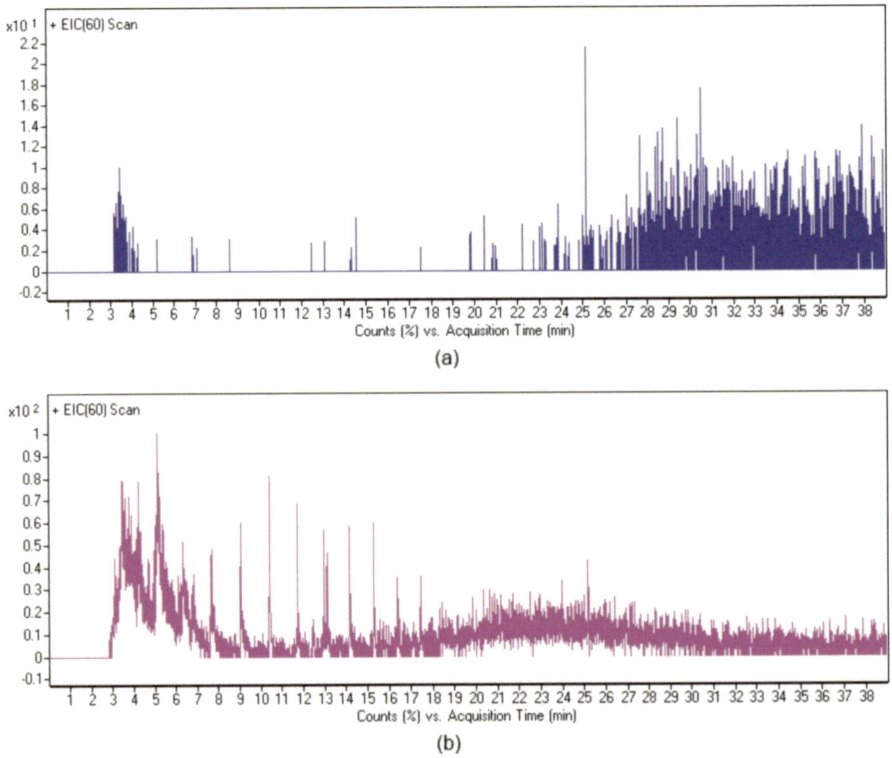

Figure 22.
Fragmentogram m/z 60 DT-GC/MS of the polyolefin samples a) 1 and b) 2.

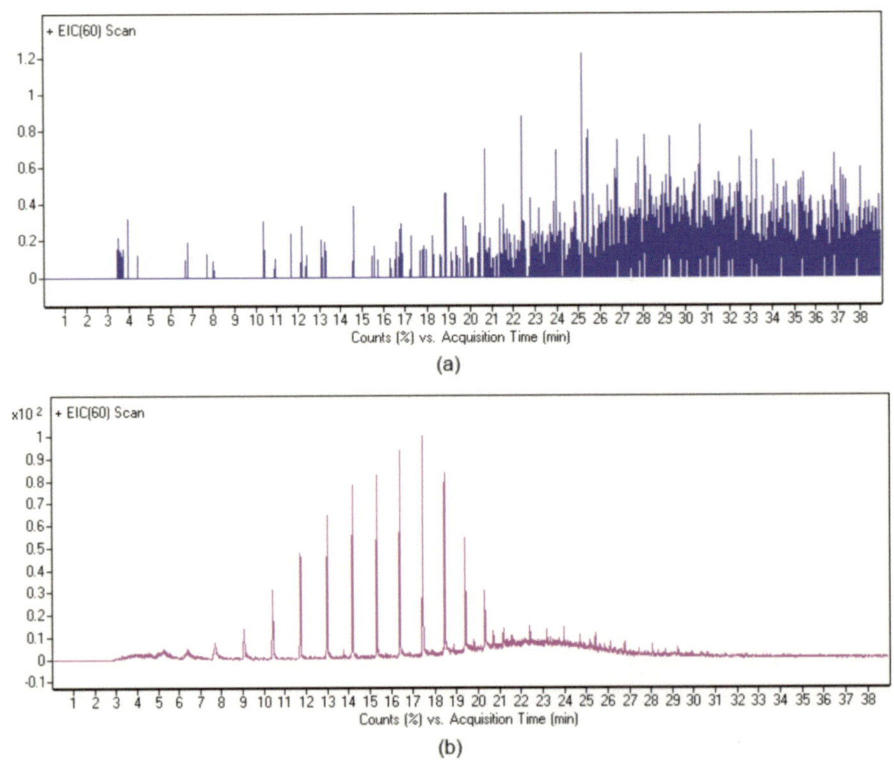

Figure 23.
Fragmentogram m/z 60 DT-GC/MS of the polyolefin samples a) 3 and b) 4.

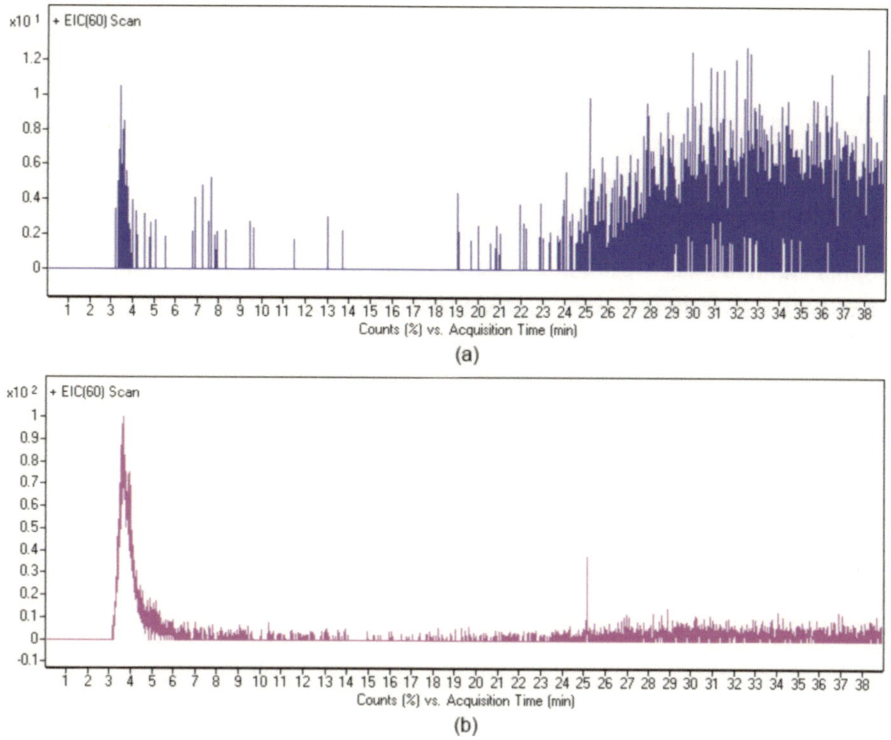

Figure 24.
Fragmentogram m/z 60 DT-GC/MS of the polyolefin samples a) 5 and b) 6.

The six samples were analyzed by TD-GC/MS and the residual of each sample is subjected to Py-GC/MS.

The following figures show the chromatograms that were obtained for the polyolefin samples.

The Polyethylenic wax and Polyethylene pyrograms consists of serial triplets, corresponding to α, ω-alkadienes, α-alkenes and n-alkanes, respectively, in the order of increasing n + 1 carbon number in the molecule. See **Figures 16b**, **17b**, **18b** and **19b**. Identification of compounds was carried out by comparison of mass spectra with data in the NIST mass spectral library.

The pyrograms of the Polypropylene samples show branched alkenes (C4 to C13) as major components. Isobutene, 2-methyl-2-butene, 3-methyl-1-pentene, dimethylpentadienes, 2,4-dimethyl-1-heptene are the most abundant branched alkenes. See **Figures 20b** and **21b**.

In the m/z 60 fragmentograms of the oxidized Polyethylene and Polyethylenic wax samples (See **Figures 22b** and **23b**) carboxylic acids are detected. In the m/z 60 fragmentogram of the oxidized Polypropylene no carboxilic acids are detected (See **Figure 24b**). In the TIC of this sample acetone is detected. However, in the pyrograms, no carboxylic acids, aldehydes or ketones are detected.

3. Conclusions

This chapter is intended for the use of both researchers and chemists who use GC/MS to analyze materials, and who need to add pyrolysis techniques to their analysis tools. Py-GC/MS is a practical, cost-effective, reliable, and flexible alternative for increasingly complex sample analyses, and can be used to analyze different kind of samples for a diversity of fields including industrial research, microbiology, forensic science, and environmental analysis.

The potential of Py-GC/MS was highlighted through different case studies, ranging from additives chemical characterization to weathered polyolefins samples.

Acknowledgements

Sonia Huertas de la Torre would like to thank Paco Martín Manzanero, who is now retired, José Luis Torres Barrera, Ángel Sánchez Fernández, and David Gómez Alda who have worked with her on the implementation of py-GC/MS technique in Repsol and in most of the work discussed in this chapter.

Conflict of interest

The author declares no conflict of interest.

Notes/thanks/other declarations

At Repsol Technology Lab we offer equipment and knowledge to apply Py-GC/MS to polymers and additives adapted to customer needs.

Author details

Sonia Huertas de la Torre
Repsol Technology Lab, Madrid, Spain

*Address all correspondence to: shuertasdlt@repsol.com

IntechOpen

© 2022 The Author(s). Licensee IntechOpen. This chapter is distributed under the terms of the Creative Commons Attribution License (http://creativecommons.org/licenses/by/3.0), which permits unrestricted use, distribution, and reproduction in any medium, provided the original work is properly cited.

References

[1] Wampler TP. Analytical pyrolysis: An overview. In: Applied Pyrolysis Handbook. 2nd ed. New York: Taylor Francis Group; 2007. p. 288

[2] Sobeih KL et al. Recent trends and developments in pyrolysis-gas chromatography. Journal of Chromatography A. 2008;**1186**:51-66

[3] White DM, Garland DS, Beyer L, Yoshikawa K. Applied Pyrolysis Handbook, 2nd ed. Florida, USA: Taylor & Francis, CRC Press; 2004

[4] Tienpont B, David F, Vanwalleghem F, Sandra P. Recent trends and developments in pyrolysis–gas chromatography. Journal of Chromatography A. 2001;**1186**(1–2): 51-66

[5] Bhatnagar MS. A Textbook of Polymer Chemistry Basic concepts, Processing and Applications, Condensation Polymers. New Delhi, India: Editor S. Chand Publishing; 2012

[6] Kusch P. Pyrolysis-Gas Chromatography/Mass Spectrometry of Polymeric Materials, Advanced Gas Chromatography-Progress in Agricultural, Biomedical and Industrial Applications. London, United Kingdom: IntechOpen; 2012. DOI: 10.5772/32323. Available from: https://www.intechopen.com/chapters/32828

[7] Hermabessiere, L. et al. Pyrolysis-GC/MS: A powerful tool for microplastics identification. In: Conference: ECsafeSEAFOOD Conference, 2017, Brussel (Belgium).

[8] Fischer M, Scholz-Bottcher BM. Simultaneous trace identification and quantification of common types of microplastics in environmental samples by pyrolysis-gas chromatography-mass spectrometry. Environmental Science & Technology. 2017;**51**:5052-5060

[9] Kusch P, Knupp G. Horizons in Polymer Research, Analysis of Synthetic Polymers and Copolymers by Pyrolysis–Gas Chromatography/Mass Spectrometry. New York, USA: Nova Science Publishers; 2007. pp. 141-191

[10] Wampler TP. Pyrolysis–capillary gas chromatography–mass spectrometry for the determination of polyvinyl chloride traces. Solid environmental samples. Journal of Chromatography A. 2007;**911**(2):235-247

[11] White DM. Pyrolysis–GC/MS fingerprinting of environmental samples. Journal of Analytical and Applied Pyrolysis. 2004;**71**(1):107-118

[12] Aguado J, Serrano DP, San Miguel G. Analysis of products generated from the thermal and catalytic degradation of pure and waste polyolefins using Py–GC/MS. Journal of Polymers and the Environment. 2007; **15**(2):107-118

[13] Campo J, Nierop KGJ, Cammeraat E, Andrei V, Rubio JL. Pyrolysis of phthalic acid esters: Their fate. The Science of the Total Environment. 2011;**36**(1):313-318

[14] Evans CJ, Evershed RP, Black HIJ, Ineson P. Application of pyrolysis–gas chromatography/mass spectrometry to study changes in the organic matter of macro- and microaggregates of a mediterranean soil uppon heating. Journal of Chromatography A. 2003; **1218**(30):4817-4827

[15] Kusch P. Application of the curie–point pyrolysis-high resolution gas chromatography for analysis of synthetic polymers and copolymers. Chemia Analityczna (Warsaw). 1996; **41**(2):241-252

[16] Languri GM, Boon JJ. Between myth and reality: mummy pigment from the

hafkenscheid collection. Studies in Conservation. 2005;**50**(3):161-178

[17] Gaston JW et al. Quantification of vehicle paint components containing polystyrene using pyrolysis-gas chromatography/mass spectrometry. Analytical Methods. 2012;**4**(7): 1989-1995

[18] Lee H, Lee D, Seo JM. Analysis of paint traces to determine the ship responsible for a collision Analysis of paint traces to determine the ship responsible for a collision. Scientific Reports. 2021;**11**(1):134

[19] Defeyt et al. Polyurethane coatings in twentieth century outdoor painted sculptures. Part II: comparative study of four systems by means of Py-GC/MS. Heritage Science. 2017;**5**:15

[20] Toapanta T et al. Influence of surface oxidation on the quantification of polypropylene microplastics by pyrolysis gas chromatography mass spectrometry. Science of The Total Environment. 2021;**796**:148835

[21] Ivleva NP et al. Chemical analysis of microplastics and nanoplastics: Challenges, advanced methods, and perspectives. Chemical Reviews. 2021; **121**(19):11886-11936

[22] Wenig P et al. Efficient analysis of Py-GC/MS data by a large scale automatic database approach: An illustration of white pitch identification in pulp and paper industry. Journal of Analytical and Applied Pyrolysis. 2010; **87**(1):85-92

[23] Reyes-Rivera J. Cheminformatics Applied to Analytical Pyrolysis of Lignocellulosic Materials. London, United Kingdom: Intechopen; 2021. DOI: 10.5772/intechopen.100147

[24] Montalbani S. Pyrolysis-gas chromatography mass spectrometry and chemometric analysis for the characterization of complex matrices [thesis]. Bologna, Italy: University of Bologna; 2012

[25] Brunow G, Lundquist K, Gellerstedt G. Lignin. In: Analytical methods in wood chemistry, pulping, and papermaking. Berlin, Heidelberg: Springer; 1999. pp. 77-124

[26] Kusch P. Introductory Chapter: Analytical Pyrolysis-Gas Chromatography/Mass Spectrometry of Polymeric Materials. London, United Kingdom: IntechOpen; 2018. DOI: 10.5772/intechopen.81596

[27] Fries E et al. Identification of polymer types and additives in marine microplastic particles using pyrolysis-GC/MS and scanning electron microscopy. Environmental Science: Processes & Impacts. 2013;**15**:1949-1956

[28] Shadkami F, Helleur RJ. Recent applications in analytical thermochemolysis. Journal of Analytical and Applied Pyrolysis. 2010;**89**(1):2-16

[29] Kleine-Benne, B. Rose B, Versatile automated pyrolysis GC combining a filament type pyrolyzer with a thermal desorption unit, GERSTEL Application Note 11/2011. Available from: www. gerstel.de/pdf/p-gc-an-2011-04. pdf.

[30] Dyer E, Newborn GEJ. Thermal degradation of carbamates of methylenebis (4-phenyl isocyanate). Journal of the American Chemical Society. 1958;**80**:5495-5498

[31] Hiltz JA. Analytical pyrolysis gas chromatography/mass spectrometry (py-GC/MS) of poly(ether urethane)s, poly(ether urea)s and poly(ether urethane-urea)s. Journal of Analytical and Applied Pyrolysis. 2015;**113**:248-258

Chapter 16

Structural, Morphological and Optical Properties of Perfume Atomizer Spray Pyrolysis CdO Thin Films: Effect of Solution Volume

Kasinathan Kasirajan, Karuppasamy Radhi Devi, Murugesan Rajini and Marimuthu Karunakaran

Abstract

CdO films have been deposited on glass substrates with different solution volume (30, 40 and 50 mL) at 200°C using perfume atomizers spray pyrolysis method. X-ray diffraction studies shows that the prepared thin film had cubic and polycrystalline nature. Scanning electron microscope shows the influence of solution volume on surface morphology of the CdO thin film. Optical studies show that in these films the electronic transition is of the direct transition type. The optical energy gap for the films of as deposited are vary from 2.12 to 2.00 eV with solution volume. Photoluminescence results analysis confirmed that the dependence of optical energy gap on solution volume. The Hall measurements were carried out and the results were discussed.

Keywords: Thin films, X-ray diffraction, Thickness, Transmittance, Photoluminescence

1. Introduction

Cadmium oxide (CdO) thin films are one of the transparent conducting oxide (TCO) materials and have been fascinated the investigators for their remarkable properties such as tunable band structure, high extinction coefficient, possible multiple exciton generation, electronic and transport properties [1–4]. CdO is a degenerate n-type semiconductor with a wide energy gap and high electrical conductivity [5]. Its high electrical conductivity is due to the increased grain size at high substrate temperature, moderate electron mobility, and high carrier concentration. The properties of CdO can be controlled to a suitable value to for the suit particular applications by controlling some of the process parameters viz. film thickness, the temperature of the substrate, pH value of the solution, Wt.% and solution flow rate etc. Though CdO is a first transparent conducting oxide film studied by Bedekar in 1907, not many investigations have been done for quite a long time. During last few years CdO is being extensively investigated in bulk and

IntechOpen

thin film form. CdO has interesting properties like large band gap, low electrical resistivity and high transmission in the visible region etc. These characteristics make it suitable for a variety of applications, including photodiodes, phototransistors, photovoltaic transparent electrodes, liquid crystal displays, IR detectors, and anti-reflection coatings [6–10].

For the last few decades various techniques such as chemical vapor deposition, sputtering, chemical bath deposition, sol–gel, SILAR, thermal evaporation, layer by layer assembly, spin coating, activated reactive evaporation, metal organic chemical vapor deposition (MOCVD), pulsed laser deposition and spray pyrolysis techniques are employed for the preparation of CdO thin film [11–29]. Among these routes, spray pyrolysis technique is a simple and low cost chemical method for the preparation of thin films with large area of coating. In this work the simplified perfume atomizer spray pyrolysis homemade setup is used for prepare a CdO thin films. The usage of a perfume atomizer offers a number of advantages over traditional spray gun assembly: low cost, no need for carrier gas, fine atomization, better wettability between sprayed micro particles, and nearly no loss of the precursor to the environment. We recently demonstrated that by effectively synthesizing CdO films, this simplified spray process can be a desirable alternative to the standard spray technique. The structure, morphology and optical behavior of the film depend on molarity, solution pH, substrate temperature, age of solution and solution volume. The precursor solution pH affects the hydrolysis and condensation behavior of the solution, which, in turn, influences the structure of the resultant film, morphological and optical behavior of the films. The role of the solution aging on the properties of thin films and interesting photo luminescent characteristics had been already reported [30–32]. In general, the properties of coated film depend on pH of the precursor solution, substrate temperature, molarity of the starting solution and solution volume. In this chapter, the effect of volume of solution on the structural, morphological and optical properties of CdO films prepared by perfume atomizer spray pyrolysis deposition method were investigated and reported in detail (Section 2).

2. Experimental

CdO films have been prepared by using cadmium acetate dihydrate as a source material of Cd and double distilled water was used as solvent. The glass substrates of 1.5 cm × 1.5 cm were used for the CdO deposition. The substrates were cleaned following a well-established methodology consisting on an initial washing step with soap and water followed by a washing procedure using acetone and isopropyl alcohol in an ultrasonically bath and a final rinsing step with distilled water [33]. For the implementation of the perfume atomizer spray pyrolysis method, homemade system was designed and implemented (**Figure 1**).

The system consists on a perfume atomizer, a power supply and a heating plate. The optimized deposition parameters such as substrate – spray nozzle distance (25 cm), spray angle (about 45°), spray time (5 s) and spray interval (30 s) were kept constant. The precursor solution was prepared at room temperature and kept on a container attached to the atomizer. Once the precursor solution was prepared, cleaned glass substrates were positioned onto the heating plate and heated until the temperature of the substrates reached to the desired value. The glass substrates were positioned at a horizontal separation distance of 25 cm from the perfume atomizer. The CdO thin films have been prepared by spraying a solution (30, 40 and 50 mL) composed of cadmium acetate dissolved in double distilled water and when the droplets of sprayed solution reached the preheated substrate, owing to the

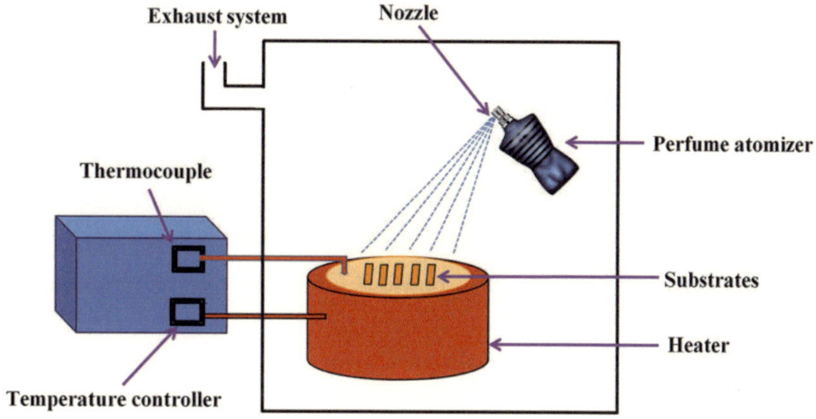

Figure 1.
Perfume atomizer spray setup.

pyrolytic decomposition of solution, a well adherent, pinhole free, uniform yellowish colored films of cadmium oxide are formed on the substrate surface according to the following reaction [34, 35].

$$Cd(CH_3COO)_2 + 3H_2O \rightarrow CdO\downarrow + CH_4\uparrow + 4H_2\uparrow + 3CO_2\uparrow \quad (1)$$

After the CdO deposition on the glass substrate, the samples were cooled down at room temperature and stored in dry conditions for further characterization. For the present study, the appropriate chemical and physical parameters were changed to study the effect of solution volume such as 30, 40 and 50 mL on structural, morphological and optical properties of CdO thin film.

3. Results and discussions

3.1 X-ray diffraction analysis

X-ray diffraction (XRD) is a versatile, non-destructive analytical technique for identification and quantitative determination of the various crystalline compounds, known as 'phases', present in solid materials and powders [36–38]. X-ray diffraction spectra were recorded using Philips X Pert PRO X-ray diffraction system (Cu $K\alpha$ radiation; K = 1.54056 Å). The average crystallite size (D) of the CdO film was calculated from Scherer's equation for the (111) plane [39].

$$D = \frac{0.9\lambda}{\beta\cos\theta} \quad (2)$$

Where, λ is the X-ray wavelength, β is the full-width at half-maximum of the peak and θ is the reflection angle. Dislocation density (δ) and strain (ε) for (111) plane was evaluated using the relations [40, 41].

$$\varepsilon = \frac{\beta\cot\theta}{4} \quad (3)$$

$$\delta = \frac{1}{D^2} \tag{4}$$

The number of crystallites per unit area is calculated from the following relation:

$$n_c = \frac{t}{D^3} \tag{5}$$

The lattice constant 'a' of the cubic phase of CdO films was determined by the following relations [42].

$$\frac{1}{d^2} = \left\{ \frac{h^2 + k^2 + l^2}{a^2} \right\} \tag{6}$$

where (hkl) and d have the same meaning as before. **Table 1** shows the estimated lattice constants, which are in good agreement with the conventional lattice constant (a) value.

Different sets of starting solutions were prepared by dissolving 0.2 M cadmium acetate with different volumes such as 30, 40 and 50 mL in doubly deionized water. The solutions thus got are sprayed manually by a perfume atomizer on preheated glass substrates at 200°C. The recorded XRD spectra of CdO thin film grown from different solvent volumes are shown in **Figure 2** and show sharp and narrow diffraction peaks at (111), (200), (220), (311) and (222) planes confirms that the films coated have good crystallinity and also it reveals that the films are polycrystalline in nature with cubic structure. It is also observed that all the films have preferential orientation along the (111) plane irrespective of the solvent volume. The intensity of the peaks (111), (200) and (220) increases as the solvent volume increases, showing the crystalline nature of the samples. The other prominent peaks are also obtained in the XRD pattern corresponds to the planes (200), (220), (311) and (222) according to the JCPDS Card No. 75-0591. The findings demonstrate that as the solution volume increased the intensity of the (111) peak is also increased significantly.

The calculated crystallite size and other crystal parameters are presented in **Table 1**. It is observed that the crystallite size increases with an increase in solution volume. This is because a low solution volume allows for the formation of a dense film structure with small crystals, whereas a high solution volume allows for a slower nucleation process, high agglomeration, and the formation of larger crystals. The deviation in the lattice parameter values of the films coated with solutions having solvent volumes 30 and 40 mL suggest that the films were under strain. The strain may be because of the oxygen vacancies and the interstitial Cd atoms incorporated in the CdO lattice. Film coated with 50 mL solvent volume has the minimum value of strain and dislocation density (δ) which strongly favors for the improved crystallinity of this film. The lesser value of δ obtained for the film coated with 50 mL solvent volume shows the degree of crystallization of the film.

Volume of the solution (mL)	Lattice constant 'a' (Å)	Film thickness (nm)	Crystallite size (nm)	Dislocation density ($\times 10^{15}$) lines.m^{-2}	Strain ($\times 10^{-3}$)	Number of crystallites ($\times 10^{16}$ m^{-2})
30	5.049	850	34	0.8800	3.59	2.2192
40	4.666	880	42	0.5633	2.88	1.1767
50	4.670	920	43	0.5633	2.88	1.2304

Table 1.
Structural parameters of CdO thin films at different solution volume.

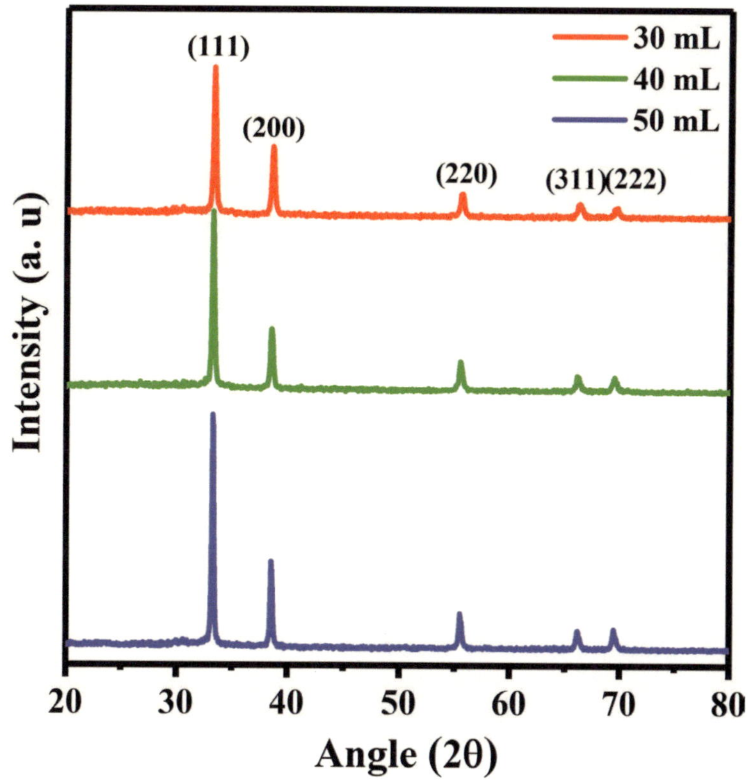

Figure 2.
XRD spectra of CdO thin film at different solution volume.

3.2 SEM and EDAX analysis

Scanning electron microscopy (SEM) is a widely used tool for surface analysis of thin films. SEM, accompanied by X-ray analysis, is considered a relatively rapid, inexpensive and basically nondestructive approach for surface analysis [43]. High resolution images of surface topography, with excellent depth of field are produced using a highly-focused, scanning (primary) electron beam [43–46]. Elemental composition of a specimen can be analyzed using energy dispersive X-ray (EDAX) technique. Other than preliminary insight peak of the examined sample, EDAX also offers quick and efficient characterization technique that acquires elemental information of the prepared samples at high resolutions [47, 48].

Figure 3(a–c) shows the SEM image of CdO thin film deposited with different solution volumes such as 30, 40 and 50 mL. The surfaces of the coated films are and densely packed and homogeneous. It is seen that there are well defined grains and it confirms the crystalline nature of the coated film. From the microscope, as the solution volume increases small sized grains are fused together and begin to for a patch of grain for 40 mL solution volume. The SEM image for 50 mL solution volume reveals that the spherically shaped grains with different size are found scattered throughout the substrate surface uniformly as reported earlier [49]. This study confirms that as solution volume increases, grain size increases, and these findings are well consistent with XRD tests. **Figure 3(d)** show the EDAX spectra and it confirms the presence of expected element in the prepared CdO films. It is found that the weight percentage of Cd and O atoms are 77.99 and 26.01% respectively.

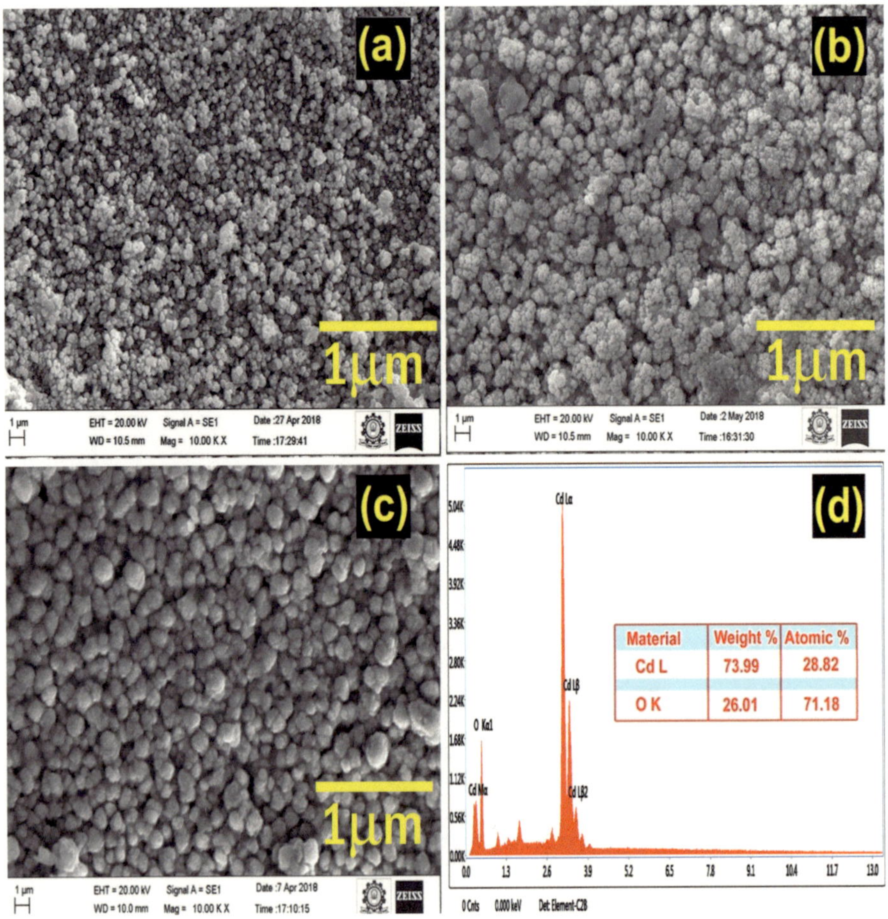

Figure 3.
SEM image of CdO thin film: (a) 30 mL, (b) 40 mL, (c) 50 mL, and (d) EDAX spectrum.

3.3 UV–Visible spectroscopic analysis

The optical transmittance spectrum in the wavelength region 300–800 nm was recorded using UV–Vis spectrometer. **Figure 4** shows the optical transmittance spectra of the CdO film deposited at different solvent volume. The variations in the solvent volume affect evidently the optical properties. Film coated with 30 mL solvent volume show a high transmittance of 68%. The transmittance decreases for the films with increase in solvent volume from 40 to 50 mL. The high thickness of the films and the optical scattering at the increased number of grain boundaries cause the decrease of transmittance. Usually in transparent metal oxides, metal or oxygen ratio in the film decides the percentage of optical transmittance. Metal rich film usually shows less transparency [50]. Also the optical transmittance is known to strongly depend on the grain size of the films. A smaller grain size marks to the high transmittance for the films coated with less solvent volume. When the grain size increases, the grain boundary scattering reduces, which causes a decrement in the transmittance. The sharp absorption edges obtained in the spectrum for all volumes clearly show the good crystallinity of the films.

The optical band gap (E_g) of the films was estimated from the transmittance data where the photon energy ($h\upsilon$) and absorption coefficient (α) are related by the equation [51].

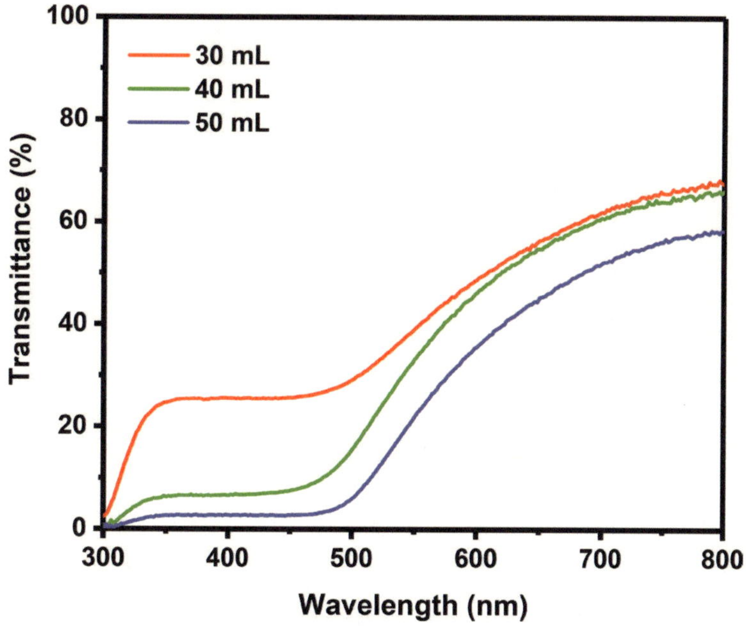

Figure 4.
Optical transmittance spectra of CdO thin film at different volume of solution.

$$\alpha h\nu = B\left(h\upsilon - E_g\right)^n \tag{7}$$

For allowed direct transition, the value of n = 1/2 and B is constant. Tauc graph is plotted between $(\alpha h\upsilon)^2$ against $h\upsilon$ to calculate the band gap of prepared CdO films. The fundamental absorption which clearly shows itself by a rapid fall in the transmittance can be used to find out the band gap of the as deposited samples to decide whether the CdO film deposited using the simplified spray technique have direct band gap $(\alpha h\upsilon)^2$ vs. $(h\upsilon)$ plot are drawn. Since better linearity is obtained in the $(\alpha h\upsilon)^2$ vs. $(h\upsilon)$ plot, the direct band gap values are determined by extrapolating the linear portion of this plot to the energy axis shown in **Figure 5**. The optical band gap (E_g) is found to be in the range of 2.12–2.00 eV for the coated films. As the solvent volume increases, the optical band gap is red shifted from 2.12 to 2.00 eV. Thus increased solvent volume reduces the optical band gap. This decreasing band gap is due Moss–Burstein effect. The Moss–Burstein effect is associated with lifting the Fermi level into the conduction band of degenerate semiconductors, which leads to decrease the band gap.

Furthermore, using the following relationships, we were able to determine the refractive index (n) and optical extinction coefficient (k) of cadmium oxide thin films [52, 53].

$$n = \frac{(1+R)}{(1-R)} + \sqrt{\frac{4R}{(1-R)^2} - k^2} \tag{8}$$

$$k = \frac{\alpha\lambda}{4\pi} \tag{9}$$

where each sign has its own meaning. In general, film transmittance is inversely proportional to refractive index (n) and extinction coefficient (k) at the same

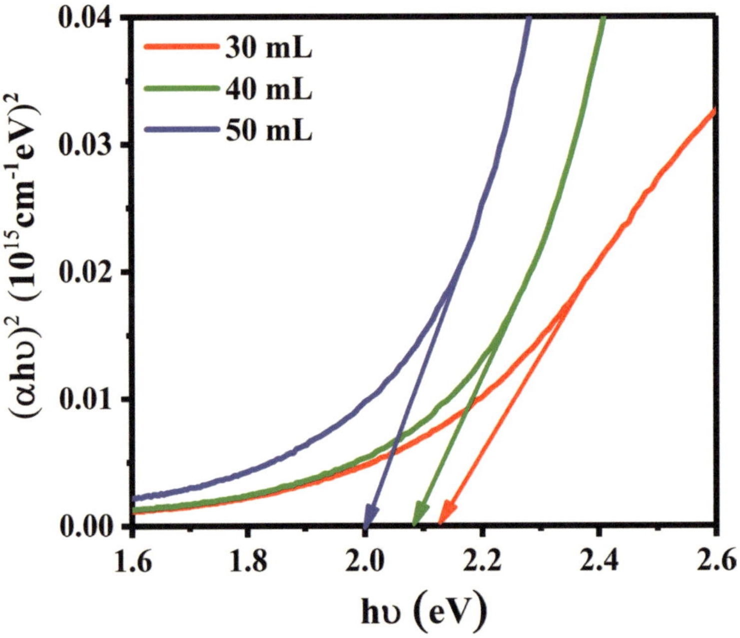

Figure 5.
Tauc plot of the CdO thin film at different volume of solution.

wavelength. Similarly, in both (n) and (k) values, we have seen an increase in parallel variation. **Table 2** shows the predicted (n) and (k) values. This data value clearly shows that both (k) and (n) values are low in all wavelength ranges, indicating that all films have good transparency and smooth surface roughness. When the solvent volume was increased from 30 mL to 50 mL, the refractive index and extinction coefficient both increased. This is due to a decrease in the optical energy gap.

Dielectric nature of CdO films were also quantified using the following equation [54].

$$\varepsilon_r = n^2 - k^2 \tag{10}$$

$$\varepsilon_i = 2nk \tag{11}$$

3.4 Photoluminescence spectroscopic analysis

PL spectroscopy is used to determine the optical quality of CdO semiconductor films and the common radiative transition in the semiconductor occurs between the states at the bottom of the conduction band and the top of the valence band in the prepared CdO thin film. Room temperature luminescence spectrum was recorded using (SHIMADZU-5301) spectrofluorometer. The PL spectra of different solution volume produced CdO thin films obtained at room temperature under the excitation of 325 nm are shown in **Figure 6**. The peak (ultraviolet light emission) about 360 nm was caused by free exciton recombination. The blue emission due to the passage of electrons from oxygen vacancies to the valence band is attributed to the peak centred at 458 nm. The blue emission peak at 494 nm is generally attributed to the radiative recombination of a photo-generated hole with an electron occupying the oxygen vacancy. This peak is attributed to the excitonic transitions which are size-dependent and excitation wavelength independent of

Volume of the solution (mL)	Transmittance (%)	Band gap (eV)	Refractive index (n)	Extinction coefficient (k)	Dielectric constant (ε_r)	Dielectric constant (ε_i)
30	68	2.12	2.130	0.08	4.56	0.32
40	66	2.08	2.134	0.06	4.55	0.56
50	58	2.00	2.136	0.05	4.55	0.81

Table 2.
Optical and dielectrical properties of CdO thin films.

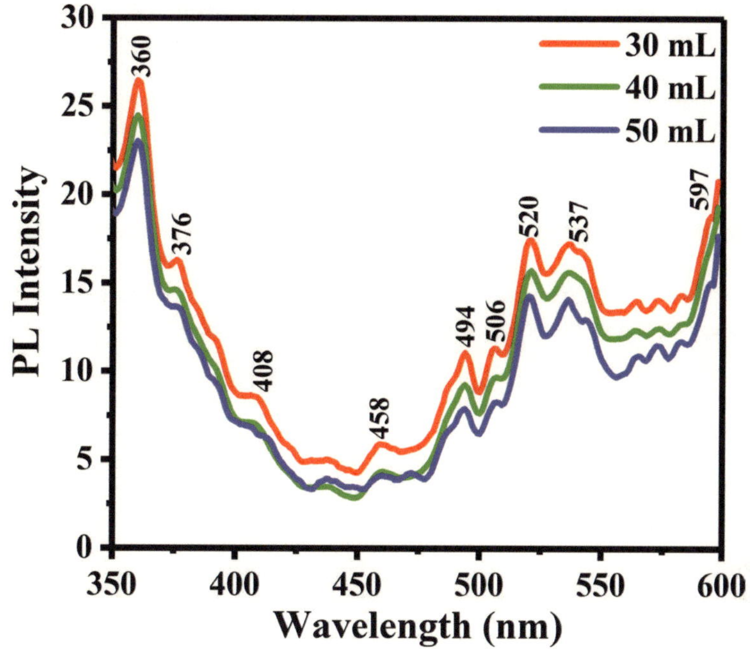

Figure 6.
Photoluminescence spectra of CdO thin film at different solution volume.

the certain wavelength range [55]. The peak at 506 nm arises from the combination of holes from the valence band and electrons from the conduction band. This peak might also be attributed to the electronic transitions from cadmium interstitial and oxygen vacancies to the top of the valence band [56]. The green emission peak at 520 nm might have originated from the electronic transitions of ionized oxygen vacancies from the deep level donor to the valence band [57]. The low intensity of this peak indicates a low concentration of surface defects in the prepared CdO thin film. The green emission peak at 595 nm might have originated from the oxygen vacancies on the surface of CdO. The intensity of the principal emission peaks likewise decreases as the solution volume increases, which could be due to the creation of more Cd and O vacancies as well as an increase in charge carrier recombination.

3.5 Electrical studies

Hall Effect measurements were carried out in a van der Pauw configuration (Ecopia HMS-3000) at room temperature. The negative sign of the Hall coefficient confirms that the conduction is n-type. The electrical parameters in the CdO films were clearly influenced by the solution volume. By increasing the volume of precursor solution, the carrier concentration (n) was increased. For 50 mL of solution, the resistivity (ρ) ($0.552 \times 10^{-3}\ \Omega$ cm) is lower and the carrier concentration ($6.165 \times 10^{-3}\ \Omega$ cm) is higher. Resistivity (ρ) for different solution volumes is 0.835×10^{-3}, 0.665×10^{-3}, and 0.552×10^{3} cm, carrier concentration (n) is 3.547×10^{20}, 4.448×10^{20}, and 5.369×10^{20} cm^{-3}, and carrier mobility (μ) is 21.11, 17.51, and 16.51, cm^{2} V^{-1} s^{-1}.

4. Conclusion

Cadmium oxide thin films have been deposited by perfume atomizer spray pyrolysis technique on glass substrate at different solution volume. In this study, the effect of volume of solution on structure, morphology and optical properties of prepared CdO thin film were reported. X-Ray diffraction studies shows that the preferred growth orientation along (111) plane with cubic crystal structure for all solution volume. It is evident that the microstructural parameters vary with volume of solution and may be useful to enhance the properties of CdO thin film for various applications. An average optical transmittance about 55% in the wavelength range 480–800 nm, was obtained. The Tauc graphs were plotted and revealed that the band gap of the CdO thin films can be altered by changing the appropriate solution volume. Intense PL emission peaks were observed at different wavelengths confirms that the presence of oxygen vacancies, defect centre and recombination of electron hole.

Conflict of interest

The authors declare no conflict of interest.

Author details

Kasinathan Kasirajan[1], Karuppasamy Radhi Devi[2], Murugesan Rajini[1] and Marimuthu Karunakaran[1]*

1 Thin Film and Nano Science Research Lab, Department of Physics, Alagappa Government Arts College, Karaikudi, Tamilnadu, India

2 Department of Physics, Sethupathy Government Arts College, Ramanathapuram, Tamilnadu, India

*Address all correspondence to: tvdkaruna@gmail.com

IntechOpen

© 2021 The Author(s). Licensee IntechOpen. This chapter is distributed under the terms of the Creative Commons Attribution License (http://creativecommons.org/licenses/by/3.0), which permits unrestricted use, distribution, and reproduction in any medium, provided the original work is properly cited.

References

[1] Chopra KL, Major S, Pandya DK. Transparent conductors—A status review. Thin Solid Films. 1983;**102**(1): 1-46

[2] Dawar AL, Joshi JC. Semiconducting transparent thin films: Their properties and applications. Journal of Materials Science. 1984;**19**(1):1-23

[3] Hartnagel HI, Dawar AX, Jain AK, Jagadish C. Chapter 2: Semiconducting Transparent Thin Films. Philadelphia: Institute of Physics; 1995 p. 22

[4] Gupta RK, Ghosh K, Patel R, Mishra SR, Kahol PK. Highly conducting and transparent tin-doped CdO thin films for optoelectronic applications. Materials Letters. 2008;**62**(25):4103-4105

[5] Murthy LCS, Rao KK. Thickness dependent electrical properties of CdO thin films prepared by spray pyrolysis method. Bulletin of Materials Science. 1999;**22**(6):953-957

[6] Saha B, Das S, Chattopadhyay KK. Electrical and optical properties of Al doped cadmium oxide thin films deposited by radio frequency magnetron sputtering. Solar Energy Materials and Solar Cells. 2007;**91**(18):1692-1697

[7] Su LM, Grote N, Schmitt F. Diffused planar InP bipolar transistor with a cadmium oxide film emitter. Electronics Letters. 1984;**20**(18):716-717

[8] Balu AR, Nagarethinam VS, Suganya M, Arunkumar N, Selvan G. Effect of solution concentration on the structural, optical and electrical properties of SILAR deposited CdO thin films. J. Electron Devices. 2012;**12**:739-749

[9] Dakhel, A. A. Influence of hydrogenation on the electrical and optical properties of CdO thin films. Semiconductor Science and Technology 2008;23(5):055017.

[10] Holland L. Vaccum Deposition of Thin Films. London: Chapman and Hall Ltd.; 1966

[11] Ueda N, Maeda H, Hosono H, Kawazoe H. Band-gap widening of CdO thin films. Journal of Applied Physics. 1998;**84**(11):6174-6177

[12] Vigil O, Cruz F, Morales-Acevedo A, Contreras-Puente G, Vaillant L, Santana G. Structural and optical properties of annealed CdO thin films prepared by spray pyrolysis. Materials Chemistry and Physics. 2001;**68**(1-3): 249-252

[13] Zhao Z, Morel DL, Ferekides CS. Electrical and optical properties of tin-doped CdO films deposited by atmospheric metalorganic chemical vapor deposition. Thin Solid Films. 2002;**413**(1-2):203-211

[14] Dakhel AA, Henari FZ. Optical characterization of thermally evaporated thin CdO films. Crystal Research and Technology: Journal of Experimental and Industrial Crystallography. 2003;**38**(11):979-985

[15] Seo DJ. Structural and optical properties of CdO films deposited by spray pyrolysis. Journal of the Korean Physical Society. 2004;**45**(6):1575

[16] Bhosale CH, Kambale AV, Kokate AV, Rajpure KY. Structural, optical and electrical properties of chemically sprayed CdO thin films. Materials Science and Engineering: B. 2005;**122**(1):67-71

[17] Saha B, Thapa R, Chattopadhyay KK. Bandgap widening in highly conducting CdO thin film by Ti incorporation through radio frequency magnetron sputtering

technique. Solid State Communications. 2008;**145**(1-2):33-37

[18] Dhawale DS, More AM, Latthe SS, Rajpure KY, Lokhande CD. Room temperature synthesis and characterization of CdO nanowires by chemical bath deposition (CBD) method. Applied Surface Science. 2008;**254**(11):3269-3273

[19] Kose S, Atay F, Bilgin V, Akyuz I. Int. J. Hyd. Energy. 2009;**34**:5260

[20] Yıldırım MA, Ateş A. Structural, optical and electrical properties of CdO/Cd $(OH)_2$ thin films grown by the SILAR method. Sensors and Actuators A: Physical. 2009;**155**(2):272-277

[21] Mishra RL, Sharma AK. S. G. Prakash. Dig. J. Nanomater. Biostruct. 2009;**4**:511

[22] Yakuphanoglu F. Nanocluster n-CdO thin film by sol–gel for solar cell applications. Applied Surface Science. 2010;**257**(5):1413-1419

[23] Lamb DA, Irvine SJC. A temperature dependant crystal orientation transition of cadmium oxide films deposited by metal organic chemical vapour deposition. Journal of Crystal Growth. 2011;**332**(1):17-20

[24] Carballeda-Galicia DM, Castanedo-Perez R, Jimenez-Sandoval O, Jimenez-Sandoval S, Torres-Delgado G, Zuniga-Romero CI. High transmittance CdO thin films obtained by the sol-gel method. Thin Solid Films. 2000;**371**(1-2): 105-108

[25] Sravani C, Reddy KR, Reddy PJ. Physical behaviour of CdO films prepared by activated reactive evaporation. Semiconductor Science and Technology. 1991;**6**(10):1036

[26] Ellis DM, Irvine SJC. MOCVD of highly conductive CdO thin films. Journal of Materials Science: Materials in Electronics. 2004;**15**(6):369-372

[27] Gupta RK, Ghosh K, Patel R, Kahol PK. Highly conducting and transparent Ti-doped CdO films by pulsed laser deposition. Applied Surface Science. 2009;**255**(12):6252-6255

[28] Caglar, M., and Yakuphanoglu, F. Fabrication and electrical characterization of flower-like CdO/ p-Si heterojunction diode. Journal of Physics D: Applied Physics 2009;42(4):045102.

[29] Henriguez, R., Grez, P., Munoz, E., Lincot, D., Dalchiele, E. A., Marotti, R., and Gomez, H. One-step pentiodynamic synthesis of polycrystalline cadmium oxide (CdO) thin films in DMSO solution. Sci. Technol. Adv. Mater 2008;9:025016.

[30] Rozati SM, Moradi S, Golshahi S, Martins R, Fortunato E. Electrical, structural and optical properties of fluorine-doped zinc oxide thin films: Effect of the solution aging time. Thin Solid Films. 2009;**518**(4):1279-1282

[31] Kumar PR, Kartha CS, Vijayakumar KP, Abe T, Kashiwaba Y, Singh F, et al. On the properties of indium doped ZnO thin films. Semiconductor Science and Technology. 2004;**20**(2):120

[32] Biswal RR, Velumani S, Babu BJ, Maldonado A, Tirado-Guerra S, Castaneda L, et al. Fluorine doped zinc oxide thin films deposited by chemical spray, starting from zinc pentanedionate and hydrofluoric acid: Effect of the aging time of the solution. Materials Science and Engineering: B. 2010;**174**(1-3):46-49

[33] Oliva AI, Solıs-Canto O, Castro-Rodrıguez R, Quintana P. Formation of the band gap energy on CdS thin films growth by two different techniques. Thin Solid Films. 2001;**391**(1):28-35

[34] Seo DJ. Structural and optical properties of CdO films deposited by spray pyrolysis. Journal of the Korean Physical Society. 2004;**45**(6):1575

[35] Uplane MD, Kshirsagar PN, Lokhande BJ, Bhosale CH. Characteristic analysis of spray deposited cadmium oxide thin films. Materials Chemistry and Physics. 2000;**64**(1):75-78

[36] Buerger MJ. X-Ray Crystallography. New York: Wiley Publication; 1942

[37] Klug HP, Alexander LE. X-Ray Diffraction Procedures. New York: Wiley Publication; 1954

[38] Cullity BD. Elements of X-Ray Diffraction. Massachusetts: Addison Wesley Publication; 1956

[39] Williamson GK, Smallman RE III. Dislocation densities in some annealed and cold-worked metals from measurements on the X-ray Debye-Scherrer spectrum. Philos. Mag. 1956;**1**:34-46. DOI: 10.1080/14786435608238074

[40] S. Maheswari, M. Karunakaran, K. Hariprasad, K. Kasirajan, L. Bruno Chandrasekar, T. Alshahrani, Mohd Shkir, and S. AlFaify. Enhanced room-temperature ammonia vapor-sensing activity of nebulizer spray pyrolysis fabricated SnO_2 thin films: An effect of Er doping. Journal of Materials Research 2021;36:657-667. DOI:10.1557/s43578-020-00033-0

[41] Thomas R, Mathavan T, Shkir M, AlFaify S, Kim HS, Kathalingam A. Influence of yttrium doping on microstructural and optical properties of FTO thin films prepared by nebulizer spray technique. Materials Today Communications. 2020;**24**:101087

[42] K. Kasirajan, A. Nancy Anna Anasthasiya, Omar M. Aldossary, Mohd Ubaidullah, M. Karunakaran. Enhanced room-temperature ammonia vapor-sensing activity of nebulizer spray pyrolysis fabricated SnO_2 thin films: An effect of Er doping. Sensors and Actuators 2021;A319:112531.

[43] Goldstein IJ et al. Scanning Electron Microscopy and X-Ray Microanalysis, 3rd Ed. New York: Plenum Press; 2003

[44] D. McMullan Oatley and K. C. A. Smith. The Development of the Scanning Electron Microscope: The Beginnings of Electron Microscopy. In: P. W. Hawkes, editor. Advances in Electronics and Electron Physics Suppl. 16. London: Academic Press; 1985. p. 443.

[45] S. Banerjee. Proc. of National Workshop on Advanced Methods for Materials Characterization (NWMMC-2004) Mat. Res. Soc., India; 2004.

[46] Reed SJB. Electron Microprobe Analysis, 2nd Ed. Cambridge: Cambridge University Press; 1993

[47] B. E. Stern and D. Lewis. Pitman Pub. Corp.; 1971. p. 205.

[48] Russ CJ. Fundamentals of Energy Dispersive X-Ray Analysis. London: Butterworths; 1984

[49] Shanmugavel G, Balu AR, Nagarethinam VS. Preparation of cadmium oxide thin films by spray technique using perfume atomizer and effect of solvent volume on their physical properties. International Journal of Chemistry and Materials Research. 2014;**2**(9):88-101

[50] Radhi Devi K, Selvan G, Karunakaran M, Rajesh Kanna G, Kasirajan K, Annalakshmi V. IJRSR. 2018;**9**(1):23606-23609

[51] Maheswari S, Karunakaran M, Annalakshmi V, Kasirajan K. Effect of gadolinium on tin oxide thin films prepared by nebulizer spray pyrolysis

method. Results in Materials. 2020;**8**:100130

[52] Arif M, Shkir M, AlFaify S, Sanger A, Vilarinho PM, Singh A. Linear and nonlinear optical investigations of N: ZnO/ITO thin films system for opto-electronic functions. Optics and Laser Technology. 2019;**112**:539-547

[53] A. Jesu Jebathew, M. Karunakaran, K. Deva Arun Kumar, S. Valanarasu, V. Ganesh, Mohd Shkir, S. AlFaify, A. Kathalingam. Effect of novel Nd3+ doping on physical properties of nebulizer spray pyrolysis fabricated ZnS thin films for optoelectronic technology. Physica B: Condensed Matter 2019;572:109-116.

[54] A. Jesu Jebathew, M. Karunakarana, K. Deva Arun Kumar, S. Valanarasu, V. Ganesh, Mohd Shkir, I. S. Yahia, H. Y. Zahran, A. Kathalingam. An effect of Gd3+ doping on core properties of ZnS thin films prepared by nebulizer spray pyrolysis (NSP) method. Physica B: Condensed Matter 2019;574:411674.

[55] W. Dong, C. Zhu. Opt. Mater. 2003;22(3):227-233.

[56] Pan Z, Zhang P, Tian X, Cheng G, Xie Y, Zhang H, et al. Properties of fluorine and tin co-doped ZnO thin films deposited by sol–gel method. Journal of Alloys and Compounds. 2013;**576**:31-37

[57] Krunks M, Dedova T, Kärber E, Mikli V, Acik IO, Grossberg M, et al. Growth and electrical properties of ZnO nanorod arrays prepared by chemical spray pyrolysis. Physica B: Condensed Matter. 2009;**404**(22):4422-4425

Chapter 17

Recent Perspectives in Biochar Production, Characterization and Applications

Asfaw Gezae Daful, Meegalla R. Chandraratne and Marie Loridon

Abstract

This chapter presents the most promising features and applications of biochar along with their optimal pyrolysis conditions. Biochars have a range of physicochemical properties depending on the feedstock and pyrolysis conditions, which greatly affect their wide applications. The biochar production and its characteristics, including the effect of feedstocks and different process-parameters on the properties and yield of biochar are thoroughly examined. The higher pyrolysis-temperature can give higher carbon-contents, pH, and surface-areas of biochars while volatiles and molar-ratios of O/C, H/C and N/C decrease with pyrolysis-temperature. Higher carbon-content and neutral-pH biochars have high affinity for organic pollutants due to high surface areas, making them attractive for adsorption and catalysis purposes. Biochars with higher-pH are preferred for soil application to correct soil-acidity. Thus, the pyrolysis temperature should be selected as per the final application of the biochar. Characterization of biochars of different feedstocks and pyrolysis conditions is reviewed and presented along with their proximate and ultimate analysis.

Keywords: biochar, pyrolysis, biomass, characterization, proximate analysis, ultimate analysis

1. Introduction

Biochar is a porous carbonaceous solid material produced by the thermal decomposition of biomass from plant or animal waste under oxygen-free or limited oxygen conditions [1–3]. The International Biochar Initiative [4] defines biochar as "a solid material obtained from the thermochemical conversion of biomass in an oxygen-limited environment". Biochars have a wide range of physicochemical properties [5], which greatly affect their potential applications in various agronomic and industrial sectors. The feedstock and the method of biochar production have a significant impact on the biochar characteristics, such as concentrations of elemental constituents, density, porosity, and pH, which collectively impact the suitability of the biochar for various applications [3]. Biochar is used to upgrade the soil quality [6, 7] in agricultural areas. It slows down the rate of decomposition and release of nutrients from the soil and hence, enhances the soil quality. In various industries, biochar is used in waste treatment [8–10] to remove organic contaminants and

 IntechOpen

heavy metals. Biochar can be used as a fuel in power generation because it contains a high carbon percentage in it.

Biomass is a very potential source of renewable energy [11] materials and chemicals [12, 13]. Agricultural residues, algal biomass, forest residues, manures, activated sludge, energy crops, digestate, etc. are the main sources of biomass [14] to be used as a raw material. Biomass can be converted into high-value products using various physical, thermochemical and biochemical processes. Thermochemical conversion processes, such as pyrolysis, gasification, torrefaction, and hydrothermal carbonization of carbonaceous biomass are used for biochar is production, at high temperatures ranging from 300–900°C and under O_2-free conditions [15]. The physicochemical and mechanical properties of biochars depend on the pyrolysis operating conditions and feedstock used [16]. The selection of a suitable kind of feedstock is usually determined by the availability of that material in areas where the biochar is likely to be produced, as this reduces the cost of transport while decreasing the carbon footprint of the biochar technology. Biochar production from the biomass depends upon the thermochemical process used and process parameters considered. The literature on the biomass pyrolysis revealed that the production of the biochar depends upon several factors such as type of biomass, moisture content, and particle size, reaction conditions (reaction temperature, reaction time, heating rate) and surrounding environment (carrier gas type, flow-rate of carrier gas) and other factors (catalyst, reactor type) [3, 6, 8, 11, 16].

The main objective of this chapter is to show the potential use of waste biomass for the production of biochar which is an important material with numerous industrial and environmental applications. The biochar characterization is presented to understand the physical and chemical properties of biochar, including variations of the biochar properties as a function of production conditions and feedstocks, and to evaluate the applicability of biochar in desired fields. This chapter exhaustively describes the possible feedstocks for biochar production, biochar production processes particulary slow pyrolysis process, and pyrolysis process conditions, the properties of biochar such as biochar characterization, proximate and ultimate analysis of biochars. Some important industrial and environmental biochar applications are discussed and finally conclusion is presented.

2. Raw materials for biochar production

The main sources of raw material for biochar production include: municipal waste, agricultural and forest residues, energy crops, and animal waste, which are grouped as lignocellulosic and non-lignocellulosic biomasses. Lignocellulosic biomass [17–19] are abundant fibrous plant parts, non-food ‘second generation’ feedstock, including agro-industrial residues, forest-industrial residues, energy crops, municipal solid waste, etc. Chemically, biomass is a complex composition of carbon, hydrogen, oxygen, sulfur, nitrogen, and small quantities of few other elements which include alkali metals, alkaline earth metals, and heavy metals, depending upon the species or type of biomass. The proportion of these elements in the biomass is a function of species of biomass, growing condition, and geographical situation of the region [20]. Lignocellulosic biomass is mainly composed of cellulose (38–50%), hemicelluloses (23–32%), lignin (15–25%) and small amounts of extractives [3, 16, 21, 22]. Among these components, cellulose and hemicelluloses are linear and chain polysaccharides respectively, while lignin is a cross-linked phenolic polymer. Biomass with varying contents of hemicellulose, cellulose, and lignin may yield biochars with distinctive physicochemical properties. The abundant biomass reserves and its renewability have been the main driving forces for research and utilization of biomass. Thus, such agricultural and animal waste disposal can be reduced and converted into value-added products such as biochars using pyrolysis processes. A review by Li and Jiang presented on non-lignocellulosic biomass characteristics, thermochemical behaviors of main components (e.g., C, O, N, P, and metals), characterization methods, conversion process, and the main applications of non-lignocellulosic biochar [23]. Song and Guo studied the quality variations of poultry litter biochar generated at different pyrolysis temperatures [24]. Unlike lignocellulosic biomass, the non-lignocellulosic biomass has a greater threat to the ecological environment because of its higher contents of heavy metals and *hetero-atom* like nitrogen, phosphorus, sulfur [25], which may dissolve in a water systems, leading to water pollution and accumulation in the food chains [26].

Poultry litter (PL), a solid waste resulting from chicken rearing, is being explored as a feedstock for biochar production and examined the effect of pyrolysis temperature on the quality PL biochar and identify the optimal pyrolysis temperature for converting PL to agricultural-use biochar [24]. Physically, PL is a mixture of bedding materials (e.g., wood shavings, sawdust, and peanut hull), bird excreta, feather, feed spills, and chemical treatments like alum and sodium bisulphate. Through pyrolysis, PL can be readily transformed into biochar [24].

3. Biochar production processes

Biomass-derived biochar production is formed via a complex process, but the reaction mechanism of biomass pyrolysis can be described as occurring mainly through three general steps, as depicted in Eq. (1):

$$\begin{gathered}\text{Biomass} \overset{heat}{\rightarrow} \text{Moisture} + \text{Dryresidues}\\ \text{Dryresidues} \overset{heat}{\rightarrow} \text{Volatile\&Gases} + \text{Pre} - \text{biochar}\\ \text{Pre} - \text{biochar} \overset{heat}{\rightarrow} \text{Volatile\&Gases} + \text{Biochar}\end{gathered} \tag{1}$$

The first step is the removal of available moisture from the biomass, which becomes dry feedstock by heating. Then pre-biochar and volatile compounds are formed. In the last step, chemical compounds in the pre-biochar rearrange and form a carbon-rich solid product known as biochar. Major thermochemical technologies for biochar production include pyrolysis, gasification, torrefaction, and hydrothermal carbonization. Pyrolysis is one of the thermochemical technologies for converting biomass into energy and chemical products consisting of liquid bio-oil, solid biochar, and pyrolytic gas [3, 11, 20, 27–29]. Depending on the heating rate, pyrolysis temperature, and residence time, biomass pyrolysis can be divided into slow, intermediate, fast and flash pyrolysis mainly aiming at maximizing either the bio-oil or biochar yields. Operating conditions of various pyrolysis processes and product distribution (biochar, bio-oil, and gas). Thus, biochar yield greatly depends on the type of pyrolysis used. Slow pyrolysis conducted at longer residence time and at a moderate temperature (350–550°C) in the absence of O_2 results in a higher yield of biochar (~30%) than the fast pyrolysis (~12%) or gasification (~10%) [3, 11, 27, 28]. Pyrolysis requires relatively dry feedstock (usually moisture content < 30 wt %, but moisture contents of ≤ 10 wt % are preferred), and grinded to different particle sizes based on the type of pyrolysis. Feedstock with high moisture content consumes more energy accounting for increasing heat of vaporization during the heating of biomass towards the pyrolysis temperature. Additionally, the gases and vapors produced in pyrolysis using a high moisture feedstock are diluted with steam

Ref.		Pyrolysis Temp. [C]	Pyrolysis Time	Characteristics			Proximate analysis (%)			Ultimate analysis (%)									Molar ratio			
			min	pH	BET $\frac{m^2}{g}$	Moisture (%)	Volatiles	Fixed carbon	Ash	C	H	O	S	N	Ca	Mg	K	Na	$\frac{H}{C}$	$\frac{O}{C}$	$\frac{(O+N)}{C}$	$\frac{(O+N+S)}{C}$
[1]	Date	RM		5.90	0.98	6.20	69.90	20.90	2.90	45.40	5.60	40.40	5.50	0.00					1.47	0.67	0.67	0.71
		400	11 hr	9.50	1.99	4.80	43.20	45.00	7.10	60.90	2.50	25.60	2.20	1.20					0.49	0.32	0.33	0.35
	Rhode	RM		6.10	1.97	7.80	66.50	11.00	14.70	2.50	5.50	28.70	5.30	1.90					1.54	0.51	0.55	0.59
	grass	400	11 hr	9.70	16.78	1.80	11.80	56.60	28.80	56.70	2.20	8.20	1.60	1.90					0.46	0.11	0.14	0.15
[31]	Date palm waste	RM				7.60	70.59	22.30	7.11	43.19	5.83	39.00	4.16	0.70	2.53	0.68	1.32	0.28	1.61	0.68	0.69	0.73
		300	240			3.29	40.08	45.49	14.42	57.99	4.08	20.82	2.14	0.54	4.85	1.53	2.18	0.40	0.84	0.27	0.28	0.29
		400				3.13	20.25	63.41	16.34	66.87	3.54	11.44	1.36	0.45	6.04	1.57	2.17	0.42	0.63	0.13	0.13	0.14
		500				2.96	9.31	71.00	19.68	72.30	2.11	4.50	1.02	0.42	5.81	1.93	2.23	0.48	0.35	0.05	0.05	0.06
		600				2.25	6.85	72.44	20.71	72.89	1.74	3.28	0.98	0.39	7.77	1.90	2.38	0.53	0.28	0.03	0.04	0.04
		700				2.12	5.47	73.49	21.05	73.42	1.14	3.19	0.85	0.35	7.65	1.92	2.69	0.50	0.19	0.03	0.04	0.04
		800				2.09	3.91	74.70	21.39	74.63	0.86	2.27	0.54	0.31	8.08	2.02	2.71	0.58	0.14	0.02	0.03	0.03
[32]	Bagasse	500		9.30	202.00	1.30	9.17	80.97	8.57	85.59	2.82	10.48		1.11	0.16	0.04	0.26	0.03				
	Cocopeat	500		10.20	13.70	2.55	14.30	67.25	15.90	84.44	2.88	11.67		1.02	0.27	0.06	2.30	1.37				
	Wood stem	500		9.50	316.00	1.46	12.79	83.47	2.28	89.31	2.57	7.34		0.78	0.17	0.02	0.20	0.01				
	Wood bark	500		9.60	13.60	0.36	18.14	68.66	12.84	84.84	3.13	10.20		1.83	1.97	0.11	0.65	0.00				
[33]	rice husk	RM		6.40		8.50	63.78	14.73	12.99	31.39	3.39	43.40		0.35								
		300	60	5.30		1.52	45.49	30.80	22.19	47.15	4.52	23.98		0.65								
		400	90	7.70		3.92	16.54	48.60	30.94	54.33	2.06	8.11		0.64								

Ref.		Pyrolysis Temp. [C]	Pyrolysis Time	Characteristics			Proximate analysis (%)			Ultimate analysis (%)				Molar ratio
		500	120	8.30		2.16	7.44	60.10	30.30	57.35	1.48	8.08		0.63
		300	120	4.20		4.18	46.67	31.87	17.29	46.16	4.51	27.19		0.68
		400	60	5.70		3.24	22.18	45.89	28.70	52.13	2.67	12.64		0.63
		500	90	7.40		2.35	12.38	53.81	31.46	58.41	1.75	5.27		0.77
		300	90	4.10		3.59	43.10	30.64	22.67	46.14	3.83	23.33		0.44
		400	120	7.20		2.52	16.97	49.49	31.02	54.56	2.25	9.16		0.50
		500	60	8.80		1.61	8.58	57.43	32.38	56.28	1.36	7.74		0.63
[27]	Geodae Uksae 1	300			0.49			53.34		66.19				
		400			3.11			73.89		74.69				
		450			21.93			78.48		78.29				
		500			180.96			82.59		79.42				
		600			293.04			88.07		83.67				
		700			368.98			91.66		85.93				
[34]	Rice straw	300		9.00	4.50		34.54	28.06	37.40	68.72	5.22	22.78		3.28
		400		10.10	21.20		18.75	35.39	45.86	75.47	4.62	16.93		2.98
		500		10.50	45.80		10.61	38.72	50.67	81.43	2.34	15.13		1.10
		600		10.60	84.80		6.89	39.87	53.24	87.52	2.11	7.48		2.89
		700		10.60	22.50		5.88	39.52	54.60	91.15	1.26	7.01		0.58
[35]	Bamboo biomass	RM				9.37	70.31	17.75	2.57	39.00	6.10	54.00	0.02	0.60
		500	60			6.50	8.10	81.50	3.90	82.10	2.72	14.60	0.00	0.54

Ref.		Pyrolysis Temp. [C]	Pyrolysis Time	Characteristics			Proximate analysis (%)			Ultimate analysis (%)					Molar ratio	
[28]	beech wood feedstock	RM							0.19	48.45	6.12	45.08	0.02	0.15		
		470								79.58	3.60	16.57		0.25		
		520								85.04	3.81	10.79		0.37		
		570								91.45	3.05	16.82		0.11		
[17]	Pea pod									39.32	4.75	53.3		2.4		
	Cauliflower leaves									31.8	3.2	59.4		4.01		
	Orange peel									40.43	4.83	52.9		1.56		
[36]	LWB[1]	RM				7.41	83.82	12.48	8.25	49.31	5.61	39.88	0.62	0.56		
		800				0.40	13.00	73.36	13.64	75.80	2.43	7.15	0.30	0.68		
[37]	hinoki cypress	RM					84.75	15.09	0.16	51.88	6.21	41.76		0.16		
		350		7.95			41.17	58.52	0.32	75.74	5.29	18.75		0.22		
		400		8.24			33.45	66.10	0.45	77.85	4.90	17.06		0.20		
		500		8.48			21.26	78.06	0.69	85.79	3.89	10.10		0.23		
		600		9.66			12.45	86.73	0.83	91.56	2.96	5.11		0.38		
[38]	chichi manure	350		9.70			336.90	11.10	52.00	31.20	1.97	10.90	0.31		0.76	0.26
		450		10.20			30.60	14.10	53.30	27.20	1.92	11.40	0.44		0.85	0.31
		750		11.70			26.50	17.00	56.40	24.70	0.67	16.30	0.29		0.32	0.49
[39]	corrugated cardboard	RM			1.50			13.10	4.00	43.24	5.80			0.12		
		350			1.20			17.20	5.00	46.84	5.70			0.11		
		400			1.60			27.80	5.20	51.33	4.80			0.16		
		450			1.60			41.80	12.30	54.17	3.70			0.18		

Ref.		Pyrolysis Temp. [C]	Pyrolysis Time	Characteristics		Proximate analysis (%)			Ultimate analysis (%)		Molar ratio
[40]	wood	300	10	4.50		78.00	22.00	0.30	54.10	5.90	1.30
			60	5.70	6.0	42.60	57.40	0.50	71.30	4.70	0.79
		450	10	6.60	4.0	21.40	78.60	1.00	82.50	3.80	0.56
			60	6.70	23.0	16.80	83.20	1.20	86.30	3.50	0.49
		600	10	6.70	196.0	8.20	91.80	1.20	90.00	2.60	0.35
			60	9.10	127.0	6.40	93.60	1.30	92.30	2.30	0.30
		750	10	10.20	128.0	2.60	97.40	1.10	92.50	1.40	0.19
			60	10.40		2.60	97.40	1.10	92.50	1.10	0.15
	green waste	300	10	7.4		74.3	25.7	3.6	53.2	6.2	1.41
			60	8.1		48.6	51.4	6.8	69.3	5.4	0.94
		450	10	9.6		25.3	74.7	11.1	78.8	4.2	0.63
			60	10.0	17.0	18.5	81.5	12.0	82.9	3.5	0.51
		600	10	10.4		11.5	88.5	13.2	87.7	2.3	0.32
			60	11.3	46.0	8.8	91.2	13.4	88.4	2.0	0.27
		750	10	11.4		3.5	96.5	13.9	87.5	1.5	0.21
			60	11.6		1.9	98.1	13.4	93.2	1.3	0.16
	dry algae	300	10	4.9		70.0	30.0	46.3	62.7	7.2	1.38
			60	7.7		55.2	44.8	55.8	69.5	6.9	1.19
		450	10	9.1		27.5	72.5	68.6	74.5	4.5	0.72
			60	9.3	14.0	19.1	80.9	71.8	78.7	4.0	0.61
		600	10	11.1		18.9	81.1	72.2	80.1	2.7	0.41
			60	11.9	19.0	15.7	84.3	73.0	83.4	2.0	0.29

Ref.	Pyrolysis Temp. [C]	Pyrolysis Time	Characteristics	Proximate analysis (%)			Ultimate analysis (%)		Molar ratio
	750	10	12.4	10.1	89.9	74.8	86.4	1.5	0.21
		60	12.5	3.9	96.1	76.4	90.6	1.4	0.19

[1]*lignocellulosic waste biomass*
RM*: raw materal and* ***BET****: Brunauer–Emmett–Teller, surface area analysis.*

Table 1.
Characterization of raw biomass samples and biochar produced: Values for proximate and ultimate analysis.

and have a lower calorific value. Wet biomass, typically with 70 wt% or more water can be converted using hydrothermal carbonization processes. The common processes [11, 27, 28] include slow and fast pyrolysis, and the most successful approach for high-yield biochar production is via slow pyrolysis [3].

Slow pyrolysis is a conventional type of pyrolysis which is operated at moderate temperatures ranging from 300–550°C, slow heating rates of $0.1\frac{°C}{s}$ up to $0.8\frac{°C}{s}$, and longer residence time of 5 to 30 min or even 25 to 35 h [3, 30] conducted at atmospheric pressure. Slow pyrolysis is commonly used to produce biochar, with bio-oil and syngas as co-products. The typical yields of biochar, bio-oil, and syngas are 35%, 30%, and 35% of the dry biomass feedstock, respectively [3, 24] by slow pyrolysis. The main purpose conducting of slow pyrolysis is to maximize the biochar yield. The longer vapor residence times in slow pyrolsis favors the secondary reactions. Biochar produced in slow pyrolysis consists of both primary and secondary chars. The slow heating rate with moderate pyrolysis temperatures also promotes the production of biochar. Biochar yield usually depends on the raw material type & properties, and pyrolysis conditions such as processing temperature, heating rate, and pyrolysis environment [30]. The final biochar yields are decreased by increasing the process temperature because more volatiles are produced from tars at higher temperatures, leading to the production of more gases and bio-oils. Biomass containing more minerals yields less biochar [3]. The overall slow pyrolysis process can generally be exothermic due to the extensive occurrence of secondary reactions. Slow pyrolysis can accept a wide range of particle sizes (5–50 mm).

Pyrolysis of biomass also produces syngas and bio-oil as co-products together with biochar. The fraction of each that is produced depends on the pyrolysis process, but slow heating rates are recommended when biochar is the main product desired. Furthermore, pyrolysis temperatures above 250°C are recommended for the conversion of lignocellulosic biomass because decomposition of hemicellulose and cellulose begins at 250°C and is maximal at around 400°C, whereas changes in lignin structure only start to occur after heating for long durations or higher temperature pyrolysis reactors. The pyrolysis conditions used for biochar productions are related to the type of biomass and biochar quality required. Variation of these reaction parameters finally results in a variety of physicochemical properties of the biochars and affects their final application types and performances. Thus several studies have been conducted to determine the suitable raw material and optimal pyrolysis condition. The challenge is to be able to predict the quality and performance of biochars produced from given biomass and a given pyrolysis process via analysis of its physicochemical properties. Thus, to produce the right type of biochar for specific applications from certain lignocellulosic biomass, elemental composition of different biomass resources and produced biochars need to be measured. A summary of proximate analysis and ultimate analysis along with elemental composition of different raw materials for biochar production are shown in **Table 1**. Conversion of raw biomass to biochars resulted in higher contents of fixed carbon and ash, and lower contents of moisture and volatiles. Fixed Carbon (FC) of biochar was calculated as the sum of moisture, ash, and volatile matter subtracted from 100, $(FC(\%) = 100 - moisture(\%) - ash(\%) - VM(\%))$ [41].

4. Biochar quality

The quality of biochar varies with feedstock used and production conditions. Some of the commonly measured quality parameters of biochar include pH, volatile compound content, ash content, bulk density, organic carbon content, nutrient content, elemental composition, surface area, porosity, surface functional groups, cation exchange capacity, iodine number, C stability, water holding capacity (WHC), moisture content, heavy metals, electrical conductivity, polycyclic aromatic hydrocarbons (PAH) and sorption properties. The European Biochar Certificate [42] defines biochar as a "heterogeneous substance rich in aromatic carbon and minerals. It is produced by pyrolysis of sustainably-obtained biomass under controlled conditions with clean technology and is used for any purpose that does not involve its rapid mineralisation to carbon dioxide and may eventually become a soil amendment." This definition differentiates biochar from other forms of carbonaceous materials such as char and charcoal considering the sources of biomass for the production of biochar need to be renewable and sustainable [3]. The International Biochar Initiative (IBI) [4] considers several parameters relevant for assessing and comparing different biochars. These include proximate analysis, elemental composition analysis, pH, porosity, and BET surface area. In this study, proximate analysis and ultimate analysis of biochars produced from different feedstock under different pyrolysis reaction conditions are presented along with other biochar characteristics and component molar ratios of its constituent elements.

The major constituent components of lignocellulosic biomass [43], cellulose, hemicellulose, and lignin play important roles for most of the physical and chemical property modifications during the pyrolysis process. The mechanisms of pyrolysis of these polymers are chemically different from biomass species to species due to the differences in their compositions. The thermal decompositions of cellulose and hemicellulose polymers occur over a narrower temperature range whereas lignin degrades over a wider temperature range compared to those of cellulose and hemicellulose. The lignocellulosic biomass may contain some minor components other than the aforementioned polymers including some inorganic compounds and organic extractives together with substantial quantities of free and bound water [3, 31, 37, 44]. The inorganic compounds of lignocellulosic biomass which constitute less than 10% by weight of biomass, form ash in the pyrolysis process. The organic extractives of biomass refer to the nonstructural components that can be extracted by polar or nonpolar solvents including fats, waxes, proteins, terpenes, simple sugars, gums, resins, starches, alkaloids, phenolics, pectins, glycosides, mucilages, saponins, and essential oils [23, 31, 37, 44].

The biochar properties that could affect its final application will depend on the type of feedstock material characteristics and pyrolysis reaction conditions used for its production [3]. Pyrolysis temperature is the main process parameter that determines the degree of devolatilisation of the biomass. Water content of the biomass, both free and bound, is the first constituent removed in heating of biomass to temperatures up to 160°C. Thermal decomposition of biomass begins with devolatilisation or decomposition of extractives at temperatures about $\leq$220°C. Hemicellulose is the least stable polymer and break down first at temperatures of 220–315°C. Cellulose has a high degree of polymerization and exhibits higher thermal stability and it decomposes in the temperature range 315–400°C. Lignin is the most difficult component to pyrolyse which decomposes in a wide temperature range from 160–900°C [45].

Biochar characterization methods are always independent of production feedstocks, methods, conditions, and properties of the final product [27]. Several chemical characterizations of biochar ranging from biochar surface analysis to elemental composition and physical properties such as the surface area, pore size, and pore volume are commonly analyzed. The quality of biochar varies with the feedstock type and pyrolysis process conditions. Pyrolysis parameters such as heating rate, residence time, and final temperature greatly influence biochar quality [11]. Pyrolysis temperature has a critical role in biochar properties such as elemental composition, particle size, specific surface area, pore size distribution, thermal capacity, and electrical conductivity.

Certain biochar quality parameters are more important than others depending on the expected final use or application of the biochar. For agricultural application in crop production, the important quality parameters of biochar include pH, volatile compound content, ash content, water holding capacity, bulk density, pore volume, and specific surface area [24]. Carbon stability of biochar is a critical quality parameter in carbon sequestration and soil fertility enhancement. The other important quality parameters in soil fertility enhancement include surface area and nutrient content. The molar ratio of hydrogen to carbon (H/C) is another important characterization parameter of biochar which is an indicator of the degree of carbonization and the stability of biochar. The higher values of molar H/C ratio greater than 0.7 indicate the lower biochar quality and pyrolysis deficiencies. Molar oxygen to carbon (O/C) ratio is also relevant for characterizing biochar and differentiating it from other carbonization products [42] with values greater than 0.4 indicating lower biochar stability. Literatures show that the molar H/C and O/C ratios of lignocellulosic biomass are approximately 1.5 and 0.7, respectively. During pyrolysis, the biomass undergoes devolatilization and the solid portion gets enriched in carbon. The H and O are preferably removed over C and the H/C and O/C ratios tend to decrease as biomass undergoes its transformation into biochar. The H/C and O/C ratios are used to assess the degree of aromaticity and maturation [46]. The characterization of biochar produced from different feedstock and their feedstock are discussed in the next section.

5. Biochar characterization

A summary of the characterization of raw biomass samples and their biochars produced at different temperatures and other pyrolysis reaction conditions along with their values for proximate and ultimate analysis is shown in **Table 1**. The carbon and ash contents of biochar increase on increasing pyrolysis temperature while the contents of volatiles decrease with temperature [1, 37]. Pyrolysis temperature influences the structure of biochar due to the release of volatiles, thus increasing the pyrolysis temperature leads to a decreased content of volatile matter. This was observed because the increasing temperature resulted in further cracking of the volatile fractions into low molecular weight liquids and gases instead of biochar [1, 31, 34, 37, 40]. The fixed carbon and elemental carbon content of biochar increase with increasing the pyrolysis temperature, as depicted in **Table 1**. Lee et al. [27] studied characteristics of biochar produced from slow pyrolysis of Geodae-Uksae and showed the increment of the carbon content of biochar at higher temperatures. The increase in elemental carbon content of biochar at higher pyrolysis temperature implies that the biochar became increasingly carbonaceous at high temperatures, releasing hydrogen and oxygen contents. A similar trend, an increase of carbon content with pyrolysis temperature, is obtained for different raw materials, date palm waste [31], rice husk [33], rice straw [34], beechwood feedstock [28], hinoki cypress [37], corrugated cardboard [39], chicken manure, Coffee husk, and sugarcane bagasse [38], pinewood, wheat straw, green waste and dry algae [40]. As one of the purposes of biochar production is to improve carbon contents in soil, thus, the high carbon content of biochar is beneficial in terms of maximizing the amount of carbon storage. Higher temperature pyrolysis is preferred for biochar production if the biochar application is to improve soil fertility. Several studies indicate that the yield of biochar is highly dependent on the pyrolysis conditions such as temperature, heating rate and heating time and is also greatly influenced by chemical, physical and biological properties of the biomass. **Table 1** shows that the temperature of pyrolysis plays an important role in the yields of the characteristic properties of biochar. The physicochemical properties of biochars depend not only on the nature of the starting biomass but also, to a very large extent, on the condition of preparation. Pyrolysis at lower temperatures would result in a large amount of biochar, indicating at high temperature large part of the biomass is lost as volatile matters.

The proximate analysis of biochar produced at different temperatures shows the fixed carbon and ash contents increase on increasing pyrolysis temperature, while the volatile contents decrease with temperature. The proximate analysis of date palm waste driven biochar [31] shows that the fixed carbon and ash contents increase from 45.49% to 74.7% and 14.42% to 21.39%, respectively while volatile contents decrease from 40.08% to 3.91% on increasing pyrolysis reaction temperatures from 300–800°C. Park et al. [34] reported the proximate analysis of biochar produced from rice straw at different temperatures ranging from 300C to 700. The volatile contents decrease from 34.54% to 5.88% upon increasing the aforementioned temperature range, while the fixed carbon and ash contents increase from 28.06% to 39.52% and 37.4% to 54.6%, respectively. A similar trend is also observed by different authors [27, 32, 33, 37, 40] using different raw material and different pyrolysis temperatures.

The ultimate analysis indicates that pyrolysis temperature is the most influential parameter to determine the elemental composition of biochar samples as shown in **Table 1**. It is observed that carbon content of date palm waste driven biochar [31] increases from 57.99% to 74.63% on increasing pyrolysis temperature from 300–800°C. On the other hand Oxygen and Hydrogen contents decrease from 20.8% to 2.27% and 4.08% to 0.86%, respectively for the same pyrolysis temperature increase. Similarly, the ultimate analysis for metallic contents of Ca, Mg, K, and Na increase from 2.53% to 8.08%, from 0.68% to 2.02%, from 1.32% to 2.71% and from 0.28% to 0.58%, respectively for the same increment of pyrolysis temperature. Vieira et. el. [33] also reported the trend of increasing the carbon contents from 47.15–56%, from 46.14% to 58.4% and from 46.16% to 57.35% on increasing temperature from 300–500°C for pyrolysis reaction times of 60 min, 90 min and 120 min respectively, for biochar produced from rice husk. Moreover, a decrease of oxygen, hydrogen, and nitrogen contents of the biochar is observed on increasing the pyrolysis temperature. Lee et al. [27] showed the increment of the carbon content from 66.19% to 85.93% on increasing temperature from 300–700°C, for biochars produced from Geodae-Uksae 1. Several other researchers also reported the increment of carbon content and decrement of hydrogen, oxygen, and nitrogen contents with temperature for biochars produced from rice straw [34], beech wood [28], and hinoki cypress [37].

The molar ratios of Hydrogen, Oxygen, nitrogen, and sulfur to carbon are observed to decrease with temperature, as more volatile components are removed at higher temperatures making the biochar rich in carbon [31, 38].

pH of biochar is a guiding parameter to define the application of biochar as fuel or as soil fertility enhancing chemical and is correlated with the formation of carbonates and the contents of inorganic alkalis. Biochar is used in the soil as an acidity-correcting agent [47], so it is recommended that the pH conditions of the biochar should be basic because it can replace CaO due to such features. Soil acidity neutralization provides the most favorable conditions for microorganism proliferation and soil fertilization [9, 48]. Thus, the pH of biochar has been associated with having a liming effect on soil acidity, thus increasing the soil pH following the addition of biochar. Biochar can also be used as fuel, the use of acid biochar as fuel can lead to corrosion in the combustion equipment. Biochar having basic pH can cause fouling due to its mineral composition and, consequently, higher ash content than the raw biomass feed. Moreover, the pH of the biochar directly impacts the adsorption process when the carbon is used in filtration process. Therefore, a neutral pH is generally preferred [49]. Most of biochar products have alkaline pH. Some studies have indicated that ash content of feedstock in conjunction with pyrolysis severity could influence the final pH of biochar samples suggested that a large proportion of the ash in high-ash feedstock contains carbonates which could cause a liming effect [15]. The pH values of biochars produced from rice husk [33] is observed to increase with temperature, ranging from 5.3 to 8.8 for temperature range from 300–500°C with a reaction time of 60 min, and from 4.2 to 8.3 for the same temperature range with a reaction time of 120 min. Yu et al. [37] reported the pH of biochars produced from hinoki cypress at temperatures ranging from 350–600°C and their pH increases from 7.95 to 9.66. Similarly, Domingues et al. [38] reported the pH of biochars produced from chicken manure at temperatures ranging from 350–750°C and their pH increases from 9.70 to 11.7. The pH of biochar produced from different feedstock is observed to increase with temperature [40].

Thus, biochar with desirable properties can be deduced from both its proximate and ultimate analysis. The lower the O/C and H/C ratios, the higher is the loss of oxygen and hydrogen during the combustion process producing a product richer in higher elemental carbon. The International Biochar Initiative (IBI) recommends a maximum value of 0.7 for the molar H/C ratio [17] to distinguish biochar from biomass that has not been or only somewhat thermochemically altered. Thus suitable working conditions and technologies must be selected in order to produce biochar of high quality. The pH of biochar from different biomass is around 10 and the microscopic surface structure of biochars ranges from around 3 $\frac{m^2}{g}$ for rice husk biochar to around 500 $\frac{m^2}{g}$ for biochar from wood [20]. Biochar produced from different feedstock showed in increasing surface areas on increasing pyrolysis temperature [27, 34, 39].

6. Biochar applications

Biochar is considered as a multifunctional material related to carbon sequestration, contaminant immobilization by adsorption, greenhouse gas reduction, soil fertilization, and waste-water and industrial effluent treatments. Biochar is widely used in heat and power generation, in soil fertility enhancement to improve the physical properties of soil, especially in soils with bad soil structure or high bulk density [3, 47, 50], in adsorption and filtration processes in different industrial effluent treatments [3] and in catalysis or as a catalyst support [51].

Biochar is an economical adsorbent [10] removing various organic contaminants such as agrochemicals, antibiotics, polycyclic aromatic hydrocarbons, polychlorinated biphenyls, volatile organic compounds, and various inorganic contaminants like heavy metals, ammonia, nitrate, phosphate, sulfide, etc. The high adsorption capacity, high specific surface area, microporosity, and ion exchange capacity of biochar are important characteristics for its applications. The feedstock types and pyrolysis conditions used during the biochar production greatly change its physicochemical properties such as surface area, polarity, atomic ratio, elemental composition and pH, giving the overall surface property of the biochar [8, 9]. These varieties in biochar qualities have significant implications on its suitability and efficacy in the remediation of targeted pollutants. Applications of biochar in soil improves the physicochemical and biological properties of the soils [5], which contribute to soil carbon sequestration and greenhouse gas emission reduction.

Biochar produced at low temperatures may be suitable for controlling fertilizer nutrients release, while high temperatures would yield material similar to activated carbon [52]. Because of the high aromaticity, the carbon in biochar is highly recalcitrant in soils with very long residence times. Thus, biochar incorporated in soil represents a potential terrestrial carbon sink and also a means of mitigating CO_2 emissions. In addition, biochar has a significant potential to mitigate greenhouse gas emissions from agriculture, both by storing carbon in soils and through mitigating N_2O emissions [48]. Furthermore, biochar can reduce the amount of fertilizer required and the emission of N_2O and CH_4 from the soil, thus the amount of carbon emissions prevented by biochar can be significant [27, 34]. Biochar application significantly reduced the leaching of applied N fertilizers. Biochar would not only enhance soil fertility but also sequester carbon from the atmosphere further research findings revealed that biochar has an affinity for organic compounds and may sorb toxic by-products from the wastewater treatment process. DAP (di-ammonia phosphate)-based fertilizer is used and studies have shown a large proportion (> 85%) of N applied as NH_4^+, N^- was lost through NH_3 volatilization within one week after application [53]. Therefore biochar application can increase nutrient retention capacity and N use efficiency [48, 54, 55]. Several researchers demonstrated the benefits of biochar for soil, for example, wood biochar applied into a Colombian savanna Oxisol increased available Ca and Mg concentrations and pH, and reduced toxicity of Al [56, 57], moreover, biochar improved soil structure [58], created a carbon sink in soil [59], and reduced CH_4 emissions [60].

In addition to being used as a soil conditioner and carbon sequestration regent, biochar has attracted much attention in wastewater treatment fields. Recent works of literature show biochar as a highly efficient, environmentally friendly, and low-cost adsorbent [61–63]. Biochar characteristic or quality plays a critical role in contaminant removal, which is usually governed by pyrolysis temperature and feedstock type. Fully carbonized biochar produced at a higher pyrolysis temperature (> 500 °C) has higher affinity for organic contaminants due to high surface area [20], microporosity, hydrophobicity carbon-to-nitrogen (C/N) ratio [17, 31, 38], and pH [9, 15, 33, 37, 38, 40]. Partially carbonized biochars that are produced at lower pyrolysis temperatures have higher content of O-bearing functional groups like hydroxyl and carboxyl compounds and lower porosity making them more appropriate for removal of inorganic pollutants [9].

Biochar is used as an electrode as well for various electrochemical devices, including lithium-ion and Li-S batteries [64], supercapacitors [65], and microbial fuel cells [66], etc. Such biochar, namely activated biochar, are found to ne more sustainable than their fuel-based counterparts owing to its high surface area and porosity, efficient electrical and thermal conductivity, high stability, low economical cost, and availability [66, 67].

The unique chemical structure of biochar with a large surface area and tailored surface functional groups can be easily prepared by activation or functionalization and shows great potential to be used as a versatile catalyst or catalyst support in many chemical processes [51, 68–70].

7. Conclusion

Biochar is considered as a multifunctional material related to carbon sequestration, contaminant immobilization by adsorption, greenhouse gas reduction, soil fertilization, and waste-water and industrial effluent treatments. The most promising feature of biochar is the fact that it represents a low cost and sustainable products with a spectrum of applications. Biochars have a tremendous range of physical and chemical properties, which greatly affect their wide applications. The feedstock and the method by which the biochar is produced has a significant impact on biochar characteristics, including concentrations of elemental constituents, density, porosity, and pH, which collectively impact the suitability of the biochar for various applications. This chapter examines in detail the production and characteristics of biochar resulting from slow pyrolysis process, including the effect of feedstock type and different pyrolysis process parameters on the properties and yield of biochar has been thoroughly studied. The selection of a specific type of feedstock is to a great extent determined by its in a place where the biochar is likely to be produced, as this reduces the cost of transport while decreasing the carbon footprint of the biochar technology. The pyrolysis temperature affects the biochar quality, higher carbon contents of biochars can be obtained at higher temperature while volatiles and molar ratios of O/C, H/C and N/C decrease with pyrolysis temperature. Biochars of higher carbon contents are preferable for most applications. Biochars produced at low pyrolysis temperature are suitable for controlling fertilizer nutrients release, while high temperatures would yield material similar to activated carbon. The pH of biochar is also another important parameter that determines its application. More basic, higher pH, biochar is preferred for soil application usually to correct soil acidity. Neutral pH biochar is also most preferable for adsorption processes for the removal of pollutants and contaminants from industrial effluents. Biochars produced at higher pyrolysis temperature have high affinity for organic pollutants due to high surface areas. In addition, neutral pH biochar is used as energy sources because acidic biochars cause corrosion and basic biochars cause fouling problems. Thus, the pyrolysis temperature should be selected as per the final application of the biochar.

Acknowledgements

The authors would like to thank the Higher Colleges Technology office of Applied Research for financial support.

Author details

Asfaw Gezae Daful*, Meegalla R. Chandraratne and Marie Loridon
Department of Chemical Engineering, Higher Colleges of Technology, Abu Dhabi, United Arab Emirates

*Address all correspondence to: agezae@hct.ac.ae; agezae@gmail.com

IntechOpen

© 2021 The Author(s). Licensee IntechOpen. This chapter is distributed under the terms of the Creative Commons Attribution License (http://creativecommons.org/licenses/by/3.0), which permits unrestricted use, distribution, and reproduction in any medium, provided the original work is properly cited. (cc) BY

References

[1] M. Jouiad, N. Al-Nofeli, N. Khalifa, F. Benyettou, and L. F. Yousef, Journal of Analytical and Applied Pyrolysis **111**, 183 (2015).

[2] A. Pandey, T. Bhaskar, M. Stöcker, and R. Sukumaran, *Recent Advances in Thermochemical Conversion of Biomass* (Elsevier, 2015).

[3] A. Gezae Daful and M. R. Chandraratne, in *Encyclopedia of Renewable and Sustainable Materials*, edited by S. Hashmi and I. A. Choudhury (Elsevier, 2020) pp. 370 – 378.

[4] IBI, "International biochar initiative," (2021).

[5] S. Gul, J. K. Whalen, B. W. Thomas, V. Sachdeva, and H. Deng, Agriculture, Ecosystems & Environment **206**, 46 (2015).

[6] D. Aller, S. Bakshi, and D. A. Laird, Journal of Analytical and Applied Pyrolysis **124**, 335 (2017).

[7] S. Werner, K. K¨atzl, M. Wichern, A. Buerkert, C. Steiner, and B. Marschner, Environmental Pollution **233**, 561 (2018).

[8] M. Uchimiya, T. Ohno, and Z. He, Journal of Analytical and Applied Pyrolysis **104**, 84 (2013).

[9] M. Ahmad, A. U. Rajapaksha, J. E. Lim, M. Zhang, N. Bolan, D. Mohan, M. Vithanage, S. S. Lee, and Y. S. Ok, Chemosphere **99**, 19 (2014).

[10] F. R. Oliveira, A. K. Patel, D. P. Jaisi, S. Adhikari, H. Lu, and S. K. Khanal, Bioresource Technology **246**, 110 (2017).

[11] M. Tripathi, J.N.Sahu, and P. Ganesan, Renewable and Sustainable Energy Reviews **55**, 467 (2016).

[12] A. C. Opia, M. K. B. A. Hamid, S. Syahrullail, A. B. A. Rahim, and C. A. Johnson, Materials Today: Proceedings (2020), https://doi.org/10.1016/j.matpr.2020.04.045.

[13] E. J. Cho, L. T. P. Trinh, Y. Song, Y. G. Lee, and H.-J. Bae, Bioresource Technology **298**, 122386 (2020).

[14] E. Jong, A. Higson, P. Walsh, and M. Wellisch, *Bio-based Chemicals: Value added products from biorefineries*, Tech. Rep. (IEA Bioenergy: Task 42 Biorefinery, 2012).

[15] J. Lehmann, M. C. Rillig, J. Thies, C. A. Masiello, W. C. Hockaday, and D. Crowley, Soil Biology and Biochemistry **43**, 1812 (2011).

[16] N. Jafri, W. Wong, V. Doshi, L. Yoon, and K. Cheah, Process Safety and Environmental Protection **118**, 152 (2018).

[17] G. S. Mary, P. Sugumaran, S. Niveditha, B. Ramalakshmi, P. Ravichandran, and S. Seshadri, International Journal of Recycling of Organic Waste in Agriculture **5**, 43 (2016).

[18] Y. Li, B. Xing, Y. Ding, X. Han, and S. Wang, Bioresource Technology **312**, 123614 (2020).

[19] F. Wang, D. Ouyang, Z. Zhou, S. J. Page, D. Liu, and X. Zhao, Journal of Energy Chemistry (2020), https://doi.org/10.1016/j.jechem.2020.08.060.

[20] T. Kan, V. Strezov, and T. J. Evans, Renewable and Sustainable Energy Reviews **57**, 1126 (2016).

[21] K. Almi, A. Benchabane, S. Lakel, and A. Kriker, Journal of Reinforced Plastics and Composites **34**, 1231 (2015), https://doi.org/10.1177/0731684415588356.

[22] A. Gezae Daful and J. F. Görgens, Chemical Engineering Science **162**, 53 (2017).

[23] D.-C. Li and H. Jiang, Bioresource Technology **246**, 57 (2017).

[24] W. Song and M. Guo, Journal of Analytical and Applied Pyrolysis **94**, 138 (2012).

[25] E. Agrafioti, G. Bouras, D. Kalderis, and E. Diamadopoulos, Journal of Analytical and Applied Pyrolysis **101**, 72 (2013).

[26] J. Wang, Z. Xiong, and Y. Kuzyakov, GCB Bioenergy **8**, 512 (2016).

[27] Y. Lee, P.-R.-B. Eum, C. Ryu, Y.-K. Park, J.-H. Jung, and S. Hyun, Bioresource Technology **130**, 345 (2013).

[28] P. F. Jens, B. W. Scott, B. V. A. V, and D. Javier, Applied Energy **188**, 595 (2017).

[29] J. Wang and S. Wang, Journal of Cleaner Production **227**, 1002 (2019).

[30] P. Roy and G. Dias, Renewable and Sustainable Energy Reviews **77**, 59 (2017).

[31] A. R. Usman, A. Abduljabbar, M. Vithanage, Y. S. Ok, M. Ahmad, M. Ahmad, J. Elfaki, S. S. Abdulazeem, and M. I. Al-Wabel, Journal of Analytical and Applied Pyrolysis **115**, 392 (2015).

[32] Y. Lee, J. Park, C. Ryu, K. S. Gang, W. Yang, Y.-K. Park, J. Jung, and S. Hyun, Bioresource Technology **148**, 196 (2013).

[33] F. R. Vieira, C. M. R. Luna, G. L. Arce, and I. Avila, Biomass and Bioenergy **132**, 105412 (2020).

[34] J. Park, Y. Lee, C. Ryu, and Y.-K. Park, Bioresource Technology **155**, 63 (2014).

[35] L. E. Hernandez-Mena, A. A. B. Pécora, and A. L. Beraldo, Chemical Engineering Transactions **37**, 115 (2014).

[36] J. Haydary, Chemical Engineering Transactions **70**, 1033 (2018).

[37] S. Yu, J. Park, M. Kim, C. Ryu, and J. Park, Bioresource Technology Reports **6**, 217 (2019).

[38] R. R. Domingues, P. F. Trugilho, C. A. Silva, I. de Melo, L. C. A. Melo, Z. M. Magriotis, and M. A. S'anchez-Monedero, Plos One **12** (2017).

[39] F. Sotoudehnia, A. Baba Rabiu, A. Alayat, and A. G. McDonald, Journal of Analytical and Applied Pyrolysis **145**, 104722 (2020).

[40] F. Ronsse, S. van Hecke, D. Dickinson, and W. Prins, GCB Bioenergy **5**, 104 (2013).

[41] L. Zhao, X. Cao, O. Mašek, and A. Zimmerman, Journal of Hazardous Materials **256-257**, 1 (2013).

[42] EBC(2012), *European Biochar Certificate - Guidelines for a Sustainable Production of Biochar'. European Biochar Foundation (EBC)*, Tech. Rep. (Arbaz, Switzerland, http://www.europeanb iochar.org/en/download, Version 6.3E of 14th August 2017,).

[43] P. Basu, *Biomass Gasification, Pyrolysis and Torrefaction Practical Design and Theory*, 2nd ed. (Elsevier Inc., 2013).

[44] F.-X. Collard and J. Blin, Renewable and Sustainable Energy Reviews **38**, 594 (2014).

[45] H. Yang, R. Yan, H. Chen, D. H. Lee, and C. Zheng, Fuel **86**, 1781 (2007).

[46] J. Lehmann and S. Joseph, *Biochar for Environmental Management: Science and Technology* (Earthscan, 2009).

[47] G. Agegnehu, A. Srivastava, and M. I. Bird, Applied Soil Ecology **119**, 156 (2017).

[48] S. Weldon, D. P. Rasse, A. Budai, O. Tomic, and P. Dörsch, Soil Biology and Biochemistry **135**, 173 (2019).

[49] *K. Qian*, A. Kumar, H. Zhang, D. Bellmer, and R. Huhnke, Renewable and Sustainable Energy Reviews **42**, 1055 (2015).

[50] Q. Zhang, M. Saleem, and C. Wang, Science of The Total Environment **671**, 52 (2019).

[51] J. Lee, K.-H. Kim, and E. E. Kwon, Renewable and Sustainable Energy Reviews **77**, 70 (2017).

[52] W. Gwenzi, N. Chaukura, F. N. Mukome, S. Machado, and B. Nyamasoka, Journal of Environmental Management **150**, 250 (2015).

[53] C. Chen, I. Phillips, L. Wei, and Z. Xu, Environ. Sci.Pollut. Res. **17**, 1098 (2010).

[54] J. Lehmann, J. Gaunt, and M. Rondon, Mitigation and Adaptation Strategies for Global Change **11**, 403 (2006).

[55] M. Esfandbod, I. R. Phillips, B. Miller, M. R. Rashti, Z. M. Lan, P. Srivastava, B. Singh, and C. R. Chen, Ecological Engineering **98**, 157 (2017).

[56] M. Julie, R. Marco, M. Diego, S. J. Riha, and L. Johannes, Plant and Soil **333**, 117 (2010).

[57] Y. Ding, S. Liu, Yunguo Liu, Z. Li, X. Tan, X. Huang, G. Zeng, L. Zhou, and B. Zheng, Agronomy for Sustainable Development **36**, 36 (2016).

[58] D. L. Jones, G. Edwards-Jones, and D. V. Murphy, Soil Biology and Biochemistry **43**, 804 (2011).

[59] S. P. Sohi, E. Krull, E. Lopez-Capel, and R. Bol, in *Advances in Agronomy*, Advances in Agronomy (Academic Press, 2010) pp. 47–82.

[60] Y. Liu, M. Yang, Y. Wu, H. Wang, Y. Chen, and W. Wu, Journal of Soils and Sediments **11**, 930 (2011).

[61] V. Gargiulo, A. Gomis-Berenguer, P. Giudicianni, C. O. Ania, R. Ragucci, and M. Alf'e, Energy and Fuels **32**, 10218 (2018).

[62] G. Chu, J. Zhao, Y. Liu, D. Lang, M. Wu, *B. Pan*, and C. E. W. Steinberg, Journal of Hazardous Materials **373**, 106 (2019).

[63] R. He, X. Yuan, Z. Huang, H. Wang, L. Jiang, J. Huang, M. Tan, and H. Li, Colloids and Surfaces A: Physicochemical and Engineering Aspects **579**, 123642 (2019).

[64] J. Li, F. Qin, L. Zhang, K. Zhang, Q. Li, Y. Lai, Z. Zhang, and J. Fang, J. Mater. Chem. A **2**, 13916 (2014).

[65] J. Chang, Z. Gao, X. Wang, D. Wu, F. Xu, X. Wang, Y. Guo, and K. Jiang, Electrochimica Acta **157**, 290 (2015).

[66] T. Huggins, H. Wang, J. Kearns, P. Jenkins, and Z. J. Ren, Bioresource Technology **157**, 114 (2014).

[67] J. L. Goldfarb, G. Dou, M. Salari, and M. W. Grinstaff, ACS Sustainable Chemistry & Engineering **5**, 3046 (2017), https://doi.org/10.1021/acssuschemeng.6b02735.

[68] L. Pierri, A. Gemenetzi, A. Mavrogiorgou, J. Borges Regitano, Y. Deligiannakis, and M. Louloudi, Molecular Catalysis **489**, 110946 (2020).

[69] A. Kumar, J. Kumar, and T. Bhaskar, Environmental Research **186**, 109533 (2020).

[70] X. Xiong, I. K. Yu, L. Cao, D. C. Tsang, S. Zhang, and Y. S. Ok, Bioresource Technology **246**, 254 (2017).

Section 3

Pyrolysis for Waste Management

Chapter 18

Mixed or Contaminated Waste Plastic Recycling through Microwave - Assisted Pyrolysis

Piero Frediani and Marco Frediani

Abstract

A single type of thermoplastic polymer is easily recycled through a mechanical process, but this way can't be followed in the presence of mixed or contaminated plastic. In this case, one of the main followed solutions is a thermochemical process and among them, microwave-assisted pyrolysis is one of the emerging technologies. This chapter offers an update of the microwave-assisted pyrolysis of mixed or contaminated waste plastic as a very promising example of chemical recycling. Furthermore, some unpublished results in this field will be reported such as the pyrolysis of waste lead containing polyethylene coming from end cycle batteries or the pyrolysis of waste polypropylene from facemasks used for covid protection. Finally, some examples of pilot plants will be described and commented as well as several industrial cooperations.

Keywords: microwave, pyrolysis, waste mixed plastic, waste contaminated plastic, waste plastic disposal

1. Introduction

The goal of realizing the conditions for achieving a truly circular economy is one of the main challenges in which scientists are involved to solve the problems concerning both global warming and the end of mineral resources. Over the last twenty years, the number of companies involved in environmental issues is continuously increasing and now almost all companies are, even in a different way, involved to solve this problem. These resources also represent a strong driving force for the development of sustainable industrial processes.

The world plastic production in 2019 was $368.0x10^6$ tons with an increase of 2.5% concerning for 2018 [1] while in the same year the European production was $57.9x10^6$ tons among thermoplastic and thermosetting polymers.

Among them, polyolefins are the most produced and employed materials for applications in everyday life for industrial, domestic, and technological applications [1]. They are thermoplastic polymers and are mainly used for packaging, whose life cycle is very short which means they may be disposed of in a short time after their production. A less important part of them is employed to realize furniture, insulating materials, automotive parts, and so on and their life cycle is considerably longer (ten years or more).

Polymeric materials are easily recycled through mechanical processes, as reported in **Figure 1**, for the production of renewed objects prolonging the life of these plastic materials, avoiding their disposal through landfilling or combustion, and reducing the use of raw materials. This is a very friendly and economic process, however, it may be applied only when a single plastic material is available or when the material is not contaminated or strongly deteriorated. In these cases, other routes such as thermochemical processes may be followed as an environmentally friendly procedure avoiding to send these materials to combustion or landfilling, so giving a contribution to realizing a circular economy and reducing the emission of greenhouse gases (GHG).

Thermochemical processes may supply fuels and chemicals using recycled or renewable feedstock as alternatives to oil-based raw materials [2]. These led to the development of new technologies able to carry out the optimal use of such resources minimizing the environmental impact and producing a limited amount of secondary waste. It means that waste disposal or contaminated plastic materials may be transformed into useful products. Among these processes, the microwaves technologies have taken large attention due to their high efficiency to supply the energy required for a plethora of industrial processes. The main performances on the use of microwave as correlated to a classical heating system are resumed in **Table 1**.

Figure 1.
Pathway to recycle thermoplastic materials.

Microwave heating		Conventional heating	
Advantages	Volumetric heating	**Detriments**	Surface heating
	Short reaction time (minutes)		Long reaction time (hours)
	High heating efficiency		Heating efficiency is usually low
	The low thermal conductivity of polymers may be overwhelmed: easy heating of polymers		Hard heating of polymers: their thermal conductivity is low
	The gas formed does not contain combustion gas		The gas formed is contaminated by combustion gas (Direct heating)
Detriments	Electrical power is required	**Advantages**	Every fuel source may be employed
	Microwave absorber is required		Additives are not required, some heating carrier are sometimes used

Table 1.
Correlation between microwave and conventional heating.

This chapter aims to offer a comprehensive description of new results reported to provide an updated description to the specialist and open this knowledge to a greater audience. So our intent is to contribute to extend the application of this technology to many other fields.

2. Previous results

Pyrolysis is the thermal decomposition of a material at a relatively elevated temperature under an inert atmosphere. It involves a change of the chemical composition of the waste polymer that will be converted into a solid, a liquid, and a gaseous product. These products will be directly used as solid, liquid, or gaseous fuels to produce energy or employed as raw materials for the synthesis of new products [3].

Pyrolysis of end-cycle plastic materials using the microwave as heating technology (Microwave-Assisted Pyrolysis: MAP) has been reviewed ten years ago [4–6]. The polymeric materials have a low efficiency to convert microwave radiation into heat (see **Table 1**), but if a microwave absorber is added, sometimes erroneously called a catalyst, the pyrolysis may run very efficiently. The absorbers are compounds showing a high efficiency to convert microwave into heat and the main types employed are silicon carbide, carbon, iron, and so on. Anyway, in the literature, it is possible to find a list of many different compounds able to convert MW into heat, each one with specific features.

In the previous review [4] was reported some preliminary results on the pyrolysis of several waste plastics such as High-Density Polyethylene (HDPE), Polypropylene (PP), Polystyrene (PS), Polyvinylchloride (PVC), Polyethylentherephtalate (PET), and end cycle tyre but these data were referred to 10 years ago.

3. Update on MAP of plastic waste

This chapter aims to report a comprehensive description of the new results obtained to provide an accessible description not only to specialized researchers but also to open this knowledge to a greater audience.

MAP of plastic wastes has emerged as a promising chemical method to displace waste plastic and generating fuels suitable to produce energy and chemical products. Several parameters such as the plastic origin and its composition, temperature, reaction time, MW power, MW absorber and eventually the catalyst present and its composition has been reported in the literature in different reviews during the last years [3, 7, 8]. The catalyst, if present, rapidly loses its activity due to the formation of char on its surface. Energy balance showed that 5 MJ of electric energy was required to process 1 kg of HDPE using a continuous MAP system (energy efficiency 89.6%). It is interesting to note that the requested energy could be generated from the gas products formed during pyrolysis (6.1 MJ), making the process self-sufficient from an energetic point of view.

Some of the main theoretical and experimental aspects of the microwave-materials interactions are reported alongside the issues related to a microwave pyrolytic process of materials [9]. Due to several parameters affecting the interactions between materials and microwaves a rigorous detection of the reaction temperature may be a hard task during MAP [10] both for its detection and uniformity of the distribution of temperature on the material during the overall pyrolytic run.

The MAP is furthermore a promising process available for other industrial applications allowing for energy recovery directly from plastic wastes. Moreover, it offers some interesting advantages, such as high energy efficiency, absence of waste production, lower reaction temperature if correlated to classical heating processes. In the following parts are reported the MAP of synthetic plastic produced through the polymerization of monomer obtained from mineral oils together with citations of MAP of biomasses (polymers from natural source) to produce bio-gas, bio-oil, and bio-char.

3.1 Disposal of a single type of plastic

Disposal of a single type of plastic waste through pyrolysis is not the best choice when they are clean and not deteriorated, because they can be easily reused through mechanical recycling as reported in **Figure 1**. Mixed plastics may be hardly sent to a mechanical process because a large amount of expensive compatibilizers are required, due to the low miscibility of different plastics. Otherwise, if the plastics, even if of a single type, are contaminated or they have lost a large part of their performances, pyrolysis represents one the best choice. Waste or contaminated polyolefins were disposed through MAP using tyres or carbonaceous char as MW absorber [3, 11–15]. The presence of the absorber mixed with the char formed in the course of the pyrolysis process is not a problem because it may be easily recycled to the pyrolysis process or(depending on the absorber used) employed together with the char formed for several applications such as solid fuel, support for catalyst, filler for fiber-reinforced composites and so on. The study of MAP on single plastic is however important to understand the mechanism of the process, the nature and composition of the products, and the interaction among intermediates when pyrolysis of composites or mixed plastics is run.

3.1.1 Polyethylene (PE)

PE is the most common plastic used today, it is produced as High-density Polyethylene (HDPE) through Ziegler-Natta catalysis or Low-Density Polyethylene (LDPE) from a radical polymerization. Other classes of intermediate products (between HDPE and LDPE) are present on the market. PE is a polymer, with production over 100.0 x 10^6 tons/y accounting for 34% of the total plastic market. It is a polymer, primarily used for packaging (plastic bags and films), geomembranes and containers including bottles, tanks, and so on). Waste or contaminated PE may be disposed of through MAP using tyres or carbonaceous char as MW absorber. HDPE was converted into waxy products (yield 80.2 -83.9 wt. %) while the yield of gas was 15.7-19.2 wt. % and that one of solid was 0.4—0.6 wt. % [9, 10, 14] when standard apparatus was employed. However HDPE was converted into a low viscosity liquid by using a very low MW power, but a long time was required to obtain a complete conversion. It is possible to use a pyrolytic apparatus containing a system able to fractionate the vapor formed, improving the residence time of the waxy products in the oven. In these conditions, a low viscosity and density liquid was obtained and the overall pyrolysis was improved. In all cases, the time of the process was strongly reduced concerning processes using classical thermal heating. The liquid fraction from HDPE contained linear alkanes and 1-alkenes with negligible formation of branched, cyclic, or aromatic hydrocarbons.

MAP of HDPE was realized also using a bed of activated carbon [11] obtaining a liquid containing hydrocarbons with a carbon chain length profile close to that one of a diesel. The chain of HDPE was broken, across all operating temperatures, and a lighter liquid product with a narrower range of chain lengths was formed if compared to the traditional pyrolysis using a bed of traditional coke. High-temperature pyrolysis of plastic has been also reported by Jang et al. [16] without any correlation with other results previously reported in the literature.

The MAP of PE using a mixed catalyst containing cracking active components (γ-Al_2O_3, amorphous aluminum silicate, fly ash, zeolite molecular sieve, natural mineral clay, and solid acid) together with an MW absorber (C-containing compounds, Si-containing compounds, Fe-containing compounds, and polymers) and coking inhibiting component (alkali metal salt, P-containing compounds, B-containing compounds, magnesium sulfate, diethylpolysiloxane and p-tert-Butyl catechol) were reported [17]. The use of in-situ or ex-situ catalysts such as NiO and HY zeolite is reported to improve the quality of the products formed [18].

3.1.2 Polypropylene (PP)

PP is a thermoplastic polymer obtained through stereospecific polymerization with properties close to those of PE, but slightly hard, and resistant to a higher temperature. It is white, non-polar, and used in a wide variety of applications. It is the second large produced plastic, after PE. PP has been pyrolyzed using tyres or carbonaceous char as MW absorber [14] obtaining a low density (0.746-0.760 g/mL) and viscosity (0.63-0.95 cP) liquid (yield 56.5-74.7 wt. %), a char (yield 9.1-12.0 wt. %) and a gaseous fraction (yield 13.3 - 34.4 wt. %) containing hydrogen and aliphatic hydrocarbons. Liquid from MAP of PP was formed as a mixture of methyl-branched alkane and alkenes, and sometimes aromatics. The composition was a function of pyrolysis conditions.

Similar results were later reported by Suriapparo et al. [19], using different MW absorbers such as graphite, aluminum, silicon carbide, activated carbon, lignin, and fly ash. The optimal conditions were an MW power of 450 W and a ratio of PP: graphite = 100: 1, but only an oil yield of 48 wt. % was reached. The oil had a heating value of 44.45 MJ/Kg and contained 50 wt. % of C_3-C_6 gaseous hydrocarbons. The reaction was repeated with 50 g of PP under the same conditions, obtaining an 83 % of energy recovery in the oil. Alkenes and cycloalkanes were the major compounds in the oil, but their relative yields varied significantly with different absorbers. The catalytic activity of the absorber was hypothesized even if the results may be explained with the different efficiency in the conversion of MW into heat and the transfer of the heat to the polymer. The MAP was also carried out in the same conditions using commercial PE and polyisoprene.

3.1.3 Polystyrene (PS)

PS is an amorphous, colorless, and transparent thermoplastic polymer synthesized from an aromatic alkene (styrene) through radical polymerization. PS can be solid or foamed, the PS used for general purpose is clear, hard, and brittle. It have a low-density and a high barrier to heat and noise when expanded (EPS) but a low barrier to oxygen and water vapor and a relatively low melting point. PS is one of the most widely used plastics, after PE and PP, it can be naturally transparent, but can be colored. It is employed as protective packaging, containers, lids, bottles, trays, tumblers, and so on. Reverse polymerization of waste PS was realized obtaining styrene and other aromatics through MAP, using tyres or carbonaceous char as MW absorber [20–25]: Tyres may be employed if the final use of the liquid is as a diesel fuel, carbon if the liquid formed (styrene) must be directly used to produce new PS. Working at atmospheric pressure a clear and low viscosity liquid containing styrene as the major product was always collected together with a low amount of char and gas. Using a MW power of 3 kW and 100 g of PS together with 47.3 g of carbon, it gave a liquid (yield 86.5 wt. %) containing a higher amount of single-ring aromatic compounds, as evaluated by Gas Chromatographic-Mass Spectrometric analysis (GC–MS) (aromatics C_6–C_{10} 93.9%, among which styrene is 66.0%) a char (yield 9.8 wt. %) and a gas (yield 3.7 wt. %). Improvements in residence time, by using low MW power or a fractionating system directly inserted over the oven and before the collecting system, allowed to obtain a liquid with low viscosity and density even if the char yield was increased to 10.0%. If the process was realized at reduced pressure (21.3 KPa) the liquid product was formed with a yield of 84.3 wt. % containing the monomeric styrene in a concentration of 71.9 % (determined by GS-MS).

Waste PS was also pyrolyzed in the presence of aluminum as MW absorber, at temperatures as high as the melting point of aluminum, obtaining styrene and other substituted aromatic compounds [22]. The rate of MAP and yield of the products was found to depend on the size, shape, and form of the aluminum. The reaction was faster using the coil, slower for strips, and negligible for the cylindrical form. The products of the pyrolysis were found to contain 88 wt. % of liquid (substituted benzene together with polycyclic aromatics and condensed ring aromatics), 9–10 wt. % of gas, and a low amount of char.

A pyrolysis process was also run in a batch reactor for MAP using activated carbon [21]. The quality of the oil from pyrolysis of PS was assessed for the possible applicability of the liquid in fuel production. The best conditions were a MW power of 450 W and a polymer/activated carbon ratio of 10:1, resulting in an oil yield of 93.0 wt. %. The liquid contained alkenes 8.4 wt. %, α-methyl styrene 1.0 wt. %, condensed ring aromatics 23.2 wt. %, and benzene derivatives 26.8 wt. %. The C_9-C_{12} aromatics were 93.0 wt. % while it was not reported the amount of styrene (C_8H_8) in the liquid.

3.1.4 Poly(vynylchloride) (PVC)

PVC is another largely used polymer obtained through radical polymerization of vinyl chloride. It is employed in two basic forms: rigid and flexible. The main uses are in construction for pipes, doors, windows, non-food packaging, food-covering sheets, and cards. It can be made softer and more flexible by the addition of plasticizers. In this form, it is used in plumbing, electrical cable insulation, imitation leather, flooring, signage, phonograph records, inflatable products, and many applications where it replaces rubber. Together with cotton or linen, it has been used in the production of canvas. According to its thermodynamic stability, the pyrolysis of PVC was obtained with the initial loss of HCl followed by cracking the hydrocarbon chain, so the gas contains HCl and low molecular weight hydrocarbons (ethene and propene) while the main compounds present in the liquid fraction were aromatic hydrocarbons such as benzene, toluene, dialkylbenzenes and so on, formed through Diels Alder reactions involving polyunsaturated compounds formed as intermediates in the course of the MAP process.

3.1.5 Polyesters

The most widespread synthetic polyester fiber is obtained by polycondensation of ethylene glycol and terephthalic acid (Polyethylentherephtalate: PET). It is also mainly used to produce bottles devoted to contain sparkling water and carbonated soft drinks due to the low permeability to CO_2. PET may be also disposed of through pyrolysis with the formation of a gaseous fraction containing carbon dioxide (22.7%) and carbon monoxide (13.3%) due to decarboxylation and decarbonylation of the ester groups [26] while the liquid product contains aromatic hydrocarbons (benzene and toluene) and oxygenated compounds (benzoic acid, acetaldehyde, and benzaldehyde). However, the yield of liquid product was lower and PET may be more efficiently recycled through microwave-assisted reverse polymerization (hydrolysis) in the presence of methanol obtaining a high yield of ethene glycol and dimethyl terephthalate the monomer employed for the synthesis of PET.

The microwave-assisted reverse polymerization (hydrolysis) of polycarbonate (PC) in the presence of NaOH using tetrahydrofuran/water as solvent has been also reported [27] obtaining bisphenol-A, the main monomer employed for the synthesis of PC.

3.1.6 Polylactic acid (PLA)

PLA is a thermoplastic polyester formally obtained by condensation of lactic acid. It can be prepared by ring-opening polymerization of lactide, the cyclic dimer of the lactic acid. PLA has become a popular material due to its production from a renewable resource (lactic acid is industrially produced by fermentation of glucose, fructose, or galactose using lactic acid bacteria). Its widespread application has been hindered by numerous physical and processing shortcomings. PLA is the most widely used plastic filament material in 3D printing.

A reverse polymerization of PLA may be realized through MAP using different MW absorbers (tyre, carbon, Fe), apparatus set-up, and MW power [28]. A liquid rich in 3,6-dimethyl-1,4-dioxane-2,5-dione (lactide, one of the starting monomer for the production of PLA) and other oxygenated organic compounds were formed. The pure form of l-lactide was separated from the collected liquid by crystallization, while the meso-form of lactide remained in the mother liquors and was identified through GC–MS analysis. Up to 27.7% of lactide was obtained and might be recovered. Furthermore, simple acids such as acetic and propionic acids (up to 17.1 %), carbonyl compounds, and fragments of PLA backbone randomly cleaved were present in the liquid collected.

When the pyrolysis of PLA was run in the presence of tyres (used as MW absorber) the products showed the presence of cross-reactions between the intermediates formed from PLA and tyres affecting both yield and characteristics of the liquid. Aromatic compounds were formed from tyres pyrolysis and they improved the solvent properties of the liquid, preventying the crystallization of lactide.

3.2 Mixed plastics

In 2017 a proposal on friendly management of waste/contaminated polymeric materials from differentiated municipal solid waste collection through microwave-assisted pyrolysis [29] was reported to enlighten the properties and possible use of the main products formed.

Zou et al. [30] in 2021 reported the pyrolysis of plastic wastes with SiC as an MW absorber and it is a new, continuous MAP process for fuel production. Higher pyrolysis temperatures promoted the cracking of wax and lighter and more stable hydrocarbons were formed. Talc as a filler in commercial polypropylene showed high cracking activity. Incorporating ZSM-5 catalysts in the waste using a space velocity of 10 h^{-1} and a pyrolysis temperature of 620°C a yield of liquid of 48.9 % was obtained and this product consisted of 73.5 % of gasoline-range hydrocarbons rich in aromatics (45.0 %) and isomerized aliphatic hydrocarbons (24.6 %). However, the catalyst rapidly lost its activity using a feedstock/catalyst ratio of 5:1.

Microwave-assisted catalytic pyrolysis of municipal solid waste was reported by Yu et al. in 2020 [31] using a catalyst able to be regenerated. The catalyst was employed in a fluidized bed pyrolysis furnace with a microwave generator. Data were not reported on the summary of the Chinese paper.

An artificial mixture of cellulose, paraffin oil, kitchen waste, and garden waste that closely mimic municipal solid wastes (MSW) was pyrolyzed at different reaction conditions [32] using MAP. Ten different MW absorbing materials such as aluminum, activated carbon, garnet, iron, silica beads, cement, SiC, TiO_2, fly ash, and graphite was tested. MAP was run up to 600°C, and the effects of MSW/MW absorber ratio and composition of the model MSW mixture were reported. The MW absorber affected the yields of oil, gas, and char, and played a catalytic role in altering the selectivity towards the various components present in the liquid. The oil contained oxygenated compounds (furans, phenolics, cyclic-oxygenates) from the biomass present in MSW, aliphatic, and aromatic hydrocarbons (mono- and polycyclics). Aromatic hydrocarbons were mainly derived from lignin decomposition while aliphatic hydrocarbons were derived from cellulose and plastic pyrolysis. The highest yield of oil (53 wt. %) was achieved with a 1/1 wt./wt. ratio of MSW/graphite. The energy in the oil corresponded to energy recovery of nearly 95 % with an 85 % deoxygenation. Using a high ratio of MW absorber/MSW monoaromatics such as benzene, toluene, xylene and styrene, and C_8-C_{20} aliphatic hydrocarbons were formed with high selectivity, while polycyclic aromatics were obtained with low selectivity. Methane, ethene, propene, isobutene, and hydrogen were the major products in the gaseous phase, whose selectivities varied with MSW composition.

MSW containing about 40 % of food was treated through MAP at different MW power [33]. The maximum oil yield of 30.2 wt. % was obtained under the optimized pyrolysis conditions: 400°C, residence time 30 min and a nitrogen flow rate of 50 mL/min at the microwave power of 450 W. Surprisingly the liquid contained oxygen, sulfur, nitrogen, and phosphorous containing compounds: methylphosphine [CH_3PH_2] (the main compound), 2-ethoxyethylenamine (CH_3CH_2-O-CH_2CH_2-NH_2), 2-methoxyethylenamine [CH_3CH_2-O-CH_2CH_2-NH_2], 2-fluoropropane [C_3H_7F], (2-hydroxyethyl) (trimethylsilylmethylen)sulfide [Me_3Si-CH_2-S-CH_2CH_2OH], and 1,3-bis(2-hydroxymethylen)urea [OC(NH-$CH_2OH)_2$] were identified and quantified by GC–MS analysis. The heating value of the oil was 23.94 MJ/kg.

3.3 Plastic composites

The MAP of some plastic composites such as end cycle tyres [34–39], multilayer packaging beverages, corn derived plastic bags, biomasses and so on have also been reported.

3.3.1 Tyre

Waste tyres are well known to have relevant disposal or reprocessing problems under environmental and economic sustainable conditions and their disposal is a challenge for industrial and academic research. Mechanical recycling to produce renewed tyres or granular tyres (employed in several fields such as athletics track, road paving, and so on) are followed but these processes use only a fraction of tyres to be disposed of. In this contest, pyrolysis represents a modern valid alternative to generate value-added products. Anyway improvements in the heat transfer technology are crucial to optimize the efficiency of the process itself. The use of MAP is one of the most promising heating technologies for their pyrolysis, due to MAP's ability to heat quickly and directly any MW absorbing material. Tyre contains a high amount of MW absorbing materials such as metal wires, metal oxides, and carbon, and they quickly absorb MW and turn it into heat [12–19]. The process was performed in a short time comparing with traditional heating techniques [2]. Typical products were a char (yield 40.6-65.0 wt. %), a liquid (yield 20.7-44.0 wt. %) and a gas (yield 9.0-27.4 wt. %). MAP variables such as MW power, reaction time (15-100 min) and tyre mass, may strongly affect the properties of the products. The char was characterized through chemical (ultimate analysis and ion coupled plasma-mass spectroscopy (ICP-MS)), morphological (BET surface area, scanning electron microscopy), and X-ray diffraction (XRD) analyses. It contains a large amount of amorphous carbon (up to 92.0%) and inorganic compounds formed from additives employed in tyres formulation. XRD analyses of crystalline phases of char showed a marked MW effect: different crystalline ZnS forms, spharelite or wurtzite were identified. The presence of these compounds suggested that tyres were heated to a temperature higher than that one usually accounted. The liquid was a low viscosity oil (<2.9 cP, with a large amount of single-ring aromatic hydrocarbons) while the gas contains light hydrocarbons, hydrogen and only traces of N_2. The three products collected had a high calorific value, respectively 34 MJ/kg for solid, 45 MJ/kg for liquid, and 46 MJ/kg for the gas fraction. The most performing conditions were achieved using an MW power of 3 kW per 0.2 kg of tyres even if these conditions were not optimized.

The char from MAP of waste tyres contains iron and carbon. The iron may be separated using an electromagnet and sold as armonic steel (iron) while the carbon can be directly reused to produce new tyres, or used in many other production processes: as a pigment for the production of plastics, or textile printing, for printer toners, etc. It was also evaluated for Oxygen Reduction Reaction (ORR) in electrocatalysis [40]. The presence of different metals together with high carbon in char may produce synergic catalytic effects which are necessary for ORR. The char obtained from microwave-assisted pyrolysis of waste tyres represents a low-cost and friendly source of metals need for the preparation of the cell cathode for ORR.

The liquid is a fuel oil with low sulfur content (< 1%), formed in about 35% of the tyre treated. It contains aliphatic and aromatic hydrocarbons and may be used as marine diesel fuel or sent to a refining treatment to obtain any kind of fuel.

The gas contains hydrogen and low molecular weight hydrocarbons and may be employed as fuel gas or directly used to produce the electricity required for the process [41].

3.3.2 Multilayer packaging beverage

Multilayer packaging beverages are containers of composite material usually manufactured using five layers of materials: LDPE, Ink, Board, Aluminiun (Al), LDPE [42]. The most famous product is Tetrapack©. At the end of their life cycle multilayer packing beverages contaminated by the liquid present in the container, may be displaced through MAP using different MW absorbers (none, chopped tyre, carbon, and iron powder) and apparatus set-ups. Board was pyrolyzed forming water containing bio-oil where alcohols, aldehydes, acids, and anhydrosugars were present, according to pyrolysis conditions. LDPE was converted as reported in chapter 3.1.1, into a high viscosity liquid (wax), solid at room temperature, except when a fractionating system was directly connected to the pyrolysis oven. In these last conditions linear alkanes, alkenes, cyclic, and aromatic hydrocarbons were formed. Al was always recovered as unscratched samples.

3.3.3 Corn derived plastic bags

Europe introduced measures on the management of waste packaging in 2005 through the revision of Directive 94/62/EC and the introduction of the norm EN13432:2000 where it was specified the characteristics of bio-materials to be employed for plastic bags. These directives were adopted in Italy in 2011 and, as a consequence, the polyethylene base shopping bags were forbidden and replaced by bio-degradable materials. Bio-plastics are materials obtained from a vegetable source such as corn and they are largely used, especially those derived from corn starch with the aim to reduce world pollution. Their production requires starch destructuring, complexation, blending with specific synthetic and/or natural polymers, and the addition of compatibilizers, plasticizers, and other additives [43, 44]. This great variability of processing methods brings different types and grades of bio-polymers with a wide range of properties which allows their use in various fields [45].

Corn-derived plastic bags (CDP) are biodegradable and they may be disposed of through anaerobic digestion, however their energy and chemical content, required for their production, will be largely lost, only biogas was sometimes recovered. Furthermore, the biodegradation process is too long with respect to the time required for the biodegradation of the biomass present inside the container, usually derived from a waste collection. For this reason CDPs must be initially separated and then separately bio-degraded. In a greener process, these wastes may be employed for energy production with the recovery of their thermal content, but the atom economy of the process is very poor because the chemical content of CDP is lost. MAP of CDP was performed to evaluate the possibility to efficiently dispose of this waste obtaining a liquid useful as a source of chemicals and/or fuels. MAP was performed using different MW power, MW absorbers (carbon or Fe), and apparatus obtaining a liquid, a char, and a gas in amount depending on the pyrolysis conditions. Liquids were always separated into three phases: upper, middle, and bottom and deeply characterized. They contained a large amount of aromatic acid, phthalates, and their derivatives in the upper fraction; water (70 wt. %), organic acids, alcohols, anhydro-sugars, and their pyrolysis products were present in this fraction; bottom fraction showed close properties and composition than the upper fractions. The same classes of compounds were present in upper and bottom phases but in the last ones, these compounds were oligomers of those present in the upper fractions. Working with more drastic conditions bottom fraction was reduced while char yield was increased. MAP converted a waste CDP, even contaminated, obtaining some chemicals or even fuel thus avoiding to send them to anaerobic digestion for their transformation into carbon dioxide, methane, water, and a residue with a very low recovery of chemicals or energy. MAP process may represent a green solution to dispose of CDP recovering a liquid useful as a source of chemicals or employed to produce fuels. Furthermore, this process avoids the contamination of the soil improvement obtained by an anaerobic digestion with the residue of plasticizers present in CDP.

3.3.4 Waste electric and electronic equipment (WEEE)

The collection of waste electric and electronic equipment (WEEE) is largely and rapidly increasing, and even if WEEE is recycled today there are still valuable residues left after recycling, ending up in landfills. MAP may be a valuable way to recycle these components. Six different fractions (from light dust to particles sized of 7-12 mm) were pyrolyzed, producing an oil, a gas, and a solid residue [46]. A mass reduction was observed as a function of process time, independently on the reaction temperature for all of the WEEE treated.

WEEE from the envelope of end-life computers was pyrolyzed using different absorbers and set-ups in a multimode batch reactor [47]. A large amount of the liquid fraction (yield 76.6 wt. %) was obtained together with a strong reduction of the solid residue (yield 14.2 wt. %). The liquid fraction was characterized using Fourier Transform-Infrared Spectroscopy-Attenuated Total Reflection (FT-IR ATR), Nuclear Magnetic Resonance Spectroscopy (^{1}H NMR), and quantitative GC–MS analysis. The liquid showed a low density, viscosity and contained a high concentration of useful chemicals such as styrene (up to 117.7 mg/mL), xylenes (up to 25.6 mg/mL of p-xylene) while halogenated compounds were absent or present in undetectable amounts.

Waste printed circuit boards (a fraction of WEEE) were also pyrolyzed using carbonaceous absorbers such as graphite or activated carbon [48]. Char was the main product (58 wt. %) and the presence of a high amount of absorber reduces the yield of tar and gas. Char contained phenolic compounds and phenylphosphates in significant amounts. CO_2 was the major component in the gas fraction, while the concentration of H_2 in the gas was increased when activated carbon was used as an absorber. The major metal-containing compounds present in char were Cu, Pb, Ti, while a lower amount of Bi, Fe, and Ca were present.

MAP of WEEE with bromine-containing compounds as flame retardant was also performed with special attention to the fate of these compounds. Conversion is increased with increasing temperature, reaching 93.3 wt. % at 650°C [49]. High pyrolysis temperature enhanced the transfer of bromine containing compounds to pyrolysis gas while the presence of K_2CO_3, Na_2CO_3 and NaOH mixed with the WEEE formed KBr and NaBr reducing the presence of HBr in the gas. Increasing heating time did not exhibit a remarkable influence on pyrolysis conversion: Working at 350°C, the main compounds in the liquid were phenols (91.1 %) while at 650°C, polycyclic and monocyclic aromatic hydrocarbons (except phenols) increased to 20.5 % and 19.0 %, respectively. Meanwhile, the non-condensable gases were composed of CO_2, CO, CH_4, and H_2 and they improved significantly with increasing temperature. Working in the presence of ZSM-5 and kaolin as catalysts, the monocyclic aromatic hydrocarbons (except phenols) and C_{11}-C_{20} compounds in the oil was increased while non-condensable gas was reduced.

A review of existing classical pyrolysis techniques for recycling of plastics in WEEE with special attention to the recovery of the products obtained (monomers, hydrocarbons, phenols, etc.) was reported [50]. Special attention was devoted to the processes to remove bromine containing compounds in the starting materials such as solvent extraction, supercritical fluid technology, and so on. Co-pyrolysis was also investigated and plastic wastes were displaced without the use of solvents or catalysts. The catalysts were employed to affect the distribution of the products and to enhance the removal of bromine containing compounds from pyrolysis oils.

4. Metal recovery from waste/contaminated plastics through MAP

Metal containing plastic has been treated with MAP to recover the metal and recycle all the other components like a solid, a liquid and a gas.

a. Several metals are present in waste or contaminated plastics, for instance, lead is present in the envelope of lead-acid battery realized using HDPE as the material of the container. At the end of life of the lead-acid battery the Pb of the electrodes and H_2SO_4 are recovered and recycled while the envelope must be disposed of as a contaminated HDPE because it contains Pb. This contaminated plastic may be pyrolyzed using a MAP process using carbon as an MW absorber. In the course of the process gas is formed (containing hydrogen and low molecular weight hydrocarbons) together with a liquid (containing a small amount of water, and aliphatic hydrocarbons close to those reported in paragraph 3.1.1 on the MAP of HDPE). The char contains the MW absorber, the low amount of carbon formed from HDPE, and the lead present as a contaminant [51] and this char may be recycled as an MW absorber. For each recycle of the char, the concentration of lead is slightly increased every time. When an appropriate concentration of lead was reached the MW absorber was sent to a process for the extraction of the lead present, for instance through acid or metallurgical treatment, then the MW absorber may be recycled or in the case of carbon, used as a solid fuel together with the carbon formed from HDPE. In **Figure 2** is reported a schematic representation of the process.

b. An analogous process was employed to recover precious metals (i.e. Ag) present in some painted materials at the end of the life cycle of painted (for instance in the painted car bodywork). The plastic painted materials were pyrolyzed through MAP collecting a liquid employed as a fuel, a gas that may be burned to produce energy while the precious metal remained in the char together with the MW absorber. The char was recycled several times, up to an appropriate concentration of the precious metals was present in the char. At this point, the precious metal was recovered through a metallurgical process.

c. Using an analogous methodology the printed circuit board, a class of WEEE, has been treated to pyrolyze the plastic material recovering an oil containing a large amount of phenols [46] and a residue where are present, in a high amount, several metal containing compounds such as Cu, Pb, Ti, and minor amounts of other metal containing compounds including Bi, Fe and Ca.

d. Another application of MAP for the recovery of metal containing plastics is represented by pyrolysis of waste tyres. In this MAP [52], as reported in paragraph 3.3.1, harmonic steel (iron) present as a reinforcing material in tyre (about 5-10% of the tyre), remains in the char, from which it was recovered with an electrical

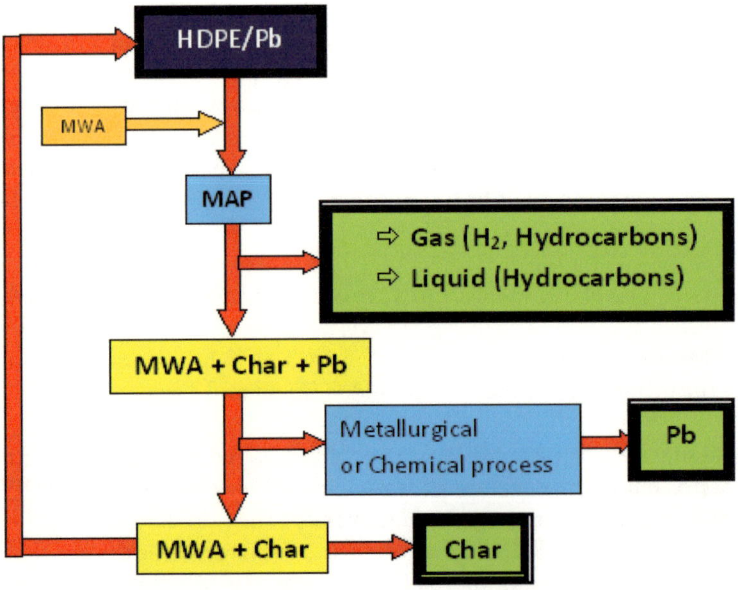

Figure 2.
A schematic representation on the MAP of lead containing HDPE to recover lead. MWA: Microwave absorber.

magnet and sent to a metallurgical process to enter in a new cycle of utilization. The char formed during the pyrolysis of tyre (recycled carbon, rCB) may be used to prepare new tyres or may be employed to prepare anode for Oxygen Reduction Reaction (ORR) [40] as previously suggested in paragraph 3.3.1.

5. Some industrial applications

a. Environmental Waste International (https://www.ewi.ca/tyre.html):

This company, since 1992, located in Whitby – Ontario (Canada), designs specialized eco-friendly waste reduction solutions and systems, using patented Reverse Polymerization processes and proprietary Microwave Delivery Systems. The recycle of scrap tyres has been a huge priority for EWI since 1994, demonstrating its capabilities. The results show the realization of a continuous feed system and its operation. EWI was able to prove the economic tyre recycling projections.

b. Molectra (https://wastemanagementreview.com.au/a-rubberised-archetype-molectra-technologies/):

This company developed a series of machines and a technique to process tyres extracting rubber in a fine powder format, which is unique (the name of the company means MOLECularTRAnsformation). The process comprises mechanical, chemical, and MW treatments. The residual carbon is pure (97.4 %) and can be crushed to form carbon black, which is used to manufacture new tyres, plastics, paints, inks, and batteries. Alternatively, the carbon can be converted into activated carbon for water purification, gold extraction, and air filtration. Another clever development is the inclusion of fertilizers and soil microbes into the char. The capacities can be 2-10 tons/h of processed tyres.

c. SBC (https://www.sbiofuel.com/downloads.html):

SBC pursues industrialization and commercialization of its proprietary and patented technology; the utilization of MAP of any material of hydrocarbon origin, being it any biomass, plastic, or rubber. The product strategies differ as a function of the markets and according to the maturity of technologies.

d. Tyrebirth s.r.l. - Italy (https://www.tyrebirth.com/en/):

Tyrebirth has developed, following proprietary patents [38, 53, 54] a very high-efficient recycling process that allows the creation of high-quality products through the MAP of end-of-life tyres. The Tyrebirth process produces solid, liquid, and gaseous products:

i. Solid - carbon black (rCB) - about 40% of the tyres treated. It has very high performances, it is dry, unpolluted and with quality characteristics 95% close to virgin carbon black. The rCB can be directly reused to produce new tyres, or used in many other production processes: as a pigment for the production of plastics, or textile printing, for printer toners, and so on.

ii. Solid - harmonic steel - about 10% of the production. This product has strong market demand. The steel is easily separated and delivered to steelwork for reuse.

iii. Liquid - Fuel Oil with low sulfur content (< 1%) - about 35% of the tyres treated. The liquid product is a low sulfur diesel marine oil and can be used in ships' diesel engines, in thermal plants for the production of electricity, and other specific installations.

iv. The gas – a mixture containing LPG gases and hydrogen. It is about 15% of the tyres treated. It may be used to produce more than 70% of the plant's electric energy required through cogeneration. The capacity of the plant is 540 Kg/h. Some pictures of the plant are reported in the following **Figure 3**.

e. Techwave s.r.l. - Italy (https://www.techwave.it/)

Techwave follows an innovative technological solution using proprietary patents of the MAP process [35, 39] to recover the material and energy content of waste plastic obtaining new materials from waste or contaminated plastics. The process is innovative, energetically sustainable, and has a low environmental impact. The innovation of the process is the use of microwave to supply the energy required by the process and a MW absorber, obtaining uniform heating of the material avoiding the problems of classical thermal heating. Another characteristic of the plant is its compactness because it may be installed in two standard containers (**Figure 4**).

f. Pyrowave - Canada (https://www.pyrowave.com/en/)

Pyrowave provides a unique technology using a MAP process [55] to regenerate post-consumer plastics by breaking them down into intermediate products that are used to make new plastics identical to virgin plastics, restoring their full value. Pyrowave paves the way to a truly circular economy of plastics. The regeneration of plastics allows for infinite recycling and avoids the Green House Gases associated with the extraction of virgin material.

g. Industrial co-operations

Some co-operation has been realized among companies and some important firms or institutions for chemical recycling of synthetic polymers, even if some of them, at this time, do not plan to use the microwave to supply the energy

Figure 3.
A picture of the Tyrebirth plant (by curtesy of Tyrebirth s.r.l.)

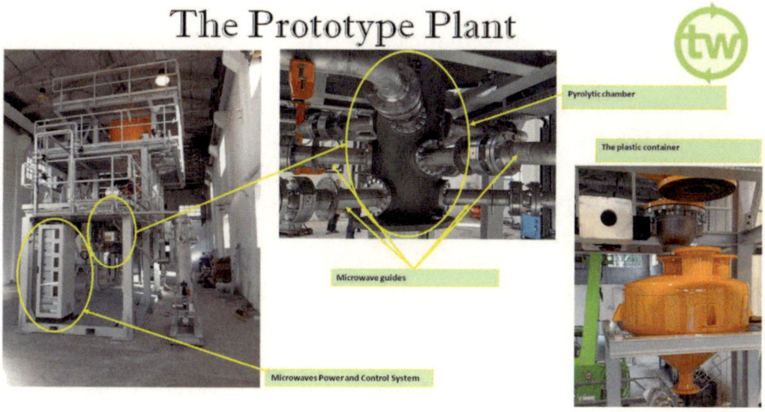

Figure 4.
Some picture of the prototype plant of Techwave s.r.l. (by curtesy of Techwave s.r.l.)

for the process, but the use of MW may lead to a better improvement of the process itself.

i. Montreal-based specialty recycling company Polystyvert (for the chemical recycling of polystyrene throw solvent dissolution) has closed a round of funding to facilitate the development of full-scale PS recycling plant [56]. Proceeds will be used to facilitate the development of a full-scale plant – a strategic milestone for the company to demonstrate the high level of purity reached by their products.

ii. AmSty, the largest PS producer in the Americas, and Agilyx Corporation a wholly-owned subsidiary of Agilyx AS (Euronext Growh – Oslo: "AGLX"), a pioneer in the advanced recycling of post-use plastics, announced an agreement to explore the development of a jointly owned advanced recycling facility of a 50–100 tons/ already in use at the parties' Regenyx joint venture where post-use PS products are converted back into virgin-equivalent styrene monomer.

iii. Versalis (Eni's chemical company) and AGR (an Italian company) have signed an agreement to develop technological innovations and new products and applications with recycled rubber [57].

iv. ReVital Polymers, Pyrowave and INEOS Styrolution are realizing a North American PS recycling consortium [58] to launch a closed-loop for recycling single-serve PS packaging using recycling technology pioneered by Pyrowave (a MAP process) obtaining styrene monomer used for manufacturing new PS products.

v. SABIC and Plastic Energy are set to start construction of world's first commercial unit to significantly upscale production of certified circular polymers derived from used plastic [59].

vi. BASF, Quantafuel, and REMONDIS have signed a Memorandum of Understanding (MoU) to jointly evaluate cooperation in chemical recycling including a joint investment into a pyrolysis plant for mixed plastic waste [60].

REMONDIS supplies suitable plastic waste to the plant and BASF uses the resulting pyrolysis oil as feedstock in its production.

vii. Neste successfully processed 400 tons of liquefied plastic waste at its refinery in Finland this fall (2021), an amount corresponding to the annual amount of plastic waste generated by 20,000 average European citizens [61]. Neste's target is to process over 1 x 10^6 ton of waste plastic annually from 2030 onwards.

viii. Techwave has evaluated the possibility to pyrolyze facemasks used for Covid19 pandemia in one of its microwave ovens using the MAP and a microwave absorber [62]. The facemasks were converted into the classical three products of a MAP (liquid, small amount of solid, and gas) having a composition close to those reported in the pyrolysis of PP. The possible contamination was absent in the products formed due to the pyrolysis conditions adopted.

ix. The Fraunhofer Institute UMSICHT, with SABIC and Procter & Gamble (P&G) start a collaboration to demonstrate the chemical recycling of used facemasks using a dedicated research pyrolysis plant [63]. The masks were first automatically shredded and then thermochemically converted to pyrolysis oil. Any residual pollutants or pathogens, such as the Coronavirus were absent in the products. The pyrolysis oil was employed by SABIC as feedstock for the production of new PP resin obtaining a high-quality PP.

x. The Techwave plant may improve the KuWert project [64] because the MAP-Techwave plant may be installed in two standard containers, so it may be installed on the ocean platform of the KuWert project or, more interestingly, on a ship devoted to collect the plastic present in the ocean [65]. In this last option, the ship may sail where a large amount of plastics are present, collect them, and immediately introduced into the pyrolysis plant present on board. In this way the waste plastics will be converted into useful materials, part of them employed to supply the energy required by the plant and the ship while the remaining fuel may be sold on the market to obtain the revenue need for the activity. After that the waste plastic may be immediately eliminated, avoiding the cost for their transport and only the final products may be sent to the market of raw materials.

xi. EEW Energy from Waste together with DSD plans to develop a new plant for processing mixed plastics and sorting residues [66] through chemical recycling. The process will be developped by the end of this year (2021) to process up to 200,000 tons/y of plastic waste.

xii. Borealis starts a new project for an increase on supply of chemically recycled feedstock for the production of more circular base chemicals and polyolefin-based products [67]. This research is carried out with project partner Stena Recycling and provided a successful feasibility study and final investment decision, operations are expected to begin in 2024.

xiii. Dow a global leader in materials science, and Mura Technology, the global pioneer of an advanced plastic recycling solution, announced a partnership to keep plastic waste out of the environment [68]. The collaboration will support the rapid scaling of Mura's new HydroPRS™ (Hydrothermal Plastic Recycling Solution). This process can recycle all forms of plastic – including multi-layer, flexible plastics with the first 20,000 tons/y line expected to be operational in 2022.

xiv. LyondellBasell successfully starts up a new pilot Molecular Recycling Facility (*MoReTec)* at its Ferrara, Italy, site [69]. The pilot plant is capable of processing between 5 and 10 kg/h of household plastic waste.

xv. Repsol, Axens, and IFPEN, have joined forces to develop the pioneering and patented Rewind™ Mix process [70]: It purifies the plastics pyrolysis oils allowing the direct and undiluted processing in existing petrochemical plants for the production of circular plastics. Repsol, Axens, a worldwide technology provider and IFPEN, the renowned French research and innovation player in the field of energy, have developed a pioneering and patented process to enhance the chemical recycling of plastic waste and boost circular materials production. Pyrolysis is one of the most promising pathways for the chemical recycling of plastic waste, which otherwise the Rewind™ Mix process removes impurities such as silicon, chlorine, diolefins, and metals from the plastics pyrolysis oils produced, allowing the direct use as undiluted feed to petrochemical units.

xvi. Chemical recycling, has been developed by ExxonMobil to molecularly convert difficult-to-recycle plastics into the virgin-quality raw material used to make a wide range of valuable new products and potentially repeat the process over and over again [71]. The facility will have an initial capacity of 25,000 tons/y of plastic waste, with plans to scale up to 33,000 tons/y shortly.

xvii. DEMETO project: a Modular, scalable and high-performance DE-polymerization by MicrowavE TechnolOgy [72]. Based on an internationally patented technology, the project foresees to bring at industrial level (through a completely functional pilot plant) an innovative process, based on applying a new microwave technology to a well-known chemical reaction: the reverse polymerization (alkaline hydrolysis depolymerization) of PET.

6. Conclusions

This review shows several MAP processes proposed for friendly and economical disposal of waste/contaminated plastics to reduce the carbon footprint and obtain valuable products useful to close the cycle of a circular economy.

MAP is a very interesting way because it does not produce waste from chemical recycling of plastics but produces three classes of product: a char, an oil and a gas every one available as fuel or the source of new materials for the synthesis of new products. The process is self-sustainable from an energetic point of view rendering the process economically sustainable.

The interest of industrial companies in this field is improving day by day, as reported in the last part of the chapter because this is an important aspect for the realization of a truly circular economy and the reduction of the Green House Gas in the atmosphere.

Acknowledgements

Authors personally thank MIUR-Italy for "Progetto Dipartimenti di Eccellenza 2018-2022" allocated to Department of Chemistry "Ugo Schiff".

Author details

Piero Frediani and Marco Frediani*
Department of Organic Chemistry, University of Florence, Florence, Italy

*Address all correspondence to: marco.frediani@unifi.it

IntechOpen

© 2021 The Author(s). Licensee IntechOpen. This chapter is distributed under the terms of the Creative Commons Attribution License (http://creativecommons.org/licenses/by/3.0), which permits unrestricted use, distribution, and reproduction in any medium, provided the original work is properly cited.

References

[1] Plasticeurope, Plastic-the fact 2019 An analysis of European plastics production, demand and waste data, Available from https://www.plasticseurope.org/en/resources/publications/4312-plastics-facts-2020 [accessed 2021-06-25].

[2] Dogu O, Pelucchi M, Van de Vijver R, Van Steenberge PHM, D'hooge DR, Cuoci A, Mehl M, Frassoldati A, Faravelli T, Van Geem KV: The chemistry of chemical recycling of solid plastic waste via pyrolysis and gasification: State-of-the-art, challenges, and future directions. Prog Energ Comb Sci, 2021; 84:1009001. DOI:10.1016/j.pecs.2020.100901.

[3] Scheirs J, Kaminsky W, Feedstock Recycling and Pyrolysis of Waste Plastics: Converting Waste Plastics into Diesel and Other Fuels, 2006, J. Wiley & Sons Ltd, ISBN:9780470021545, DOI:10.1002/0470021543.

[4] Undri A, Rosi L, Frediani M, Frediani P: Microwave Pyrolysis of Polymeric Materials, in Microwave heating, U. Chandra ed., InTech, 2011. 207-226.

[5] Motasemi F, Afzal MT, A review on the microwave-assisted pyrolysis technique, Ren Sust Energ Rev 2013; 28:317-330. DOI: 10.1016/j.rser.2013.08.008.

[6] Fernandez Y, Arenillas A., Menendez JA: Microwave Heating Applied to Pyrolysis, in Advances in Induction and Microwave Heating of Mineral and Organic Materials, Grundas S ed, Chapt 7, 723-752. DOI: 10.5772/13548.

[7] Zhou N, Dai L, Lyu Y, Li H, Deng W, Guo F, Chen P, Lei H, Ruan R, Catalytic pyrolysis of plastic wastes in a continuous microwave assisted pyrolysis system for fuel production", Chem Eng J, 2021;418:129412. DOI:10.1016/j.cej.2021.129412.

[8] Arshad H, Sulaiman SA, Hussain Z, Naz Y, Basrawi F: Microwave assisted pyrolysis of plastic waste for production of fuels: a review, MATEC Web of Conferences 2017, 131:02005/1-02005/7. DOI:10.1051/matecconf/201713102005.

[9] Chandra U. Microwave Heating, InTech, 2011. ISBN 978-953-307-573-0.

[10] Bartoli M, Frediani M, Briens C, Berruti F, Rosi L: An overview of temperature issues in microwave-assisted pyrolysis, Processes, 2019;7:658. DOI:10.3390/pr7100658.

[11] Wong SL, Ngadi N, Abdullah TAT, Inuwa IM, Current state and future prospects of plastic waste as source of fuel: A review, Renew Sust Energ Rev, 2015;50:1167-1180. DOI:10.1016/j.rser.2015.04.063.

[12] Estahbanati MRK, Kong XY, Soo HS, Eslami A, Current Developments in the Chemical Upcycling of Waste Plastics Using Alternative Energy Sources, ChemSusChem 2021. doi.org/10.1002/cssc.202100874.

[13] Ludlow-Palafox C, Chase HA Microwave pyrolysis of plastic wastes, in Scheirs J, Kaminsky W ed., Feedstock Recycling and Pyrolysis of Waste Plastics 2006;569-594. DOI:10.1002/0470021543.ch21.

[14] Undri A, Rosi L, Frediani M, Frediani P: Efficient Disposal of Waste Polyolefins through Microwave Assisted Pyrolysis", Fuel, 2014;116:662-671. DOI:10.1016/j.fuel.2013.08.037.

[15] Russell AD, Antreou EI, Lam SS, Ludlow-Palafox C, Chase HA: Microwave-assisted pyrolysis of HDPE using an activated carbon bed, RSC Advances 2012;2:6756.

[16] Jiang H, Liu W, Zhang X, Qiao J, Liu W, Qiao J, Chemical Recycling of

Plastics by Microwave-Assisted High-Temperature Pyrolysis, Global chall 2020,4:1900074, Hoboken, NJ.

[17] Wang W, Wang J, Sun G, Catalyst used for microwave pyrolysis of waste plastics, and preparation method and application method thereof, Faming Zhuanli Shenqing 2013, CN 103252226 A 20130821.

[18] Ding K, Liu S, Zhou N, Peng P, Wang Y, Chen P, Ruan R: Catalytic microwave-assisted pyrolysis of plastic waste over NiO and HY for gasoline-range hydrocarbons production, Energ Conv Manag, 2019;196:1316-1325. DOI:10.1016/j.enconman.2019.07.001.

[19] Suriapparao DV, Vinu R, Resource recovery from synthetic polymers via microwave pyrolysis using different susceptors, J Anal Appl Pyrol, 2015;113:701-712. DOI:10.1016/j.jaap.2015.04.021.

[20] Undri A, Frediani M, Rosi L, Frediani P: Reverse Polymerization of Waste Polystyrene through Microwave Assisted Pyrolysis, J Anal Appl Pyrol, 2014;105:35-42. DOI: 10.1016/j.jaap.2013.10.001.

[21] Hussain Z, Khan KM, Hussain K: Microwave-metal interaction pyrolysis of polystyrene, J Anal Appl Pyrol, 2010;89:39-43.

[22] Hussain Z, Khan KM, Perveen S, Hussain K, Voelter W: The conversion of waste polystyrene into usefull hydrocarbons by microwave –metal interaction pyrolysis, Fuel Process Techn, 2012;94:145-150.

[23] Bartoli M, Rosi L, Frediani M, Undri A, Frediani P: Depolymerization of polystyrene at reduced pressure through a microwave assisted pyrolysis, J Anal Appl Pyrol, 2015;113:281-287. DOI:10.1016/j.jaap.2015.01.026.

[24] Prathiba R, Shruthi M, Miranda LR: Pyrolysis of polystyrene waste in the presence of activated carbon in conventional and microwave heating using modified thermocouple, Waste Manag, 2018;76:528-536.

[25] Frediani P, Undri A, Rosi L, Frediani M: Waste/Contaminated Polystyrene Recycling through Reverse Polymerization" in Polystyrene Synthesis, Characteristics and Applications, Cole Lynwood ed., Novapublisher, 2014. ISBN: 978-1-63321-371-5.

[26] Williams PT, Williams EA: Interaction of plastic in Mixed Plastic Pyrolysis, Energ Fuels, 1999;13:188-196.

[27] Rosi L, Bartoli M, Undri A, Frediani M, Frediani P: Synthesis of Dianols or BPA through catalytic hydrolyisis/glycolysis of waste polycarbonates using a microwave heating. J Mol Catal, A, 205;408: 278-286. DOI: 10.1016/j.molcata.2015.07.027.

[28] Undri A, Rosi L, Frediani M, Frediani P: Conversion of poly(lactic acid) to lactide via microwave assisted pyrolysis J Anal Appl Pyrol, 2014;110:55-65. DOI:10.1016/j.jaap.2014.08.003.

[29] Rosi L, Frediani M, Undri A, Bartoli M, Frediani P: A Friendly Management of Waste/Contaminated Polymeric Materials from Differentiated Waste Collection through Microwave Pyrolysis" chapt. 4 in Municipal Solid Waste: Management Strategies, Challenges and Future Directions, Nikolaos Tzortzakis ed. Novapublisher, 2017:83-125, ISBN: 978-1-53611-865-0.

[30] Zhou N, Dai L, Lyu Y. Li H, Deng W, Guo F, Chen P, Lei H, Ruan R: Catalytic pyrolysis of plastic wastes in a continuous microwave assisted pyrolysis system for fuel production, Chem Eng J, 2021;418:129412. DOI:10.1016/j.cej.2021.129412.

[31] Yu Z, Zhang Y, Zhang X, Ma X: A kind of catalyst capable of regeneration

of microwave assisted catalytic pyrolysis of municipal solid waste device for rapid, Shiyong Xinxing Zhuanli Shuomingshu, 2020, CN 210595928 U 20200522.

[32] Suriapparao DV, Vinu R: Bio-oil production *via* catalytic microwave pyrolysis of model municipal solid waste component mixtures, RSC Adv, 2015;5:57619-57631. DOI:10.1039/C5RA08666C.

[33] Kadlimatti HM, Raj Mohan B, Saidutta MB: Bio-oil from microwave assisted pyrolysis of food waste-optimization using response surface methodology, Biomass Bioenerg, 2019;123:25-33. DOI:10.1016/j.biombioe.2019.01.014.

[34] Undri A, Meini S, Rosi L, Frediani M, Frediani P: Microwave Pyrolysis of Polymeric Materials: Waste Tires Treatment and Characterization of the Value-Added Products, J Anal Appl Pyrol, 2013;103:149-158. DOI: 10.1016/j.jaap.2012.11.011.

[35] Frediani P, Rosi L, Frediani M, Undri A, Occhialini S, Production of hydrocarbons from copyrolysis of plastic and tire material with microwave heating, 2012: WO2012110990, PCT/IB2012/050747.

[36] Undri A, Sacchi B, Cantisani E, Toccafondi N, Rosi L, Frediani M, Frediani P: Carbon from Microwave Assisted Pyrolysis of Waste Tires, J Anal Appl Pyro, 2013;104:396-404. "http://dx.doi.org/doi:10.1016/j.jaap.2013.06.006"doi:10.1016/j.jaap.2013.06.006.

[37] Undri A, Rosi L, Frediani M, Frediani P: Upgraded Fuel from Microwave Assisted Pyrolysis of Waste Tire, Fuel, 2014;115:600-608. DOI: 10.1016/j.fuel.2013.07.058.

[38] Frediani P, Rosi L, Frediani M, Meini S, Undri A, Occhialini S: Produzione di idrocarburi da Pirolisi di gomme, Italian Pat, 2011, 1404132B1.

[39] Frediani P, Rosi L, Frediani M, Undri A, Occhialini S: Produzione di idrocarburi da Co-Pirolisi di Materiali Plastici e Pneumatici con riscaldamento a Microonde, Italian Pat, 2011, 1404131B1.

[40] Passaponti M, Rosi L, Frediani M, Salvietti E, DeLuca, A, Giaccherini A, Innocenti, M: Microwave assisted pyrolysis of waste tires: study and design of half-cells SOFCs with low environmental impact, ECS Trans, 2017; 78:1933-1940. DOI:10.1149/07801.1933ecst.

[41] Scandinavian Biofuel company: Advanced and sustainable biofuels, Available from https://www.sbiofuel.com/downloads.html (accessed 2021-06-28).

[42] Dyson J: Packaging materials, 9. Multilayer packaging for food beverage, in ILSI Europe report series, Available from https://www.pac.gr/bcm/uploads/9-multilayer-packaging-for-food-and-beverages.pdf (accessed 2021-06-28).

[43] Bastioli C: Properties and applications of Mater-Bi starch-based materials, Polym Degrad Stabil, 1998;59:263-272.

[44] Haque MM-U, Alvarez V, Paci M, Pracella M, Processing, compatibilizationand properties of ternary composites of Mater-Bi with polyolefins and hempfibres, Compos A: Appl S, 2011;42:2060-2069.

[45] La Mantia FP, Morreale M, Green composites: a brief review, Compos A: Appl, 2011;42:579-588.

[46] Andersson M, Knutson Wedel M, Forsgren C, Christeen, J, Microwave assisted pyrolysis of residual fractions of waste electrical and electronics equipment, Min Eng, 2012;29:105-111. DOI:10.1016/j.mineng.2011.09.005.

[47] Rosi L, Bartoli M, Frediani M, Microwave assisted pyrolysis of halogenated plastics recovered from waste computers, Waste Manag, 2018;73:511-522. DOI:10.1016/j.wasman.2017.04.037.

[48] Suriapparao DV, Batchu SP, Jayasurya S, Vinu R: Selective production of phenolics from waste printed circuit boards via microwave assisted pyrolysis, J Clean Prod, 2018;197:525-533. DOI:10.1016/j.jclepro.2018.06.203.

[49] Zhang T, Mao X, Siyal Asif A, Fu J, Jiang Z, Deng Z, Song Y, Polina C, Qu J, Wang D, et al.: Microwave-assisted catalytic pyrolysis of waste printed circuit boards, and migration and distribution of bromine, J Hazard Mat, 2021;402123749.

[50] Charitopoulou MA, Achilias DS, Kalogiannis KG, Lappas AA, Novel trends in the thermo-chemical recycling of plastics from WEEE containing brominated flame retardants, Environm Sci Poll Res Int, 2020.

[51] Frediani M, unpublished results / Research convention between University of Florence and Piombo Leghe srl-It.

[52] Undri A, Rosi L, Frediani M, Frediani P, Nuove Possibilità per la Valorizzazione di Pneumatici Usati: La Pirolisi a Microonde, Chim Ind (Milan), 2012;146-148.

[53] Innocenti G, Simoni R, D'Alessandro G, Andreini V, Frediani P, Sonego A, Occhialini S: Dispositivo di pirolisi, It Patent n. 102018000003163 (01/03/2018), rif. P020275IT-01/sf.

[54] Frediani P, Rosi L, Frediani M, Undri A, Occhialini S, Meini S: Production of hydrocarbons from pyrolysis of tyres, 2012: WO2012110991, PCT/IB2012/050748.

[55] Douchet J, Chaouki J, Sobhy A: Catalyst for distributed batch microwave pyrolysis, system and process thereof, WO2012097448A1 and reference therein cited.

[56] Available at http://www.mrcplast.com/news-news_open-389610.html [accessed 2021-07-15]; Polimerica, 2021/07/05.

[57] Available at https://www.eni.com/en-IT/media/press-release/2020/11/eni-versalis-agreement-agr.html [accessed 2021-07-17]; Polimerica 2020/11/17.

[58] Available at https://www.ineos-styrolution.com/news/revital-polymers-pyrowave-and-ineos-styrolution-partner-to-launch-closed-loop-north-american-polystyrene-recycling-consortium [accessed 2021-07-15].

[59] Available at https://www.sabic.com/en/news/26247-sabic-and-plastic-energy-set-to-start-construction-of-pioneering-advanced-recycling-unit [accessed 2021-07-15].

[60] Available at https://www.basf.com/global/en/who-we-are/sustainability/whats-new/sustainability-news/2021/basf-quantafuel-and-remondis-want-to-cooperate-on-chemical-recycling-of-plastic-waste.html [accessed 2021-07-15].

[61] Available at https://www.neste.com/releases-and-news/plastics/neste-successfully-completed-its-first-industrial-scale-processing-run-liquefied-waste-plastic [accessed 2021-07-15]; https://www.polimerica.it/articolo.asp?id=25563.

[62] Techwave private communication.

[63] Available at https://www.fraunhofer.de/en/press/research-news/2021/june-2021/closed-loop-recycling-pilot-project-for-single-use-face-masks.html; https://www.polimerica.it/articolo.asp?id=26080.

[64] Available at http://www.kuwert.hs-bremen.de/downloads/2019-08-13-KuWert_Grafik-LQ.pdf [Accessed 2021-07-17].

[65] suggestion of the authors.

[66] Available at https://www.bkv-gmbh.de/news-reader-4/dsd-enters-chemical-recycling.html [Accessed 2021-07-21]; https://www.polimerica.it/articolo.asp?id=26287.

[67] Available at https://www.borealisgroup.com/news/true-to-its-everminds-mind-set-borealis-drives-collaborative-project-in-sweden-to-increase-supply-of-chemically-recycled-feedstock-for-the-manufacture-of-more-circular-base-chemicals-and-plastic-products [Accessed 2021-08-07].

[68] Available at https://corporate.dow.com/en-us/news/press-releases/dow-and-mura-technology-announce-partnership-to-scale-game-chang.html [Accessed 2021-07-21].

[69] Available at https://www.lyondellbasell.com/en/news-events/corporate-financial-news/lyondellbasell-successfully-starts-up-new-pilot-molecular-recycling-facility/ [Accessed 2021-07-21].

[70] Available at https://www.ifpenergiesnouvelles.com/article/repsol-axens-and-ifpen-develop-new-process-boost-circular-materials-production [Accessed 2021-07-21].

[71] Available at https://energyfactor.exxonmobil.com/insights/partners/steps-up-advanced-recycling-plastic-waste/#:~:text=Advanced%20recycling%2C%20also%20called%20chemical,that%20process%20over%20and%20over [Accessed 2021-07-21].

[72] Available at https://cordis.europa.eu/project/id/76857 [Accessed 2021-08-12].

Chapter 19

Basic Morphological, Thermal and Physicochemical Properties of Sewage Sludge for Its Sustainable Energy and Material Use in the Circular Economy

Barbara Novosel, Vesna Mislej and Viktor Grilc

Abstract

The treated sewage sludge is a smart material that provides sustainable use of energy and material that can be adapted to the needs of modern human life. Pyrolysis is one of the waste recovery operation that enables the generation of various useful groups of substances from treated sludge, which can then be utilized separately by final energy and material recovery processes: (i) solid residue (pyrogenic material), (ii) liquid fraction (bio-oil), and (iii) non-condensable gas. Those fractions are the basis for fertilizers and alternative fuel or new chemicals production. The chapter combines data on the quality of treated sludge as a case study of an urban wastewater treatment plant (UWWTP) with the capacity of 360,000 population equivalents (PE). The analytical results of dried anaerobically treated sludge over a 10-year period (from 2010 to 2020) are evaluated with respect to the limit values from directives, resolutions, guidelines, and national legislation related to specific processing procedures and quality of the final product. The data provide a comprehensive insight into the basic properties of treated sludge and allow thoughtful use of this smart material for WtE processes, organic matter material conversion, and agricultural use in terms of protecting the environment and human health.

Keywords: circular economy, fertilizer, incineration, pyrogenic material, pyrolysis, sewage sludge, solid residue, urban wastewater treatment plant, waste to energy

1. Introduction

The trend for continual improvement in the collection and treatment of wastewater in Europe's cities and towns is evident, but full compliance with the Urban waste water Directive (UWWTD) [1] has not been achieved yet [2]. The level and intensity of wastewater treatment depend on the sensitivity of the receiving surface waters according to the UWWTD [1]. In Europe, 3.1% of the load is treated at the primary level, 28.5% at the secondary level, and 68.4% at the tertiary level [3]. The EU Commission has also launched an impact assessment to evaluate different policy options with the aim of modernizing the UWWTD [2]. Among others, specifically, the UWWTD requires controls of sludge disposal and reuse [4]. Article 14 of the UWWTD specifies [1]: (i) sludge arising from wastewater treatment shall be reused whenever appropriate, and (ii) disposal routes shall minimize the adverse effects on the environment. While approximately 20–25 kg of sludge dry matter (DM) are continuously produced annually per person per year all across Europe as a result of the wastewater treatment process, the wastewater operators render the valuable resources found in sewage sludge to be reusable [5].

The term"sewage sludge"does not clearly define the status of its quality. In the literature and regulations, the term"sewage sludge" refers to untreated excess sludge, partially treated sludge, digestate, and pelletized sludge. By Sewage Sludge Framework Directive (SSD) [6], the term"sludge" refers to untreated sludge (residual sludge from sewage plants treating domestic or urban waste waters and from other sewage plants treating waste waters of a composition similar to domestic and urban waste waters, and residual sludge from septic tanks and other similar installations for the treatment of sewage), while the term"treated sludge" refers to sludge which has undergone biological, chemical or heat treatment, long-term storage or any other appropriate process so as significantly to reduce its fermentability and the health hazards resulting from its use. Using the appropriate term in the professional literature helps us to distinguish between the quality of sludge obtained at different levels of processing.

On the European scale, the most common recovery procedures are agriculture and incineration [3]. In the latter procedure, it is not clear whether it is incineration for the purpose of waste to energy (WtE) processes (R 1 recovery operation), or it is just a matter of reducing the amount of sewage sludge (D 10 disposal operation). According to the Waste Framework Directive (WFD) the recovery procedure R 1 takes precedence over the disposal operation D 10 [7].

Quality of municipal infrastructure of urban agglomerations and appropriate pre-treated industrial wastewater has high importance on the stable operation of municipal wastewater treatment plants, and properties of treated sludge. Continuous quality control of treated sludge is the starting point for its efficient recovery processing as a source of raw materials to follow the strategy in the new Circular Economy Action Plan [8]. Some technical solutions enable the recovery of materials from treated sludge like phosphorus, nitrogen, biopolymers, biogas, biochar, biofuel, struvite, and even recovered products from ashes [9]. A market is not always available for these innovative materials because the recovery costs are often high compared to primary raw materials. This is because fossil oil based material or extraction techniques and transportation are cheaper [9]. The chapter will address three main areas of sludge recovery operations [7], namely the use of treated sludge for: (i) energy production—waste to energy (WtE), recovery operation R 1, (ii): the organic matter recovery with pyrolysis, recovery operation R 3, and (iii) agricultural use, recovery operation R 10. All end-use utilizations have advantages and disadvantages and are suitable for recovery processes from a different points of view and engineering approaches, depending on the UWWTP organic load capacity, level of raw wastewater treatment, and excess sludge pre-treatment stage (e.g. aerobic or anaerobic stabilization, dehydration, hygienization, and pelletization). Important are also others conditions such as the UWWTP geographical site, transport infrastructure, and available facilities for sewage sludge utilization. National legislation, the national strategic program for the management program of this specific waste stream, and the national level of public awareness of the usefulness of the sludge as a raw material for the CE are also important.

A case study of the composition of treated sewage sludge, generated at Central wastewater treatment plant Ljubljana (CWWTPL) has been performed. The CWWTPL treats the municipal wastewater of the capital city Ljubljana and is the largest UWWTP in Slovenia [10]. The purpose of the analysis and the research was to evaluate the feasibility of sewage sludge pre-treatment at the site of origin, anaerobic stabilization of excess sludge and biogas production, seasonal fluctuations in the quality of treated sludge, and the comparability of quality of the annual representative samples. The purpose is also to evaluate the sewage sludge as a smart material. The obtained results are the basis for sustainable and predictable use of treated sludge in the CE [11], both in the field of WtE and material utilization. The most important legal requirements for specific purposes of sewage sludge use are given. The chapter combines specifications, guidelines, and limit values regarding WtE processes, pyrolysis, and fertilizing products with real data of treated sludge quality and its potential. Quality data of the CWWTPL sludge were collected over a ten-year period of operation, from 2010 to 2020. The last year is also the starting year for a number of forecasts on improving the environment and preventing climate change.

1.1 Pyrolysis: recovery operation R 3, thermal conversion of organic matter into new products

According to WFD [7] the recovery operation R 3 means recycling/reclamation of organic substances which are not used as solvents (including composting and other biological transformation processes), gasification and pyrolysis using the components as chemicals, waste preparing for reuse, and recovery of organic materials in the form of backfilling.

Pyrolysis is a thermochemical anaerobic decomposition of biomass into a range of useful products. The process is typically carried out in a temperature range from 300°C through 650°C [12]. During pyrolysis, large, complex hydrocarbon molecules of biomass break down into relatively smaller and simpler molecules, forming gaseous (non-condensable volatile compounds) and liquid products (bio-oil), and solid residue (biochar). Pyrolysis offers the potential of material utilization of UWWTS. The densified pyrolytic fraction of sewage sludge is suitable for energy utilization, and on the other hand, the resulting solid residue is potential for material utilization on fields or forests. Biochar is by mass the major pyrolytic product or resideous material. Its characteristics make it promising for application in agriculture due to its high immobilization affinity of heavy metals [13]. Biochar retains the main part of the carbon in stable solid form [12]. In addition, sewage sludge biochar contains significant content of phosphorus, potassium, and nitrogen, which further demonstrates the great potential of its use as a fertilizer [12].

There is no special EU legislation regarding the waste pyrolysis process. European Biochar Foundation (EBC) has launched the guidelines for specification and certification procedure for biochar evaluation [14]. As a specification for pyrolysis technology is prescribed: (i) the use of waste heat or the use of liquid and gaseous pyrolysis products, and (ii) compliance with nationally defined emission limit values. For the characterization of biochar properties, the limit values in **Table 1** must be observed. The biochar for all application classes must be analyzed at least according to the EBC Basic Analysis Package ([14], Annex 1). The latter reference also indicates which types of biomass are permissible for each application class.

As both biochar properties and the environmental footprint of its production are largely dependent on the pyrolysis parameter and the type of feedstock to be used, a secure control and assessment system for its production and analysis had to be introduced. According to reference [14], biochar is defined by its quality

EBC—label, application classes	Parameter	EBC—feed	EBC—AgroBio	EBC—Agro	EBC—material
EBC—class		Class I	Class II	Class III	Class IV
Ultimate analysis		TC, C_{org}, H, N, O, S, ash			
Indicator of the degree of carbonization and stability	H/C_{org}	<0.7			
	O/C_{org}	<0.4			
Physical properties	Water content, dry matter (DM), bulk density, specific surface area (BET), pH, salt content				
Thermogravimetric analysis					
Nutrients		At least N, P, K, Mg, Ca			
Potencially toxic metals (PTM)	Pb	$10\ g\ t^{-1}$ ($88\%_{DM}$)	$45\ g\ t^{-1}{}_{DM}$	$150\ g\ t^{-1}{}_{DM}$	$250\ g\ t^{-1}{}_{DM}$
	Cd	$0.8\ g\ t^{-1}$ ($88\%_{DM}$)	$0.7\ g\ t^{-1}{}_{DM}$	$1.5\ g\ t^{-1}{}_{DM}$	$5\ g\ t^{-1}{}_{DM}$
	Cu	$70\ g\ t^{-1}{}_{DM}$		$100\ g\ t^{-1}{}_{DM}$	$250\ g\ t^{-1}{}_{DM}$
	Ni	$25\ g\ t^{-1}{}_{DM}$		$50\ g\ t^{-1}{}_{DM}$	$250\ g\ t^{-1}{}_{DM}$
	Hg	$0.1\ g\ t^{-1}$ ($88\%_{DM}$)	$0.4\ g\ t^{-1}{}_{DM}$	$1\ g\ t^{1}{}_{DM}$	
	Zn	$200\ g\ t^{-1}{}_{DM}$		$400\ g\ t^{-1}{}_{DM}$	$750\ g\ t^{-1}{}_{DM}$
	Cr	$70\ g\ t^{-1}{}_{DM}$		$90\ g\ t^{-1}{}_{DM}$	$250\ g\ t^{-1}{}_{DM}$
	As	$2\ g\ t^{-1}$ ($88\%_{DM}$)	$13\ g\ t^{-1}{}_{DM}$		$15\ g\ t^{-1}{}_{DM}$
Organic contaminants	16 PAH according to EPA priority list	$4 \pm 2\ g\ t^{-1}{}_{DM}$		$6.0 \pm 2.2\ g\ t^{-1}{}_{DM}$	$30\ g\ t^{-1}{}_{DM}$
	Benzo(a)pyren	$25\ mg\ t^{-1}$ ($88\%_{DM}$)	—		
	PCB, PCDD/F	([14], chapter 9)	Once per pyrolysis unit for the first production batch. For PCB: $0.2\ mg\ kg^{-1}{}_{DM}$, for PCDD/F: $20\ ng\ kg^{-1}$ (I-TEQ OMS), respectively.		

Table 1.
Limit values for biochar properties.

characteristics, the raw materials used, its sustainable production, and the end-use (**Table 1**).

The pyrolysis of non-plant biomasses such as treated sludge, livestock manure, certain digests or bones may also produce valuable raw materials and could be used in the interests of the bio-economy and climate protection, but these raw materials have not yet been included in the EBC feedstock list and are therefore not subject of the EBC guidelines [14].

According to the declaration of biochar properties, the organic carbon or TOC (C_{org}) must be defined. By the last version of EBC guidelines [14] the lower limit value of 50% m/m_{DM} has been reconsidered. All solid pyrolysis products below this limit value were considered only as pyrogenic carbonaceous materials (PCM).

The EBC certificate guarantees that only climate-positive biochar production technology is used and does not release unburned pyrolysis gases into the atmosphere. Reference [14] specified the conditions for the safe and harmless operation of pyrolysis units. Regarding the environment protection and preventing climate change the most important conditions are: (i) with the exception of the preheating of the pyrolysis reactor, the use of fossil fuel for heating is prohibited, (ii) if the pyrolysis reactor is electrically heated, the use of renewable energy source or the use of surplus electricity is recommended, (iii) the non-condensable pyrolysis gas must be burned or it can be trapped and used for the chemical industry, (iv) the bio-oil can also be stored and used for other energy and material purposes, (v) syngas combustion must comply with national emission thresholds, and (vi) the heat produced by pyrolysis process must be used.

1.2 EU legislation regarding the energy recovery from waste

The IED Directive [15] shall not apply to pyrolysis and gasification plants if the gases resulting from this thermal treatment of waste are purified to such an extent that they are no longer a waste prior to their incineration and they can cause emissions no higher than those resulting from the burning of natural gas.

In the EU regulation on waste incineration and co-incineration, the limit values for polluting substances, related to substances in sludge are presented. There is a possibility of linking the characteristics of the sludge with environmental pollution and harm to human health, so the energy recovery from treated sludge must be provided safely and harmlessly. Data provision and assessment must be made from a holistic point of view to consider WtE processes. These processes must be evaluated from an economic as well as from the environment point of view regarding the emissions into the air and into the surface water, and last but not least also regarding the generated residues.

1.2.1 Technical provisions to waste incineration plants and waste co-incineration plants

Technical provisions to the new waste incineration plants and waste co-incineration plants, according to IED Directive [15] are considered. In **Table 2** presented emission limit values of polluting substances into the air according to the standardized oxygen atmosphere. Limit values for waste incineration plants in **Table 2**, footnote a, apply for facilities: (i) that were in operation and have a permit in accordance with applicable Union law before 28 December 2002, and (ii) that were authorized or registered for waste incineration and have a permit granted before 28 December 2002 in accordance with applicable Union law, provided that the plant was put into operation no later than 28 December 2003. In the view of the competent authority, these facilities were the subject of a full request for authorization before 28 December 2002.

Limit value for NOx in **Table 2**, footnote b, applies for existing waste incineration plants with a nominal capacity exceeding 6 tonnes per hour or new waste incineration plants. The limit value for NOx in **Table 2**, footnote c, applies for existing waste incineration plants with a nominal capacity of 6 tonnes per hour or less. A new waste incineration plant means any waste incineration plant not covered by the existing plant.

Table 3 presents the limits values for polluting substances causing the emissions into the air according to transitional provisions of the IED Directive. The limit values shall be applied as follows: (i) in **Table 3**, footnote a, to combustion plants that co-incinerate waste from 1 January 2016 referred to the reference ([15], Article 30 (2)) and (ii) in **Table 3**, footnote b, to combustion plants that co-incinerate waste from 7 January 2013 referred to the reference ([15], Article 30 (3)).

Polluting substances/standard oxygen concentration in waste gas	Waste incineration plant[a]	Cement kilns co-incinerating waste
	Sampling period of waste gas	
	daily average	
	11% v/v	10% v/v
Total dust	10	30
TOC	10 (gaseous and vaporous organic substances)	
CO	50	
HCl	10	10
HF	1	1
SO_2	50	
NO_x as NO_2	200[b]	500
	400[c]	
Heavy metals and their compounds	A minimum of 30 min and a maximum of 8 h	
Cd + Tl	$\sum_{tot}$; 0.05	
Hg	0.05	
Sb, As, Pb, Cr, Co, Cu, Mn, Ni, V	$\sum_{tot}$; 0.5	
Organic pollutants	A minimum of 6 h and a maximum of 8 h	
Dioxins and furans	0.1	

The significance for footnotes a, b and c are provided in the main text of sub-subsection 1.2.1.

Table 2.
Emission limit values for the following polluting substances in mg/Nm³ for heavy metals and in ng/Nm³ in combination with toxic equivalence factor (TEF) for dioxins and furans in their gaseous and vapor forms.

1.2.2 Technical provisions to technological wastewater from incineration plants and waste co-incineration plants

In flue gas cleaning technological wastewater is generated, which captures dust of fly ash and condensed volatile compounds of heavy metals. The generated process water must be treated in accordance with the limit values for pollutants in the generated wastewater, which are uniformly determined for all waste incineration and co-incineration plants [15]. Limit values for total suspended solids (TSS) in pre-treated technological wastewater as defined in Annex I of UWWTD [1] are: (i) 30 mg L^{-1} by 95% of all results or (ii) 40 mg L^{-1} by 100% of all results.

In terms of sustainable energy utilization of sewage sludge, in order to prevent the circulation of already removed polluting substances back to nature, it is necessary to assess the quality of treated technological wastewater and prevent deterioration of the chemical and ecological status of the receiving surface water. **Table 4** shows a comparison of the limit values for the discharge of treated technological wastewater from the incineration and co-incineration plant [15] with the limit values for good chemical status and very good and good ecological status of receiving surface water. The chemical status of waters is assessed by the content and concentration of Environmental quality standards (EQS), expressed as priority substances (PS) or priority hazardous substances (PHS) [16, 17]. The ecological status is assessed, among other conditions, by the presence and concentration of non-synthetic special pollutants (SP) [16]. Since the dilution factor of some polluting substances in the treated wastewater from the incineration and co-incineration

Polluting substances/standard oxygen concentration in waste gas	**Total rated thermal input, MWth**			
	<50	**50–100**	**100–300**	**>300**
	Sampling period of waste gas			
	daily average			
	6% v/v			
Total dust[a]	50	30	25	20
Total dust[b]		20		10
SO_2[a]	—	400	200	
SO_2[b]			200	150
NO_x as NO_2[a]		300	200	
NO_x as NO_2[b]			200	150
Heavy metals and their compounds	A minimum of 30 min and a maximum of 8 h			
Cd + Tl	$\sum_{tot}$; 0.05			
Hg	0.05			
Sb, As, Pb, Cr, Co, Cu, Mn, Ni, V	$\sum_{tot}$; 0.5			
Organic pollutants	A minimum of 6 h and a maximum of 8 h			
Dioxins and furans	0.1			

The significance for footnotes a and b are provided in the main text of sub-subsection 1.2.1.

Table 3.
Emission limit values for combustion plants co-incinerating solid fuels with the exception of biomass for the following polluting substances in mg/Nm³ and in ng/Nm³ in combination with toxic equivalence factor (TEF) for dioxins and furans in their gaseous and vapor forms ([15], Articles 30(2) and 30(3)).

Polluting substances	**IED Directive [15]**
	Emission limit values for unfiltered samples (mg L^{-1} except for dioxins and furans)
Hg	0.03
Cd	0.05
Pb	0.2
Ni	0.5
Dioxins and furans, TEQ	0.3 ng L^{-1}
As	0.14
Tl	0.05
Cr	0.5
Cu	0.5
Zn	1.5

Table 4.
Emission limit values for discharges of wastewater from the cleaning of waste gases.

plant into the environment must be around 1000, it is necessary to pay full attention to this sustainability aspect of the WtE process.

EU Member States may, when assessing the monitoring results against the relevant EQS, take into account: (i) natural background (NB) concentrations for metals and their compounds where such concentrations prevent compliance with the relevant EQS, (ii) hardness, pH, dissolved organic carbon or (iii) other water quality parameters that affect the bioavailability of metals, the bioavailable concentrations being determined using appropriate bioavailability modeling.

Mercury is an atmospheric deposit and is behaving like a ubiquitous persistent, bioaccumulative and toxic substance (PBT). Due to its volatility mercury is often present in the environment. Limiting the circulation of mercury in the atmosphere as much as possible is an important task in establishing the sustainability of sewage sludge utilization in the CE.

Establishing the sustainable utilization of sewage sludge with WtE processes is based on a holistic assessment of the environment in which the facility is located. Sewage sludge is a specific waste stream with a complex composition and only a good knowledge of all aspects of environmental impact and properties of sewage sludge allows the design of sustainable WtE technologies from treated sludge.

1.3 EU legislation regarding material recovery from treated waste in the agriculture

The framework European Sewage sludge Directive (SSD) [6] regulates the use of sewage sludge in agriculture for the EU Member States. According to reported quantities on recovery procedures for sewage sludge for the period 2013–2015 in Europe, 48.5% m/m sewage sludge is used for agriculture [3]. The data in Ref. [3] show that in many EU Member States this process is still the most topical.

The selection of polluting substances, for which a limit value is set, include heavy metals, which are potentially toxic and are essential in a small amount for various biochemical and physiological functions within the plants, animals, and human [18] (**Table 5**). A pH limit is additionally prescribed for the quality of the soil. In adapting its national legislation EU Member States may permit the limit values they fix to be exceeded in the case of the use of soil and sludge for cultivation of commercial food crops exclusively for animal consumption (**Table 5**, footnote a).

Treated sludge and soil properties	Soil[a]	Treated sludge[a]	The maximum annual quantities of PTM introduced into soil[c]
pH	6–7	Not relevant	—
PTMs	mg kg^{-1}_{DM}		kg ha^{-1} y^{-1}
Cd	1–3	20–40	0.15
Cr	Limit value is not defined yet		
Cu	50–140[b]	1000–1750	12
Hg	1–1.5	16–25	0.1
Ni	30–75[b]	300–400	3
Pb	50–300	750–1200	15
Zn	150–300[b]	2500–4000	30

The significance for footnotes a, b and c are provided in the main text of the subsection 1.3.

Table 5.
Limit values for use of sewage sludge in agriculture according to SSD [6].

EU Member States also may permit the limit values they fix to be exceeded in respect of these parameters on soil with a pH consistently higher than 7 (**Table 5**, footnote b). The maximum authorized concentrations of PTMs may not exceed those values by more than 50%. The EU Member States must also seek to ensure that there is no resulting hazard to human health or the environment and in particular to ground water. The amount of PTMs annually spread on soil is calculated on a 10-year average (**Table 5**, footnote c).

National legislation of the EU Member States regarding the use of sludge in agriculture is not uniform, there are differences in the limit values for the key parameters, and there is also a difference in the legislation demands between USA and EU [19].

Given the strategic objectives of the CE philosophy and the reduction of waste, it is necessary to give priority to the use of recovered waste over the depletion of raw materials. In order to include as much recovered waste as possible in the raw material cycle, the EU has adopted a Fertilizing Resolution (FR) [20] for the use of recovered waste in fertilizer, which aims to unify the European fertilizer market. According to the vision of FR, the 'EU fertilizing product' means a fertilizing product which is labeled with the logo CE ("conformité européenne") when made available on the market. It includes fertilizers produced from recovered or organic materials. The FR, which will not enter full force until 16 July 2022, will not affect the implementation of SSD [6]. The possibility of using sewage sludge as a fertilizer product is still ambiguous. From the conditions for the selection of possible treated waste for use as a fertilizer, it can be understood that this can be all wastes that are composted or anaerobically stabilized [20].

FR does not prescribe the extent of plant response when applying a particular fertilizer. Slovenian legislation is more restrictive in this area and prescribes the quantity and quality of the digestate used for the purpose of a fertilizer (**Table 6**) [21].

1.3.1 Designation of component material categories for the potential EU solid fertilizer

According to FR sewage sludge could be used as input material for the component materials categories (CMC) for fertilizer as a blending compost (CMC 3) and treated digestate (CMC 5), if it does not contain more than 6 mg PAH kg_{DM}^{-1}. **Tables 7** and **8** provide the basic starting conditions for the quality of the recovered waste that can be used as input material for the production of a fertilizer.

Biological criteria	**Unit**	**Quality class 1**	**Quality class 2**
Plant (Chinese cabbage in pot) germination 15% m/m or 25% (v/v) granules: fresh vegetable mass (FVM)	[% m/m]; [% v/v]	≥ 100% of control substrate; germination: ≥95%, germination delay: 0 days	No limits
Germination rate (Cress in Petri dish) 30% m/m or 50% v/v granules: FVM		≥90% of control substrate; germination: ≥90%, germination delay: 0 days	
Green mass (Chinese cabbage), granules: peat moss 20% m/m	%	No limit	
Determination of the content of unwanted weed seeds and plant propagules in growing substrates and soil improvers			
Weed seeds and plant propagules, granules: peat moss 20% m/m	N° L^{-1}	<2	

Table 6.
Determination of the plant response. Digestate with >20% m/m_{DM}: Characterization of the effect of digestate granules as soil improvers and growth substrates on plant germination and growth [21].

Parameter/ criteria	Unit	Compost (CMC 3)	Digestate other than fresh crop digestate (CMC 5)
		Macroscopic impurities (MI); Glass, metal, plastics	
Particles >5 mm	% m/m_{DM} [21]	^{1th}Class, ^{2sd}Class: < 5	^{1th}Class, ^{2sd}Class: < 5
Particles >2 mm[a]		^{1th}Class: < 0.5 ^{2sd}Class: < 2	^{1th}Class, ^{2sd}Class: < 2
	$g\ kg^{-1}_{DM}$ [20]	3	3
MI_{total}		5	5
		POPs	
$PAHs_{16}$[b]	$mg\ kg^{-1}_{DM}$	6	6

The significance for footnotes a and b are provided in the main text of sub-subsection 1.3.1.

Table 7.
Designation of component material categories (CMCs) for the potential EU solid fertilizer incorporating sewage sludge [20, 21].

Parameter/criteria	Unit	Compost (CMC 3)	Digestate other than fresh crop digestate (CMC 5)
	Stability criteria [20]		
Oxygen uptake rate	mmol $O_2\ kg^{-1}_{OM}\ h^{-1}$	Max 25	Max 25
Residual biogas potential	L biogas g^{-1} VS[a]	—	0.25
	Stability criteria [21]		
(AT_4)	mg $O_2\ g^{-1}_{DM}$	< 15	No limit
Acetic acid	$mg\ L^{-1}$ (mass/volume)	< 100	< 300
Propionic acid			
	Micro-organisms to be tested/pathogens		
Salmonella spp.	CFU in 25 g		Absence
Escherichia colli or Enterococcaceae	CFU in 1 g		1000[b]

The significance for footnotes a and b are provided in the main text of sub-subsection 1.3.1.

Table 8.
Designation of component material categories (CMCs) for the potential EU solid fertilizer incorporating sewage sludge [20, 21].

Conditions for use include the content of total MI, polycyclic aromatic hydrocarbons (PAHs) as persistent organic pollutant (POPs), adequate stability and content of pathogens as evidence of hygienic integrity.

According to FR stricter limit values will apply in the future for the presence of plastics above 2 mm (**Table** 7, footnote a): (i) from 16 July 2026 the maximum limit value point shall be no more than 2.5 g/kg dry matter, and (ii) by 16 July 2029 this limit value shall be re-assessed in order to take into account the progress made with regards to separate collection of bio-waste.

According to FR, sixteen PAHs have been identified as polluting substances indicative of POPs contamination (**Table** 7, footnote b): naphthalene, acenaphthylene, acenaphthene, fluorene, phenanthrene, anthracene, fluoranthene, pyrene, benzo[a] anthracene, chrysene, benzo[b]fluoranthene, benzo[k]fluoranthene, benzo[a]pyrene, indeno[1,2,3-cd]pyrene, dibenzo[a,h]anthracene and benzo[ghi]perylene.

Residual biogas potential of digestate (**Table 8**, footnote a) is an indicator of mineralization efficiency of excess sludge into the digestate and is expressed as a produced biogas per unit of material that is lost on ignition of the dry solids at 550° C. Loss on ignition is expressed as volatile solids (VS). The treatment process of the digestate must include the hygenization process. It must be free of *Salmonella* spp. and the maximum number of *Escherichia coli* or *Enterococcaceae* expressed in colony-forming units (CFU) is 1000 (**Table 8**, footnote b).

1.3.2 Limit values and declaration for product function categories of the EU fertilizing products

According to FR the product function categories are: (i) liquid or solid fertilizer (organic or inorganic straight or compound macro and micronutrients), (ii) liming material, (iii) organic or inorganic soil improver, (iv) growing medium, (v) non-microbial plant biostimulant, and (vi) fertilizing product blend [20]. In a solid organo-mineral fertilizer, each physical unit shall contain organic carbon (C_{org}) and all the nutrients in their declared content. A physical unit refers to one of the component pieces of a product, such as granules or pellets. Inorganic micronutrient fertilizers shall be made available to the end-user only in packaged form.

If sewage sludge is pre-treated at the site of fertilizer production, the final product line must be separate from the processing of other input materials, and there must be a strictly prevented physical contact between the input and output material during storage.

An organic fertilizer shall contain organic carbon (C_{org}) and primary nutrients (PN) (P, K, Ca, Mg, Na, S) of solely biological origin, but no other material which is fossilized or embedded in geological formations. It shall be derived from plant residues or livestock manure (**Tables 9** and **10**).

Table 9 provides the technical specification of product function categories (PFC) of the EU fertilizing products, which could potentially contain the sewage sludge.

In a solid organo-mineral fertilizer, each physical unit shall contain C_{org} and all PN in their declared content (**Table 9**, footnote a). Limit values shall not apply where copper or zinc has been intentionally added to an organo-mineral fertilizer for the purpose of correcting a soil micronutrients (MICN) deficiency and is declared in accordance with Annex III of FR (**Table 9**, footnote b).

Straight or compound inorganic fertilizer shall contain one or more primary macro-nutrients (P MACN) and one or more secondary macronutrients (S MACN). **Table 10** presents the minimal content for nutrients for those two groups of fertilizer: (i) content of only one P MACN (**Table 10**, footnote a), (ii) content of only one P MACN and one or more S MACN (**Table 10**, footnote b), (iii) content of more than one P MACN (**Table 10**, footnote c), and (iv) content of no P MACN and more than one S MACN (**Table 10**, footnote d). The total sodium oxide (Na_2O) content shall not exceed 40% by mass (**Table 10**, footnote e). If an inorganic fertilizer contains more than 1% by mass of C_{org} derivated from chelating or complexing agents, nitrification inhibitors, denitrification inhibitors or urease inhibitors, coating agents, urea (CH_4N_2O), or calcium cyanamide ($CaCN_2$) (**Table 10**, footnote f), shall meet pathogen requirements set out in **Table 8**. Limit values shall not apply where copper or zinc has been intentionally added to an organo-mineral fertilizer for the purpose of correcting a soil micronutrients deficiency and is declared in accordance with Annex III of FR (**Table 10**, footnote g).

Declared minimum content of the total sum of (MICN) boron, cobalt, copper, iron, manganese, molybdenum, and zinc for compound solid inorganic micronutrient fertilizer, PFCs 1(C)(II)(b) shall be 5% by mass. That type of fertilizer shall be made available to the end-user only in packaged form. Contaminants in an inorganic micronutrient fertilizer must not exceed the limit values presented in **Table 11**.

Parameter	Unit	Organic fertilizer, PFC 1(A)(I)		Organo-mineral fertilizer, PFC 1 (B)(I)	
		Content of only one PN	Content of more than one PN	Content of only one PN	Content of more than one PN
Declaration					
Primary nutrients (PN), maximum content					
NH_4NO_3	% N	—		<16	
Primary nutrients (PN), minimal content					
N_{org}	% N	—		1	0.5
N_{tot}	% N	2.5	1	2.5	2
P_{tot}	% P_2O_5	2	1	2	
K_{tot}	% K_2O	2	1	2	
Sum of N, P, K	%	—	4	—	8
Organic carbon					
C_{org}	% C	15		7.5[a]	
Limit values					
Polluting substances					
Cd	mg kg_{DM}^{-1}	1.5		—	
Cd (<5% P_2O_5)		—		3	
Cd (>5% P_2O_5)	mg Cd kg^{-1} of P_2O_5			60	
Cr^{VI}	mg kg_{DM}^{-1}	2			
Hg		1			
Ni		50			
Pb		120			
As (in inorganic compounds)		40			
Biuret ($C_2H_5N_3O_2$)	g kg_{DM}^{-1}	Not present		12	
Potentially toxic metals					
Cu	mg kg_{DM}^{-1}	300		600[b]	
Zn		800		1500[b]	

The significance for footnotes a and b are provided in the main text of sub-subsection 1.3.2.

Table 9.
Limit values and declaration for solid PFCs of EU fertilizing products.

Table 12 presents the limit values and declaration of nutrients for organic soil improver, inorganic soil improver, growing medium, and non-microbial plant biostimulant. Declared organic soil improver may contain peat, leonardite, and lignite, but no other material which is fossilized or embedded in geological formations. Growing medium could be solid or liquid materials/substrate in which plants (including algae) grow (**Table 12**, footnote a).

The plant biostimulant shall be an EU fertilizing product which stimulates plant nutrition processes independently of the product's nutrient content with the sole aim of improving one or more of the following characteristics of the plant or the

Parameter	Unit	Straight inorganic fertilizer, PFC 1(C)(I)(a)(i)		Compound inorganic fertilizer, PFC 1(C)(I)(a)(ii)	
		1[a]	2[b]	3[c]	4[d]
Declaration (% by m/m in the form as available on the market)					
Primary macronutrients (P MACN), minimal content					
N_{tot}	% N	10		3	—
P_{tot}	% P_2O_5	12			—
K_{tot}	% K_2O	6			—
Secondary macronutrients (S MACN), minimal content					
Ca	% CaO	12	1.5	—	1.5
Mg	% MgO	5	1.5	—	1.5
Na[e]	% Na_2O	1		—	1
S	% SO_3	10	1.5	—	1.5
Sum of P MACN and S MACN	%	18			
Organic carbon					
C_{org}[f]	% C	<1			
Limit values					
Polluting substances, maximal content					
Cd (>5% P_2O_5)	mg kg^{-1} of P_2O_5	60			
Cd (<5% P_2O_5)	mg kg_{DM}^{-1}	3			
Cr^{VI}		2			
Hg		1			
Ni		100			
Pb		120			
As		40			
Biuret ($C_2H_5N_3O_2$)	g kg_{DM}^{-1}	12			
Perchlorate (ClO_4^-)	mg kg_{DM}^{-1}	50			
Potentially toxic metals, maximal content					
Cu	mg kg_{DM}^{-1}	600[g]			
Zn		1500[g]			

The significance for footnotes a, b, c, d, e, f and g are provided in the main text of sub-subsection 1.3.2.

Table 10.
Limit values and declaration of nutrients for solid inorganic fertilizer declared as a PFCs 1(C)(I).

plant rhizosphere: (i) nutrient use efficiency, (ii) tolerance to abiotic stress, (iii) quality traits, or iv) availability of confined nutrients in the soil or rhizosphere (**Table 12**, footnote b). The plant biostimulant shall have the effects that are claimed on the label for the plants specified thereon (**Table 12**, footnote c).

Pathogens content of EU fertilizing products declared in **Table 12** must not exceed the limits set out in **Table 8**. Evaluation of the sewage sludge, its solid pyrolytic residue (PCM), and its mixture with other treated waste as a fertilizer according to FR will be a complex procedure. In addition to the physicochemical evaluation, the EU fertilizer products label will require a number of other administrative procedures - from packaging to sale.

Polluting substance	Fertilizer with more than one MICN, expressed in mg/kg of total micronutrients content
Cd	200
Hg	100
Ni	2000
Pb	600
As	1000

Table 11.
Upper limit values of polluting substances for compound solid inorganic micronutrient fertilizer, PFCs 1(C)(II)(b) of EU fertilizing products.

Parameter	Unit	Organic soil improver, PFC 3 (A)	Inorganic soil improver, PFC 3 (B)	Growing medium, PFC 4[a]	Non-microbial plant biostimulant[b;c]
Declaration of nutrients (% by m/m in the form as available on the market)					
Biological origin	%	>95	—	—	—
Dry matter		≥20	—	—	—
Organic carbon, minimum content					
C_{org}	% C	7.5	—	—	—
Limit values					
Polluting substances					
Cd	mg kg_{DM}^{-1}	2	1.5	1.5	1.5
Cr^{VI}		2	2	2	2
Hg		1	1	1	1
Ni		50	100	50	50
Pb		120	120	120	120
As (in inorganic compounds)		40	40	40	40
Potentially toxic metals					
Cu	mg kg_{DM}^{-1}	300	300	200	600
Zn		800	800	500	1500

The significance for footnotes a, b and c are provided in the main text of sub-subsection 1.3.2.

Table 12.
Limit values and declaration of nutrients for PFCs 3(A), PFCs 3(B), PFCs 4 and non-microbial plant biostimulant (PFCs 6(B)) of EU fertilizing products.

2. Materials and methods

In terms of reviewing basic data on morphological, thermal, and physicochemical properties of sewage sludge, a case study for the CWWTPL was conducted [10]. It is an urban one-stage mechanical-biological wastewater treatment plant aimed at conventional secondary treatment of municipal wastewater, a small amount of industrial wastewater, and providing the public service for acceptance and treatment of excess sludge from the small WWTPs and the contents of septic tanks.

The findings of long-term morphology, physicochemical investigations, and thermal analyses of pelletized dehydrated digestate (granules), produced at CWWTPL, which were carried out in the period from 2010 to 2020, are presented. Predominantly accredited methods for physicochemical characterization and non-accredited methods for morphological properties (bulk density, particle size distribution, and specific surface area determination) and thermal properties were applied. Chemical elements that are the basis for WtE processes, the elements that are important for the use of residues after thermal treatment (combustion or pyrolysis), and elements that are important for the use of granules as a fertilizer or as a material for soil remediation and improvement of the ecological condition were determined. Additional research of thermal properties was done with the conventional accredited methods for organic matter content, loss on ignition, and volatile matter. Enabling more extensive information in the change in weight of granules under the thermal load was derived with simultaneous thermal analysis of granules and mass spectroscopy of the released gases.

Confidence in the results of research into the properties of sewage sludge is based on an appropriate quality control system (including sampling), targeted end-use, and a list of required parameters. Only high-quality performed analysis enable credible results and the setting of optimal guidelines for the successful CE.

2.1 Materials

After thickening by mechanical equipment, along with the addition of a strong cationic polyelectrolyte, the surplus sludge is anaerobically stabilized under mesophilic conditions at 37–40°C and pH of 7.4 in a digester by a hydraulic retention time of at least 20 days. The produced digestate is a thick black-gray suspension of anaerobic biomass with 3.6% m/m of dry matter. Anaerobic stabilization is followed by centrifugal dehydration (cake production) and granulation in a drum dryer at 90°C [19]. The drying process is conducted in batch mode. Simultaneously the digestate cake gets hygenized and granulated. The selection and execution of the cake drying are in accordance with Technical Standard (TS) from the group CEN/TC 308, Characterization of sludges, CEN/TR 15473:2007, Good practice for sludge drying.

Prepared and characterized were also laboratory samples of the basic solid residues generated after controlled thermal loading of granules in order to simulate pyrolytic biochar and incineration bottom ash.

2.2 Methods and techniques

Inflow and outflow of water at the CWWTPL are measured continuously with the Khafagi type venturi meter and an in-line flow meter. The wastewater samples are prepared, cooled, and stored in accordance with the standards from the group water quality—sampling: (i) ISO 5667-10:1992, Part 10: Guidance on sampling of waste waters, and (ii) ISO 5667-3: 2012, Part 3: Preservation and handling of water samples. The COD in raw and treated wastewater is monitored by measurements of daily representative 24 h time-proportional composite water samples using method from the group of standards ISO/TC 14/SC 2 "Physical, chemical and biochemical methods".

From the composite daily sub-samples, mass-proportional representative batch and monthly composite samples of dried and pelletized digestate (in granules) are made, tested, and stored at 5 ± 1°C. Similarly, mass-proportional representative annual composite samples according to monthly delivered amount were prepared out of the refrigerated monthly samples. For the granules sampling procedure, a quality system that includes a manual time-proportionate daily sampling in accordance to the standard ISO 5667-13:2011, water quality, sampling, Part 13: Guidance on the sampling of sludge from sewage and water treatment works, and EN 15002:2015, characterization of waste, preparation of test portions from the laboratory sample, was established, as well as a routine control of produced granules as they were shipped to the stakeholder (**Table 13**). The granules sampling plan takes into consideration the relevant technical standards (TS): (i) EN 14899:2005, Characterization of waste, sampling of waste materials, framework for the preparation and application of a sampling plan, (ii) CEN/TS 15442:2006, Solid recovered fuels, methods for sampling and (iii) CEN/TR 15310–3:2006, Characterization of waste, sampling of waste materials, Part 3: Guidance on procedures for sub-sampling in the field.

Most of the results for the annual representative samples are obtained from official evaluation prepared for the purpose of conducting a public tender for the transport and final treatment of granules from the place of origin. The evaluations were prepared by an authorized contractor [22] and are part of the tender documentation for the implementation of large value public contracts. The methods are selected according to the legal requirements depending on the sample matrix and the scope of the sample under consideration. Slovenian national legal requirements prescribe the quality of the sewage sludge used as a bioresource or as a material for the WtE, and the range of methods and techniques for quality parameters determination. Standard analytical methods for characterization of granules as a potential fertilizer (PF) and as a solid recovered fuel (SRF) were applied as specified by TSs: (i) CEN/TC 223 "Soil improvers and growing media", (ii) CEN/TC 292 "Characterization of waste", (iii) CEN/TC 308 "Characteritation of sludge", and (iv) CEN/TS 343 "Solid recovered fuels". In accordance with the CEN/TS 343, TS EN 15402:2011, Determination of the content of volatile matter (VM), the proportion of VM that causes the granules mass loss in a covered crucible at a thermal load of 900°C over a period of 7 min is determined.

2.2.1 Sampling pattern for representative samples of granules

Periodicaly, every 3 h, 200 mL samples of granules were manually collected from the conveyor belt. Preparation of representative composite daily sample for each shipment is provided by homogenizing and quartering the three-hour sub-samples. Daily samples are stored in plastic bottles of 1 L at 5°C ± 1°C. The sampling period runs from 1 January to 31 December. At the end of each batch of drying process preparation of the composite batch sample (sample increment of 400 mL in volume from each daily sample) is executed by homogenizing and quartering of sub-samples. Representative batch samples are stored in glass bottles of 0.75 L at 5°C ± 1°C. At the end of the calendar month preparation of a composite monthly sample (sample increment of 400 mL) from each daily sample is provided. Prepared are four composite samples (4 × 1 L) by homogenizing and quartering the daily sub-samples. The monthly samples are stored at 5°C ± 1°C in brown glass bottles. At the end of the calendar year preparation of a composite mass-proportional annual sample is prepared from each monthly sample. Four annual composite samples (4 × 1 L) are obtained by homogenizing and quartering the monthly sub-samples. Storage is done at 5°C ± 1°C in brown glass bottles.

All annual representative samples are still kept properly and are available for further investigation.

2.2.2 Granules characterization

At CWWTPL the quality control of granules is provided by performing regular quality control at all levels of sludge pre-treatment. To ensure the stable granules quality, dry matter is regularly checked on an hourly and daily basis with moisture analyzers with a halogen heater. In addition, the quality of representative samples of batch, monthly and annual samples are checked in the laboratory using standard analytical methods for characterization of granules: (i) the oven dry method for moisture content at 103°C (CEN/TC 223, EN 13040:2007, Sample preparation for chemical and physical tests, determination of dry matter content, moisture content and laboratory compacted bulk density) and at 105°C (CEN/TC 444, Environmental characterization, Sludge, treated biowaste, soil and waste, EN 15934:2012, Calculation of dry matter fraction after determination of dry residue or water content), (ii) laboratory furnace (LF) for organic matter content (OM) at 450°C (LF) (CEN/TC 223, EN 13039:2011, Determination of organic matter content and ash), (iii) LF for loss on ignition (LOI) at 550°C (CEN/TC 292, EN 15169:2017, Determination of loss on ignition in waste, sludge and sediments), (iv) LF for ash content at 900°C in LF (CEN/TS 343, EN 15403:2011, Determination of ash content, modified), and (v) standard laboratory equipment for bulk density (as received) (CEN/TS 343, CEN/TS 15401:2010, Determination of bulk density, modified). The content of carbonates (inorganic source of carbon and CO_2) is determined as a mass difference between LOI at 550°C and LOI at 900°C due to the thermal decomposition of magnesium- and calcium carbonates, predominately caused by the latter.

Measurements of the specific surface area of granules by gas adsorption are performed using a Micromeritics ASAP 2020 analyzer according to ISO 9277:2010. The adsorptive gas was nitrogen, and the specific surface area (SSA) was calculated using the Brunauer-Emmett-Teller method (BET). A simple method for determining the ignition temperature in the granules layer was provided by a non-standardized method in an open LF with flues and with the isothermal temperature program.

To determine the chemical properties of granules the annual composite granular sample was milled to <1 mm using the Retsch SK1 hammer mill and, when necessary, down to <0.5 mm in the Retsch ZM 200 mill. Microwave-assisted digestion procedure for preparation of the PF to determine the PTMs content was done with aqua regia, while microwave-assisted digestion of granules as a SRF was done with a more invasive acid mixture of concentrated hydrofluoric (HF), nitric (HNO_3) and hydrochloric acid (HCl), and the results obtained are shown in separate tables.

The majority of the methods used for wastewater and granules characterization are accredited according to the Technical Standard SIST EN ISO/IEC 17025:2017, General requirements for the competence of testing and calibration laboratories and performed by authorized contractors.

Simultaneous thermal analysis (STA) with thermogravimetric analysis (TGA), derivative thermogravimetric analysis (DTG), and differential thermal analysis (DTA) of granules and mass spectroscopy of released gases are performed in an oxidative and in an inert atmosphere. STA apparatus (Netzsch's products were used) allowed the simultaneous acquisition of mass loss (TGA) and thermal effects (DTA) during thermal analysis. The DGT curve is obtained by calculating the derivative of the TG curve. The DTA measurement gives combustion curve or thermal change curve as a function of thermal load and information on the heat balance for each stage of thermal decomposition of granules. The DTG analysis enables a "fingerprint" of the thermal behavior and a thermal decomposition profile curve ($\Delta m/\Delta t$). Data manipulation and transformation are performed by Netzsch Proteus 6.1.0 software.

Fingerprint of thermal decomposition of granules for determination of their behavior at thermal load is carried out on the two representative annual samples (**Table 13**): (i) for the year 2010 and (ii) for the year 2012. In 2011 the annual sample 2010 (labeled as 2010/2011) was characterized with the TGA/DTG/DTA

Year of operation	2010	2011	2012	2013	2014	2015
N° of delivered shipment	165	141	203	165	145	167
N° of batch samples	45	60	46	45	42	45
N° of monthly samples			12			
Annual sample, time period			Jan–Oct			no s.
Year of operation	2016	2017	2018	2019	2020	
N° of delivered shipment	176	154	194	194	194	
N° of batch samples	40	40	37	37	37	
N° of monthly samples			12			
Annual sample, time period	Jan–Oct	no s.	Jan - Dec	no s.		

no s.—the annual sample was not prepared.

Table 13.
Sampling procedure for granules.

analysis in both atmospheres: (i) the inert atmosphere of Ar (purity 99.999%) and in (ii) the oxidizing atmosphere (80% v/v of Ar and 20% v/v of O_2, purity 99.999%). The findings are presented as a short review of the study in Ref. [23]. In 2012 TGA analysis for the annual sample 2012 is performed in both atmospheres (labeled as 2012/2012), and additional the TGA analysis in both atmospheres for the sample 2010 (labeled as 2010/2012) for the purpose of comparison was repeated. In 2016 the TGA analysis (inert atmosphere) was repeated on the sample 2012 (labeled as 2012/2016), and again in 2018 on both samples: (i) the sample 2010 (labeled as 2010/2018), and (b) the sample 2012 (labeled as 2012/2018). Both representative samples, 2010 and 2012, are still properly stored (closed packaging, in a dark and cool room). Each time, 100 mg of finelly ground granules were weighed into the TG/DTA crucible (0.3 mL, Al_2O_3) and exposed to heating from T_{room} to 1500°C at the heating rate of 10 K min^{-1}. Protective and purge gas flows were set to 30 and 50 mL min^{-1}, respectively.

In the year 2017, using a method described by CEN/TS 343, CEN/TR 15716:2008, Determination of combustion behavior (Combustion behavior), the proximate analysis was performed on the representative annual sample 2012 (labeled as 2012/2017) with the TGA technique in the combination with: (i) isothermal and non-isothermal temperature program in the temperature range from room temperature (T_{room}) to 900°C in the inert atmosphere, and (ii) the isothermal temperature program in the oxidative atmosphere at 900°C. The proximate analysis was performed in Netzsch STA 449 F3 Jupiter. 50.0 ± 0.5 mg of the milled granules were used to provide the analysis. The sample was weighted into the TG/DTA crucible (0.3 mL, Al_2O_3). The contents of moisture, volatile matter (VM), fixed carbon (C_{fix}), and ash were determined with this single analysis of the granules consisting of several stages. In the first stage, moisture is determined by sample heating to 110°C with 20 K min^{-1}, keeping the temperature constant for 15 min in an inert Ar atmosphere (purity 99.999%). VM is determined by consecutive heating to 900°C at 20 K min^{-1}, followed by an isothermal hold for 15 min. In the last stage, the purge gas is changed to the oxidizing atmosphere (80% v/v of Ar and 20% v/v of O_2, purity 99.999%), and the temperature is kept constant at 900°C for 120 min. The mass loss in this stage is attributed to C_{fix}, and the residual mass is ascribed as ash.

The year 2012 is chosen also as the reference year for further investigations of representative annual samples of CWWTPL and comparisons of results obtained in future research.

2.2.3 Characterization of solid residues after thermally treated granules

Two types of techniques for solid residue preparation were used: (i) pyrogenic residue, prepared with TGA technique using inert atmosphere (pyrolysis, Ar, 99.999 vol%)—2000 mg of granules were isothermally heated for 60 min at 450°C, in TG crucible 3.4 mL, Al_2O_3; the procedure was repeated until enough amount of sample has been achieved, and (ii) simulation of four type of bottom ash - the ignition (incineration) of granules in the laboratory heating furnace (LF) at 450°C (240 min), 550°C (180 min), 700°C (120 min) and 900°C (60 min). In the resulting residues were determined the content of nutrients (TOC, N, P, K, and Mg), the proportion of volatile part of nutrients due to the thermal treatment, and the proportion of water-soluble nutrients with the extraction method in accordance with the TS CEN/TC 223, EN 13652:2002, Extraction of water-soluble nutrients.

3. Results and discussion

3.1 Sewage sludge treatment at the CWWTPL

Because of the mixed sewer system, the amount of treated wastewater over a 10-year period varies between 1.5 million and 4.045 million m^3 per month and 32.6 million m^3 in the year 2010 to 24.6 million m^3 in 2020. The removed COD varies between 550 t of O_2 to 1770 t of O_2 per month and is lower at the higher quantity of inflow in the period of wet weather.

The average daily amount of excess sludge over a 10-year period is about 17.5 t_{DM} or 51 g $PE^{-1}{}_{COD}$ (18.6 kg $PE^{-1}{}_{COD}$ $year^{-1}$). The produced amount of excess sludge is lower than that reported in the literature (20–25 kg of sludge per one PE $year^{-1}$) [5]. Due to the fluctuations of the raw municipal wastewater inflow and related removed COD (**Figure 1**), the amount of generated granules, and also their useful matter content delivered to the stakeholder, varied accordingly (**Figure 2**). On average, the daily amount of granules production over a 10-year period is 10.6 t_{DM} or 31 g $PE^{-1}{}_{COD}$ (11.3 kg $PE^{-1}{}_{COD}$ $year^{-1}$). Over the year, these values fluctuate more markedly due to the influence of wastewater temperature on the properties of activated sludge and consequently on the final quantity and quality of granules.

Dynamics at CWWTPL is very important for the granules stakeholder to be able to make a risk assessment and adjust his procedures to the quantities and energy

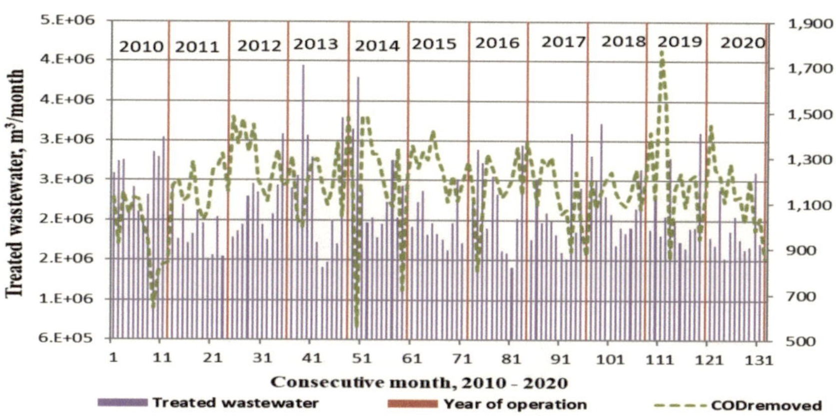

Figure 1.
Dynamics of the amount of treated wastewater and COD removal rate.

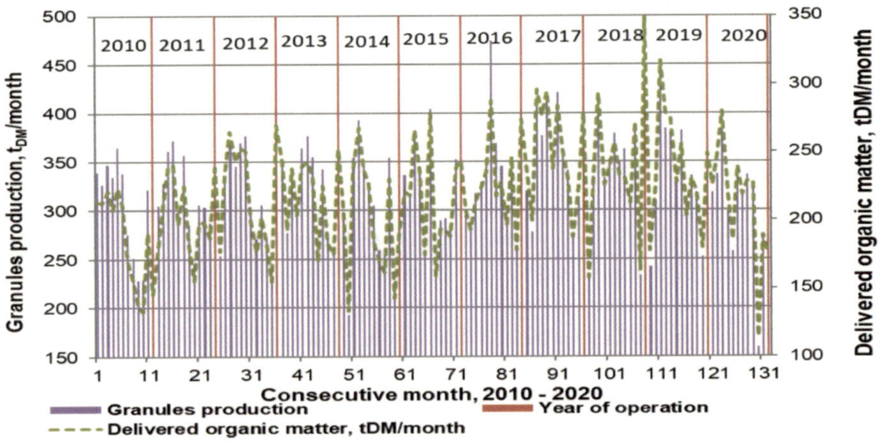

Figure 2.
Dynamics of granules production in their net calorific value (NCV) for the consecutive months in the 10 years time period.

content of granules (**Figure 2**). The problem for the sustainable use of granules in agriculture is the seasonal demand for fertilizer, while the granules are generated over the whole year with the highest production rate in the summer [19]. On average, 6.380 t_{DM} of excess sludge is produced annually (**Figure 3a**) with content of organic matter of 73.0% m/m_{DM}. The anaerobic stabilization reduces the amount of sludge for 39% m/m_{DM}, yielding about 3880 t_{DM} to be transformed into granules having average organic matter content of 67.4% m/m_{DM}. By taking over the granules, the stakeholder gained an average of 55,300 GJ of energy per year in the ten years period (**Figure 3b**).

Biogas production in digestor contributes to a double economic benefit. The amount of sludge is reduced and fuel is obtained for site heat supply (heating the digester and drying the cake). Biogas production has its seasonal fingerprint as well. Due to summer conditions at the biological stage of wastewater treatment, the biogas production (**Figure 4a**) and the granules NCV (**Figure 4b**) are lowest. On average, 17.2 L of biogas per one PE_{COD} was produced at CWWTPL over the 10 years period.

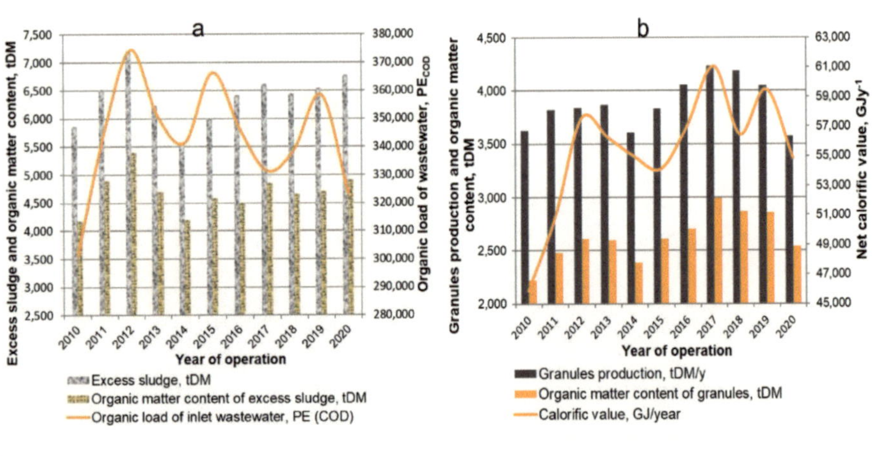

Figure 3.
Dynamics of 10 years period: (a) excess sludge production and an organic load of inlet wastewater and of excess sludge, and (b) annual dynamics of granules production, their organic matter content, and calorific value.

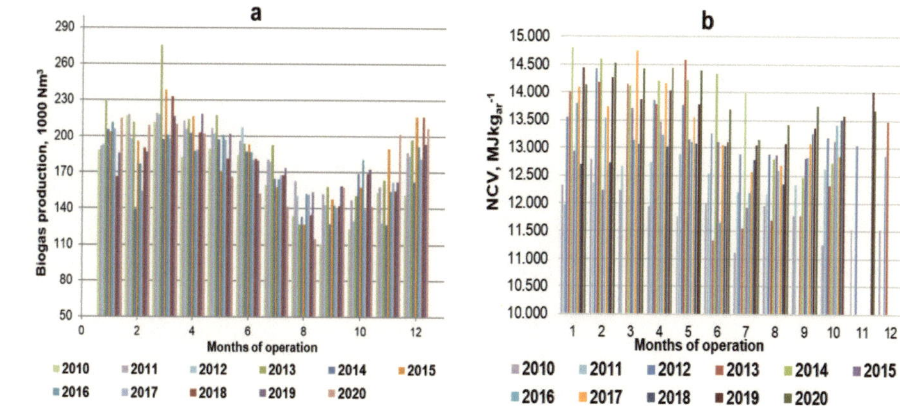

Figure 4.
Seasonal dynamics over the 10 years period: (a) biogas production and (b) NCV_{ar} (as received) of granules.

3.2 Granules characterization

3.2.1 Morphologic properties

Table 14 presents data for the morphological characteristics of annual representative granules samples. Granules are a dry, homogeneous non-hazardous hygienized fine-grained waste with a grain size distribution of 2–4 mm (**Figure 5a**). No foreign solid particles are observed with the eye. Granules do not contain macroscopic impurities or solid particles of glass, plastic, and metals exceeding 2 mm in diameter. Also, no other mineral particles exceeding 5 mm in diameter are present. Granules' response to mechanical stress expressed as mechanical durability is 95.6%. Mechanical durability is a measure of the resistance of compressed fuels toward shocks and/or abrasion during handling and transportation.

Seasonal fluctuations in bulk density in the period from 2017 to 2020 are not significant, the bulk density of granules is highest in early spring. There is also no significant fluctuation in bulk density at the annual level. Only the value for the annual sample 2010 differs significantly (**Table 14**), and that is also determined for other quality parameters of that sample, which will be presented below. The volume of 1m^3 granules for the period 2017–2020 has a mass of about 640 kg (**Figure 5b**).

Granules are slightly basic. Their electrical conductivity is 2.38 mS cm^{-1}, which means that they contain a small proportion of salts. The granules are compacted, difficult to break, which is evident in their low value for the specific surface area (BET) from 0.9 m^2 g^{-1} to 1.4 m^2 g^{-1} (**Table 14**).

Parameter	Unit	2010	2011	2012	2014	2016	2018
Electroconductivity$_{25°C}$	mS cm^{-1}	No a.	3.81	2.38	No a.	2.48	2.71
Specific surface area, BET	m^2 g^{-1}		0.954	0.934	1.406	No a.	
pH	—	No a.	7.1	7.9	No a.	8.0	7.3
Bulk density	kg m^{-3}	704	656	641	683	646	628
Compacted bulk density		No a.	683	No a.		702	809

Table 14.
Basic morphologic and physical properties of granules.

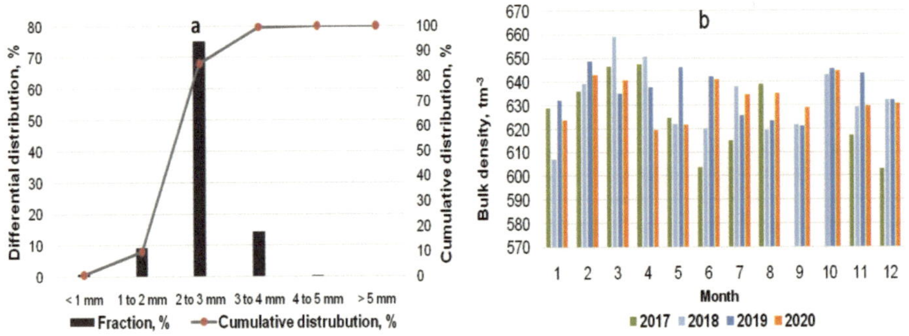

Figure 5.
(a) Granules size distribution, and (b) dynamics of granules bulk density.

3.2.2 Basic thermal properties of granules: determination with conventional laboratory methods

Organic matter is the main component of granules, which releases energy during combustion, rendering steam and CO_2. In general, sewage sludge is semi-biomass, its main constituents being water, organic matter, which is a mixture of residual decomposition products of excess aerobic biomass and anaerobic decomposition products, which are mostly denatured proteins (due to cationic polymer addition, dehydration, and drying of the digestate), carbohydrates, lipids, fatty acids and inorganic substances (salts and complex natural minerals) are also present. Granules behavior under thermal load with conventional laboratory methods in the oven dryer, LF, or combustion chamber (determination of NCV) is an indicator of their proximate properties: moisture, volatile and organic matter (VM, OM), carbonates, and ash content (**Table 15**). The organic fraction or biomass content that can burn in the presence of oxygen at 450°C is 66.9% m/m_{DM}, the further fraction that can burn at 550°C is 0.7% m/m_{DM} higher. The share of VM is 53.9% m/m_{DM}, which means that a high volume of gaseous products is formed during incineration (**Table 15**).

Parameter	Unit	2010	2011	2012	2013	2014	2016	2018
DM, 103°C[1]	% m/m	No a.	No a.	91.3	90.8	91.3	90.5	89.9
DM, 105°C[2]		91.3	90.7	91.5	90.7	91.5	91.0	90.4
OM, 450°C[1]	% m/m_{DM}	No a.	62.7	66.9	65.7	65.7	65.6	66.6
LOI, 550°C[2]		61.1	67.6	67.6	66.3	67.4	66.3	68.6
Ash, 900°C, modif.[3]		33.2	31.8	28.5	29.6	29.4	29.8	28.1
VM, 900°C[2]		No a.	No a.	53.9	54.0	53.3	53.0	No a.
$T_{ignition}$	°C	540	560			No a.		
Heat content	GJ/y	46.58	51.67	55.61	54.50	54.13	57.83	59.27
NCV	MJ/kg_{DM}	12.91	13.68	14.90	13.31	14.43	14.16	14.48

[1]*Temperature prescribed for biodegradable waste or biomass.*
[2]*Temperature prescribed for treated waste into solid recovered fuel.*
[3]*Prescribed temperature is 550°C, but the determination of the residue at 900°C also includes the thermal decomposition of carbonates.*

Table 15.
Basic thermal properties of granules.

The granules ignition point is determined by a non-standardized method and apparatus. It means the temperature at which, when heated in an oxidizing atmosphere, the granules spontaneously ignite and immediately continue to burn. The granules temperature of the ignition ($T_{ignition}$) is at 550°C on average (**Table 15**).

Tables 14 and **15** provide the results, which are taken from expert assessments of granules, which were published in the period from 2010 to 2018 in expert reports prepared by an authorized waste assessment contractor. Only the results for the year 2010 differ significantly due to extensive floods on Ljubljana city area and the inflow of background water to the CWWTPL.

On average, the granules' dry matter content is at least 90% w/w. Fluctuations in the results for dry matter content are due to the drier operating conditions, the choice of analytical method, and determination temperature. It is characteristic of the granules that they lose most of their mass even at a low temperature of the thermal load, namely at a temperature of 450°C. With further heating, the weight loss is smaller.

The loss on ignition is often used as an estimate for the content of organic matter in the sample. Inorganic decomposition products (e.g. H_2O, CO_2, SO_2) are released and some inorganic substances are volatile under the reaction conditions. Determination of LOI at 550°C and 900°C allows us to determine the proportion of organic matter (at 550°C) as well as the proportion of CO_2 (at 900°C) resulting from the thermal decomposition of naturally occurring carbonates that cause CO_2 emissions for which it is not necessary to purchase emission coupons (6a). In 2010, the contamination of raw wastewater with inorganic impurities led to an increase in the ash content in the granules (**Table 15**) and especially an increase in the carbonate content, which continued in 2011 (**Table 15**, **Figure 6a**). **Figure 6b** also shows the fluctuation of the calorific value on an annual basis, which is not always logically related to the fluctuation of the LOI and carbonates.

Granules production at CWWTPL is not a continuous process. Due to the specific layout of facilities and machine capacity, this process is performed in batch mode. The quality of representative samples of individual batch series for the year 2020 (**Table 13**) fluctuates markedly seasonally (**Figure 7**). The contents of the biomass (LOI at 450°C) are characterized by the lowest content over the summer. At that time, the mineral content, determined as the difference between LOI at 550°C and LOI at 450°C, is also the lowest (**Figure 7**). The latter mass loss is probably due to the thermal decomposition of $MgCO_3$ or minerals containing magnesium, phosphorus, and carbonates.

The summer period is characterized by higher wastewater temperatures, less precipitation, lower concentration of activated sludge because of more intensive

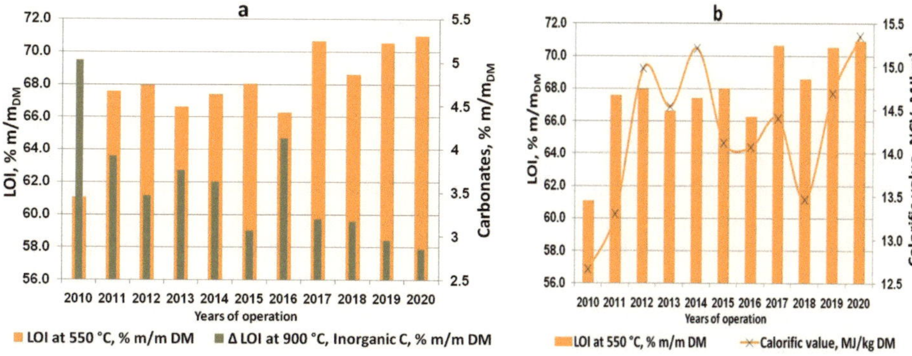

Figure 6.
(a) Dynamics of LOI of granules at 550°C and carbonates content, and b) comparison of annual dynamics of LOI of granules at 550°C and NCV.

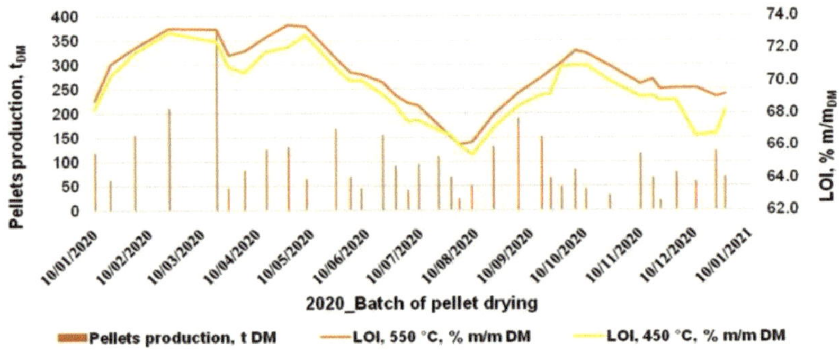

Figure 7.
*Seasonal dynamics of OM content and LOI of the representative batch samples (**Table 13**) of the drying process in the year 2020.*

removal of excess sludge, and consequently higher granules production, which is consequently reflected also in their final quality.

At CWWTPL the biogas production is also the lowest during sommer and early autumn (**Figure 4a**), the same is true for NCV_{ar} (**Figure 4b**).

3.2.3 Basic thermal properties of granules: determination with the advanced techniques

According to TS Combustion behavior solid fuel combustion consists of three relatively distinct but overlapping phases: (i) heating phase (time to thermal decomposition), (ii) pyrolysis, and gas phase combustion (time of gas-phase burning), and (iii) char combustion (time to char burnout). The STA is the most used technique in thermal analysis to give us sufficient information about the thermal behavior of solid fuel.

The results of two techniques of the STA method are provided on granules: (i) proximate analysis and (ii) fingerprint of thermal decomposition of granules in the oxidative and inert atmosphere provided by the non-isothermal temperature program. The proximate analysis of solid granular fuel executed with the TGA technique is a standard procedure for the determination of behavior regarding volatile release and combustion. This technique is comparable with the method as specified in TS EN 15402:2011 for volatile matter (VM) determination. The proximate analysis was performed according to the prescribed procedure (TS Combustion behavior) on the annual sample 2012/2017.

The obtained results of the performed technique (**Figure 8**), moisture (8.12% m/m), VM (51.9% m/m_{DM}), and ash (30.8% m/m_{DM}), are comparable to those obtained with the conventional methods (**Figure 6, Table 15**). The proximate analysis performed with a TGA gives us an additional parameter of solid fuel properties, and that is a combustible carbon content or fixed carbon (C_{fix} = 11.8% m/m_{DM}). When switching an inert atmosphere of thermal loading at 900°C to an oxidizing atmosphere, further mass loss can be attributed to the carbon remaining in the sample as elementary carbon, which was bound in substances that were depolymerized, and thermochemically converted to carbon. 88.2% m/m_{DM} of granules carbon content is bound in substances which, in an inert atmosphere when heated to 900°C, are volatile or are thermo-chemically converted into volatile gaseous compounds. The content of TOC for the annual sample 2012 is 38.4% m/m_{DM} (**Table 15**), and only 30.7% m/m_{DM} of carbon as a part of TOC is presented as a non-volatile matter and is burnout in the third phase of combustion.

The proximate analysis was performed in six stages (**Table 16, Figure 8**). The moisture was determined in the first stage, when the sample was heated to 110°C with

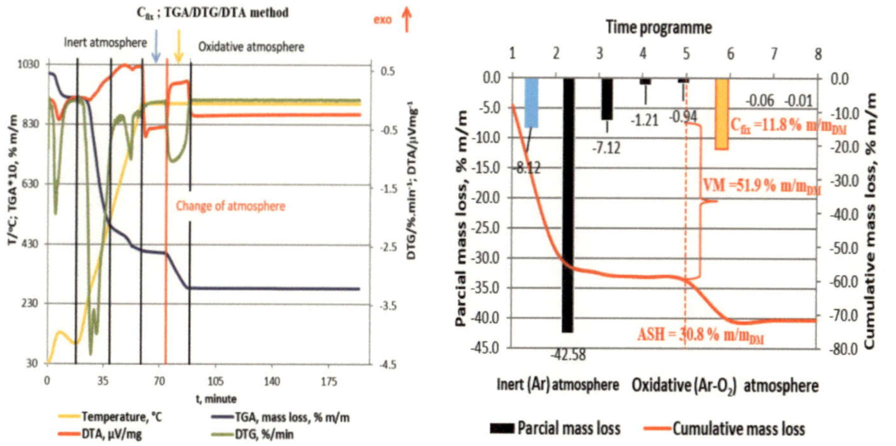

Figure 8.
The proximate analysis of granules (the annual sample 2012/2017) with TGA/DTG/DTA technique: C_{fix}) the fingerprint of granules decomposition in an inert atmosphere combined with an oxidative atmosphere in a sequential mode, and time program) determination of the stages of granules mass loss.

20 K min^{-1}, keeping the temperature constant for 15 min in an inert Ar atmosphere. At the time of 4.8 min, the signal peak for DTG was −1.95% min^{-1} and the heat balance was endothermic (**Figure 8**, C_{fix}). VM was determined by consecutive heating to 900° C at 20 K min^{-1}. This part of proximate analysis consists of three stages: (i) second stage with two signal peaks for DTG (at 27.0 min the first signal peak is −4.34% min^{-1} and at 30.8 min the second signal peak for DTG is −3.98% min^{-1}), (ii) third stage with one signal peak for DTG at 51.8 min (−1.01% min^{-1}), and (iii) stage four followed by the isothermal hold at 900°C in an inert atmosphere for 15 min. In the fifth stage, the purge gas was changed to the oxidizing atmosphere (80% v/v of Ar and 20% v/v of O_2), and the temperature was kept constant at 900°C for 120 min. In this stage the carbonized sample was burnout (C_{fix}), the signal peak for DTG is −1.01 at 80 min, and the heat balance is exothermic. In the sixth stage, only insignificant mass change is recorded. The residual mass is ascribed as high-temperature ash, and the cumulative mass loss is 71.8% m/m (**Figure 8**, Time program). The mass loss followed in 8 steps of the time program is shown in **Figure 8b** and **Table 16**.

The TGA/DTG/DTA granules analysis gives us a useful picture and data on their behavior, with which we can predict or estimate the course of incineration and

Time program, step	Time period, minute	Stage	Temperature at the end of time step, °C	Cumulative mass loss, % m/m
1	0.0–18.0	1	98.5	−8.12
2	18.0–39.0	2	515.1	−50.7
3	39.0–53.4	3	801.8	−57.8
4	53.4–58.4		902.1	−59.0
5	58.4–74.0	4	900.2	−60.0
6	74.0–88.7	5	900.0	−71.7
7	88.7–133.3	6		−71.8
8	133.3–193.3			−71.8

Table 16.
*Stages of proximate analysis of the sample 2012/2017—overview of the time program and cumulative mass loss (**Figure 8**, time program).*

pyrolysis, the formation of emissions and residues on the full scale [23]. Using the TGA/DTG/DTA/EGA techniques, we can estimate the temperature range when substances characteristic of the decomposition of biological macromolecules take place, formation of the main decomposition gaseous products H_2O and CO_2, and determination of the temperature range when biologically inert substances react to thermal load. The latter allows us to determine the temperature range of CO_2 release resulting from naturally occurring carbonates that cause CO_2 emissions for which no emission coupons need be purchased. Reference [23] conducted a STA study of the representative annual sample 2010/2011 in an inert (pyrolysis) and oxidative (incineration) atmosphere. The study shows that in a comparable temperature range from room temperature to 1000°C the decomposition in an oxidative atmosphere takes place in at least in two stages, while in an inert atmosphere it takes place in three stages. Due to the continuous decrease in mass by isothermal temperature program in an inert atmosphere, the STA was prolonged to 1200°C. In the last temperature range of the inert atmosphere, the residue mass was further decreased by 4.5% m/m, which means, that the decomposition in an inert atmosphere takes place in four stages. The study [23] reveals that the first stage of granules decomposition in the temperature range from room temperature to 185°C takes place equally regardless of the atmosphere, with the evaporation of H_2O. The mass loss in this temperature profile is 6.1% m/m, while the result for the mass loss for the annual sample 2010, determined with the conventional oven-dry method (non-isothermal temperature profile), is 8.7% m/m (**Table 15**). In both atmospheres, the granules' thermal behavior in the temperature range from 185 to 420°C is very dynamic. The highest mass losses are observed: (i) in an inert atmosphere 32.5% m/m, the signal peak of DTG is $-1.97\%\ min^{-1}$ at 324°C, and (ii) in an oxidative atmosphere 30.8% m/m, the signal peak of DTG is $-2.14\%\ min^{-1}$ at 252°C. This stage in the thermal behavior of granules in both atmospheres presents the fingerprint for that type of material. It is a result of the decomposition of a major part of the granules' biomass. A comparison of the TGA curves profiles in both atmospheres shows that the TGA curves behave similarly to the temperature 486°C. At that temperature, the residue mass was 56.0% in the oxidative atmosphere and 56.4% by mass in the inert atmosphere. The further course of the TG curves was different.

Slightly different results were obtained in repeated analysis on the newer aparatous Netzsch STA 449 F3 Jupiter in the year 2012. For the sample 2010/2012 (**Figure 9a** and **b**) TGA curves behave similarly to the temperature of 346°C, whereby the residue mass was 70.1% m/m in the oxidative atmosphere and 71.2% m/m in the inert atmosphere. To the temperature of 486°C both TGA curves had a similar slope, with the residue mass being 58.0% m/m in the oxidative atmosphere and 55.2% m/m in the inert atmosphere. After that temperature TGA curves begin to behave significantly different.

This means that the presence of oxygen influences mass loss to differ significantly only above 486°C, while the dynamics of mass loss (DTG) and changes in heat balance (DTA) in different atmospheres begin to differ at temperature 185°C, immediately after the loss of free and bound water [23] (**Figure 9**). The determined temperature limit is close to $T_{ignition}$ of granules, which is 550°C (**Table 15**). The mass loss for the annual sample 2010 due to decomposition of carbonates determined with the combination of the two conventional methods in LF (non-isothermal temperature profile at 550°C and at 900°C) is 5.3% m/m_{DM} (**Figure 6a**). Carbonates content evaluated on the basis of the fingerprint for thermal behavior (isothermal temperature profile) for the sample 2010/2011 is 6.5% m/m, and for the sample 2010/2012 9.2% m/m (**Figure 9a**, pyrolysis mass loss from 550 to 900°C). Similar behavior was confirmed in the sample 2012/2012 analysis (**Figure 9c** and **d**), which had at 346°C the residue mass of 67.1% m/m in the oxidative atmosphere

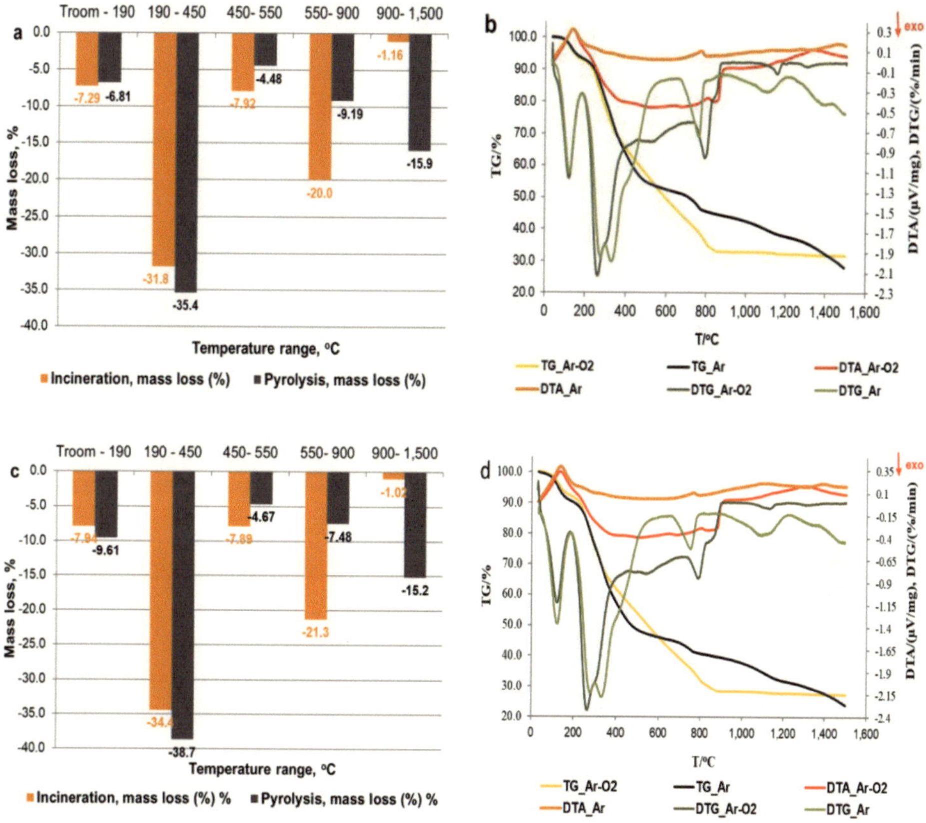

Figure 9.
Comparison of the fingerprint of thermal decomposition in the oxidative and the inert atmosphere for (a) mass loss for each stage for sample 2010/2012, (b) thermogram for sample 2010/2012, (c) mass loss for each stage for sample 2012/2012, and d) thermogram for sample 2012/2012.

and 66.1% m/m in the inert atmosphere, and at 486°C the residue mass was 54.8% in the oxidative atmosphere and 49.1% m/m in the inert atmosphere.

In order to determine the impact on the stability of annual representative samples due to longer storage and in order to determine the repeatability of the TGA/DTA/DTG techniques, repeated thermogravimetric analysis in an inert atmosphere were performed on annual samples 2010 and 2012 (pyrolysis) (**Figure 10**). Compared to the analysis in the year 2011 [23] when apparatus Netzsch STA 409 was used with the temperature range from room temperature to 1200°C, aparatous Netzsch STA 449 F3 Jupiter was used, which allows the temperature range from room temperature to 1500°C, and the heating rate was the same. Repeated analysis were performed in the years (**Figure 10**): (i) 2012 (sample 2010/2012 and sample 2012/2012), (ii) 2016 (2012/2016), and 2018 (2010/2018 and 2012/2018). The partial mass loss of samples at individual characteristic temperatures is repeated in the same way in all samples (**Figure 11a**), with only a slight upward trend at 400°C. The same is true for the peaks of the DTG signals (**Figure 11b**). There were no significant biological-chemical transformations or a significant change observed in the quality of granules after the storage period. Some substances became slightly more thermally stable, suggesting a smaller temperature lag for the DTG signal peaks toward higher temperatures (**Table 17**).

Pyrolysis is an anaerobic thermo-chemical decomposition process that enables the generation of various useful groups of substances from the treated sludge, which can then be utilized separately in various energy and material recovery

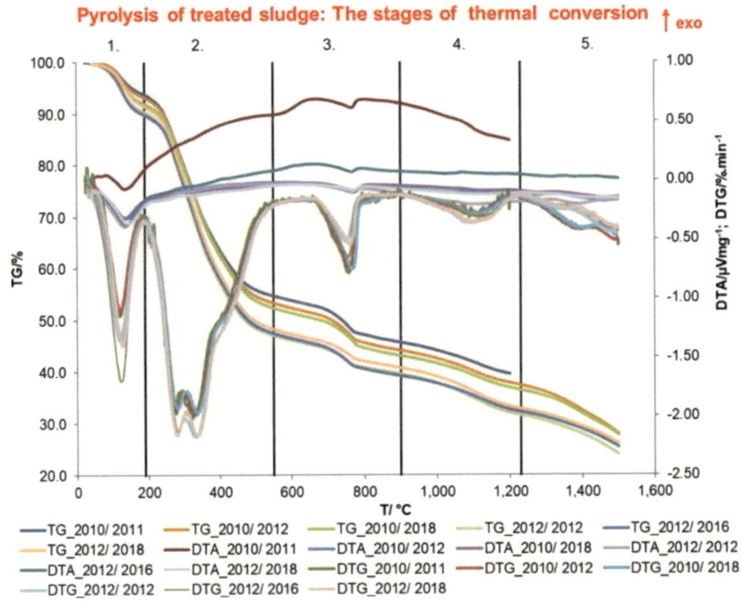

Figure 10.
Fingerprint of thermal decomposition in an inert atmosphere for annual samples 2010 and 2012 - analyzed in 2011, 2012, 2016, and 2018.

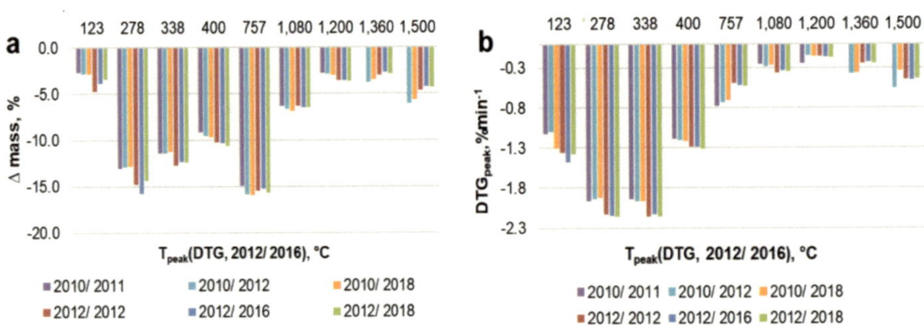

Figure 11.
*Comparison of the fingerprint of thermal decomposition in the inert atmosphere (5 stages of mass loss) for the annual samples 2010 and 2012 (analyzed in the years 2011, 2012, 2016 in 2018) at the temperature of the DTG_{peak} for the sample 2012/2016 (**Figure 10**, **Table 17**): (a) Δmass loss (% m/m) and (b) DTG (% min^{-1}).*

Stage of thermal conversion, °C	1st Troom—190	2nd 191–550		3rd 551–900	4th 901–1230	5th 1231–1500
Sample/year of analysis			T_{peak}(DTG), °C			
2010/2011	121	269	324	758	1114	—
2010/2012	120	277	329	759	1099	1383
2010/2018	121	282	337	768	1116	1393
2012/2012	123	279	332	753	1093	1345
2012/2016	123	278	338	757	1080	1360
2012/2018	127	279	339	763	1094	1366

Table 17.
Comparison of T_{peak}(DTG) of the samples 2010 and 2012 analyzed in 2011 (only sample 2010), 2012, 2016 (only sample 2012), and 2018.

processes. Due to carbonization, part of the carbon remains in the residue (PCM) as bound carbon, so pyrolysis is useful as a technology that reduces the carbon footprint as opposed to the incineration procedure that generates greenhouse gases. The use of PCM in agriculture actually means carbon sequestration and thus a double bonus for the environment.

Further pyrolytic experiments were performed on the sample 2012 [13, 24]. Under pyrolytic conditions two main processes were taken place—volatilization and carbonization. Despite the high temperature (up to 1.500°C) that was reached in the apparatus Netzsch STA 449 F3 Jupiter (**Figure 9d** and **10**) the sample during pyrolysis remained solid. Melting of ashes occurs during granules incineration above temperature 1100°C. The pyrolytic products are [24]: (i) solid residue (pyrogenic material), (ii) aqueous light oily fraction (water condensate), (iii) pyrolysis oil as a heavy fraction (bio-oil), and (iv) non-condensable gas (syngas). The latter product was continuously withdrawn from the reactor and was co-fired (potential recovery procedure R 1). The semi-pilot scale experiment yielded [24]: 15.1% m/m syngas, 17.8% m/m water condensate, 16.8% m/m bio-oil and 50.3% m/m of PCM. Most of the water condensate originating from moisture, chemically or crystalline bound water, and water produced during thermal decomposition of the sludge was collected in the light oily fraction and contains a high concentration of condensed water-soluble substances. The yield of the most valuable product bio-oil was low due to the low content of macromolecules and their poor quality (oxidized and degraded organic matter). The bio-oil must be further refined and completely free of water to make it useful as a fuel. The produced PCM contained 32.4% of carbon by weight and had a calorific value of 11.1 MJ kg^{-1}.

3.2.4 Characterization of granules according to energy and material recovery operations

Despite anaerobic stabilization and biogas production (**Figure 4a**), there is still enough organic matter left in the granules that can be used for WtE process (**Figure 6**, **Tables 15** and **18**). **Table 18** provides the chemical properties of representative annual samples for the period from 2010 to 2018. In addition to energy recovery, the WtE processes enable also the material recovery of many inorganic compounds or elements present in the granules. They also contain some interesting elements that remain in the ash. These are primarily elements of the lithosphere (silicon, calcium, sodium, magnesium, aluminum, potassium, and iron, all in the form of oxides). They are in addition to carbon, nitrogen, hydrogen, sulfur and phosphorus the major elements in treated sludge (**Table 18**). Some other elements presented in **Table 18** are also on the list of critical raw materials [25, 26].

Untreated ash may be used in the construction industry. All procedures of material recovery must be used sensibly because during combustion granules substances are thermo-chemically converted, most of the elements are converted into oxides that are insoluble in water and their bioavailability is highly questionable.

Recovery procedures regarding ash utilization on the full scale, which purpose is to obtain only a certain element, e.g. extraction of phosphorus or other critical raw materials, are economically still unprofitable. WtE processes have economic benefits, but the facilities are also the source of emission of produced volatile compounds and particles into the air (**Tables 2–4**). **Table 18** presents the parameters, which are arranged in different groups: (i) elements relevant to the economic benefits, but are also the source of emission into the air (C, H, N, and S), (ii) halogens, which can cause corrosion of mechanical equipment (F and Cl), (iii) the major elements, which are important for material recovery of the ash (P), and (iv) volatile metals and their compounds, that are harmful to human health (Cd, Hg). Based on the given values for all the listed elements (**Table 18**), it is not possible to accurately

Parameter	Unit	2010*	2011*	2012	2013	2014	2016	2018
Economic benefits, source of green CO_2 of biological origin, and natural source of inorganic CO_2/emission into the atmosphere [15]								
TC/C_{tot}	$\%\ m/m_{DM}$	No a.	No a.	40.4	41.2	39.5	43.0	42.6
TOC/C_{org}		29.04	33.6	38.4	37.6	39.5	33.1	35.7
H		4.57	5.58	4.38	4.21	5.08	No a.	
N	$\%\ m/m_{DM}\ NO_2$	16.1	17.9	20.2	19.4	18.3	18.9	19.7
S	$\%\ m/m_{DM}\ SO_2$	1.52	1.78	2.30	1.86	2.52	2.12	1.60
Corosion potential								
F	$\%\ m/m_{DM}$ HF	0.011	0.012	0.013	0.012	0.015	No a.	
Cl	$\%\ m/m_{DM}$ HCl	0.011	0.066	0.088	0.098	0.113	0.123	0.076
Potential resource recovery of inorganic compounds/ash components								
TIC	$\%\ m/m_{DM}$	No a.	No a.	2.0	3.6	<0.1	9.9	6.9
P		1.07	2.18	2.22*	1.92	1.77	1.62*	1.45*
Fe		1.2	1.1	0.92	1.1	1.2	No a.	
Al		1.0	0.97	0.88	0.78	0.88	No a.	
SiO_2	$mg\ kg_{DM}^{-1}$ Si	510	1700	1510	no a.	4700	No a.	
Si_{tot}	$\%\ m/m_{DM}$	No a.	No a.	3.3	5.6	6.1	9.4	7.6
Ca		5.2	5.1	4.6*	4.8	4.7	3.65*	6.4*
Mg		1.0	1.1	0.96*	0.42		No a.	
Ti		No a.	No a.	0.14	0.15			
K_{tot}		0.22	0.23	0.30*	0.36	0.39	0.28*	0.32*
Na		No a.	No a.	0.12	0.14		No a.	
Environmental impact/volatile metals and their compounds (VMs)/emission into the atmosphere [8, 15].								
Sb	$mg\ kg_{DM}^{-1}$	< 20	< 20	4.3	3.7	2.6	3.5**	2.7
As		< 20	< 20	3.5	3.9	3.8	3.6**	3.3
Co		9.0	9.6	17	8.4	6.7	9.7**	6.8
Cu		310	300	350	410	540	230*	350
Mn		270	280	310	270	300	No a.	250
Cd		1.0	1.1	1.2	1.2	1.1	1.1**	0.80
Cr		84	100	120	110	110	130**	100
Hg		2.6	2.7	1.8	1.7	2.3	1.7**	1.14
Ni		54	60	86	82	80	73**	75
Pb		84	85	93	80	78	85**	66
V		<20	<20	17	18	19	No a.	20
Tl		<10	<10	<0.5	<0.5	<0.5		0.13
$\sum$ VMs		<885	<838	<1000	<985	<1264	—	874
Scan of additional parameters with economic or environmental significance								
Ba	$mg\ kg_{DM}^{-1}$	260	260	320	260	260	No a.	
Be		No a.	<0.50	<0.5	<0.5	<0.5		
Br		<50	30	<21	<37	62		

Parameter	Unit	2010*	2011*	2012	2013	2014	2016	2018
Se		<20	<20	3.4	3.1	4.4		
Zn		920	970	1000*	930	920	720*	790*

*Digestion for subsequent determination of aqua regia soluble portion of elements.
**Average of ten representative monthly samples (Jan-Oct.).

Table 18.
Basic physicochemical properties of granules, interesting for WtE processes.

predict whether the WtE process will exceed the permitted emissions into the environment. The latter will certainly occur if the flue gas cleaning system is inadequate. For the sustainable energy use of granules, in addition to reducing their amount and economic benefits offered by this recovery procedure, it is very important that advanced flue gas cleaning and volatile metal removal techniques prevent the polluting substances their circulation in nature as much as possible. The granules produced at CWWTPL are a sustainable material for SRF is revealed by reference [27]. The latter is a case study of the establishment of an integral quality system for wastewater treatment and sludge management at CWWTPL according to TS CEN/TS 343, EN 15358:2011, Quality management system, Particular requirements for their application to the production of SRFs. For the period from 2012 to 2020, a classification of the granules defines their conformity as an SRF with Class code NCV 3–4; Cl 1; Hg 3. In Ref. [27] is also pointed out that mercury is one of the most problematic volatile metals regarding the quality of cleaned flue gases. It is transmitted to the environment through flue gas emissions [28]. By effectively cleaning the flue gases generated by WtE processes, we achieve the interruption of the circulation of substances that cause emissions into the air.

3.2.5 Granules and their solid residues after thermally loading evaluation as a fertilizer

In the EU the use of sewage sludge in agriculture is regulated by the framework directive SSD [6], which is summarized differently in each EU Member State. The limit values for soil and sewage sludge quality prescribed by SSD are presented in **Table 5**. The set of parameters for which the SSD prescribes limit values is short. It also does not prescribe a declaration for the content of C_{org}, primary nutrients, secondary nutrients, macronutrients, and micronutrients as does the Fertilizing Resolution (FR) (**Tables 9, 10** and **12**). In general, the existing European market for fertilizers, which are produced from treated waste, is non-uniform. The latter will be addressed by (FR) [20], which will take full effect from 16 July 2022 (**Tables 7–12**). The use of granules in agriculture should be viewed in terms of soil quality (heavy metal content, pH, organic content) and in terms of granules quality. For their application, it is necessary to take into account the seasonal time, the type of plantation, the purpose of the crop usage, and the period of the plant growth.

Metals in the environment are usually in the form of compounds. Soil pH plays an important role in the bioavailability of heavy metals into the plant. If the soil is alkaline, soluble metal ions precipitate into less soluble compounds, e.g. hydroxides, which are less bioavailable and therefore less harmful. Elements in granules can have the function of nutrients if they are water-soluble or if they are bioavailable. This is especially important for phosphorus and potassium.

Table 19 presents the biological properties of representative annual samples for the years 2016 and 2018. Granules are anaerobically stabilized and hygienized waste, which therefore has limited biological properties and, depending on the degree of stabilization, still has the possibility of further self-degradation. In terms

Parameter	Unit	2016	2018
Stability criteria			
(AT_4)	mg O_2 g^{-1}_{DM}	43	35
Acetic acid	mg L^{-1} (mass/volume)	236	530
Propionic acid		57.9	38
Soil improvers and growing media/determination of plant response			
The effect of granules as a soil improver and growth substrate on plant germination and growth			
Plant germination and growth (Chinese cabbage)	% m/m	73[1]	87[4]
Germination rate (Cress)		53[2]	47[5]
Green mass (Chinese cabbage)		0[3]	79[6]
Determination of the content of unwanted weed seeds and plant propagules in granules as a growing substrate and soil improver			
Weed seeds and plant propagules, granules: peat moss 20% m/m	N° L^{-1}	<2	<2

[1]*Granules: peat moss 30% m/m.*
[2]*Granules: peat moss 32% m/m.*
[3]*The green part of the plant did not develop.*
[4]*Granules: peat moss 15% m/m.*
[5]*Granules: peat moss 33% m/m.*
[6]*Granules: peat moss 15% m/m.*

Table 19.
Biological parameters/stability criteria of CWWTPL granules [21].

of the "stability" parameter, the limit value for short-chain fatty acids for 2018 was slightly exceeded (**Tables 8** and **19**). There is no limit value for parameter AT4 for granules as a digestate. The biological properties of the granules are suitable. By preparing an appropriate substrate, they enable germination and plant growth and do not introduce weeds into the plantation (**Table 19**).

In addition to C_{org}, the granules contain a wide range of other elements, which are declared as either primary nutrients, secondary nutrients, or macro-and micro-elements. **Table 20** presents the chemical composition of representative annual samples for the period from 2010 to 2018. The values of the elements are expressed in units, as prescribed for the declaration and limit values of the elements or their relevant compounds according to FR [20] (**Tables 9–12**). Among the polluting substances in the granules, the most critical is mercury, the value of which exceeds the limit values for fertilizers (**Tables 9, 10** and **12**). Mercury is an environmental pollutant from atmospheric deposition, which enters with precipitations to a municipal treatment plant. The effect of biological removal of mercury from wastewater at CWWTPL is 84.6% on average with standard deviation of $\pm$ 17.0% [19, 27], so municipal wastewater treatment plants play an important role in preventing the circulation of mercury in the environment and protecting human health. The phenomenon of Hg content in Slovenia is historically linked to the operation of a mercury mine in Idrija, where cinnabarite ore mining was supposed to begin in the last decade of the 15th century, and ceased in 1977. It was the second-largest cinnabarite mine in the world. In the period from 2010 to 2018, a decrease in Hg content in granules produced at CWWTPL is observed (**Tables 18** and **20**). It is likely that the Hg content will continue to decline.

The general finding regarding the possible use of granules as compost or digestate is that these two treated forms of treated sludge are useful primarily as an additive to other processed waste, which then forms a mixture corresponding to the quality declaration and limit values set by FR.

Parameter	Unit	2010	2011	2012	2014	2016	2018
			Declaration				
			PN (**Table 9**)				
C_{tot}	% C	No a.	No a.	36.6	36.1	35.9	38.5
C_{org}	% C	26.51	30.4.	34.8	36.1	30.1	32.3
NH_4-N	% N	No a.		0.63		No a.	
	$mg\ kg^{-1}_{DM}$, soluble in water		No a.		2271	2260	2600
NO_3-N					No a.	<2.5	16
N_{org}	% N	No a.	No a.	4.8		No a.	
N_{tot}	% N	4.17	4.94	5.57	5.07	5.20	5.49
	$kg\ N\ m^{-3}$	26.7	31.7	35.7	34.8	33.6	34.4
P	% P_2O_5	2.25	4.99	4.62	3.72*	3.36	3.01
K	% K_2O	0.244	0.255	0.326	4.30*	3.09	3.44
			S MACN (**Table 10**)				
Ca	CaO, %	6.69	6.50	5.80	6.02*	4.64	8.10
Mg	MgO, %	1.53	1.64	1.49	1.67*	No a.	
Na	Na_2O, %	No a.	No a.	0.147*	0.109*		
S	SO_3, %	1.72	2.02	2.60	2.88	2.41	1.86
			MICN (**Table 11**)				
Fe	%	1.10	1.00	0.836*	1.10*	No a.	
Mn		0.0248	0.0255	0.0281*	0.0274*	No a.	0.0226*
Cu		0.0285	0.0273	0.0281	0.0494*	0.0209	0.0325
Zn		0.0846	0.0884	0.0906	0.0842*	0.0655	0.0714
B	$mg\ kg^{-1}$				No a.		
Co		8.3	8.7	15.4*	6.1*	8.8*	6.2*
Mo		<50	<50	7.5	8.2*	No a.	
			Environmental aspects				
			Polluting substances, PTMs				
Cd	$mg\ kg_{DM}^{-1}$	1.0	1.1	1.1	No a.	0.82	0.88
Cr		84	100	93		90	110
Hg		2.6	2.7	1.6		0.83	1.4
Ni		54	60	73		49	72
Pb		84	85	85		61	66
As		<20	<20	3.5*	3.8*	3.6*	3.3*

**Microwave-assisted digestion with hydrofluoric (HF), nitric (HNO_3), and hydrochloric (HCl) acid mixture for subsequent determination of elements.*

Table 20.
Basic physicochemical properties of granules, relevant for fertilizing.

Solid residues generated at thermally treated sludge are an interesting substitute for phosphorus-potassium fertilizers and, in addition, a good conditioner for acidic, mineral-poor soils. At the laboratory level, the study was carried out regarding the water solubility of nutrients in resulting residues from the sample 2012 after its exposure to inert and oxidative thermal treatment (**Table 21**). PCM is by mass the major pyrolytic product with the organic carbon content of 29.4% m/m_{DM}, BET surface area of 6.82 $m^2\ g^{-1}$, and bulk density of 820 kg m^{-3}. The study on the water solubility of generated PCM shows that pyrolysis offers a much greater opportunity for material recovery from sewage sludge than from incineration ashes. It was found that the water solubility of phosphorus in treated sludge's PCM produced at 450°C is higher compared to the residues from the oxidizing atmosphere; in contrast, the water solubility of K and Mg is lower (**Table 21**). According to ashes properties, it was found that the highest water solubility and the highest BET

Parameter	Unit	Granules 2012	Technique for residues preparation; temperature of thermal load (°C)				
			Pyrolysis, TGA	Incineration, LF			
			450	450	550	700	900
			PCM 2012	ASH 450	ASH 550	ASH 700	ASH 900
Solid fraction							
Shape	—	Granules					
Mass loss by preparation							
Mass loss	% $Volatile_{DM}$	—	57.3	55.4	68.0	71.5	71.5
$P_{volatile}$	% m/m_{DM} of P_{total}		No a.	<0.001			
$Mg_{volatile}$	% m/m_{DM} of Mg_{total}			<0.02			77.3
$K_{volatile}$	% m/m_{DM} of K_{total}			<0.02	16.4	No a.	16.2
Physicochemical properties							
BET area	$m^2\ g^{-1}$	0.934	6.82	52.4	31.6	18.2	8.49
Bulk density	kg m^{-3}	641	820	590	360	440	530
TOC	% m/m_{DM}	38.4	29.4	No a.			
TC		40.4	29.4				
N_{tot}		6.15	4.12				
P	mg kg^{-1}_{DM}	22,253	40,870	69,150	70,517	76,050	83,155
Mg		9949	18,985	30,270	32,233	34,500	7700
K		2989	4085	6205	7801	7810	8800
Water-soluble nutrients							
pH	—	7.8	8.0	6.8	9.1	12.4	12.4
P	% m/m_{DM} of P_{total}	5.76	5.37	0.16	<0.01		
	mg kg^{-1}_{DM}	1282	2195	112	<10		
Mg	% m/m_{DM} of Mg_{total}	1.91	1.30	4.66	4.64	<0.002	
	mg kg^{-1}_{DM}	184	246	1411	1496	<2	
K	% m/m_{DM} of K_{total}	23.6	3.82	26.14	12.1	9.47	3.20
	mg kg^{-1}_{DM}	706	156	1622	943	740	282

Table 21.
Fraction characterization of granules and their solid residues after thermal treatment.

surface area has the ash generated at 450°C. At higher temperatures of oxidative thermal treatment, the nutrients are chemically converted to a form that is no longer water-soluble, reducing the possibility of nutrient recovery of the residue (**Table 21**) [29]. The advantage of extracting phosphorus from PCM is also that pyrolysis does not cause sintering or melting, as happens with incineration at temperatures above 1000°C. According to the guidelines for sustainable production of biochar [14], it is potentially useful as a fertilizer for non-agricultural land, for reclamation of degraded land, and for the preparation of artificial soils. It could be used to amend depleted soil to increase fertility, porosity, and water retention. However, the concentration of copper and nickel increases due to the volatile and decomposed fraction of the pellets, while the concentration of mercury decreases [13, 19].

To increase the useful bioavailable components new economically viable technologies need to be developed to liberate the nutrients from sewage sludge ash. Some materials recovered from the UWWTS are still considered waste even if they are of good quality and have a market value [9]. At the moment, European legislation does not favor the recycling of these materials in the economy. The example of the FR, explicitly defining the conditions or processes under which struvite or ashes produced from UWWTS could cease to be a waste and become a product, is a good starting point [9]. Due to the impoverishment of natural sources of phosphorus, it is especially important to find opportunities for recycling it.

4. Conclusions

Municipal wastewater is a product of human life. With the help of natural processes, man has created suitable conditions for its biological treatment. Generated sewage sludge is of biogenic origin, its production is dynamic, and that is reflected in seasonal fluctuations of its quality and quantity. The results of dried sludge granules analysis over a 10-year period of operation at the CWWTPL are presented, as well as the year 2020, which is the starting year for the upcoming new ten-year period. 2020 is also a control year for many forecasts of improving the environment and preventing climate change. The data provide a comprehensive insight into the properties and potentials of treated sludge and allow thoughtful studies for use of this smart material in WtE, critical raw material recovery, organic matter recovery with pyrolysis, and agricultural use of the solid pyrolytic product or as raw granules.

The sewage sludge composition during the period 2010–2020 is similar, only 2010 differs markedly, which was characterized by heavy floods. They did not spare the capital of Slovenia, when a lot of urban area were under the water. Normal changes of anthropogenic or meteorological behavior do not have any special effect on the average annual values of granules. Within a year, these values fluctuate more markedly due to the influence of wastewater temperature on the properties of activated sludge and consequently on the final quality of granules.

Knowledge of the sewage sludge properties enables quick adaptation to the requirements of the revised legislation and, above all, prudent use. The chapter brings together and comments on existing legislation and also legislation in preparation. The data make it possible to compare the quality of granules with other UWWTPs and confirm the feasibility of the drying technology of dehydrated digestate.

For energy recovery, sewage sludge must be used in technically advanced facilities to prevent harmful emissions of substances into the air and prevent the circulation of harmful substances in nature.

For the use of sewage sludge in agriculture, it is important to emphasize that this is a material that would make sense to add to the mixture of components for fertilizer and not just as a stand-alone substrate, while controlling the quality of the soil and that the application to the soil is in the form of a dynamic approach. Annual application amount, soil quality control, seasonal application period, and type of crop must be planned. With a dynamic approach, it is necessary to prevent the limit values for polluting substances per hectare of agricultural land from being exceeded.

Pyrolysis is a specific thermochemical decomposition of biomass into a range of useful products without oxygen. Pyrogenic carbonaceous materials are by mass the major pyrolytic product of treated sludge, but it has not yet been included in the EBC feedstock list and is therefore not subject to the EBC guidelines.

The study on the water solubility of generated residues after thermal treatment of granules shows that pyrolysis offers a much greater opportunity for material recovery from sewage sludge. It was found that the water solubility of phosphorus in biochar produced at 450°C is higher compared to the residues from the oxidizing atmosphere; in contrast, the water solubility of K and Mg is lower.

Without good sludge quality control, a sustainable strategy for the use of treated sludge in the CE cannot be achieved. To have confidence in the results of research into the properties of sewage sludge, it is necessary to have a good sampling procedure, targeted end-use, and a list of required parameters. Only high-quality analysis enables confidence in the results and the setting of optimal guidelines for a successful CE. Such an approach to sewage sludge quality control improves the public acceptance of this material in the CE.

The scope of analysis to determine the properties of sewage sludge can be reduced if the quality of sewage sludge is well known and the technology of municipal wastewater treatment and its quality does not change significantly.

Author details

Barbara Novosel[1], Vesna Mislej[2]* and Viktor Grilc[3]

1 UL, Faculty of Chemistry and Chemical Technology, Ljubljana, Slovenia

2 JP VODOVOD KANALIZACIJA SNAGA d.o.o., Ljubljana, Slovenia

3 Environmental Protection College, Velenje, Slovenia

*Address all correspondence to: vesna.mislej@vokasnaga.si

IntechOpen

© 2022 The Author(s). Licensee IntechOpen. This chapter is distributed under the terms of the Creative Commons Attribution License (http://creativecommons.org/licenses/by/3.0), which permits unrestricted use, distribution, and reproduction in any medium, provided the original work is properly cited.

References

[1] Council Directive 91/271/EEC of 21 May 1991 concerning urban waste-water treatment OJ L 135, 30.5. 1991. pp. 40-51

[2] Urban waste water: Improvement in collection and treatment across the EU helps reducing pollution in the environment [Internet]. 2020. Available from: Urban waste water (europa.eu) [Accessed: July 06, 2021]

[3] Europe's water in figures, an overview of the European drinking water and waste water sectors, 2017 edition, EurEau [Internet]. 2017. Available from: eureau_water_in_figures.pdf (danva.dk) [Accessed: July 06, 2021]

[4] Urban Waste Water Directive Overview [Internet]. 2020. Available from: ec.europa.eu/environment/water/water-urbanwaste/index_en.html [Accessed: July 06, 2021]

[5] Waste water treatment—Sludge management. Briefing note. EurEau [Internet]. 2021. Available from: file (eureau.org) [Accessed: July 06, 2021]

[6] Directive 86/278/EEC on the protection of the environment, and in particular of the soil, when sewage sludge is used in agriculture. OJ L 181, 4.7.1986. 1986. pp. 6-12

[7] WFD. Directive 2008/98/EC of the European Parliament and of the Council of 19 November 2008 on waste and repealing certain Directives (Text with EEA relevance). OJ L 312, 22.11.2008. pp. 3-30

[8] Circular Economy Action Plan [Internet]. 2020. Available from: ec.europa.eu/environment/strategy/circular-economy-action-plan_en [Accessed: July 06, 2021]

[9] Position Paper, Enabling the Circular Potential of Sewage Sludge within the EU Legislative Framework, A Critical Analysis of the Current Urban Waste Water Treatment Sludge Legislation with Respect to the Circular Economy. EurEau [Internet]. 2021. Available from: file (eureau.org) [Accessed: August 06, 2021]

[10] Central Wastewater Treatment Plant Ljubljana [Internet]. 2006. Available from: Centralna čistilna naprava Ljubljana | JP VOKA SNAGA [Accessed: July 06, 2021]

[11] Communication from the Commission to the European Parliament, the Council, the European economic and social Committee and the Committee of the Regions, The role of waste-to-energy in the circular economy, 2017, COM (2017) 34 final

[12] Biomass PB. Gasification, Pyrolysis, and Torrefaction, Practical design and Theory. 2nd ed. London: Academic Press: Elsevier; 2013. DOI: 10.1016/C2011-0-07564-6

[13] Mislej V, Novosel B, Klinar D, Zalar, Serjun V, Zupan K, et al. Pyrolysis of dried sewage sludge (In Slovenian "Piroliza posušenega blata"). In: Kaučič V et al., editors. Proceedings of the Slovenian Chemical Days 2015. Ljubljana: Slovenian Chemical Society; 2015. pp. 1-9

[14] EBC. 2012 European Biochar Certificate—Guidelines for s Sustainable Production of Biochar. 9.5E of 1st August 2021. Arbaz, Switzerland: European Biochar Foundation (EBC). 2021. Available from: european-biochat.org.Version [Accessed: June 11, 2021]

[15] Directive 2010/75/EU of the European Parliament and of the Council of 24 November 2010 on Industrial Emissions (Integrated Pollution Prevention and Control) Text with EEA Relevance OJ L 334. 2010. pp. 17-119

[16] Directive 2000/60/EC of the European Parliament and of the Council of 23 October 2000 Establishing a Framework for Community Action in the Field of Water Policy OJ L 327, 22.12.2000. pp. 1-73

[17] Directive 2013/39/EU of the European Parliament and of the Council of 12 August 2013 Amending Directives 2000/60/EC and 2008/105/EC as Regards Priority Substances in the Field of Water Policy Text with EEA Relevance OJ L 226 24.8.2013. pp. 1-17

[18] Prakash BO. Health risks of potentially toxic metals contaminated water. In: Kanayochukwu Nduka J, Nageeb Rashed M, editors. Heavy Metal Toxicity in Public Health. London, UK: IntechOpen; 2020. DOI: 10.5772/intechopen.92141

[19] Mislej V, Grilc V, Novosel B, Mladenovič A, Zalar SV. Sewage sludge management at the central wastewater treatment plant Ljubljana—Characterization of its properties, possibilities for sustainable utilization and legislative obstacles. The Holistic Approach to Environment. 2022;**12**(1): 9-25. DOI: org/10.33765/thate.12.1.2

[20] Regulation (EU) 2019/1009 of the European Parliament and of the Council of 5 June 2019 Laying Down Rules on the Making Available on the Market of EU Fertilising Products and Amending Regulations (EC) No 1069/2009 and (EC) No 1107/2009 and Repealing Regulation (EC) No 2003/2003, PE/76/2018/REV/1, OJ L 170, 25.6.2019. pp. 1-114

[21] Decree on the treatment of biodegradable waste and the use of compost or digestate, OJ of Republic of Slovenia, N° 99/13, 56/15 and 56/18

[22] National Laboratory of Health, Environment and Food. Department for Waste and Soil [Internet]. Available from: Waste and soil - NLZOH [Accessed: July 06, 2021]

[23] Mislej V, Novosel B, Vuk T, Grilc V, Mlakar E. Combustion behaviour and products of dried sewage sludge—Prediction by thermogravimetric analysis and monitoring the co-incineration process in a cement factory. Chemical Engineering Transactions. 2012;**29**: 685-690. DOI: 10.3303/CET1229115

[24] Mislej V, Klinar D, Zalar Serjun V, Grilc V, Žgajnar Gotvajn A, Novosel B, et al. Pyrolysis—The initial stage of an integrated approach to optimal material recovery of sewage sludge. In: Proceedings of the 2nd Conference on Holistic Sludge Management (HSM'16); 7–9 June 2016, Malmö. London: IWA; 2016. pp. 1-10

[25] Communication from the Commission to the European Parliament, the Council, the European Economic and Social Committee of the Regions on the 2017 list of Critical Raw Materials for the EU, Brussels, 13.9.2017, COM(2017), 490 final

[26] Critical raw materials. Internal Market, Industry, Entrerpreneurship and SMEs [Internet]. 2012. Available from: Critical raw materials | Internal Market, Industry, Entrepreneurship and SMEs (europa.eu) [Accessed: August 06, 2021]

[27] Mislej V, Babič R, Nastran S, Grilc V, Novosel B. Sewage sludge—A sustainable material for the production of solid recovered fuel. Chemical Engineering Transactions. 2021;**88**(1057):1062. DOI: 10.3303/CET2188176

[28] Chemical, Wastes and Climate change. Interlinkages and Potential for Coordinated Action [Internet]. 2021. Available from: Chemicals, Wastes and Climate Change: Interlinkages and Pote ntial for Coordinated Action | UNEP - UN Environment Programme [Accessed: July 06, 2021]

[29] Mislej V, Novosel B, Serjun VZ, Grilc V, Marinšek M, Zupan K.

Characterization of residues, resulting from the thermal treatment of the sewage sludge, with the purpose of nutrient recovery. In: Kljun J, Klemenčič M, Drvarič T, Talian S, editors. Book of abstracts of the conference Cutting edge (Ce 2015). Ljubljana: Faculty of Chemistry and Chemical Technology: Cutting Edge; 2015. p. 17

Chapter 20

Activated Carbons from Waste Tyre Pyrolysis: Application

Mzukisi Matandabuzo and Delford Dovorogwa

Abstract

The development of better and efficient methods of consuming less and/or wasting little resource materials is becoming more important. In this study, pyrolytic waste tyre carbon black residue and commercial grade activated carbon were characterized and evaluated against adsorption of mercury vapor. The performance of the raw carbon black residue and the activated carbon against mercury vapor generated in the laboratory was determined using a designed reactor system. The adsorption of Hg^{+} was investigated at temperatures ranging from 200 to 280°C for 6 hours. Batch experiments were conducted for the different carbon residue samples and characterization analysis were done before and after adsorption using the spectroscopic, microscopic, and structural techniques to elucidate the structural arrangements and properties of the carbonaceous materials. Spectroscopic analysis of these carbonaceous residues showed a C=C stretching vibration attributed to the lignocellulose aromatic ring at 1657–2000 cm^{-1}. Comparatively, it was also observed that the Infrared spectrum of raw carbon black exhibits less functional groups as compared to the H_2SO_4-AC and H_2O_2-AC carbonaceous residues prepared.

Keywords: Tyre waste, pyrolysis, activated carbon, mercury vapor, adsorption

1. Introduction

The continually growing global economies inevitably results in a steady increase of the automobile industry, which unavoidably produces huge amounts of end-of-life waste tyres. It is estimated that about 1.5 billion waste tyres are produced annually worldwide [1]. In South Africa, the majority of these waste tyres are landfilled and do not readily decompose, contributing to both land and air pollution [2]. The carbon black currently produced via pyrolysis has not been put to any significant use as to justify its economic/environmental production from the plant operations. On the hand, mercury (Hg) is a known neurotoxin and can be dangerous once it is released into the environment [3]. It is a toxic heavy metal that bio accumulates in organisms and causes brain and liver damage if digested or inhaled by human beings. The efficiency and sustainability of control and management of mercury vapor release from compact flourescent lamp waste in South Africa, however is not adequately documented and therefore specific studies aimed at developing cheaper and locally available technology to control this are necessary.

The phenomenon of mercury, (Hg vapor) Hg-vapor sorption is well recognized and accepted. However, vapor sorption of many vapors is a well-recognized physical principle. Gaseous materials such as Hg vapor diffuse into a solid host material

IntechOpen

because of differences in concentration. This driving force for adsorption continues until equilibrium conditions are established between the vapor phase and the solid phase. The solubility of one material in another is generally governed by the rule "like dissolves like." Mercury vapor exists as a nonpolar molecule that would likely be most soluble in nonpolar substances such as other metals and nonpolar hydrocarbons [4]. At first glance the new energy efficient lighting technology appears to be the better way to reduce energy consumption and protecting the environment. However, the serious mercury management concerns associated with use of CFL lamps needs to be considered and addressed. Depending on the type of lamp, a longitudinal compact fluorescent lamp may contain mercury from greater than 0.1 up to 100 milligrams (mg). Although cleaning and recovery methods have been developed and are implementation stages, efficient levels of the containment of the released mercury vapor during the recycling processes are not clearly known. Therefore there is a need for the development of a cheaper and locally available mercury recovery adsorbent material from waste tyres pyrolytic carbonaceous residues for use at recycling facilities in South Africa. This study therefore seeks to investigate the use of carbon black residue from waste tyre pyrolysis as an adsorption agent for the removal of mercury vapor from fluorescent lamp waste recycling, yet another waste hurdle, the country is grappling on.

2. Materials and methods

2.1 Preparation of waste tyre carbon materials

Two forms of carbon residues were considered during this research study on the characterization and development of a sustainable adsorption material for mercury vapor recovery. One form of the carbon residue was sourced from waste tyre pyrolysis by-product, the carbon black residue by RECO, a waste to energy Solution Company based in Pretoria, Gauteng province. The other carbon residue sample was a commercially available activated carbon collected from RECLITE, a fluorescent lamp waste recycling company based in Germiston, Johannesburg, Gauteng Province in South Africa. The company, RECLITE, has imported the fluorescent lamp recycling plant, including the activated carbon filter cartridges from Sweden. The commercial grade Activated Carbon was produced and supplied by MRT Systems, Swedish based company specializing in technology and equipment for the safe disposal/recycling of mercury containing waste products. All chemicals and reagents were obtained from Sigma-Aldrich Merck and used as received unless stated otherwise. Sulfuric acid (95–99% H_2SO_4), hydrogen hydroxide (H_2O_2), and 32% hydrochloric acid (HCl). The experimental materials include the following: Electric Oven (0–300°C), 100 mL volumetric flasks, de-ionized water, desiccator, vacuum filter, drying oven (Temperature 90–120°C for drying Carbon black C\char), and the Carbon black (CB) was collected RECOR. The carbon black was taken into the oven (Scientific Oven Series 2000 Labotec SA) at 120°C to dry. After drying, it was crushed and sieved into smaller pieces of about 0.85–0.1 mm internal diameter (**Figure 1A** and **B**). The as-prepared sieved carbon black material was stored in an Oven at 100°C prior to activation.

2.2 Activation of carbon black

Activated carbon (AC) was prepared from waste tyre pyrolytic carbon black using one-step chemical activation method [5]. The two different chemical activating agents, Sulfuric acid (H_2SO_4) and hydrogen peroxide (H_2O_2) were used in this study, in a 1:4 ratio (carbon black: agent).

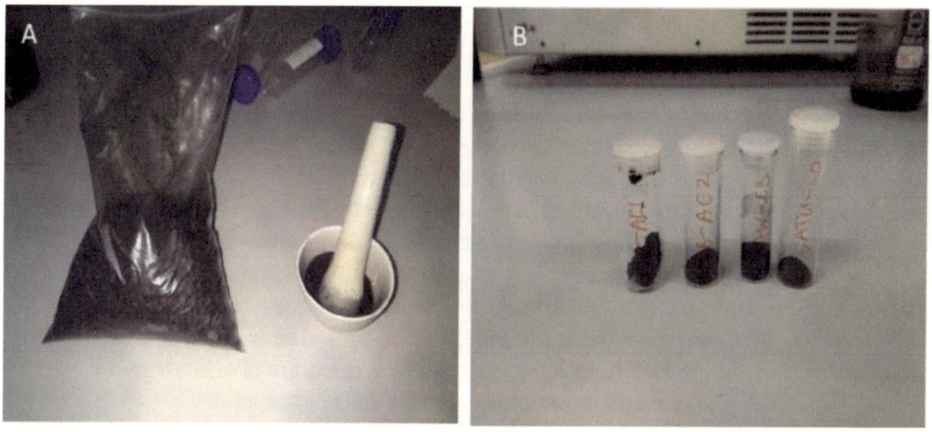

Figure 1.
(A) Grinding of commercial grade AC; (B) prepared raw and commercial AC samples ready for characterization.

2.3 Activation of carbon black (CB) using sulfuric acid and hydrogen peroxide

In a typical experiment, approximately 15 g of dried-sieved carbon black was impregnated with 100 mL of H_2SO_4 and the impregnation mixture was kept for 12 hours in an oven at 100°C as shown in **Figure 2**. Therefore, the mixture was further heated in an oven for 6 hours at temperatures ranging from 200 to 280°C. Consequently, a black bubbles-like structure of activated carbon was produced and was allowed to cool. Upon cooling, the activated carbon was rinsed with hot distilled water to remove acid from the activated carbon until the pH of the resulting solution was neutral, that is, pH = 7. The resultant activated carbon was further dried in an oven at 120°C overnight. The resultant product was thus named: Sulfuric acid-based carbon black-activated carbon (CB-AC 1). On the other hand, the hydrogen peroxide-based activated carbon was rinsed with cold distilled water to remove hydrogen peroxide residue from the activated carbon. The resultant activated

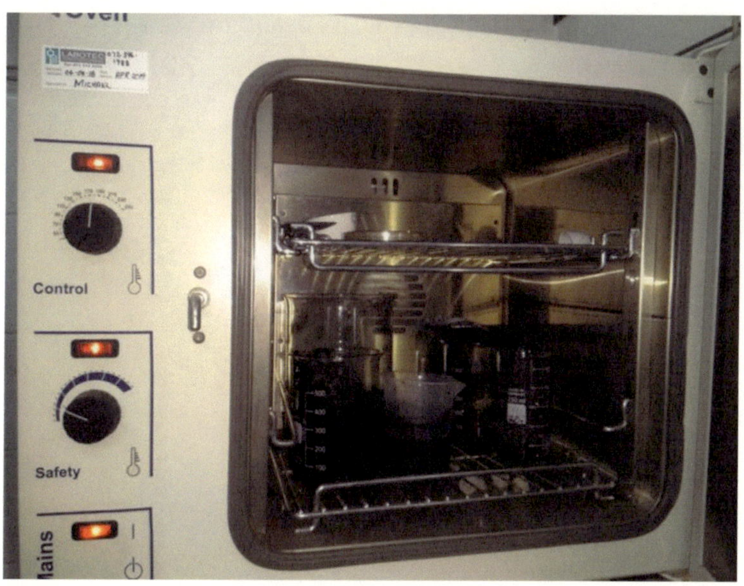

Figure 2.
Samples of raw carbon black in an electric oven for chemical activation.

carbon was further dried in an oven at 120°C overnight. The resultant product was thus named: Hydrogen peroxide-based carbon black-activated carbon (CB-AC 2).

2.4 Characterization of carbon residues

Spectroscopic, microscopic and structural characterization methods were applied to determine the qualitative and structural properties of the carbonaceous materials in this study. The analytical methods employed have been chosen as the most suitable ones considering the instrument, material availability, and cost. The selected methods and techniques will then be adapted to enhance the accuracy and precision of analysis of the material as far as possible. The as-prepared activated carbons were characterized using Fourier transform infrared spectroscopy (Perkin Elmer, System 2000 FT-IR, USA), scanning electron microscopy/energy dispersive X-ray spectroscopy (Jeol, JSM-6390 LV SEM with Noran system Six software, USA), powder X-ray diffraction (Brker X8 Proteum pXRD), and Brunnauer Emmett–Teller methods.

2.5 Adsorption of Hg using carbon black residues

A sample of liquid mercury (4.89 g) in a 250 mL flask was placed in a water bath with controlled water temperature of 50–100°C at atmospheric conditions. The determination of the mercury vapor adsorption of the carbon residues was done through an experimental set up where the mercury vapor was generated through placing liquid mercury contained in a flask on temperature-controlled water bath. The flask was connected to a reactor glass column filter packed with the carbon residue samples and connected to an active peristaltic air vacuum pump. The experiment was run for 6 hours, **Figure 3**. This was repeated for all the 4 samples of the carbon residues (raw CB; saturated commercial AC; sulfuric acid activated CB; hydrogen peroxide activated CB). The same experiment set up was repeated using an electric hotplate as a heat source (0–300°C) replacing the water bath. This experiment set up was placed and run under an active laboratory fume hood cupboard to ensure safe extraction and emission of all potential fugitive mercury vapor.

Figure 3.
Experimental set up for the extraction generation and adsorption of mercury vapor on the developed AC from waste Tyre pyrolysis. (a peristaltic vacuum pump connected to the filter unit to extract the mercury vapor under the fume wood cupboard).

3. Results and discussion

3.1 Characterization of carbon black materials

The results of FT-IR analysis (FT-IR spectra) of the carbon black and its activated carbons are shown in **Figure 4** (A-B carbon black-based activated carbons), (C Raw carbon black or tyre waste), and (D Saturated commercial carbon black). In **Figure 4A**, absorption bands around 1100 cm^{-1} with an overtone at 1430 cm^{-1} corresponding to Si-O-Si asymmetric stretch [6]. The C=C stretching vibration attributed to the lignocellulose aromatic structure appeared at 1657–2000 cm^{-1} in carbon black and in activated carbons [6, 7]. Another vibration bands at 2345 cm^{-1} appeared in both carbon black and activated carbons can be assigned to the alkyne groups [8]. However, it was also observed that the IR spectrum of raw carbon black showed less functional groups as compared to the sulfuric acid and hydrogen peroxide-based activated carbons [9]. The latter is due to the fact that, during/after activation, various functional groups are formed. The C-O-C stretching vibrations in ether, phenol and esters groups appeared around 1041–1248 cm^{-1}. In their study, [7] also reported acyclic C-O-C groups around 1300–1000 cm^{-1} coupled with aromatic structures (C=C). The stretching vibration band at 1000 cm^{-1} attributed to C-O group is distinct in sulfuric acid-based activated carbon. It has been reported that chemical activation with hydrogen peroxide oxidant normally yield carbonaceous structures with oxygen-functionalities on the surface [7]. In this study, oxygen-functionalities were observed in the surface of the activated carbon and are believed to have participated in the activated carbon-Hg interaction, and subsequently influenced the Hg uptake and adsorption rate of Hg by activated carbons.

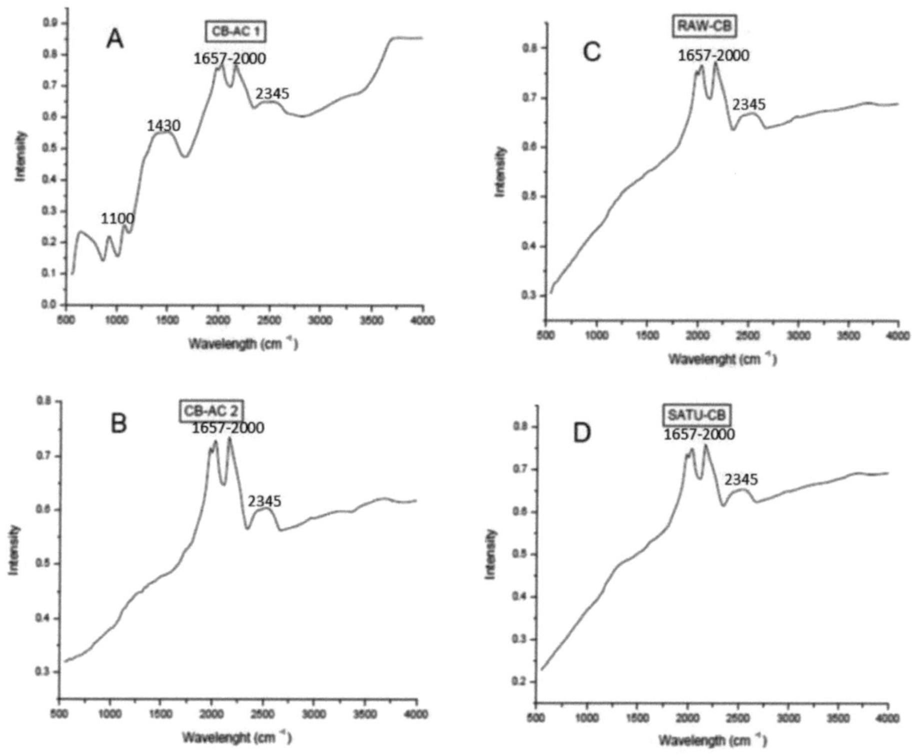

Figure 4.
The FT-IR spectra of carbonaceous materials. A, B: Sulfuric acid and hydrogen peroxide-based activated carbons, respectively. C-raw carbon black, and D-saturated commercial carbon black.

BET analysis results revealed that activated carbon prepared via impregnation with hydrogen peroxide exhibits high surface area, high micropore volume, and average pore volume diameter. However, both activated carbons (CB-AC 1 and CB-AC 2) prepared via chemical activation in this study showed improved surface area and micropore volume. It is believed that the activating agents were good enough to percolates the interior of the waste tyre carbon black, breaks the lateral bonds in the cellulose molecules (increases Inter and Intra miscelle voids) and able to influence the surface properties [9]. The BET results obtained confirmed the successfulness of the chemical activation method and the influence of the chemical activating agents on the surface of carbon black. Based on the surface area and micropore volume results obtained, raw carbon black and the two as-prepared activated carbons were further used for the adsorption of Hg vapor. It has been reported that the uptake of an gaseous or liquid materials by an adsorbent normally rely on the BET surface area and pore volume, which all indicate the degree of microporosity in an carbonaceous material [7, 9].

Pyrolytic carbon black and as-prepared activated carbons were also studied under pXRD to check the crystalline arrangement of the carbonaceous materials. In the untreated carbon black (**Figure 5**), peaks at $2\theta = 27$ and 58° corresponds to the reflections from (111) and (222) planes for cubic ZnS. Additionally, peaks at $2\theta = 51$ and 64° corresponding to the reflections (102) and (110) planes, respectively, are of the wurtzite phase of ZnO (F.A. [10]). It has been reported that Zinc Oxides are normally added as vulcanization agents/catalysts during tire manufacturing and they decompose to different forms of ZnS during pyrolysis [6]. However, as shown in **Figure 5**, the diffraction patterns depict broad bands centered around $2\theta = 24$, 25 and 45°, associated or attributed to (002) and (100 and/or 101) planes, respectively [11]. These denote the stacking height Lc and the lateral size of the crystallites La in the carbonaceous material [12]. The presence of a predominantly porous structure, especially in H_2O_2-AC, is confirmed by the sharp peak around $2\theta = 25°$ [12]. Naturally, carbon black is a lignocellulosic material (containing interlinked cellulose, hemicellulose, and lignin structures). However, the introduction/use of an

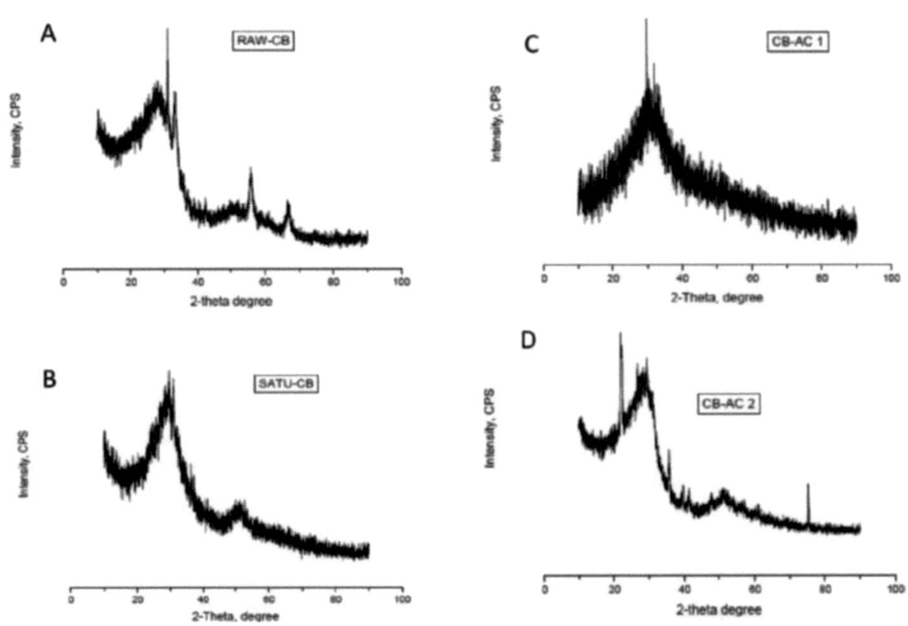

Figure 5.
Powder XRD patterns of pyrolytic carbon black and activated carbons.

activant (i.e. H_2O_2) normally brings the breakdown and alteration of the 3D linkages between the chemical components of carbon black (cellulose, hemicellulose, lignin), thereby forming skeletons of fully disorganized carbonaceous materials (activated carbon). Furthermore, the X-ray diffraction patterns of saturated carbonaceous material (SATU-CB, **Figure 5C**) shows relatively broad, weak and low intensity peaks as compared to raw carbon black or H_2O_2-AC, confirming the interaction of carbonaceous material with adsorbate. In comparison, [6] also studied the crystallographic structure of activated carbon prepared via KOH activation. A broad hump peak was observed around 2θ = 22°, which was attributed to the amorphous nature of the carbon material before activation. After KOH treatment, sharp peaks around 2θ = 23° were observed and assigned to the turbostratic structure and corresponds to the (002) reflection.

The surface morphology and elemental content of the carbon black and as-prepared activated carbons were determined using field emission scanning electron microscopy-coupled with energy-dispersive X-ray spectroscopy (FESEM/EDX). **Figure 6A** and **B** shows the SEM image and the EDX of carbon black used in this study, respectively. According to **Figure 6A**, carbon black exhibits a dense and rough porous structure with very limited pores. EDX confirms a highly carbon-dominated material (carbonaceous), having high carbon content and minimal presence of other elements such as sodium (Na), silicon (Si), and Sulfur (S). Upon activation with hydrogen peroxide and sulfuric acid, the as-prepared activated carbons were also analyzed with FESEM/EDX. **Figure 6A** and A1 shows the FESEM/EDX results of the activated carbon prepared by hydrogen peroxide activation before adsorption of Hg. As shown in the image (**Figure 6A**), the surface morphology of the resultant activated carbon is dominated by increased and improved irregular cracks and pores with many shapes and sizes [9]. This type of porous morphology confirms that activation with hydrogen peroxide was successful and has produced carbonaceous material with improved porous structure (as shown in the BET analysis). The advantage of field-emission scanning electron microscope (FESEM) is that, it studies the structures of carbonaceous materials to microporous level. After adsorption it was expected that some Hg will be spotted and noticed on the surface of the activated carbon. As shown in **Figure 7B** and **B1**, the surface of the activated carbon is covered by dense and light materials of Hg. To that effect, the light spots of Hg infused in the pores of the activated carbon were also deeply checked and focused with high resolution of the microscopy and EDX (**Figure 7**). **Figure 8** illustrates the round-ball like structure of Hg found inside the pores of the activated carbons. Furthermore, the backscatter electron technique was applied to further analyze the surface of the materials against Hg. Light and heavy Hg was seen and confirmed by EDX analysis (**Figure 7C** and **D**, respectively). Mercury is known to show dense and lighter spots when screened by the backscatter electrons (**Figure 9**). In the study by [13], it

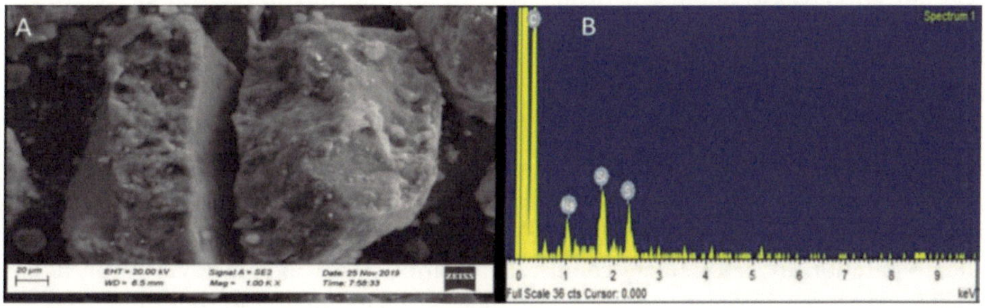

Figure 6.
FESEM/EDX of pyrolytic carbon black.

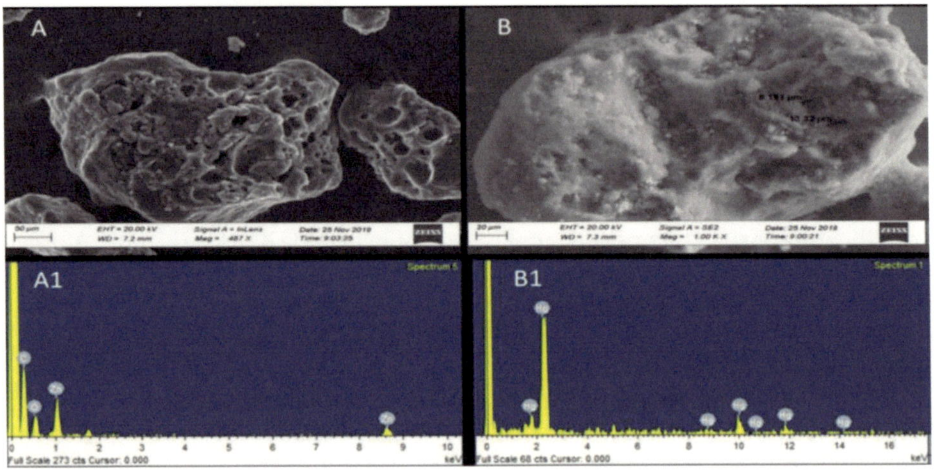

Figure 7.
FESEM/EDX of CB-AC 2 before and after adsorption of Hg.

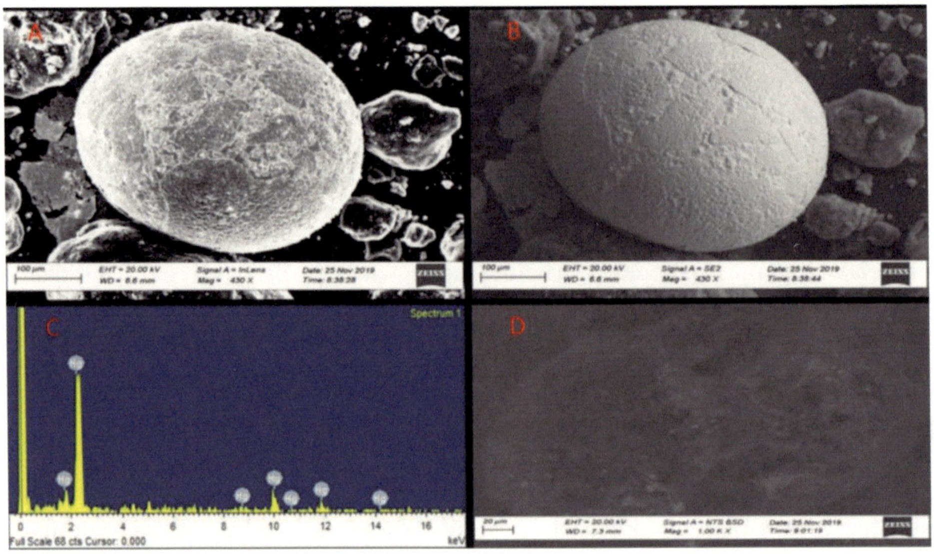

Figure 8.
FESEM/EDX of Hg in the activated carbons prepared.

was revealed through the SEM/EDX analysis that high elementary mercury vapor concentration was found in synchronicity with high sulfur content in a mercury-saturated carbon material as very white spots in the SEM images [13]. In this study, it can been concluded that the interactions between Hg and the activated carbons were influenced and facilitated by heterogeneous surface (oxygen-functionalities) of the adsorbent and its narrow micropores (**Figure 9**). Due to the heterogeneous surface nature of the activated carbons and variety of oxygen-functionalities, the adsorption of Hg is believed to have involved both the physisorption and chemisorption valence forces in both multilayer and monolayer fashion. Recently, reported the preparation and subsequent chemical activation of carbon black via pyrolysis of waste scrap tyres. In order to achieve complete activated material, they employed pyrolysis temperature of 600°C, and further KOH chemical activation at 800°C.

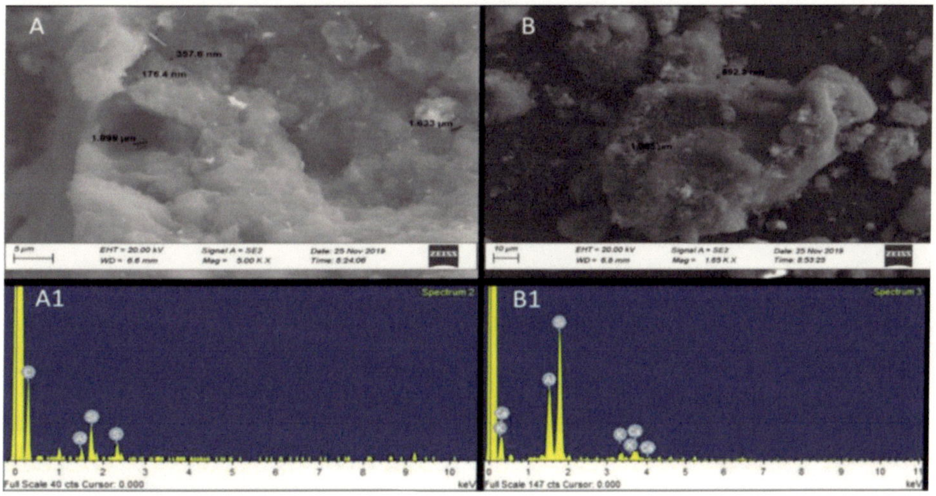

Figure 9.
FESEM/EDX of CB-AC 1 before and after adsorption of Hg.

It was observed that KOH solution chemical activated predominantly affected the porous structure of the resultant activated material.

4. Conclusion

The current study presents chemical preparation, characterization and performance evaluation of pyrolytic carbon black and its as-prepared activated carbons as potential adsorbents for mercury (Hg) vapor. FTIR, BET, SEM/EDX and pXRD were employed to elucidate the structural arrangements and properties of the carbonaceous materials. FTIR analysis of these carbonaceous materials showed a C=C stretching vibration attributed to the lignocellulose aromatic ring at 1657–2000 cm^{-1}. Comparatively, it was also observed that the IR spectrum of raw carbon black exhibits less functional groups as compared to the H_2SO_4-AC and H_2O_2-AC carbonaceous materials prepared. BET analysis confirmed the effectiveness of the chemical activation method and the influence of the chemical activants on the surface of carbon black.

It was observed that the porous morphology of activated carbons prepared via hydrogen peroxide activation showed carbonaceous material with improved porous structure and was complimented by BET analysis results obtained. However, after adsorption it was expected that some Hg will be spotted and noticed on the surface of the activated carbon. However, backscatter electron analysis was employed and round-ball like structures of Hg were found inside the pores of the activated carbons. Additionally, light and heavy Hg was seen and confirmed by EDX analysis. Furthermore, the pXRD patterns confirmed that the use of an activant such as H_2O_2 normally alter the 3D linkages between the chemical components of carbon black (cellulose, hemicellulose, lignin), thereby forming skeletons of fully disorganized carbonaceous materials, thus resulting in increased porosity of the carbon residue.

Acknowledgements

The authors would like to acknowledge RECOR and RECLITE companies for providing the pyrolytic Waste Tyre and a commercial grade activated carbon, respectively.

Author details

Mzukisi Matandabuzo[1*] and Delford Dovorogwa[2]

1 Qualitative and Quantitative Analysis, Animal Production Institute, Agricultural Research Council, Irene, South Africa

2 EnviroFact Consulting CC, Johannesburg, South Africa

*Address all correspondence to: matandabuzom@arc.agric.za

IntechOpen

© 2022 The Author(s). Licensee IntechOpen. This chapter is distributed under the terms of the Creative Commons Attribution License (http://creativecommons.org/licenses/by/3.0), which permits unrestricted use, distribution, and reproduction in any medium, provided the original work is properly cited.

References

[1] Danon B, der Gryp V, Schwarz P, Görgens CE. A review of dipentene (DL-limonene) production from waste tire pyrolysis. J. Anal. Appl. Pyr. **2015**;*112*(1-13)

[2] Park C, Kim B, Yu Y. The regeneration of waste foundry sand and residue stabilization using coal refuse. J. Hazard. Mater. **2012**;*203*:176-182

[3] Agency for Toxic Substances and Disease Registry (ASTDR - USA), **1999**. Toxicological Profile for Mercury.

[4] Dewing HH, Schluter RB. Characterization and recovery of mercury from electrical manufacturing Wastesby thermal desorption. USA Report of Investigations. **1994**:9513

[5] Matandabuzo M, Ajibade PA. Synthesis and surface functionalization of multi-walled carbon nanotubes with imidazolium and pyridinium-based ionic liquids: Thermal stability, dispersibility and hydrophobicity characteristics. J. Mol. Liq. **2018**;*268*:284-293

[6] Shilpa, Kumar, R., Sharma, A, **2018**. Morphologically tailored activated carbon derived from waste tires as high-performance anode for Li-ion battery. *J. Appl. Electrochem*. *48*, 1-13.

[7] Heo YJ, Park SJ. H_2O_2/steam activation as an eco-friendly and efficient top-down approach to enhancing porosity on carbonaceous materials: The effect of inevitable oxygen functionalities on CO2 capture. Green Chem. **2018**;*20*:5224-5234

[8] Saka C. BET, TG–DTG, FT-IR, SEM, iodine number analysis and preparation of activated carbon from acorn shell by chemical activation with $ZnCl_2$. J. Anal. Appl. Pyrolysis. 2012;*95*:21-24

[9] Dimpe KM, Ngila JC, Philiswa N, Nomngongo PN. Application of waste Tyre-based activated carbon for the removal of heavy metals in wastewater. Cogent Eng. **2017**;*4*:1330912. DOI: 10.1080/23311916.2017.1330912

[10] López FA, Centeno TA, Rodríguez O, Alguacil FJ. Preparation and characterization of activated carbon from the char produced in the thermolysis of granulated scrap tyres. J Air Waste Manag Assoc. **2013**;*63*: 534-544

[11] Omri A, Benzina M. Characterization of activated carbon prepared from a new raw lignocellulostic material: Ziziphus spina-christi seeds. J. de la Société Chimique de Tunisie. **2012;175183**

[12] Badie SG, Yassin M, Mostafa T, Gadelrab M, Abdullah ID. X-ray diffraction patterns of activated carbons prepared under various conditions. Carbon. Sci. **2007**;(*2*):75-100

[13] Musmarra D, Karatza D, Lancia A, Prisciandaro M, di Celso M. Adsorption of elemental mercury vapors from synthetic exhaust combustion gas onto HGR carbon. J. Air Waste Manag. Assoc. **2016**;*66*:698-706

Chapter 21

Conceptualization and Design of a Small Pyrolysis Plant for the Sustainable Production of Flexible Bricks and Bituminous Concrete from Polyethylene Terephthalate Waste

Ngonidzashe L. Shangwa, Wilson R. Nyemba, Simon Chinguwa and Tien-Chien Jen

Abstract

Polyethylene terephthalate is majorly used for packaging of various products because of its resistance to chemical attack and environmental degradation, but the proper disposal of this non-biodegradable material has been a major challenge. Pyrolysis is the melting of plastic in the absence of oxygen. Currently pyrolysis of polyethylene terephthalate is considered as a viable recycling method since it only requires 5% of the calorific value of polyethylene terephthalate. This research was aimed at designing a pyrolysis plant for the production of construction materials with acceptable mechanical properties such as compressive strength and water absorption. Sustainable, eco-friendly road construction from bituminous concrete with waste polyethylene terephthalate has the capability of reducing carbon emissions. The polyethylene terephthalate bituminous composite has the flexibility of plastic but strength of concrete. The bricks have a maximum compressive strength of 10 N/mm^2 which is within the standard range and have less water absorbing tendencies hence have a longer lifespan. Value addition is equally important in the pyrolysis plant so as to contribute to sustainable development. This book chapter reviews the different products such as polyethylene terephthalate composite bricks and flexible pavements which can effectively use polyethylene terephthalate waste as a raw material.

Keywords: pyrolysis, bricks, waste, polyethylene terephthalate, recycling, bituminous

1. Introduction

The word plastic refers to any material which is made of synthetic or organic compound which has malleable properties. Polyethylene Terephthalate (PET) is being used for beverage and food packages all over the world, but the problem with

 IntechOpen

the disposal of this non-biodegradable hydrocarbon materials has been a major challenge to the world. The major features which play a big role in the usage of plastic are malleable, light weight, impervious to water and low cost [1]. Recycling of plastic into construction materials can help in providing a cleaner environment. Much of the waste generated from this PET has ended up in water drains and eventually into the ocean. It is the belief of some school of thoughts that by 2050, there will be more plastics than fish in the ocean of the world if this trend is not quickly addressed [2]. Concrete bricks are a type of brick that is commonly used in low and medium cost housing development in developing countries since it is cheap and easy to produce. The conversion of the solid residue remaining after pyrolysis to bricks will add to sustainable value addition recycling of waste PET.

During pyrolysis the hydrocarbon bonds are broken down due to the heat applied. **Figure 1** shows the most common but crude way in which people have been disposing of plastic waste in Zimbabwe.

Table 1 shows the different types of plastics and their symbols. The different plastic types have different uses affected by their chemical structure.

Waste utilization has become an attractive alternative to disposal given the scarcity of space for landfilling and its ever increasing cost; utilization of waste PET in concrete bricks is viable. The use of waste products in concrete not only makes it economical, but also helps in reducing disposal problems [3]. The three types of pyrolysis namely conventional/slow pyrolysis, fast pyrolysis and flash pyrolysis differ majorly in terms of temperature, residence time and products made. **Table 2** summarizes the differences in order to understand the process.

Figure 1.
Disposal of waste in Harare, Zimbabwe.

Symbol	Type of plastic	Application
1	Polyethylene terephthalate (PET)	Clear transparent plastic
2	High-density polyethylene (HDPE)	Opaque plastic bottles
3	Polyvinyl chloride (PVC)	Water service pipes, cable insulation, footwear packaging
4	Low density polyethylene (LDPE)	Grocery bags, packaging film
5	Polypropylene (PP)	Margarine tubs, microwaveable meat trays.
6	Polystyrene	Yoghurt pots, fish trays, egg cartons, vending cups, plastic cutlery and toys
7	Other	Plastics that do not fall into any of the above categories. e.g., melamine.

Table 1.
Types of plastics and typical uses.

Method	Temperature (°C)	Residence time	Heating rate (°C/s)	Major products
Conventional pyrolysis	400–500	5–30 min	Low 10	Gases, char, bio-oil (tar)
Fast pyrolysis	400–650	0.5–2 s	High 100	Bio-oil (thinner), gases, char
Flash pyrolysis	700–100	< 0.5 s	Very high >500	Gases, bio oil

Table 2.
Differences between types of pyrolysis [4].

Heating rate	Time	Temperature (°C)	Char	Liquid	Gas
Slow (<10^3 W/m^2)	1000s	~500°C	25%	35%	40%
Medium (>10^4 W/m^2)	100s	~500°C	17%	58%	35%
Fast (>10^5 W/m^2)	1s	~500°C	15%	65%	20%
Flash (>10^6 W/m^2)	0.01s	~500°C	10%	70%	20%
Slow (<10^3 W/m^2)	1000s	~1000°C	40%	35%	25%
Medium (>10^4 W/m^2)	100s	~1000°C	20%	37%	43%
Fast (>10^5 W/m^2)	1s	~1000°C	15%	20%	65%
Flash (>10^6 W/m^2)	0.01s	~1000°C	0%	15%	95%

Figure 2.
Illustration of pyrolysis properties on end products [4].

Hence from the table, as the temperature increases, the residence time decreases and the solid products become less. The amount of solid residue produced is determined by the type of pyrolysis and mainly affected by the heating rate. **Figure 2** clearly outlines the effect of heating rate and other pyrolysis factors on the products.

Hence from the figure it can be deduced that in order to obtain a higher yield of the solid residue or char, the heating rate should be slow.

2. Background and literature

Plastic pyrolysis is a tertiary recycling method for disposing of waste and repurposing it in a sustainable manner [5]. Incineration and landfilling are two traditional methods for disposing of plastic waste in developing countries, both of which have shown to be unsustainable. Plastic waste is primarily disposed of in landfills or incinerated. Utilization of PET in bricks and flexible pavements has gained momentum over the past decade. The general failures linked with basic bituminous asphaltic concrete (BAC), which are commonly used in surfacing courses in bricks and flexible pavements include rutting, bleeding, polish, cracking and potholes. Repurposing of the waste PET plastic in raw BAC used in bricks and pavements will be a more sustainable waste management option since flexible pavements constitute over 90% of paved roads around the world [6]. Due to various environmental issues such as air pollution and soil contamination, as well as economic resistance due to space waste and disposal costs, these approaches are encountering significant popular resistance [7]. The conversion of PET to bricks would be a more sustainable method of recycling plastic waste.

2.1 Incineration and landfilling (pyrolysis processes waste disposal)

Incineration is a waste treatment process which includes the combustion of waste for recovering energy [8]. Landfills occupy space and are therefore unsustainable. Pyrolysis is a tertiary recycling technology because chemical degradation results in the creation of products with a higher added value than the raw material that undergoes pyrolysis. Pyrolysis is defined as the cracking process in the absence or in poor content of oxygen. By mass, 67 percent of plastic waste may be converted into fuel, gasoline oil, or diesel [8].

2.2 Advantages of plastic pyrolysis over other recycling methods

Pyrolysis of waste PET plastic is a more sustainable way of waste management as it creates value. The residue from the tertiary recycling method can be further treated to produce secondary raw material. The advantages of pyrolysis are:

- It is an inexpensive method of processing waste and can be used with a variety of feedstock.
- Minimizes the amount of non-biodegradable waste that goes to landfills and incinerators, thereby reducing greenhouse gas emission.
- Reduced risk of water pollution.
- Creates jobs for low income earners in developing countries based on the waste generated, they collect the waste for recycling.
- In comparison with combustion it has lower process temperatures and has a lower emission of pollutants.
- Hence pyrolysis of waste PET plastics will be a more sustainable method of waste management rather than the traditional methods.

Figure 3 shows the different stages in the chemical degradation process. It starts with the initiation step which involves homolytic breaking down of the carbon to carbon bonds either by random or end chain scission. The result is two radicals and it leads to propagation which is the hydrogen transfer reaction either by intra or intermolecular forming and unzipping leading to unstable free radicals. Then the final step termination of the free radicals forming unreactive materials.

3. Methodology

A sustainable alternative was essential in the design due to the issues with non-biodegradable plastic waste. In the last few decades, the recycling of PET as building material has been gaining ground, the properties of the plastic make it most suitable [5]. The goal of the study was to simplify existing technologies for value addition pyrolysis. Three different options were considered. Careful consideration and analysis were conducted in order to determine which of the three concepts provided the best solution to the Product Design Specification (PDS). The plant's primary selection criteria were the heat exchanger and the source of heat. Use of the binary dominance matrix was essential because it allowed for simultaneous analysis of all concepts. SolidWorks was used to model three design possibilities, as illustrated in **Figure 3**. The heat exchanger in Concept (a), the Double Pipe, was laminar for fluid flow hence the rate of heat flow was lower than in the other concepts. Since Liquefied Petroleum Gas (LPG) is used as a heat source in Concept (b), the Shell and Tube, the target market without electricity may choose LPG. Furthermore, a Shell and Tube heat exchanger could be more expensive than a helical coil heat exchanger. Due to its helical coiled heat exchanger and coiled heater on the reactor, Concept (c), the Helical Pipe Heat Exchanger, dominated the weighted objectives. The selection was heavily influenced by the ease of assembly and fabrication. The different concepts are shown in **Figure 4**.

Due to its great strength, formability, and corrosion resistance, stainless steel was chosen over mild steel. **Table 3** shows the binary dominance matrix that was used to choose the most optimal concept based on functionality, efficiency, ease of maintenance, reliability, simplicity, cost, quality, and ergonomics. Concept (c) was chosen as a result.

Due to the helical heater on the pyrolysis tank there is even heating and the molten PET is then combined with the bituminous asphaltic concrete (BAC). This mixture is then put into casting bricks and also used in making concrete [8]. Recycling of PET and using it in construction will be a great stride since the PET is non-biodegradable and its chemical properties will further make the BAC durable.

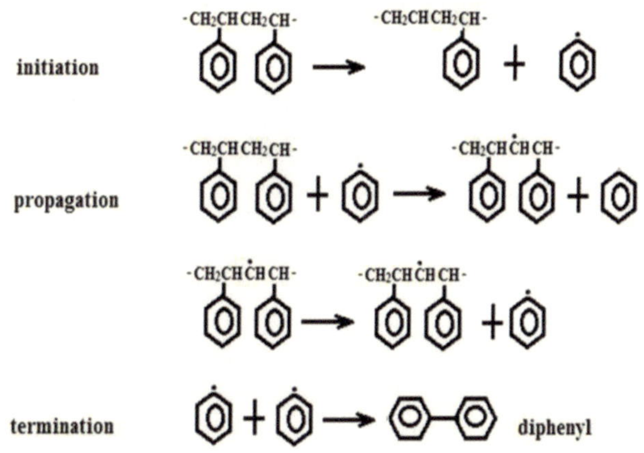

Figure 3.
Stages of radical degradation reaction.

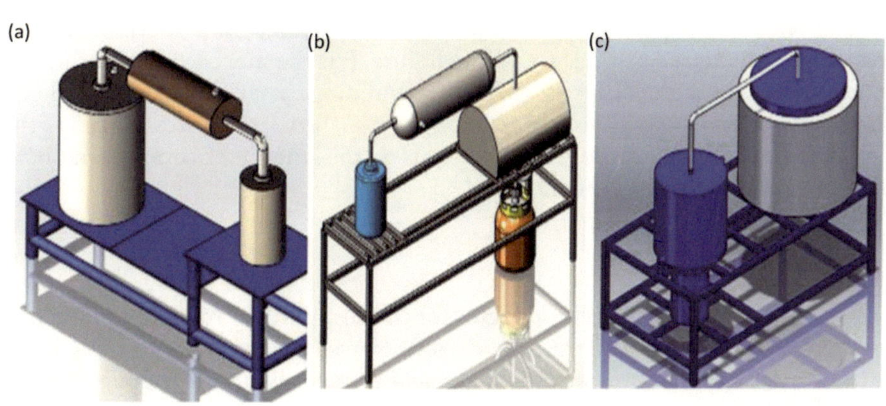

Figure 4.
Concepts (a) double pipe; (b) shell and tube; (c) helical pipe heat exchanger.

Criteria	Weight	Concept rating/10			Concept weighting/100%		
		(a)	(b)	(c)	(a)	(b)	(c)
Function	10	7	5	9	70	50	90
Efficiency	9	6	6	8	54	54	72
Cost	8	4	6	8	32	48	64
Ease of maintenance	6	7	7	9	42	42	54
Reliability	6	5	5	8	30	30	48
Ease of manufacture	4	6	7	9	24	28	36
Lifespan	4	7	8	9	28	32	36
Simplicity of layout	2	5	7	8	10	14	16
Quality	1	8	6	9	8	6	9
Weight	1	8	5	9	8	5	9
Ergonomics	1	6	6	8	6	6	8
Total score					312	315	443

Table 3.
Concept selection using the binary dominance matrix.

3.1 Experimental design development

The pulverized plastic wastes used in this work were collected from disposed plastic waste used for packaging water and fruit juices at Petreco Zimbabwe. The PET waste (**Figure 5**) was collected and cleaned with tap water to remove any form of contaminants and deleterious materials before sun drying for a minimum of three days.

The waste PET was added into the pyrolysis tank with a helical heating coil for even heat distribution. Oxygen was eliminated from the chamber by use of oxygen scavengers. Weight batching was used to be able to eliminate errors due to size and void in the material. Whilst molten the PET was tapped out of the pyrolysis tank into the brick cast mold. Mechanical tests were key to ascertain the properties of the brick. Hence compression test, water absorption test and flexural strength test were carried out. The different mechanical tests are described below.

3.1.1 Compressive strength test

Compressive strength is a mechanical test that measures the amount of compressive load a material can sustain before failure. It is the most accurate indicator for measuring the engineering properties of a building material because a higher compressive strength indicates a good and strong material. Compressive strength in bricks and flexible pavements greatly depends on the amount of compaction, heat treatment and the concrete mix ratio.

3.1.2 Water absorption test

Absorption of water is a critical factor for the durability of bricks and flexible pavements. When water penetrates brick, it decreases the strength of bricks. Hence the brick internal structure must be adequately dense to avoid the leaking of water. To increase density and decrease water absorption of clay bricks, the firing temperature must be raised [7]. Generally, the water absorption values decreased with

Figure 5.
Sample pulverized PET.

increasing temperature, and decreased with increasing amounts of waste plastic in the mixtures. Hence PET plastic would reduce the water absorption of the pavements and bricks.

3.1.3 Flexural strength

The flexural strength of brick is its ability to resist bending through internal stresses generated within the brick or flexible pavement. Elastic stresses are developed within the brick and pavement material to resist the external load. Resistant tensile stress that develop, keep the bricks in shape until the failure of the structure. Whenever the material is overloaded the structure will rupture, and flexural cracks are developed at the point of highest tensile stress on the brick or flexible pavement [5]. It's an integral test in determining properties of the brick.

4. Results and discussion

The model of choice was the Helical Heat Exchanger. In order to enhance mobility and temperature monitoring, wheels and a thermometer also included in the frame as shown in **Figure 6**. In this model electrical energy is used to melt the plastic making use of an oxygen scavenger to remove oxygen from the reactor. In order to optimize the intended operation, the heat exchanger's dimension was to be calculated. Using Eq. (1) below to calculate the required length of wire, ten turns of the helical copper pipe were chosen.

$$L = N\sqrt{(2\pi r)^2 + p^2} \tag{1}$$

$$\mathrm{L} = 10\sqrt{2\pi(7.5)^2 + 20^2}$$

Using Eq. (2), the helical pipe volume was found.

$$V_e = \frac{\pi}{4}(d_o)^2 L \tag{2}$$

$$V_e = \frac{\pi}{4}(15)^2 \times 512 = 90.4 / \mathrm{mm}^3$$

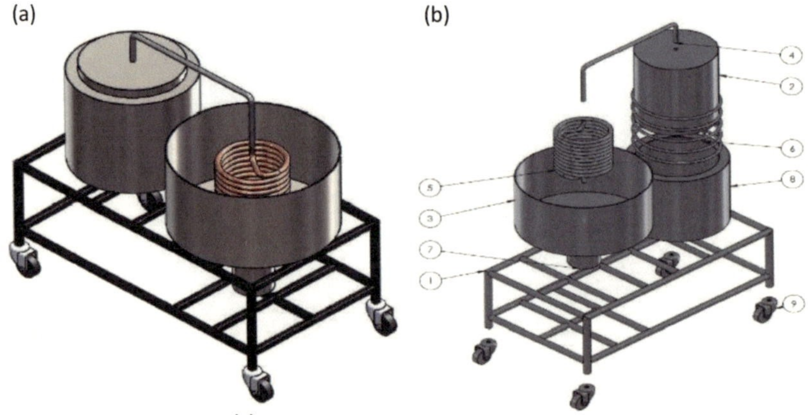

(c)

ITEM NO.	PART NUMBER	QTY.
9	ROULETTE 3,8	4
8	reactor cassing	1
7	Collector	1
6	heater	1
5	Helix heat exchanger	1
4	Pipe	1
3	Heat exchanger tank	1
2	Tank	1
1	Stand	1

Figure 6.
SolidWorks model of the heat exchanger (a) assembled heat exchanger (b) exploded view (c) parts list.

4.1 Flexible pavements and bricks experimental results

The pulverized PET was melted in the pyrolysis chamber in the absence of oxygen Different amounts of PET were used in the mixture with bituminous asphaltic concrete to optimize the best properties from the materials. After testing the compressive strength of the bricks at different percentages of PET the results in **Figure 7** were obtained.

Figure 7 shows that the compressive strength of flexible pavements decreased with increase in the percentage of the PET. The results showed a maximum

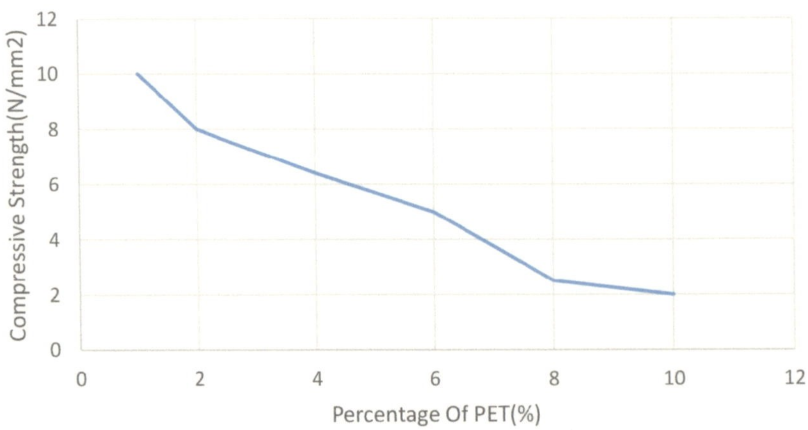

Figure 7.
Compressive strength results.

compressive strength of 10 N/mm^2. However, the water absorption properties decreased with increase in percentage content of PET that is the brick absorbed less water with an increase in the amount of PET.

Since coarse aggregates account for more than 60% of the aggregates, the BAC can be defined as coarse dense-graded and it is suited for all pavement layers and traffic conditions. The PET reduced the flexible pavement and brick compressive strength whilst increasing its fatigue resistance. In addition, it created better mixing between asphalt and the aggregates. During the compressive test the PET bricks had a maximum compressive strength of 10 N/mm^2 which was within the standard range and had less water absorbing tendencies and hence a longer lifespan. Generally bigger aggregate sizes give better mechanical properties, as well as good permeability and traction, and are typically focused at achieving high surface drainage and durability [9]. The flexible pavements with PET as a filler also exhibited increased workability and durability, as well as resistance to irreversible deformation, stress, and low-temperature cracking, and moisture damage [8]. The filler's purpose was to improve stiffness, workability, moisture resistance, and aging properties. The use of waste PET as a flexible pavement is one of the possible uses of the plastic waste.

Globally there is growing demand for high-performance pavement due to increase in traffic and failure of ordinary pavements. In a bid to increase durability of roads especially in developing countries where the roads are deplorable, use of PET waste with BAC and cement as a binder. PET wastes in flexible pavement, brick construction and road rehabilitations would essentially utilize several million metric tons of PET wastes from the waste stream which would have good sustainable effects on the environment such as minimization of pollution and greenhouse gas emissions. It also encourages ecosystem balance by the limiting of non-biodegradable PET wastes from the ecosystem.

5. Conclusion

The concept of employing composite brick and pavement was investigated in this research. PET materials melted during the pyrolysis process due to their low melting point of 250°C. This research brought about the purpose of eco-friendly construction material production in the drive toward achieving sustainable development in the fourth industrial revolution. It enlightens on the concept of eco-friendly road construction by use of flexible pavements and bricks by utilizing waste, non-biodegradable PET plastic. The properties of plastic meant it would not be affected by water absorption which weakens the bricks and flexible pavements. Further research can be done on exploiting the properties of PET in the manufacture of construction materials, especially its light weight and temperature regulation. As a reference for future work the residue char from pyrolysis can be further investigated for use as an activated carbon.

Author details

Ngonidzashe L. Shangwa[1*], Wilson R. Nyemba[1], Simon Chinguwa[1] and Tien-Chien Jen[2]

1 University of Zimbabwe, Harare, Zimbabwe

2 University of Johannesburg, Johannesburg, South Africa

*Address all correspondence to: ngonidzashe.shangwa@yahoo.co.uk

IntechOpen

© 2021 The Author(s). Licensee IntechOpen. This chapter is distributed under the terms of the Creative Commons Attribution License (http://creativecommons.org/licenses/by/3.0), which permits unrestricted use, distribution, and reproduction in any medium, provided the original work is properly cited.

References

[1] Rinku Verma, K. S. Vinoda, M. Papireddy, A.N.S Gowda. Toxic pollutants from waste plastic – A review. Procedia Environmental Sciences 35 (2016) 701-708.

[2] R. Harrington, By 2050, the oceans could have more plastic than fish. Business Insider. Jan 26, 2017. www.businessinsider.com

[3] Siddique, R., Khatib, J., and Kaur, I. (2008). Use of recycled plastic in concrete: A review. Waste Management Vol. 28, pp. 1835-1852.

[4] Ancheyta J, Kumar S, Singh RK. Thermolysis of high-density polyethylene to petroleum products. Journal of Petroleum Engineering. Available: https://www.hindawi.com/journals/jpe/2013/987568/. Accessed: June 2018, 2013:1-7.

[5] A.M. Azhdarpour, M.R. Nikoudel, M. Taheri. The effect of using polyethylene terephthalate particles on physical and strength-related properties of concrete: A laboratory evaluation. Constr. Build. Mater. 109 (2016) 55-62.

[6] Appiah, J. K. (2013). Modification of bitumen with waste plastics for road construction (MSc dissertation). Kwame Nkrumah University of Science and Technology, Kumasi.

[7] P. Nonthaphong, S. Kanyakam, P. Chindaprasirt. Utilisation of waste glass to enhance physical-mechanical properties of fired clay bricks. J. Clean. Prod. (2016).

[8] Kar, D. (2012). A laboratory study of bituminous mixes using a natural fibre [MSc dissertation]. National Institute of Technology, Rourkela.

[9] Qian, Z., and Lu, Q. (2015). Design and laboratory evaluation of small particle porous epoxy asphalt surface mixture for roadway pavements. Construction and Building Materials, 77, 110-116.